Fifth Edition

UNIVERSITY PHYSICS

Part I

Fifth Edition

UNIVERSITY PHYSICS
Part I

Francis W. Sears
Professor Emeritus
Dartmouth College

Mark W. Zemansky
Professor Emeritus
The City College of
The City University of New York

Hugh D. Young
Associate Professor of Physics
Carnegie-Mellon University

ADDISON-WESLEY PUBLISHING COMPANY

Reading, Massachusetts · Menlo Park, California
London · Amsterdam · Don Mills, Ontario · Sydney

This book is in the
ADDISON-WESLEY SERIES IN PHYSICS

Fourth printing, July 1978

ISBN 0-201-06937-7
ABCDEFGHIJK-DO-798

Preface

University Physics is intended for students of science and engineering who are taking an introductory calculus course concurrently. The complete text may be taught in an intensive two- or three-semester course and is also adaptable to a variety of shorter courses. Primary emphasis is on physical principles and problem solving; historical background and specialized practical applications have been given a place of secondary importance. Many worked-out examples and an extensive collection of problems are included with each chapter. *University Physics* is available as a single volume or as two separate parts. Part I includes mechanics, heat, and sound, and Part II includes electricity and magnetism, optics, and atomic and nuclear physics.

In this new edition, the basic philosophy and outline and the balance between depth of treatment and breadth of subject-matter coverage are unchanged from previous editions. We have tried to preserve those features that users of previous editions have found desirable, while incorporating a number of changes that should enhance the book's usefulness. Here are the most important changes.

1. The mks system of units, with the conventions and nomenclature of the Système Internationale, has become the principal unit system in the book. In this system the joule is the fundamental unit of energy of all forms, including heat. In the first half of the book, however, some examples and problems using English units have been retained.

2. The material on atomic and nuclear physics (Chapters 44 through 46) has been completely rewritten and expanded into three chapters, now including an elementary discussion of physics of solids, high-energy physics, and elementary particles.

3. A new chapter on relativistic mechanics has been added. Positioned somewhat arbitrarily at the end of the mechanics material, this chapter can be taken up earlier or later, or may be omitted completely if desired.

4. The chapter on electromagnetic waves has been completely rewritten, to exhibit more clearly and in simpler context the relation of wave propagation to the basic principles of electromagnetism.

5. Several sections have been added to broaden subject coverage. Among these are: 16–7 Examples (of calorimetric calculations), 19–25 (practical aspects of) Energy Conversion, 23–5 Musical Intervals and Scales, 23–7 Applications of Acoustic Phenomena, 28–10 Physiological Effects of Electric Currents, 38–7 Absorption (of light), 38–8 Illumination, 41–10 Defects of Vision, 42–1 Coherent Sources (of light), 42–13 Holography, 45–7 Semiconductors, 45–8 Semiconductor Devices, and 46–11 Radiation and the Life Sciences.

6. Some material has been reorganized. The material on surface tension has been shortened and incorporated into the hydrostatics chapter; the treatment of thermoelectricity has been shortened to a single section. The chapter on electromagnetic in-

duction has been rearranged to exhibit more clearly the various applications of Faraday's law to moving conductors and to stationary conductors in varying fields. The material on inductance and associated problems has been removed from this chapter and placed in a separate chapter.

7. Many sections have been completely rewritten for improved clarity and pedagogical effectiveness. These include the beginning of Chapter 7 (Work and Energy), Chapter 8 (Impulse and Momentum), Chapter 11 (Harmonic Motion); Section 25–4 (Gauss's Law); Chapter 26 (Potential); Section 27–7 (Polarization and Electric Displacement); Section 29–7 (R–C Circuits); Section 32–6 (Ampère's Law); and Chapter 35 (Magnetic Properties of Matter), to cite only a few examples. In a few cases, such as the opening sections of Chapters 7 and 8, the rewriting may create the illusion of de-emphasizing the use of calculus. Not so; the treatment is just as rigorous as in previous editions, but has been rearranged to follow the pedagogical principle of moving from the simple to the complex.

8. About 300 new problems have been added, bringing the total to over 1400. The added problems provide greater variety and also broader subject coverage than in previous editions. The authors have resisted the temptation to key problems to specific sections of text. Learning to select the principles appropriate for a specific problem is, after all, part of learning to solve problems. In addition, many problems require material from more than one section.

9. In every case where material has been rewritten, sound pedagogical principles and the authors' own teaching experience have guided the revision. In some instances we have shifted from a sequence in which a principle or concept is presented initially in its full generality to one that begins with special cases and then progresses to the more general statement. We hope thus to help the student attain the same final level of sophistication as previously by climbing a less steep slope.

The text is adaptable to a wide variety of course outlines. The entire text can be used for an intensive course two or three semesters in length. For a less intensive course, many instructors will want to omit certain chapters or sections to tailor the book to their individual needs. The format of this edition facilitates this kind of flexibility. For example, any or all of the chapters on relativity, hydrostatics, hydrodynamics, acoustics, magnetic properties of matter, electromagnetic waves, optical instruments, and several others can be omitted without loss of continuity.

Conversely, however, many topics that were regarded a few years ago as of peripheral importance and were purged from introductory courses have now come to the fore again in the life sciences, earth and space sciences, and environmental problems. An instructor who wishes to stress these kinds of applications will find this text a useful source for discussion of the appropriate principles.

In any case, it should be emphasized that instructors should not feel constrained to work straight through the book from cover to cover. Many chapters are, of course, inherently sequential in nature, but within this general limitation instructors should be encouraged to select among the contents those chapters that fit their needs, omitting material that is not relevant for the objectives of a particular course.

Again, we wish to thank our many colleagues who have contributed suggestions for this new edition. In particular, Prof. Robert Folk, Prof. Shelden H. Radin, and Prof. Charles W. Smith have read the entire manuscript, and their critical and constructive comments are greatly appreciated.

Hanover, New York,   F.W.S.
and Pittsburgh   M.W.Z.
November 1975   H.D.Y.

Contents

PART I
MECHANICS, HEAT, AND SOUND

1 Composition and resolution of vectors 1

1-1 The fundamental indefinables of mechanics 1
1-2 Standards and units 1
1-3 Symbols for physical quantities 4
1-4 Force 4
1-5 Graphical representation of forces. Vectors 5
1-6 Vector addition. Resultant of a set of forces 6
1-7 Components of a vector 8
1-8 Resultant by rectangular resolution 10
1-9 Vector difference 11
1-10 A word about problems 12

2 Equilibrium of a particle 15

2-1 Introduction 15
2-2 Equilibrium. Newton's first law 15
2-3 Discussion of Newton's first law of motion 17
2-4 Newton's third law of motion 18
2-5 Equilibrium of a particle 20
2-6 Friction 26

3 Equilibrium. Moment of a force 35

3-1 Moment of a force 35
3-2 The second condition of equilibrium 36
3-3 Center of gravity 39
3-4 Couples 42

4 Rectilinear motion 49

4-1 Motion 49
4-2 Average velocity 49
4-3 Instantaneous velocity 50
4-4 Average and instantaneous acceleration 51
4-5 Rectilinear motion with constant acceleration 53
4-6 Velocity and coordinate by integration 55
4-7 Freely falling bodies 57
4-8 Rectilinear motion with variable acceleration 59
4-9 Velocity components. Relative velocity 60

5 Newton's second law. Gravitation 69

5-1 Introduction 69
5-2 Newton's second law. Mass 69
5-3 Systems of units 71
5-4 Newton's law of universal gravitation 72
5-5 Mass and weight 74
5-6 Applications of Newton's second law 75

6 Motion in a plane 89

6-1 Motion in a plane 89
6-2 Average and instantaneous velocity 89
6-3 Average and instantaneous acceleration 90
6-4 Components of acceleration 91
6-5 Motion of a projectile 93
6-6 Circular motion 97
6-7 Centripetal force 99

6–8   Motion in a vertical circle   102
6–9   Motion of a satellite   104
6–10   Effect of the earth's rotation on g 106

7   *Work and energy*   *113*

7–1   Work   113
7–2   Work done by a varying force   115
7–3   Work and kinetic energy   117
7–4   Gravitational potential energy   119
7–5   Elastic potential energy   124
7–6   Conservative and dissipative forces   125
7–7   Internal work   126
7–8   Internal potential energy   127
7–9   Power   128
7–10   Power and velocity   129
7–11   Mass and energy   130

8   *Impulse and momentum*   *139*

8–1   Impulse and momentum   139
8–2   Conservation of momentum   142
8–3   Collisions   143
8–4   Inelastic collisions   143
8–5   Elastic collisions   145
8–6   Recoil   146
8–7   Rocket propulsion   148
8–8   Generalizations   149

9   *Rotation*   *157*

9–1   Introduction   157
9–2   Angular velocity   157
9–3   Angular acceleration   158
9–4   Rotation with constant angular acceleration   159
9–5   Relation between angular and linear velocity and acceleration   160
9–6   Torque and angular acceleration. Moment of inertia   161
9–7   Calculation of moments of inertia   163
9–8   Kinetic energy, work, and power   166
9–9   Angular momentum   167
9–10   Conservation of angular momentum   168
9–11   Vector representation of angular quantities   170
9–12   The top and the gyroscope   173

10   *Elasticity*   *183*

10–1   Stress   183
10–2   Strain   186
10–3   Elasticity and plasticity   186

10–4   Elastic modulus   188
10–5   The force constant   191

11   *Harmonic motion*   *195*

11–1   Introduction   195
11–2   Elastic restoring forces   195
11–3   Definitions   196
11–4   Equations of simple harmonic motion   197
11–5   Motion of a body suspended from a coil spring   202
11–6   The simple pendulum   203
11–7   Angular harmonic motion   204
11–8   The physical pendulum   204
11–9   Center of oscillation   205

12   *Hydrostatics*   *213*

12–1   Introduction   213
12–2   Pressure in a fluid   214
12–3   The hydrostatic "paradox"   215
12–4   Pressure gauges   216
12–5   Pumps   217
12–6   Archimedes' principle   218
12–7   Forces against a dam   221
12–8   Surface tension   221
12–9   Pressure difference across a surface film   224
12–10 Contact angle and capillarity   225

13   *Hydrodynamics and viscosity*   *233*

13–1   Introduction   233
13–2   The equation of continuity   234
13–3   Bernoulli's equation   235
13–4   Applications of Bernoulli's equation   236
13–5   Viscosity   241
13–6   Poiseuille's law   243
13–7   Stokes' law   245
13–8   Reynolds number   245

14   *Relativistic mechanics*   *251*

14–1   Invariance of physical laws   251
14–2   Relative nature of simultaneity   253
14–3   Relativity of time   253
14–4   Relativity of length   255
14–5   The Lorentz transformation   257
14–6   Momentum   258
14–7   Work and energy   259
14–8   Relativity and newtonian mechanics   261

15   Temperature and expansion   265

15–1   Concept of temperature   265
15–2   Thermometers   267
15–3   The establishment of a temperature scale   269
15–4   The Celsius, Rankine, and Fahrenheit scales   271
15–5   Expansion of solids and liquids   272
15–6   Thermal stresses   274

16   Heat and heat measurements   279

16–1   Heat transfer   279
16–2   Quantity of heat   281
16–3   Heat capacity   282
16–4   The measurement of heat capacity   283
16–5   Experimental values of heat capacity   284
16–6   Change of phase   286
16–7   Examples   288

17   Transfer of heat   293

17–1   Conduction   293
17–2   Radial heat flow in a sphere or cylinder   295
17–3   Convection   296
17–4   Radiation   298
17–5   Stefan's law   299
17–6   The ideal radiator   300

18   Thermal properties of matter   305

18–1   Equations of state   305
18–2   The ideal gas   305
18–3 pVT-surface for an ideal gas   308
18–4 pVT-surface for a real substance   309
18–5   Critical point and triple point   312
18–6   Effect of dissolved substances on freezing and boiling points   316
18–7   Humidity   317
18–8   The cloud chamber and the bubble chamber   318

19   The laws of thermodynamics   323

19–1   Energy and work in thermodynamics   323
19–2   Work in volume changes   323
19–3   Heat in volume changes   325
19–4   The first law of thermodynamics   326
19–5   Adiabatic process   327
19–6   Isochoric process   327
19–7   Isothermal process   327
19–8   Isobaric process   327
19–9   Throttling process   328
19–10   Differential form of the first law   329
19–11   Internal energy of an ideal gas   329

19–12   Heat capacities of an ideal gas   330
19–13   Adiabatic process of an ideal gas   331
19–14   Heat engines   333
19–15   The gasoline engine   335
19–16   The Diesel engine   336
19–17   The steam engine   336
19–18   The second law of thermodynamics   337
19–19   The refrigerator   338
19–20   The Carnot cycle   339
19–21   The Kelvin temperature scale   340
19–22   Absolute zero   341
19–23   Entropy   342
19–24   The principle of the increase of entropy   343
19–25   Energy conversion   344

20   Molecular properties of matter   349

20–1   Molecular theory of matter   349
20–2   Avogadro's number   351
20–3   Properties of matter   352
20–4   Kinetic theory of an ideal gas   353
20–5   Molar heat capacity of a gas   356
20–6   Measurement of molecular speeds   357
20–7   Crystals   357
20–8   Heat capacity of a crystal   359

21   Traveling waves   363

21–1   Introduction   363
21–2   Periodic waves   364
21–3   Mathematical representation of a traveling wave   365
21–4   Speed of a transverse wave   368
21–5   Speed of a longitudinal wave   369
21–6   Adiabatic character of a longitudinal wave   370
21–7   Water waves   372

22   Vibrating bodies   377

22–1   Boundary conditions for a string   377
22–2   Standing waves in a string   379
22–3   Vibration of a string fixed at both ends   380
22–4   Resonance   382
22–5   Interference of longitudinal waves   382
22–6   Longitudinal standing waves   383
22–7   Vibrations of organ pipes   384
22–8   Vibrations of rods and plates   386

23   Acoustic phenomena   389

23–1   Sound waves   389
23–2   Intensity   390

23–3   Intensity level and loudness   391
23–4   Quality and pitch   393
23–5   Musical intervals and scales   394
23–6   Radiation from a piston   395
23–7   Applications of acoustic phenomena   398
23–8   Beats   399
23–9   The Doppler effect   400

Answers to odd-numbered problems   A–1

Appendix   T–1

Index   I–1

Abridged
Contents

PART II

24 Coulomb's law   407

25 The electric field. Gauss's law   417

26 Potential   439

27 Capacitance. Properties of dielectrics   459

28 Current, resistance, and electromotive force   475

29 Direct-current circuits and instruments   501

30 The magnetic field   519

31 Magnetic forces on current-carrying conductors   535

32 Magnetic field of a current   547

33 Induced electromotive force   561

34 Inductance   579

35 Magnetic properties of matter   589

36 Alternating currents   609

37 Electromagnetic waves   627

38 The nature and propagation of light   639

39 Reflection and refraction at plane surfaces   653

40 Images formed by a single reflection or refraction   665

41 Lenses and optical instruments   683

42 Interference and diffraction   707

43 Polarization   737

44 Photons, electrons, and atoms   753

45 Atoms, molecules and solids   775

46 Nuclear physics   791

PART I

Mechanics, Heat, and Sound

Chapter 1

Composition and Resolution of Vectors

1–1 THE FUNDAMENTAL INDEFINABLES OF MECHANICS

Physics has been called the science of measurement. To quote from Lord Kelvin (1824–1907), "I often say that when you can measure what you are speaking about, and express it in numbers, you know something about it; but when you cannot express it in numbers, your knowledge is of a meagre and unsatisfactory kind; it may be the beginning of knowledge, but you have scarcely, in your thoughts, advanced to the state of *Science*, whatever the matter may be."

A definition of a quantity in physics must provide a set of rules for calculating it in terms of other quantities that can be measured. Thus, when momentum is defined as the product of "mass" and "velocity," the rule for calculating momentum is contained within the definition, and all that is necessary is to know how to measure mass and velocity. The definition of velocity is given in terms of length and time, but there are no simpler or more fundamental quantities in terms of which length and time may be expressed. *Length and time are two of the indefinables of mechanics.* It has been found possible to express all the quantities of mechanics in terms of only three indefinables. The third may, with equal justification, be taken to be "mass" or "force." *We shall choose mass as the third indefinable of mechanics.*

In geometry, the fundamental indefinable is the "point." The geometer asks his disciple to build any picture of a point in his mind, provided the picture is consistent with what the geometer says about the point. In physics, the situation is not so subtle. Physicists from all over the world staff international committees at whose meetings the rules of measurement of the indefinables are adopted. The rule for *measuring* an indefinable takes the place of a definition, and such a rule is sometimes called an *operational definition*.

1–2 STANDARDS AND UNITS

The set of rules for measuring the indefinables of mechanics is determined by an international committee called the *General Conference on Weights and Measures*, to which all the major countries send delegates. One of the chief functions of the Conference is to decide on a standard for each indefinable. A standard may be an actual object, in which case its main characteristic must be *durability*. Thus, in 1889 a meter bar of platinum-iridium alloy was chosen as the *standard of length*, because this alloy is particularly stable in its chemical structure. However, the preservation of a bar of this material as a world standard entails a number of cumbersome provisions, such as making a large number of replicas for

1

(a)

(b)

Fig. 1–1 (a) Etalon and (b) Michelson interferometer for use in measuring the distance x in terms of the wavelength of light.

all the major countries and comparing these replicas with the world standard at periodic intervals. On October 14, 1960, the General Conference changed the standard of length to an *atomic constant, namely, the wavelength of the orange-red light emitted by the individual atoms of krypton-86* in a tube filled with krypton gas in which an electrical discharge is main-

Table 1–1   STANDARDS AND UNITS AS OF 1969

	Standard	Measuring device	Unit
Length	Wavelength of orange-red light from krypton-86	Optical interferometer	1 meter = 1,650,763.73 wavelengths
Mass	Platinum-iridium cylinder, 1 kilogram	Equal-arm balance	1 kilogram
Time	Periodic time associated with a transition between two energy levels of cesium-133 atom	Atomic clock	1 second = 9,192,631,770 cesium periods

tained. Such a standard is much more readily reproducible than one based on a specific material object.

The *standard of mass* is the mass of a cylinder of platinum-iridium, designated as *one kilogram* and kept at the International Bureau of Weights and Measures at Sèvres, near Paris.

Before 1960, the *standard of time* was the interval of time between successive appearances of the sun overhead, averaged over a year, and called the *mean solar day*. Between 1960 and 1967 it was changed to the *tropical year 1900*, that is, the time it took the sun to move from a certain point in the heavens, known as the *vernal equinox*, back to the same point in 1900. In October 1967, the standard was changed again to the *periodic time of the radiation corresponding to the transition between the two hyperfine energy levels of the fundamental state of the atom of cesium-133*.

The three standards are listed in Table 1–1.

After the choice of a standard, the next step is to decide upon an instrument and a technique for comparing the standard with an unknown. Consider, for example, the distance x between two mirrors, A and B, of the device called an *etalon*, shown in Fig. 1–1(a). To find the number of wavelengths of orange-red light of krypton-86 in the distance x requires the use of an *optical interferometer*, one type of which (due to Michelson) is shown in Fig. 1–1(b). A mova-

Table 1–2 PREFIXES FOR POWERS OF TEN

Power of ten	10^{-12}	10^{-9}	10^{-6}	10^{-3}	10^{-2}	10^{3}	10^{6}	10^{9}	10^{12}
Prefix	pico-	nano-	micro-	milli-	centi-	kilo-	mega-	giga-	tera-
Abbreviation	p	n	μ	m	c	k	M	G	T

ble mirror M_2 on the Michelson interferometer is first made to coincide in position with A on the etalon. Then the mirror is moved slowly until it coincides with B, during which time gradations of orange and black, known as *interference fringes*, move past the cross hair in the field of view of a telescope and are counted. The motion of one complete fringe corresponds to a motion of mirror M_2 of exactly one-half wavelength. A length known as *one meter* is defined in this way as:

1 meter = 1,650,763.73 wavelengths of orange-red light of krypton-86.

The metric system of units is used exclusively in defining the standards of mass, length, and time. Most nations other than the United States and Great Britain also use the metric system exclusively for commerce and industry as well; Britain is making long-range plans to convert to the metric system, and many well-informed people believe it would be very advantageous for the United States to convert. One advantage of the metric system is that the various units for a quantity are always related by factors of ten. Thus, some units of length in common use in science and technology are:

1 angstrom unit = 1 Å = 10^{-10}m
(used by spectroscopists),

1 nanometer = 1 nm = 10^{-9}m
(used by optical designers),

1 micrometer = 1 μm = 10^{-6}m
(used commonly in biology),

1 millimeter = 1 mm = 10^{-3}m and
1 centimeter = 1 cm = 10^{-2}m
(used most often),

1 kilometer = 1 km = 10^{3}m
(a common European unit of distance).

The words "nanometer," "micrometer," and "kilometer" are all accented on the *first* syllable, *not* the second, just like the words "millimeter" and "centimeter." The prefix "nano" is pronounced "nanno." A common set of prefixes is used with all units. These and their standard abbreviations are shown in Table 1–2. Thus,

1 kilometer	= 1 km	= 10^{3} meter	= 10^{3}m,
1 kilogram	= 1 kg	= 10^{3} grams	= 10^{3}g,
1 kilowatt	= 1 kW	= 10^{3} watts	= 10^{3}W.

It is convenient to memorize Table 1–2, to have the information available when needed.

Units of length used in everyday life and in engineering in both the United States and the United Kingdom are defined as follows:

$$1 \text{ inch} = 1 \text{ in.} = \begin{cases} 41{,}929.399 \text{ wavelengths of} \\ \text{Kr light, or } \textit{exactly} \text{ 2.54 cm,} \end{cases}$$

1 foot = 1 ft = 12 in.,

1 yard = 1 yd = 3 ft,

1 mile = 1 mi = 5280 ft.

The device used to subdivide the standard of mass, the kilogram, into equal submasses is the *equal-arm balance*, which will be discussed in Chapter 5. Frequently used units of mass are:

1 microgram	= 1 μg	= 10^{-9}kg,
1 milligram	= 1 mg	= 10^{-6}kg,
1 gram	= 1 g	= 10^{-3}kg,
1 pound mass	= 1 lbm	= 0.45359237 kg.

The clock used to define the standard time interval is the *cesium clock*, a large, complex, and expensive laboratory instrument. It is extraordinarily precise and maintains its frequency constant to one part in one hundred billion (10^{11}) or better. Furthermore, it may be compared with other high-precision clocks in an hour or so, instead of the years required for comparison with the old astronomical standard. In the atomic clock, a beam of cesium-133 atoms passes through a long metal cylinder and interacts with microwaves brought in by a wave guide from a generator controlled by a quartz oscillator. The *unit of time* used throughout the world is called the *second* and is defined to be

$$1 \text{ second} = 1 \text{ s}$$
$$= 9,192,631,770 \text{ Cs periods}.$$

Other common units of time are:

1 nanosecond	= 1 ns	= 10^{-9} s,
1 microsecond	= 1 μs	= 10^{-6} s,
1 millisecond	= 1 ms	= 10^{-3} s,
1 minute	= 1 min	= 60 s,
1 hour	= 1 hr	= 3600 s,
1 day	= 1 day	= 86,400 s.

1-3 SYMBOLS FOR PHYSICAL QUANTITIES

We shall adopt the convention that an algebraic symbol representing a physical quantity, such as F, p, or v, stands for both a *number* and a *unit*. For example, F might represent a force of 10 N (where N stands for newton), p a pressure of 15 N m^{-2}, and v a velocity of 15 m s^{-1}.

When we write

$$x = v_0 t + \tfrac{1}{2} a t^2,$$

if x is in meters, then the terms $v_0 t$ and $\tfrac{1}{2} a t^2$ must be in meters also. Suppose t is in seconds. Then the units of v_0 must be m s^{-1} and those of a must be m s^{-2}. (The factor $\tfrac{1}{2}$ is a *pure number*, without units.) The units of v_0 could be written as m/s rather than m s^{-1}, but the negative-exponent form is usually more con-

venient and will be used in all such expressions in this book.

As a numerical example, let $v_0 = 10$ m s^{-1}, $a = 4$ m s^{-2}, $t = 10$ s. Then the preceding equation would be written

$$x = (10 \text{ m s}^{-1}) \cdot (10 \text{ s}) + \tfrac{1}{2} \cdot (4 \text{ m s}^{-2}) \cdot (10 \text{ s})^2.$$

The units are treated like algebraic symbols. The s's cancel in the first term and the s^2's in the second, and

$$x = 100 \text{ m} + 200 \text{ m} = 300 \text{ m}.$$

The beginning student will do well to include the units of all physical quantities, as well as their magnitudes, in all his calculations. This will be done consistently in the numerical examples throughout this book.

1-4 FORCE

Mechanics is the branch of physics which deals with the motion of material bodies and with the forces that bring about the motion. We shall postpone a discussion of motion until Chapter 4, and start with a study of forces.

When we push or pull on a body, we are said to exert a *force* on it. Forces can also be exerted by inanimate objects: a stretched spring exerts forces on the bodies to which its ends are attached; compressed air exerts a force on the walls of its container; a locomotive exerts a force on the train it is pulling or pushing. The force of which we are most aware in our daily lives is the force of gravitational attraction exerted on every physical body by the earth, called the *weight* of the body. Gravitational forces (and electrical and magnetic forces also) can act through empty space without contact. A force on an object resulting from direct contact with another object is called a *contact force*; viewed on an atomic scale, contact forces arise chiefly from electrical attraction and repulsion of the electrons and nuclei making up the atoms of material.

To describe a force, we need to describe the *direction* in which it acts, as well as its *magnitude*, which is a quantitative description of "how much" or "how hard" the force pushes or pulls, in terms of a standard unit of force. In Chapter 5 we shall see how a unit of force can be defined in terms of the units of mass, length, and time. In the meter-kilogram-second (mks) system, this unit is the *newton*, abbreviated N. A more familiar unit is the *pound*, which can be defined as the force with which the earth attracts a standard body (i.e., its weight) with a mass of 1 pound-mass as defined in Section 1–2. A particular location on the earth's surface must be specified, since the attraction of the earth for a given body varies by as much as 0.5% from one point to another. If great precision is not required, it suffices to take any point at sea level and 45° latitude.

In order for an unknown force to be compared with the force unit and thereby *measured*, some observable effect produced by a force must be used. One such effect is to alter the dimensions or shape of a body on which the force is exerted; another is to alter the state of motion of the body. Both of these effects can be used in the measurement of forces. In this chapter we shall consider only the former; the latter will be discussed in Chapter 5.

An instrument commonly used to measure forces is the spring balance, which consists of a coil spring enclosed in a case for protection and carrying at one end a pointer that moves over a scale. A force exerted on the balance changes the length of the spring, and the change can be read on the scale. The balance can be calibrated as follows. The standard pound is first suspended from the balance at sea level and 45° latitude and the position of the pointer is marked 1 lb. Any number of duplicates of the standard can then be prepared by suspending a body from the balance and adding or removing material until the index again stands at 1 lb. Then when two, three, or more of these are suspended simultaneously from the balance, the force stretching it is 2 lb, 3 lb, etc., and the corresponding positions of the pointer can be labeled 2 lb, 3 lb, etc. This procedure makes no assumption about the elastic properties of the spring except that the force exerted on it is always the same when the pointer stands at the same position. The calibrated balance can then be used to measure the magnitude of an unknown force. An analogous procedure can be used to calibrate a spring balance in newtons

(a)   (b)

Fig. 1–2

1–5 GRAPHICAL REPRESENTATION OF FORCES. VECTORS

Suppose we are to slide a box along the floor by pulling it with a string or pushing it with a stick, as in Fig. 1–2. That is, we are to slide it by exerting a force on it. The point of view which we now adopt is that the motion of the box is caused not by the *objects* which push or pull on it, but by the *forces* which these exert. For concreteness, assume the magnitude of the push or pull to be 10 N. To write "10 N" on the diagram would not completely describe the force, since it would not indicate the *direction* in which the force acts. One might write "10 N, 30° above horizontal to the right," or "10 N, 45° below horizontal to the right," but all the above information may be conveyed more briefly if we adopt the convention of representing a force by an arrow. The length of the arrow, to some chosen scale, indicates the size or *magnitude* of the force, and the direction in which the arrow points indicates the *direction* of the force. Thus, Fig. 1–3 is the force diagram corresponding to Fig. 1–2. (There are other forces acting on the box, but these are not shown in the figure.)

(a)   (b)

Fig. 1–3

Force is not the only physical quantity which requires the specification of a direction in space as well as a magnitude. For example, the velocity of an aircraft is not completely specified by stating that it is 300 miles per hour; we need to know the direction also. The concept of volume, on the other hand, has no direction associated with it.

Quantities such as volume, which involve a magnitude only, are called *scalar quantities*. Those such as force and velocity, which involve both a magnitude and a direction in space, are called *vector quantities*. Any vector quantity can be represented by an arrow, and this arrow is called a vector (or, if a more specific statement is needed, a force vector or a velocity vector).

Some vector quantities, of which force is one, are not *completely* specified by their magnitude and direction alone. Thus, the effect of a force depends also on its *line of action* and its *point of application*. (The line of action is a line of indefinite length, of which the force vector is a segment.) For example, if one is pushing horizontally against a door, the effectiveness of a force of given magnitude and direction depends on the perpendicular distance of its line of action from the hinges. If a body is deformable, as all are to some extent, the deformation depends upon the point of application of the force. However, since many actual objects are deformed only very slightly by the forces acting on them, we shall assume for the present that all objects considered are perfectly rigid. The point of application of a given force acting on a rigid body may be transferred to any other point on the line of action without altering the effect of the force. Thus a *force applied to a rigid body may be regarded as acting anywhere along its line of action.*

A vector quantity is represented by a letter in boldface type. The same letter in ordinary type represents the magnitude of the quantity. Thus the magnitude of a force F is represented by F.

Fig. 1-4 *The vectors **A**, **B**, and **C** are mathematically equal.*

1–6 VECTOR ADDITION. RESULTANT OF A SET OF FORCES

Two vector quantities are said to be equal if they have the same magnitude and direction. In Fig. 1–4 the vectors **A**, **B**, and **C**, which may represent physical quantities, are all equal, and we may write symbolically

$$A = B = C.$$

Two vector quantities which are equal need not have the same physical effect. For example, as already pointed out, two forces with the same magnitude and direction may have different points of application, and a complete description must include the point of application in addition to the magnitude and direction. Thus vector equality has a rather specialized meaning, and in this text boldface "equals" signs will be used as a reminder of this meaning.

The *vector sum* of two vector quantities is defined as follows. Let **A** and **B** in Fig. 1–5(a) be two given vectors. Draw the vectors as in (b) at any convenient point, with the initial point of **B** at the endpoint of **A**. The vector sum **C** is then defined as the vector from the initial point of **A** to the endpoint of **B**. This relationship may be expressed symbolically as:

$$C = A + B.$$

Clearly, vector addition is not the same operation as addition of ordinary numbers; in this book a

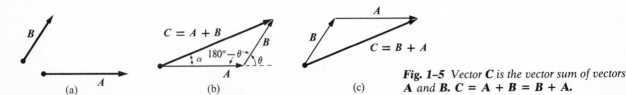

Fig. 1–5 *Vector **C** is the vector sum of vectors **A** and **B**. **C** = **A** + **B** = **B** + **A**.*

Fig. 1–6 *Parallelogram method for obtaining the vector sum of two vectors.*

boldface "plus" sign will be used to denote vector addition.

An alternative procedure is to draw the vectors as in Fig. 1–5(c), with the initial point of **A** at the endpoint of **B**. The vector **C** has the same magnitude and direction as in (b), and hence the two vector sums are mathematically equal. The order in which the vectors are added is therefore immaterial, and vector addition obeys the same *commutative law* as algebraic addition:

$$A + B = B + A.$$

The magnitude and direction of the vector sum **C** can be found from measurements on a carefully drawn diagram. They can also be computed by the methods of trigonometry. Thus, if θ represents the angle between vectors **A** and **B,** as in Fig. 1–5(b), the magnitude of **C** is given by the law of cosines:

$$C^2 = A^2 + B^2 - 2AB \cos(180° - \theta).$$

Since, for any angle θ, $\cos(180° - \theta) = -\cos \theta$, this may also be written

$$C^2 = A^2 + B^2 + 2AB \cos \theta.$$

The angle α between **C** and **A** can be found from the law of sines:

$$\frac{\sin \alpha}{B} = \frac{\sin(180° - \theta)}{C} = \frac{\sin \theta}{C}.$$

Another useful method of finding the sum of two vectors is shown in Fig. 1–6, where vectors **A** and **B** are both drawn from a common point. The vector sum **C** is the diagonal of a parallelogram of which the given vectors form two sides.

Figure 1–7 illustrates a special case in which two vectors are parallel, as in (a), or antiparallel, as in (b). If they are parallel, the magnitude of the vector sum **C** equals the sum of the magnitudes of **A** and **B.** If they are antiparallel, the magnitude of the vector sum equals the *difference* of the magnitudes of **A** and **B.** The vectors in Fig. 1–7 have been displaced slightly sidewise to show them more clearly, but they actually lie along the same geometrical line.

When more than two vectors are to be added, we may first find the vector sum of any two, add this vectorially to the third, and so on. This process is illustrated in Fig. 1–8, which shows in part (a) four vectors **A, B, C,** and **D.** In Fig. 1–8(b), vectors **A** and **B** are first added by the triangle method, giving a

Fig. 1–7 *Vector sum of (a) two parallel vectors, (b) two antiparallel vectors.*

(a)   (b)

Fig. 1–8 *Polygon method of vector addition.*

(a)   (b)   (c)

Fig. 1–9 *A force represented by the vector **R**, equal to the vector sum of **F**₁ and **F**₂, produces the same effect as the forces **F**₁ and **F**₂ acting simultaneously.*

vector sum **E**; vectors **E** and **C** are then added by the same process, to obtain the vector sum **F**; finally, **F** and **D** are added, to obtain the vector sum

$$G = A + B + C + D.$$

Evidently the vectors **E** and **F** need not have been drawn; we need only draw the given vectors in succession, with the tail of each at the head of the one preceding it, and complete the polygon by a vector **G** from the tail of the first to the head of the last vector. The order in which the vectors are drawn makes no difference, as shown in Fig. 1–8(c); the reader may wish to try other possibilities.

Now consider the following physical problem. Two forces, represented by the vectors **F**₁ and **F**₂ in Fig. 1–9, are applied simultaneously at the same point **A** of a body. Is it possible to produce the same effect by applying a *single* force at **A**, and if so, what should be its magnitude and direction? The question can be answered only by experiment; investigation shows that a single force, represented in magnitude, direction, and line of action by the vector sum **R** of the original forces, is in all respects equivalent to them. This single force is called the *resultant* of the

original forces. Hence the mathematical process of *vector addition* of two force vectors corresponds to the physical operation of finding the *resultant of two forces*, simultaneously applied at a given point.

The vector sum of **F**₁ and **F**₂ can be obtained by the construction of Fig. 1–5, where the vectors are drawn tail-to-head at any convenient point. The vector sum then has the same magnitude and direction as the resultant **R**, but not necessarily the same line of action. That is, the line of action of **R** passes through point **A**. This again illustrates that although a mathematical vector may be displaced in any way (retaining its original magnitude and direction), a force acting on a rigid body can be displaced only along its action line.

1–7 COMPONENTS OF A VECTOR

Any two vectors whose vector sum equals a given vector are called the *components* of that vector. In Fig. 1–6, for example, vectors **A** and **B** are components of the vector **C**. Evidently, a given vector has an infinite number of pairs of possible components. If the *directions* of the components are specified, however, the problem of finding the components, or of *resolving* the vector into components, has a unique solution. Thus suppose we are given the vector **A** in Fig. 1–10(a), and we wish to represent it in terms of components in the direction of the lines *Op* and *Oq*, that is, to *resolve* it into components in these directions. From the tip of vector **A** draw the dotted construction lines parallel to *Op* and *Oq*, forming a parallelogram. Vectors **A**ₚ and **A**_q, from *O* to the points of intersection of the construction lines with *Op* and *Oq*, are then the desired components, since they are in the specified directions and the given vector is their vector sum.

(a)

(b)

Fig. 1–10 *(a) Vectors **A**ₚ and **A**_q are the components of **A** in the directions Op and Oq. (b) Vectors **A**ₓ and **A**_y are the rectangular components of **A** in the directions of the x- and y-axes.*

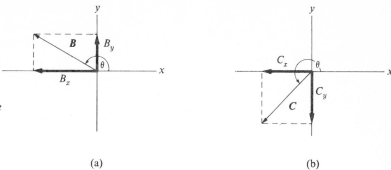

(a)   (b)

Fig. 1–11 *Components of a vector may be positive or negative numbers.*

The special case in which the specified directions are at right angles to each other is of particular importance. In Fig. 1–10(b), lines Ox and Oy are the axes of a rectangular coordinate system. The parallelogram obtained by drawing the dotted construction lines from the tip of vector $\mathbf{A}$ then becomes a rectangle, and the components $\mathbf{A}_x$ and $\mathbf{A}_y$ are called the *rectangular components* of $\mathbf{A}$.

The magnitudes of the rectangular components of a vector are easily computed. If θ is the angle which the vector $\mathbf{A}$ makes with the x-axis, then

$$A_x = A \cos \theta, \qquad A_y = A \sin \theta,$$

where A, A_x, and A_y are the magnitudes of the corresponding vectors.

The components of a vector are themselves *scalar* quantities, and they may be positive or negative. In Fig. 1–11, the component B_x is negative, since its direction is opposite to that of the positive x-axis, and also since the cosine of an angle in the second quadrant is negative; B_y is positive, but both C_x and C_y are negative.

The application of the above concepts to a physical problem is illustrated in Fig. 1–12, where a force $\mathbf{F}$ is exerted on a body at point O. The rectangular components of $\mathbf{F}$ in the directions Ox and Oy are $\mathbf{F}_x$ and $\mathbf{F}_y$, and it is found that simultaneous application of the forces F_x and $\mathbf{F}_y$, as in Fig. 1–12(b), is equivalent in all respects to the effect of the original force. *Any force can be replaced by its rectangular components, acting at the same point.*

As a numerical example, let

$$F = 10 \text{ N}, \qquad \theta = 30°.$$

Then,

$$F_x = F \cos \theta = (10 \text{ N})(0.866) = 8.66 \text{ N},$$

$$F_y = F \sin \theta = (10 \text{ N})(0.500) = 5.00 \text{ N},$$

and the effect of the original 10-N force is equivalent to the simultaneous application of a horizontal force, to the right, of 8.66 N, and a lifting force of 5.00 N.

The axes used to obtain rectangular components of a vector need not be vertical and horizontal. For

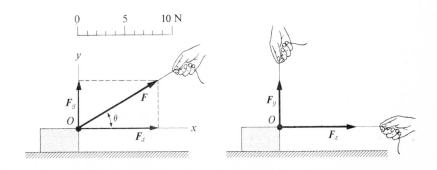

Fig. 1–12 *The inclined force* $\mathbf{F}$ *may be replaced by its rectangular components* $\mathbf{F}_x$ *and* $\mathbf{F}_y$. $F_x = F \cos \theta$, $F_y = F \sin \theta$.

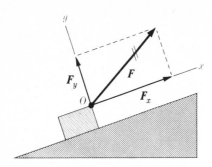

Fig. 1–13 F_x *and* F_y *are the rectangular components of* **F**, *parallel and perpendicular to the sloping surface of the inclined plane.*

example, Fig. 1–13 shows a block being pulled up an inclined plane by a force **F**, represented by its components F_x and F_y, parallel and perpendicular to the sloping surface of the plane. This choice of axes simplifies the analysis of this particular problem.

1–8 RESULTANT BY RECTANGULAR RESOLUTION

Although in principle the polygon method is a satisfactory way to find the resultant of a number of forces, it is awkward for computation because one must, in general, solve a number of oblique triangles. Therefore the usual analytical method of finding the resultant is first to resolve all forces into rectangular components along any convenient pair of axes and then combine these into a single resultant. This makes it possible to work with *right* triangles only.

Figure 1–14(a) shows three concurrent forces F_1, F_2, and F_3, whose resultant we wish to find. Let a pair of rectangular axes be constructed in any arbitrary direction. Simplification results if one axis coincides with one of the forces, which is always possible. In Fig. 1–14(b), the x-axis coincides with F_1. Let us first resolve each of the given forces into x- and y-components. According to the usual conventions of analytic geometry, x-components toward the right are considered positive and those toward the left, negative. Upward y-components are positive and downward y-components are negative.

Force F_1 lies along the x-axis and need not be resolved. The components of F_2 are

$$F_{2x} = F_2 \cos \theta, \qquad F_{2y} = F_2 \sin \theta.$$

Both of these are positive, and F_{2x} has been slightly displaced upward to show it more clearly. The numerical values of the components of F_3 are

$$F_{3x} = F_3 \cos \phi, \qquad F_{3y} = F_3 \sin \phi.$$

Both of these are negative.

We now imagine F_2 and F_3 to be removed and replaced by their rectangular components. To indicate this, the vectors F_2 and F_3 are crossed out lightly. All of the x-components can now be combined into a single force R_x whose magnitude equals the algebraic sum of the magnitudes of the x-components, and all of the y-components can be combined into a single force R_y. The Greek letter $\sum$ is often used to

Fig. 1–14 *Vector* **R**, *the resultant of* F_1, F_2, *and* F_3, *is obtained by the method of rectangular resolution. The magnitudes of the rectangular components of* **R** *are* $R_x = \sum F_x$, $R_y = \sum F_y$.

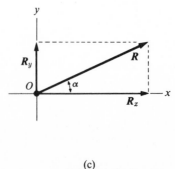

(a)   (b)   (c)

denote a sum, and $\sum F_x$ is read as *sum of the x-components of the forces.* Using the notation, we write

$$R_x = \sum F_x, \qquad R_y = \sum F_y.$$

Finally, these can be combined, as in part (c) of the figure, to form the resultant R whose magnitude is:

$$R = \sqrt{R_x{}^2 + R_y{}^2},$$

since R_x and R_y are perpendicular to each other.

The angle α between R and the x-axis can now be found from any one of its trigonometric functions. For example, $\tan \alpha = R_y/R_x$. As with the individual vectors, the components R_x and R_y may be positive or negative, and the angle α in Fig. 1–14 may be in any of the four quadrants.

Example In Fig. 1–14, let $F_1 = 120$ N, $F_2 = 200$ N, $F_3 = 150$ N, $\theta = 60°$, and $\phi = 45°$. The computations can be arranged systematically as in the following table:

Force	Angle	x-component	y-component
$F_1 = 120$ N	0	+120 N	0
$F_2 = 200$ N	60°	+100 N	+173 N
$F_3 = 150$ N	45°	−106 N	−106 N
		$\sum F_x = +114$ N	$\sum F_y = +67$ N

$$R = \sqrt{(114\text{ N})^2 + (67\text{ N})^2} = 132\text{ N},$$

$$\alpha = \tan^{-1} \frac{67\text{ N}}{114\text{ N}} = \tan^{-1} 0.588 = 30.4°.$$

1–9 VECTOR DIFFERENCE

It is sometimes necessary to subtract one vector from another. The process of subtracting one *algebraic* quantity from another is equivalent to adding the negative of the quantity to be subtracted, that is,

$$a - b = a + (-b).$$

Similarly, the negative $-\boldsymbol{B}$ of a *vector* quantity $\boldsymbol{B}$ is defined as a vector of the same length (or magnitude) but opposite in direction. Then the vector operation $\boldsymbol{A} - \boldsymbol{B}$ is defined to be the vector sum of $\boldsymbol{A}$ and $(-\boldsymbol{B})$. That is,

$$\boldsymbol{A} - \boldsymbol{B} = \boldsymbol{A} + (-\boldsymbol{B}).$$

The boldface $\boldsymbol{+}$, $\boldsymbol{-}$, and $\boldsymbol{=}$ signs remind us of the vector nature of these operations.

Vector subtraction is illustrated in Fig. 1–15. The given vectors are shown in (a). In (b), the vector sum of $\boldsymbol{A}$ and $-\boldsymbol{B}$, or the vector difference $\boldsymbol{A} - \boldsymbol{B}$, is found by the parallelogram method. Part (c) shows a second method; vectors $\boldsymbol{A}$ and $\boldsymbol{B}$ are drawn from a common origin, and the vector difference $\boldsymbol{A} - \boldsymbol{B}$ is the vector from the tip of $\boldsymbol{B}$ to the tip of $\boldsymbol{A}$. The vector difference $\boldsymbol{A} - \boldsymbol{B}$ is therefore seen to be the vector that must be added to $\boldsymbol{B}$ in order to give $\boldsymbol{A}$, since

$$\boldsymbol{B} + (\boldsymbol{A} - \boldsymbol{B}) = \boldsymbol{A}.$$

Vector differences may also be found by the method of rectangular resolution. Both vectors are resolved into x- and y-components. The difference between the x-components is the x-component of the desired vector difference, and the difference between the y-components is the y-component of the vector difference.

Vector subtraction is not often used when dealing with forces, but we shall use it frequently in connection with velocities and accelerations.

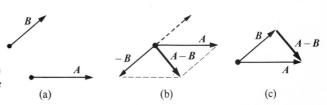

Fig. 1–15 *The vector difference* $\boldsymbol{A} - \boldsymbol{B}$ *is found in part (b) by the parallelogram method and in part (c) by the triangle method.*

(a)   (b)   (c)

1–10 A WORD ABOUT PROBLEMS

To obtain a benefit commensurate with the time expended, problem-solving should be considered as much more than merely substituting numbers for the symbols in a formula, or fitting together the pieces of a jigsaw puzzle. To merely thumb through the book until you find a formula that seems to fit, or a worked-out example that resembles the problem, is a waste of time and effort. One purpose which problems can serve is to enable you to find out for yourself whether or not you understand the assigned material, after having listened to a lecture and studied the text. Do your studying *before* you tackle the problems, instead of beginning with the problems and not referring to your book or lecture notes until you find yourself stymied. No problem assignment of reasonable length can hope to cover *every* important point, and you will miss a great deal if you read only enough to enable you to "do" your problems.

Every problem involves one or more general physical laws or definitions. After reading the statement of a problem, ask yourself what these laws or definitions are and be sure that you know them. This means that you should be able to state them to yourself clearly and explicitly, and not be satisfied with the comforting feeling that you "understand" Newton's second law although you can't quite put it in so many words.

Nearly every quantity of physical interest is expressed in terms of a unit of that quantity (some quantities are pure numbers). No answer is complete unless the proper units are given.

Unless otherwise stated, the numerical data given in the problems are to be assumed correct to three significant figures (for example, 2 m implies 2.00 m). Numerical answers should therefore be rounded off to three significant figures. That is, if long division leads to the result $6.575938 \cdots$ m s^{-1}, give the answer as 6.58 m s^{-1}. (This is automatically taken care of if you use a slide rule.) Answers are provided to about half the problems, and are rounded off to two or three significant figures.

One of the best ways of making sure you understand the principles covered by a particular problem is to work it backwards. That is, if the problem gives x and asks you to compute y, then make up another problem in which y is given and x is to be found. Another helpful procedure is to ask yourself how the result would be altered if the given conditions had been somewhat different. Suppose the friction force had been 10 N instead of 5 N? Suppose the slope of the plane had been 60° instead of 30°?

Helpful advice on the techniques of studying and of solving problems will be found in *How to Study Physics* by Chapman and *How to Study, How to Solve* by Dadourian (published by Addison-Wesley).

PROBLEMS

1–1 Find graphically the magnitude and direction of the resultant of the three forces in Fig. 1–16. Use the polygon method.

Fig. 1–16

1–2 Two men and a boy want to push a crate in the direction marked x in Fig. 1–17. The two men push with forces F_1 and F_2, whose magnitudes and directions are indicated in the figure. Find the magnitude and direction of the smallest force which the boy should exert.

Fig. 1–17

1–3 Find graphically the resultant of two 10-N forces applied at the same point:

a) when the angle between the forces is 30°;

b) when the angle between them is 130°. Use any convenient scale.

1–4 Two men pull horizontally on ropes attached to a post, the angle between the ropes being 45°. If man A exerts a force of 300 N and man B a force of 200 N, find the magnitude of the resultant force and the angle it makes with A's pull. Solve:

a) graphically, by the parallelogram method, and

b) graphically, by the triangle method. Let 1 in. = 100 N in (a) and (b).

1–5 Find graphically the vector sum $A + B$ and the vector difference $A - B$ in Fig. 1–18.

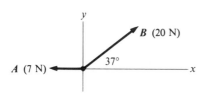

Fig. 1–18

1–6 Vector A is 2 in. long and is 60° above the x-axis in the first quadrant. Vector B is 2 in. long and is 60° below the x-axis in the fourth quadrant. Find graphically (a) the vector sum $A + B$, and (b) the vector differences $A - B$ and $B - A$.

1–7 a) Find graphically the horizontal and vertical components of a 40-N force the direction of which is 50° above the horizontal to the right. Let $\frac{1}{16}$ in. = 1 N. b) Check your results by calculating the components.

1–8 A box is pushed along the floor as in Fig. 1–2 by a force of 40 N making an angle of 30° with the horizontal. Using a scale of 1 in. = 10 N, find the horizontal and vertical components of the force by the graphical method. Check your results by calculating the components.

1–9 A block is dragged up an inclined plane of slope angle 20° by a force F making an angle of 30° with the plane. (a) How large a force F is necessary in order that the component F_x parallel to the plane shall be 16 N? (b) How large will the component F_y then be? Solve graphically, letting 1 in. = 8 N.

1–10 The three forces shown in Fig. 1–16 act on a body located at the origin. (a) Find the x- and y-components of each of the three forces. (b) Use the method of rectangular resolution to find the components, magnitude, and direction of the resultant of the forces. (c) Find the magnitude and direction of a fourth force which must be added to make the resultant force zero. Indicate the fourth force by a diagram.

1–11 Use the method of rectangular resolution to find the resultant of the following set of forces and the angle it makes with the horizontal: 200 N, along the x-axis toward the right; 300 N, 60° above the x-axis to the right; 100 N, 45° above the x-axis to the left; 200 N, vertically down.

1–12 A vector A of length 10 units makes an angle of 30° with a vector B of length 6 units. Find the magnitude of the vector difference $A - B$ and the angle it makes with vector A: (a) by the parallelogram method; (b) by the triangle method; (c) by the method of rectangular resolution.

1–13 Two forces, F_1 and F_2, act at a point. The magnitude of F_1 is 8 N and its direction is 60° above the x-axis in the first quadrant. The magnitude of F_2 is 5 N and its direction is 53° below the x-axis in the fourth quadrant. (a) What are the horizontal and vertical components of the resultant force? (b) What is the magnitude of the resultant? (c) What is the magnitude of the vector difference $F_1 - F_2$?

1–14 Two forces, F_1 and F_2, act upon a body in such a manner that the resultant force R has a magnitude equal to that of F_1 and makes an angle of 90° with F_1. Let $F_1 = R = 10$ N. Find the magnitude of the second force, and its direction (relative to F_1).

1–15 The resultant of four forces is 1000 N in the direction 30° west of north. Three of the forces are 400 N, 60° north of east; 200 N, south; and 400 N, 53° west of south. Find the rectangular components of the fourth force.

1–16 Vector A has components $A_x = 2$ cm, $A_y = 3$ cm, and vector B has components $B_x = 4$ cm, $B_y = -2$ cm. Find (a) the components of the vector sum $A + B$; (b) the magnitude and direction of $A + B$; (c) the components of the vector difference $A - B$; (d) the magnitude and direction of $A - B$.

1–17 A car drives 5 mi east, then 4 mi south, then 2 mi west. Find the magnitude and direction of the resultant displacement.

1–18 A sailboat sails 2 km east, then 4 km southeast, then an additional distance in an unknown direction. Its final position is 5 km directly east of the starting point. Find the magnitude and direction of the third leg of the journey.

1–19 Vector M, of magnitude 5 cm, is at 36.9° counterclockwise from the $+x$-axis. It is added to vector N, and the resultant is a vector of magnitude 5 cm, at 53.1° counterclockwise from the $+x$-axis. Find (a) the components of N; (b) the magnitude and direction of N.

1–20 The velocity of an airplane relative to the surface of the earth, v_{PE}, equals the vector sum of its velocity relative to the air, v_{PA}, and the velocity of the air relative to the earth, v_{AE}. Thus

$$v_{PE} = v_{PA} + v_{AE}.$$

a) Find graphically the magnitude and direction of the velocity v_{PE} if the velocity v_{PA} is 100 mi hr^{-1} due north, and if the wind velocity v_{AE} is 40 mi hr^{-1} from east to west. Let 1 cm = 20 mi hr^{-1}.

b) Find the direction of the velocity v_{PA} if its magnitude is 100 mi hr^{-1}, if the direction of v_{PE} is due north, and if v_{AE} is 40 mi hr^{-1} from east to west. What is then the magnitude of v_{PE}?

Chapter 2

Equilibrium of a Particle

2-1 INTRODUCTION

The science of mechanics is based on three natural laws which were clearly stated for the first time by Sir Isaac Newton (1642–1727) and were published in 1686 in his *Philosophiae Naturalis Principia Mathematica* (The Mathematical Principles of Natural Science). It should not be inferred, however, that the science of mechanics began with Newton. Many men had preceded him in his field, the most outstanding being Galileo Galilei (1564–1642), who in his studies of accelerated motion had laid much of the groundwork for Newton's three laws.

In this chapter, we shall make use of only two of Newton's laws, the first and the third. Newton's second law will be discussed in Chapter 5.

2-2 EQUILIBRIUM. NEWTON'S FIRST LAW

One effect of a force is to alter the dimensions or shape of a body on which the force acts; another is to alter the state of motion of the body.

The motion of a rigid body can be considered as made up of its motion as a whole, or its *translational* motion, together with any *rotational* motion the body may have. In the most general case, a single force acting on a body produces a change in both its translational and rotational motion. However, when several forces act on a body simultaneously, their effects can compensate one another, with the result that there is no change in either the translational or rotational motion. When this is the case, the body is said to be in *equilibrium*. This means (1) that the body as a whole either remains at rest or moves in a straight line with constant speed, and (2) that the body is either not rotating at all or is rotating at a constant rate.

Let us consider some (idealized) experiments from which the laws of equilibrium can be deduced. Figure 2–1 represents a flat, rigid object of arbitrary shape on a level surface having negligible friction. If a single force F_1 acts on the body, as in Fig. 2–1(a), and if the body is originally at rest, it at once starts to move and to rotate clockwise. If originally in motion, the effect of the force is to change the translational motion of the body in magnitude or direction (or both) and to increase or decrease its rate of rotation. In either case, the body does not remain in equilibrium.

Equilibrium can be maintained, however, by the application of a second force F_2 as in Fig. 2–1(b), provided that F_2 is *equal in magnitude* to F_1, is *opposite in direction* to F_1, and has the same *line of action* as F_1. The resultant of F_1 and F_2 is then zero. If the lines of action of the two forces are not the same, as in Fig. 2–1(c), the body will be in transla-

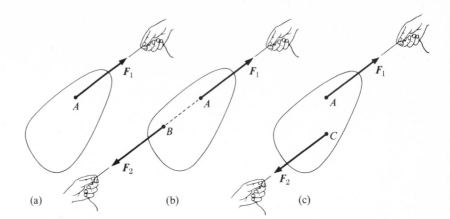

Fig. 2–1 *A rigid body acted on by two forces is in equilibrium if the forces are equal in magnitude, opposite in direction, and have the same line of action, as in part (b).*

tional but not in rotational equilibrium. (Such a combination of two forces is called a *couple*; see Section 3–4.)

If the forces F_1 and F_2 are equal in magnitude and opposite in direction, then one equals the negative of the other. That is,

$$F_2 = -F_1.$$

Then if R represents the resultant of F_1 and F_2,

$$R = F_1 + F_2 = F_1 - F_1 = 0.$$

For brevity, we shall often speak of two forces as simply being "equal and opposite," meaning that their magnitudes are equal and that one is the negative of the other.

In Fig. 2–2(a), a body in equilibrium is acted on by three nonparallel coplanar forces, F_1, F_2, and F_3. Any force applied to a rigid body may be regarded as acting anywhere along its line of action. Therefore, let any two of the force vectors, say F_1 and F_2,

be transferred to the point of intersection of their lines of action and their resultant R obtained, as in Fig. 2–2(b). The forces are now reduced to two, R and F_3; for equilibrium these must (1) be equal in magnitude, (2) be opposite in direction, and (3) have the same line of action. It follows from the first two conditions that the resultant of the three forces is zero. The third condition can be fulfilled only if the line of action of F_3 passes through the point of intersection of the lines of action of F_1 and F_2. When the lines of action of several forces pass through a common point, the forces are said to be *concurrent*; the body in Fig. 2–2 can be in equilibrium only when the three forces are concurrent.

Fig. 2–2 *When a body acted on by three nonparallel coplanar forces is in equilibrium, the forces are concurrent and the resultant of any two is equal in absolute magnitude and opposite in direction (or sign) to the third.*

(a)

(b)

The construction in Fig. 2–2 provides a satisfactory graphical method for the solution of problems in equilibrium. For an analytical solution, it is usually simpler to deal with the rectangular components of the forces. We have shown that the magnitudes of the rectangular components of the resultant **R** of any set of coplanar forces are

$$R_x = \sum F_x, \qquad R_y = \sum F_y.$$

When a body is in equilibrium, the resultant of all the forces acting on it is zero. Both rectangular components are then zero and hence, for a body in equilibrium,

$$\boxed{\boldsymbol{R} = \boldsymbol{0}, \qquad \text{or} \qquad \sum F_x = 0, \qquad \sum F_y = 0.}$$

These equations are called the first condition of equilibrium.

The *second condition for equilibrium*, to be developed more fully in Chapter 3, is a mathematical statement of the fact that the forces must have no tendency to rotate the body. In particular, when only two forces act they must have the same line of action, and when three forces act they must be concurrent, in order for the body to be in equilibrium. There is an exceptional case, three forces with parallel lines of action, which will be discussed in Chapter 3.

The first condition of equilibrium ensures that a body shall be in *translational* equilibrium; the second, that it be in *rotational* equilibrium. The statement that a body is in complete equilibrium when both conditions are satisfied is the essence of *Newton's first law of motion*. Newton did not state his first law in exactly these words. His original statement (translated from the Latin in which the *Principia* was written) reads:

"*Every body continues in its state of rest, or of uniform motion in a straight line, unless it is compelled to change that state by forces impressed on it.*"

Although rotational motion was not explicitly mentioned by Newton, it is clear from his work that he fully understood the conditions that the forces must satisfy when the rotation is zero or is constant.

2–3 DISCUSSION OF NEWTON'S FIRST LAW OF MOTION

Newton's first law of motion is not as self-evident as it may seem. In the first place, this law asserts that in the absence of any applied force a body either remains at rest or moves uniformly in a straight line. It follows that *once a body has been set in motion it is no longer necessary to exert a force on it to maintain it in motion.*

This assertion appears to be contradicted by everyday experience. Suppose we exert a force with the hand to move a book along a level table top. After the book has left our hand, and we are no longer exerting a force on it, it does *not* continue to move indefinitely, but slows down and eventually comes to rest. If we wish to keep it moving uniformly we must continue to exert some forward force on it. But this force is required only because a frictional force is exerted on the sliding book by the table top, in a direction opposite to the motion of the book. The smoother the surfaces of the book and table, the smaller the frictional force and the smaller the force we must exert to keep the book moving. The first law asserts that if the frictional force could be eliminated completely, no forward force at all would be required to keep the book moving, once it had been set in motion. More than that, however, the law implies that if the *resultant* force on the book is zero, as it is when the frictional force is balanced by an equal forward force, the book also continues to move uniformly. In other words, *zero resultant force is equivalent to no force at all.*

Furthermore, the first law defines by implication what is known as an *inertial reference system*. To understand what is meant by this term we must recognize that the motion of a given body can be specified only with respect to, or relative to, some other body. Its motion relative to one body may be very different from that relative to another. Thus a passenger in an aircraft which is making its take-off run may continue at rest relative to the aircraft but be moving faster and faster relative to the earth.

A *reference system* means a set of coordinate axes attached to (or moving with) some specified body or bodies. Suppose we consider a reference system attached to the aircraft referred to above. Everyone knows that during the take-off run, while the aircraft is going faster and faster, a passenger feels the back of his seat pushing him forward, although he remains at rest relative to a reference system attached to the aircraft. Newton's first law does not therefore correctly describe the situation; a forward force *does* act on the passenger, but nevertheless (relative to the aircraft) he remains at rest.

Suppose, on the other hand, that the passenger is standing in the aisle on roller skates. He then starts to move *backward*, relative to the aircraft, when the take-off run begins, even though no backward force acts on him. Newton's first law again does not correctly describe the facts.

We can now define an inertial reference system as one relative to which a body *does* remain at rest or move uniformly in a straight line when no force (or no resultant force) acts on it. That is, *an inertial reference system is one in which Newton's first law correctly describes the motion of a body not acted on by any force.*

An aircraft gaining speed during take-off is evidently *not* an inertial system. For many purposes a reference system attached to the earth can be considered an inertial system, although it is not precisely so, due to effects related to the earth's rotation and other motions. But if a frame of reference *is* inertial, then a second frame moving uniformly (i.e., with constant speed in a straight line, without rotation) relative to it is *also* inertial, since any body in uniform motion as seen by an observer in the first frame will also appear to an observer in the second frame to be in uniform motion. The speed and direction appear different to the two observers, but both observe that Newton's first law is obeyed.

Thus there is no single, unique inertial frame of reference, there are infinitely many; but the motion of any one relative to any other is uniform in the above sense. It follows that the concepts of "absolute rest" and "absolute motion" have no physical meaning. Experiments show that a frame of reference at rest relative to the so-called *fixed stars* is inertial, within the limits of experimental error, and so every frame in uniform motion relative to this one is also

inertial. An aircraft in steady flight over the earth is just as suitable an inertial frame of reference as the earth itself.

Finally, Newton's first law contains a qualitative definition of the concept of *force*, or at least of one aspect of the force concept, as "that which changes the state of motion of a body." (Of course, forces also produce other effects, such as changing the length of a coil spring.) When a body at rest relative to the earth is observed to start moving, or when a moving body speeds up, slows down, or changes its direction, we can conclude that force is acting on it. This effect of a force can be used to define the ratio of two forces and to define a *unit* of force, and we shall show in Chapter 5 how this is done.

2–4 NEWTON'S THIRD LAW OF MOTION

Any given force is but one aspect of a mutual interaction between *two* bodies. It is found that *whenever one body exerts a force on another, the second always exerts on the first a force which is equal in magnitude, is opposite in direction, and has the same line of action.*[*] A single isolated force is therefore an impossibility.

The two forces involved in every interaction between two bodies are often called an "action" and a "reaction," but this does not imply any difference in their nature, or that one force is the "cause" and the other its "effect." *Either* force may be considered the "action," and the other the "reaction" to it.

This property of forces was stated by Newton in his *third law of motion.* In his words,

"*To every action there is always opposed an equal reaction: or, the mutual actions of two bodies upon each other are always equal, and directed to contrary parts.*"

As an example, suppose that a man pulls on one end of a rope attached to a block, as in Fig. 2–3. The weight of the block and the force exerted on it by the surface are not shown. The block may or may not be in equilibrium. The resulting action–reaction pairs of

---

[*]There are instances, such as the motion of electrically charged particles, where this law is not obeyed, but it is correct for all the macroscopic forces we shall encounter in mechanics.

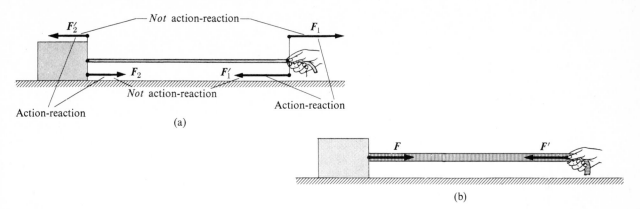

Fig. 2–3 *(a) Forces F_1 and F_1' form one action–reaction pair, and forces F_2 and F_2' another. F_1 is always equal to F_1', F_2 is always equal to F_2'. F_1 and F_2' are equal only if the rope is in equilibrium, and force F_2' is not the reaction to F_1. (Actually all forces lie along the rope.) (b) If the rope is in equilibrium, it can be considered to transmit a force from the man to the block, and vice versa.*

forces are indicated in the figure. (The lines of action of all the forces lie along the rope; the force vectors have been offset from this line to show them more clearly.) Vector F_1 represents the force exerted on the rope by the man. Its reaction is the equal and opposite force F_1' exerted on the man by the rope. Vector F_2 represents the force exerted on the block by the rope. The reaction to it is the equal and opposite force F_2', exerted on the rope by the block:

$$F_1' = -F_1, \qquad F_2' = -F_2.$$

It is very important to realize that the forces F_1 and F_2', although they are opposite in direction and have the same line of action, do *not* constitute an action–reaction pair. For one thing, both of these forces act on the *same* body (the rope) while an action and its reaction necessarily act on *different* bodies. Furthermore, the forces F_1 and F_2' are not necessarily equal in magnitude. If the block and rope are moving to the right with increasing speed, the rope is not in equilibrium and F_1 is greater than F_2'. Only in the special case when the rope remains at rest or moves with constant speed are the forces F_1 and F_2'

equal in magnitude, but this is an example of Newton's *first* law, not his *third*. Even when the speed of the rope is changing, however, the action–reaction forces F_1 and F_1' are equal in magnitude to *each other*, and the action–reaction forces F_2 and F_2' are equal in magnitude to *each other*, although then F_1 is not equal in magnitude to F_2'.

In the special case when the rope is in equilibrium, and when no forces act on it except those at its ends, F_2' equals F_1 by Newton's *first* law. Since F_2 *always* equals F_2', by Newton's *third* law, then in this special case F_2 also equals F_1 and the force exerted on the block by the rope is equal to the force exerted on the rope by the man. The rope can therefore be considered to "transmit" to the block, without change, the force exerted on it by the man. This point of view is often useful, but it is important to remember that it applies only under the restricted conditions as stated.

If we adopt this point of view, the rope itself need not be considered and we have the simpler force diagram of Fig. 2–3(b), where the man is considered to exert a force F directly on the block.

The reaction is the force F' exerted by the block on the man. The only effect of the rope is to transmit these forces from one body to the other.

A body, like the rope in Fig. 2–3, which is subjected to pulls at its ends is said to be in *tension*. The tension at any point equals the force exerted at that point. Thus in Fig. 2–3(a) the tension at the righthand end of the rope equals the magnitude of F_1 (or of F_1') and the tension at the lefthand end equals the magnitude of F_2 (or of F_2'). If the rope is in equilibrium and if no forces act except at its ends, as in Fig. 2–3(b), the tension is the same at both ends. If, for example, in Fig. 2–3(b) the magnitudes of F and F' are each 50 N, the tension in the rope is 50 N (*not* 100 N).

2–5 EQUILIBRIUM OF A PARTICLE

The materials found in nature and the processes which occur are rarely simple. As a first step in dealing with a problem of nature it is often necessary to *idealize* the situation and to make *simplifying assumptions* concerning the processes. Suppose, for example, a baseball is thrown into the air and it is desired to calculate where it lands and with what velocity. The first step in treating this problem consists in idealizing the baseball by ignoring the details of its surface and all departures from sphericity during its motion. In other words, we replace the baseball by an ideal object, i.e., a rigid smooth sphere. The next step is to ignore the rotation of the baseball and the forces brought into play by the air dragged around by the rotating ball. A further step is to regard as negligible the buoyancy and the resistance of the air. We are then left with a problem quite different from the original one; in fact we may be accused of having robbed the problem of almost all of its reality. This is true, but what remains approximates the original problem at low velocities, and has the virtue that it is amenable to simple mathematical treatment, whereas the original problem at high velocities would require the most advanced methods.

In general, the forces acting on a rigid body do not all pass through one point (they are *nonconcur-*

rent) and, as a result, the rigid body undergoes rotational as well as translational motion. There are, however, many situations of great interest where the rotation of a body is of only minor consequence and is not pertinent in the solution of the problem. The planetary motion, for example, of the earth about the sun taking place under the action of the gravitational force between the two bodies may be studied alone without regard to the earth's rotation. *A body whose rotation is ignored as irrelevant is called a particle. A particle may be so small that it is an approximation to a point, or it may be of any size, provided that the action lines of all the forces acting on it intersect in one point.* Thus a particle serves as an *idealized model* for a body whose rotational motion is of no consequence in a given situation.

In the remainder of this chapter, we shall limit ourselves to examples and problems involving the equilibrium of particles only. It is surprising how many situations of interest and importance in engineering, in the life and earth sciences, and in everyday life involve the equilibrium of particles. In many of these, it is important to know how to calculate one or two of the forces acting on a particle when the others are given. In order to do this, it is best to adhere scrupulously to the following rules.

1. Make a simple line sketch of the apparatus or structure, showing dimensions and angles.

2. Choose some object (a knot in a rope, for example) as the particle in equilibrium. Draw a separate diagram of this object and show by arrows (use a colored pencil) *all* of the forces exerted *on* it by other bodies. This is called the *force diagram* or *free-body diagram*. When a system is composed of several particles, it may be necessary to construct a separate free-body diagram for each one. Do *not* show, in the free-body diagram of a chosen particle, any of the forces exerted *by* it. These forces (which are the reactions to the forces acting *on* the chosen particle) all act *on other bodies* and appear in the free-body diagrams of those bodies.

3. Construct a set of rectangular axes and resolve any inclined forces into rectangular components. Cross out lightly those forces which have been resolved.

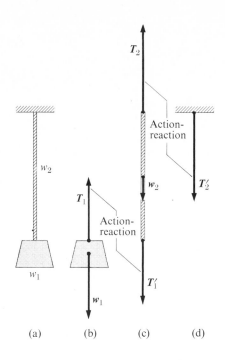

Fig. 2–4 *(a) Block hanging at rest from vertical cord. (b) The block is isolated and all forces acting on it are shown. (c) Forces on the cord. (d) Downward force on the ceiling. Lines connect action–reaction pairs.*

4. Set the algebraic sum of all *x*-forces (or force components) equal to zero, and the algebraic sum of all *y*-forces (or components) equal to zero. This provides two independent equations which can be solved simultaneously for two unknown quantities (which may be forces, angles, distances, etc.)

A force which will be encountered in many problems is the *weight* of a body, that is, the force of gravitational attraction exerted on the body by the earth. We shall show in the next chapter that the line of action of this force always passes through a point called the *center of gravity* of the body.

The force of gravitational attraction exerted on a body by the earth is but one aspect of a mutual interaction between the earth and the body. At the same time, the body attracts the earth with a force equal in magnitude and opposite in direction to the force the earth exerts on the body. Thus if a body weighs 10 N, the earth pulls down on it with a force of 10 N, and the body pulls *up* on the earth with a force of 10 N, another example of an action–reaction pair.

Example 1 To begin with a simple example, consider the body in Fig. 2–4(a), hanging at rest from the ceiling by a vertical cord. Part (b) of the figure is the free-body diagram for the body. The forces on it are its weight w_1 and the upward force T_1 exerted on it by the cord. If we take the *x*-axis horizontal and the *y*-axis vertical, there are no *x*-components of force, and the *y*-components are the forces w_1 and T_1. Then from the condition that $\sum F_y = 0$, we have

$$\sum F_y = T_1 - w_1 = 0,$$

$$T_1 = w_1. \quad \text{(First Law)}$$

In order that both forces have the same line of action, the center of gravity of the body must lie vertically below the point of attachment of the cord.

Let us emphasize again that the forces w_1 and T_1 are *not* an action–reaction pair, although they are equal in magnitude, opposite in direction, and have the same line of action. The weight w_1 is a force of attraction exerted on the body by the earth. Its

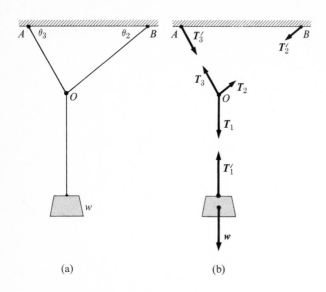

Fig. 2–5 (a) Block hanging in equilibrium. (b) Forces acting on the block, on the knot, and on the ceiling.

reaction is an equal and opposite force of attraction exerted on the earth by the body. This reaction is one of the set of forces acting *on the earth*, and therefore it does not appear in the free-body diagram of the suspended block.

The reaction to the force T_1 is an equal downward force, T_1',

$$T_1 = T_1'. \qquad \text{(Third Law)}$$

The force T_1' is shown in part (c), which is the free-body diagram of the *cord*. The other forces on the cord are its own weight w_2 and the upward force T_2 exerted on its upper end by the ceiling. Since the cord is also in equilibrium,

$$\sum F_y = T_2 - w_2 - T_1' = 0$$

$$T_2 = w_2 + T_1'. \qquad \text{(First Law)}$$

The reaction to T_2 is the downward force T_2' in part (d), exerted on the ceiling by the cord:

$$T_2 = T_2'. \qquad \text{(Third Law)}$$

As a numerical example, let the body weigh 20 N and the cord weigh 1 N. Then

$$T_1 = w_1 = 20 \text{ N},$$

$$T_1' = T_1 = 20 \text{ N},$$

$$T_2 = w_2 + T_1' = 1 \text{ N} + 20 \text{ N} = 21 \text{ N},$$

$$T_2' = T_2 = 21 \text{ N}.$$

If the weight of the cord were so small as to be negligible, then in effect no forces would act on it except at its ends. The forces T_2 and T_2' would then each equal 20 N and, as explained earlier, the cord could be considered to transmit a 20-N force from one end to the other without change. We could then consider the upward pull of the cord on the block as an "action" and the downward pull on the ceiling as its "reaction." The tension in the cord would then be 20 N.

Example 2 In Fig. 2–5(a), a block of weight w hangs from a cord which is knotted at O to two other cords fastened to the ceiling. We wish to find the tensions in these three cords. The weights of the cords are negligible.

In order to use the conditions of equilibrium to compute an unknown force, we must consider some body which is in equilibrium and on which the desired forces act. The hanging block is one such body and, as shown in the preceding example, the tension in the vertical cord supporting the block is equal to the weight of the block. The inclined cords do not exert forces on the block, but they do act on the knot at O. Hence we consider the *knot* as a particle in equilibrium whose own weight is negligible.

The free-body diagrams for the block and the knot are shown in Fig. 2–5(b), where T_1, T_2, and T_3 represent the forces exerted *on the knot* by the three cords and T_1', T_2', and T_3' are the reactions to these forces.

Consider first the hanging block. Since it is in equilibrium,

$$T_1' = w. \qquad \text{(First Law)}$$

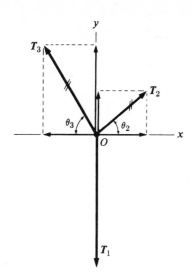

Fig. 2–6 *Forces on the knot in Fig. 2–5, resolved in x- and y-components.*

Since T_1 and T_1' form an action–reaction pair,

$$T_1' = T_1. \quad \text{(Third Law)}$$

Hence,

$$T_1 = w.$$

To find the forces T_2 and T_3 we resolve these forces (see Fig. 2–6) into rectangular components, noting that components upward or to the right (the positive axis directions) are positive, while those downward or to the left (negative axis directions) are negative.

$$\sum F_x = T_2 \cos \theta_2 - T_3 \cos \theta_3 = 0,$$
$$\sum F_y = T_2 \sin \theta_2 + T_3 \sin \theta_3 - T_1 = 0.$$

Since w is known, T_1 is also known, and these are two simultaneous equations which may be solved for T_2 and T_3.

As a numerical example, let

$$w = 50 \text{ N}, \qquad \theta_2 = 30°, \qquad \theta_3 = 60°.$$

Then $T_1 = 50$ N, and the two preceding equations become

$$T_2\left(\frac{\sqrt{3}}{2}\right) - T_3\left(\frac{1}{2}\right) = 0,$$

$$T_2\left(\frac{1}{2}\right) + T_3\left(\frac{\sqrt{3}}{2}\right) = 50 \text{ N}.$$

Simultaneous solution yields the results

$$T_2 = 25 \text{ N}, \qquad T_3 = 43.3 \text{ N}.$$

The reader may wish to show that the general solutions are given by

$$T_2 = \frac{T_1 \cos \theta_3}{\sin (\theta_2 + \theta_3)}, \qquad T_3 = \frac{T_1 \cos \theta_2}{\sin (\theta_2 + \theta_3)}.$$

These forms have the advantage that it is easy to see how the results are affected by changing the angles. When either θ_2 or θ_3 decreases, both tensions increase, as might be expected intuitively. Working out the general results has the added advantage of decreasing the amount of numerical computation needed to obtain values for T_2 and T_3.

Finally, we know from Newton's *third law* that the inclined cords exert on the ceiling the forces T_2' and T_3', respectively.

Example 3   Figure 2–7(a) shows a strut AB, pivoted at end A, attached to a wall by a cable, and carrying a load w at end B. The weights of the strut and of the cable are negligible.

Figure 2–7(b) shows the forces acting on the strut: T_1 is the force exerted by the vertical cable, T_2 is the force exerted by the inclined cable, and C is the force exerted by the pivot. Force T_1 is known both in magnitude and in direction, force T_2 is known in direction only, and neither the magnitude nor the direction of C is known. However, the forces T_1 and T_2 intersect at the outer end of the strut, and since the strut is in equilibrium under three forces, the line of action of force C must also pass through the outer end of the strut. In other words, the direction of force C is along the line of the strut. (This is not true in general. If the weight of the strut is included, or if additional forces are present, the forces are not all

Fig. 2–7

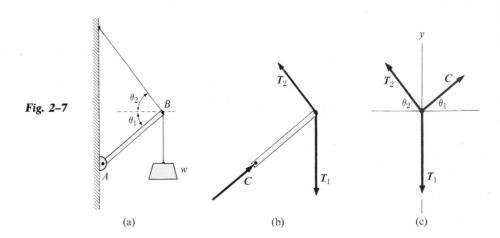

(a)   (b)   (c)

concurrent, and the force on the strut at point A is in general *not* along the line of the strut.)

Hence the resultant of T_1 and T_2 is also along this line and the strut, in effect, is acted on by forces at its ends, directed toward each other along the line of the strut. The effect of these forces is to compress the strut, and it is said to be in *compression*.

If the forces acting on a strut are not all applied at its ends, the direction of the resultant force at the ends is not along the line of the strut. This is illustrated in the next example.

In Fig. 2–7(c), the force C has been transferred along its line of action to the point of intersection B of the three forces. The force diagram for a particle at B is exactly like that of Fig. 2–6, and the problem

is solved in the same way. The conditions of equilibrium provide two independent equations among the five quantities T_1, T_2, C, θ_1, and θ_2. Hence if any three are known, the other two may be calculated.

Example 4   In Fig. 2–8 a ladder in equilibrium leans against a vertical frictionless wall. The forces on the ladder are (1) its weight w, (2) the force F_1 exerted on the ladder by the vertical wall, which is perpendicular to the wall if there is no friction, and (3) the force

Fig. 2–8 *Forces on a ladder leaning against a vertical frictionless wall.*

(a)   (b)

Fig. 2–9 *Forces on a block in equilibrium on a frictionless inclined plane.*

F_2 exerted by the ground on the base of the ladder. The force **w** is known in magnitude and in direction, the force F_1 is known in direction only, and the force F_2 is unknown in both magnitude and direction. As in the preceding example, the ladder is in equilibrium under three forces, *which must be concurrent.* Since the lines of action of F_1 and **w** are known, their point of intersection (point O) can be located. The line of action of F_2 must then pass through this point also. Note that neither the direction of F_1 nor that of F_2 lies along the line of the ladder. In part (b) the forces have been transferred to the point of intersection of their lines of action and, applying the equations for the equilibrium of a particle at this point, we get

$$\sum F_x = F_2 \cos \theta - F_1 = 0, \qquad (2\text{–}1)$$

$$\sum F_y = F_2 \sin \theta - w = 0. \qquad (2\text{–}2)$$

As a numerical example, suppose the ladder weighs 400 N, is 10 m long, has its center of gravity at its center, and makes an angle $\phi = 53°$ with the ground. We wish to find the angle θ and the forces F_1 and F_2. To calculate θ, we first find the lengths of AB and BO. From the right triangle ABC, we have

$$\overline{AB} = \overline{AC}\cos \phi = (5 \text{ m})(0.60) = 3.0 \text{ m},$$

and from the right triangle AED,

$$\overline{DE} = \overline{AD}\sin \phi = (10 \text{ m})(0.80) = 8 \text{ m}.$$

Then, from the right triangle AOB, since $\overline{OB} = \overline{DE}$,

we have

$$\tan \theta = \frac{\overline{OB}}{\overline{AB}} = \frac{8 \text{ m}}{3 \text{ m}} = 2.67.$$

Then

$$\theta = 69.5°, \qquad \sin \theta = 0.937, \qquad \cos \theta = 0.350.$$

$$F_2 = \frac{w}{\sin \theta} = \frac{400 \text{ N}}{0.937} = 427 \text{ N},$$

and from Eq. (2–1),

$$F_1 = F_2 \cos \theta = (427 \text{ N})(0.350) = 150 \text{ N}.$$

The ladder presses against the wall and the ground with forces which are equal and opposite to F_1 and F_2, respectively.

In the next chapter we shall explain another method of solving this problem, using the concept of the *moment* of a force.

Example 5   In Fig. 2–9, block A of weight w_1 rests on a frictionless inclined plane of slope angle θ. The center of gravity of the block is at its center. A flexible cord is attached to the center of the right face of the block, passes over a frictionless pulley, and is attached to a second block B of weight w_2. The weight of the cord and friction in the pulley are negligible. If w_1 and θ are given, find the weight w_2 for which the system is in equilibrium, that is, for which it remains at rest or moves in either direction at constant speed.

The free-body diagrams for the two blocks are shown at the left and right. The forces on block B are its weight w_2 and the force T exerted on it by the cord. Since it is in equilibrium,

$$T = w_2. \quad \text{(First Law)} \qquad (2\text{–}3)$$

Block A is acted on by its weight w_1, the force T exerted on it by the cord, and the force N exerted on it by the plane. We can use the same symbol T for the force exerted on each block by the cord because, as explained in Section 2–4, these forces are equivalent to an action–reaction pair and have the same magnitude. The force N, if there is no friction, is perpendicular or *normal* to the surface of the plane. Since the lines of action of w_1 and T intersect at the center of gravity of the block, the line of action of N passes through this point also. It is simplest to choose x- and y-axes parallel and perpendicular to the surface of the plane, because then only the weight w_1 needs to be resolved into components. The conditions of equilibrium give

$$\left.\begin{array}{l} \sum F_x = T - w_1 \sin \theta = 0, \\[4pt] \sum F_y = N - w_1 \cos \theta = 0. \end{array}\right\} \text{(First Law)} \quad \begin{array}{r}(2\text{–}4)\\[4pt](2\text{–}5)\end{array}$$

Thus if $w_1 = 100$ N and $\theta = 30°$, we have, from Eqs. (2–3) and (2–4),

$$w_2 = T = w_1 \sin \theta = (100\text{ N})(0.500) = 50\text{ N},$$

and from Eq. (2–5),

$$N = w_1 \cos \theta = (100\text{ N})(0.866) = 86.6\text{ N}.$$

Note carefully that, in the *absence of friction*, the same weight w_2 of 50 N is required whether the system remains at rest or moves with constant speed in *either* direction. This is not the case when friction is present.

2–6 FRICTION

Whenever the surface of one body slides over that of another, *each* body exerts a frictional force on the other, parallel to the surfaces. The force *on* each body is opposite to the direction of its motion relative to the other. Thus when a block slides from left to right along the surface of a table, a frictional force to the left acts on the block and an equal force toward the right acts on the table. Frictional forces may also act when there is no relative motion. A horizontal force on a heavy packing case resting on the floor may not be enough to set the case in motion, because of an equal and opposite frictional force exerted on the case by the floor.

The origin of these frictional forces is not fully understood, and the study of them is an important field of research. When one unlubricated metal slides over another, there appears to be an actual momentary welding of the metals together at the "high spots" where they make contact. The observed friction force is the force required to break these tiny welds. The mechanism of the friction force between two blocks of wood, or between two bricks, must be quite different.

In Fig. 2–10(a) a block is at rest on a horizontal surface, in equilibrium under the action of its weight w and the upward force P exerted on it by the surface. The lines of action of w and P have been displaced slightly to show these forces more clearly.

Suppose now that a cord is attached to the block as in Fig. 2–10(b) and the tension T in the cord is gradually increased. Provided the tension is not too great, the block remains at rest. The force P exerted on the block by the surface is inclined toward the left as shown, since the sum of the three forces P, w, and T must be zero. The component of P parallel to the surface is called the *force of static friction*, f_s. The other component, N, is the *normal* force exerted on the block by the surface. From the conditions of equilibrium, the force of static friction f_s is equal in magnitude and opposite in direction to the force T, and the normal force N is equal and opposite to the weight w.

As the force T is increased further, a limiting value is reached at which the block breaks away from the surface and starts to move. In other words, there is a certain *maximum* value which the force of static friction f_s can have. Figure 2–10(c) is the force diagram when T is just below its critical value. If the force T exceeds this value, the block is no longer in equilibrium.

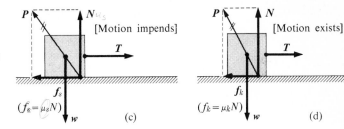

Fig. 2–10 *The magnitude of the friction force f is less than or equal to $\mu_s N$ when there is no relative motion, and is equal to $\mu_k N$ when motion exists.*

Experiment shows that for a given pair of surfaces, the magnitude of the maximum value of f_s depends on that of the normal force N; when the normal force increases, a greater force is required to make the block slide. In general the details of this relationship are quite complex, but we may often find it useful to assume simply that the maximum value of f_s is *directly proportional* to N, with proportionality factor μ_s, called the *coefficient of static friction*. The actual force of static friction can have any magnitude between zero (when there is no applied force parallel to the surface) and a maximum value given by $\mu_s N$. Thus,

$$f_s \leq \mu_s N. \qquad (2\text{–}6)$$

The equality sign holds only when the applied force T, parallel to the surface, has such a value that motion is about to start (Fig. 2–10(c)). When T is less than this value (Fig. 2–10(b)), the inequality sign holds and the magnitude of the friction force must be computed from the conditions of equilibrium.

As soon as sliding begins, it is found that the friction force *decreases*. This new friction force, for a given pair of surfaces, also depends on the magnitude of the normal force, and again it is convenient, though often rather imprecise, to represent the relation as a proportionality. The proportionality factor,

μ_k is called the *coefficient of sliding friction* or *kinetic* friction. Thus, when the block is in motion, the force of sliding or kinetic friction is given by

$$f_k = \mu_k N. \qquad (2\text{–}7)$$

This is illustrated in Fig. 2–10(d).

The coefficients of static and sliding friction depend primarily on the nature of the surfaces in contact, being relatively large if the surfaces are rough, and small if they are smooth. The coefficient of sliding friction varies somewhat with the relative velocity, but for simplicity we shall assume it to be independent of velocity. It is also nearly independent of the contact area. However, since two real surfaces actually touch each other only at a relatively small number of high spots, the true contact area is very different from the overall area. Equations (2–6) and (2–7) are useful empirical relations, but do not represent fundamental physical laws like Newton's laws. Typical numerical values are given in Table 2–1.

Table 2–1 lists the coefficients of friction of solids only. Liquids and gases show frictional effects also, but the simple equation $f = \mu N$ does not hold. It will be shown in Chapter 13 that there exists a property of liquids and gases called *viscosity* which determines the friction force between two surfaces

Table 2–1 COEFFICIENTS OF FRICTION

Materials	Static, μ_s	Kinetic, μ_k
Steel on steel	0.74	0.57
Aluminum on steel	0.61	0.47
Copper on steel	0.53	0.36
Brass on steel	0.51	0.44
Zinc on cast iron	0.85	0.21
Copper on cast iron	1.05	0.29
Glass on glass	0.94	0.4
Copper on glass	0.68	0.53
Teflon on Teflon	0.04	0.04
Teflon on steel	0.04	0.04

(a)

sliding over each other with a layer of liquid gas between them. Gases have the lowest viscosities of all materials at normal temperatures, and therefore, to reduce friction to a value close to zero, it is convenient to have an object slide on a layer of gas.

This is the principle of the *hovercraft*, a type of vehicle containing powerful blowers that maintain a cushion of air between the vehicle and the ground so that it literally floats on a layer of air. A familiar laboratory example of the principle is the linear air track, conceived originally by H. V. Neher and R. B. Leighton at the California Institute of Technology, and developed further by John Stull of Alfred University. Elastic bumpers are provided for the inverted V-shaped riders. (See Fig. 2–11(a).) When one rider only is placed on the air track and is given a push, it will hit the stationary bumper and then proceed to move back and forth many times before coming to rest. The frictional force is velocity-dependent, but at typical speeds the effective coefficient of friction is of the order of 0.001.

A similar device is the frictionless air table shown in Fig. 2–11(b), developed by Harold A. Daw of New Mexico State University. The pucks, made of plastic, are supported by over 1600 tiny air jets about an inch apart. Two-dimensional collisions, both elastic and inelastic, may be demonstrated on this table.

Fig. 2–11 *(a) The Ealing–Stull linear air track. Inverted Y-shaped sliders ride on a layer of air streaming through many fine holes in the inverted V-shaped surface. (b) The Ealing–Daw two-dimensional air table. Plastic pucks slide on a cushion of air issuing from more than a thousand minute holes in the tabletop.* [*Courtesy of the Ealing Corporation*]

(b)

Example 1 In Fig. 2–10, suppose that the block weighs 20 N, that the tension T can be increased to 8 N before the block starts to slide, and that a force of 4 N will keep the block moving at constant speed once it has been set in motion. Find the coefficients of static and kinetic friction.

From Fig. 2–10(c) and the data above, we have

$$\left. \begin{array}{l} \sum F_y = N - w = N - 20\,\text{N} = 0, \\ \sum F_x = T - f_s = 8\,\text{N} - f_s = 0, \end{array} \right\} \text{(First Law)}$$
$$f_s = \mu_s N \quad \text{(motion impends)}.$$

Hence we have

$$\mu_s = \frac{f_s}{N} = \frac{8\,\text{N}}{20\,\text{N}} = 0.40.$$

From Fig. 2–10(d), we have

$$\left. \begin{array}{l} \sum F_y = N - w = N - 20\,\text{N} = 0, \\ \sum F_x = T - f_k = 4\,\text{N} - f_k = 0, \end{array} \right\} \text{(First Law)}$$
$$f_k = \mu_k N \quad \text{(motion exists)}.$$

Hence

$$\mu_k = \frac{f_k}{N} = \frac{4\,\text{N}}{20\,\text{N}} = 0.20.$$

Example 2 What is the friction force if the block in Fig. 2–10(b) is at rest on the surface and a horizontal force of 5 N is exerted on it?

We have

$$\sum F_x = T - f_s = 5\,\text{N} - f_s = 0, \quad \text{(First Law)}$$
$$f_s = 5\,\text{N}.$$

Note that in this case $f_s < \mu_s N$.

Example 3 What force T, at an angle of 30° above the horizontal, is required to drag a block weighing 20 N to the right at constant speed, as in Fig. 2–12, if the coefficient of kinetic friction between block and surface is 0.20?

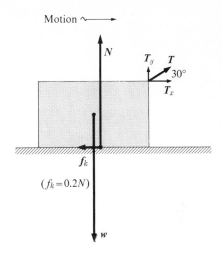

Fig. 2–12 *Forces on a block being dragged to the right on a level surface at constant speed.*

The forces on the block are shown in the diagram. From the first condition of equilibrium,

$$\sum F_x = T\cos 30° - 0.2N = 0,$$
$$\sum F_y = T\sin 30° + N - 20\,\text{N} = 0.$$

Simultaneous solution gives

$$T = 4.15\,\text{N},$$
$$N = 17.9\,\text{N}.$$

Note that in this example the normal force N is not equal to the weight of the block, but is less than the weight by the vertical component of the force T.

Example 4 In Fig. 2–13, a block has been placed on an inclined plane and the slope angle θ of the plane has been adjusted until the block slides down the plane at constant speed, once it has been set in motion. Find the angle θ.

The forces on the block are its weight w and the normal and frictional components of the force exerted by the plane. Take axes perpendicular and parallel

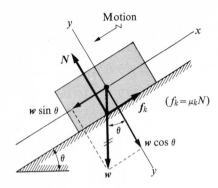

Fig. 2–13 *Forces on a block sliding down an inclined plane (with friction) at constant speed.*

to the surface of the plane. Then

$$\left.\begin{array}{l} \sum F_x = \mu_k N - w \sin \theta = 0, \\ \sum F_y = N - w \cos \theta = 0. \end{array}\right\} \text{ (First Law)}$$

Hence

$$\mu_k N = w \sin \theta, \qquad N = w \cos \theta.$$

Dividing the former by the latter, we get

$$\mu_k = \tan \theta.$$

It follows that a block, regardless of its weight, slides down an inclined plane with constant speed if the tangent of the slope angle of the plane equals the coefficient of kinetic friction. Measurement of this angle then provides a simple experimental method for determining the coefficient of kinetic friction.

PROBLEMS

Section 2–5 should be read carefully before you begin the solution of problems in this chapter. If difficulties arise, they will not result from any complicated mathematics or from a failure to remember the "formula"—after all, the only "formulas" are

$$\sum F_x = 0, \qquad \sum F_y = 0,$$

—but rather from (1) a failure to *select some one body to talk about*, and (2) *a failure to recognize precisely what forces*

are exerted on *the selected body.* Once the forces acting on the body have been clearly and correctly shown in a force diagram, the physics of the problem is finished; the rest of the computations might be (and in practice often are) turned over to a machine.

2–1 Imagine that you are holding a book weighing 4 N at rest on the palm of your hand. Complete the following sentences.

a) A downward force of magnitude 4 N is exerted on the book by _____.

b) An upward force of magnitude _____ is exerted on _____ by the hand.

c) Is the upward force (b) the reaction to the downward force (a)?

d) The reaction to force (a) is a force of magnitude _____, exerted on _____ by _____. Its direction is _____.

e) The reaction to force (b) is a force of magnitude _____, exerted on _____ by _____. Its direction is _____.

f) That the forces (a) and (b) are equal and opposite is an example of Newton's _____ law.

g) That forces (b) and (e) are equal and opposite is an example of Newton's _____ law.

Suppose now that you exert an upward force of magnitude 5 N on the book.

h) Does the book remain in equilibrium?

i) Is the force exerted on the book by the hand equal and opposite to the force exerted on the book by the earth?

j) Is the force exerted on the book by the earth equal and opposite to the force exerted on the earth by the book?

k) Is the force exerted on the book by the hand equal and opposite to the force exerted on the hand by the book?

Finally, suppose that you snatch your hand away while the book is moving upward.

l) How many forces then act on the book?

m) Is the book in equilibrium?

n) What balances the downward force exerted on the book by the earth?

2–2 A block is given a push along a table top, and slides off the edge of the table.

a) What forces are exerted on it while it is falling from the table to the floor?

b) What is the reaction to each force, that is, on what body and by what body is the reaction exerted? Neglect air resistance.

2–3 Two 10-N weights are suspended at opposite ends of a rope which passes over a light frictionless pulley. The pulley is attached to a chain which goes to the ceiling.

a) What is the tension in the rope?

b) What is the tension in the chain?

2–4 In Fig. 2–5, let the weight of the hanging block be 50 N. Find the tensions T_2 and T_3 (a) if $\theta_2 = \theta_3 = 60°$, (b) if $\theta_2 = \theta_3 = 10°$, (c) if $\theta_2 = 60°$, $\theta_3 = 0$, and (d) if $AB = 5$ m, $AO = 3$ m, $OB = 4$ m.

2–5 Find the tension in each cord in Fig. 2–14 if the weight of the suspended body is 200 N.

Fig. 2–14

Fig. 2–15

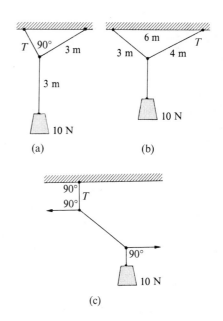

Fig. 2–16

2–6 Find the tension T in the cable, and the magnitude and direction of the force C exerted on the strut by the pivot, in each of the arrangements in Fig. 2–15. Let the weight of the suspended object in each case be 1000 N. Neglect the weight of the strut.

2–7

a) In which of the arrangements in Fig. 2–16 can the tension T be computed, if the only quantities known are those explicitly given?

b) For each case in which insufficient information is given, state one additional quantity, a knowledge of which would permit solution.

2–8 A horizontal boom 4 m long is hinged to a vertical wall at one end, and a 500-N body hangs from its outer end. The boom is supported by a guy wire from its outer end to a point on the wall directly above the boom.

a) If the tension in this wire is not to exceed 1000 N, what is the minimum height above the boom at which it may be fastened to the wall?

b) By how many newtons would the tension be increased if the wire were fastened 0.5 m below this point, the boom remaining horizontal? Neglect the weight of the boom.

2–9 One end of a rope 25 m long is attached to an automobile. The other end is fastened to a tree. A man exerts a force of 500 N at the midpoint of the rope, pulling it 1 m to the side. What is the force exerted on the automobile?

2–10 Find the largest weight w which can be supported by the structure in Fig. 2–17 if the maximum tension the upper rope can withstand is 1000 N and the maximum compression the strut can withstand is 2000 N. The vertical rope is strong enough to carry any load required. Neglect the weight of the strut.

Fig. 2–17

2–11

a) Block A in Fig. 2–18 weighs 100 N. The coefficient of static friction between the block and the surface on which it rests is 0.30. The weight w is 20 N and the system is in equilibrium. Find the friction force exerted on block A.

Fig. 2–18

b) Find the maximum weight w for which the system will remain in equilibrium.

2–12 A block hangs from a cord 10 m long. A second cord is tied to the midpoint of the first, and a horizontal pull equal to half the weight of the block is exerted on it, the second cord being always kept horizontal.

a) How far will the block be pulled to one side?

b) How far will it be lifted?

2–13 A flexible chain of weight w hangs between two hooks at the same height, as shown in Fig. 2–19. At each end the chain makes an angle θ with the horizontal.

a) What is the magnitude and direction of the force **F** exerted by the chain on the hook at the left?

b) What is the tension T in the chain at its lowest point?

Fig. 2–19

2–14 A 30-N block is pulled at constant speed up a frictionless inclined plane by a weight of 10 N hanging from a cord attached to the block and passing over a frictionless pulley at the top of the plane. (See Fig. 2–9.) Find (a) the slope angle of the plane, (b) the tension in the cord, and (c) the normal force exerted on the block by the plane.

2–15

a) A block rests upon a rough horizontal surface. A horizontal force **T** is applied to the block and is slowly increased from zero. Draw a graph with T along the x-axis and the friction force f along the y-axis, starting at T = 0 and showing the region of no motion, the point where motion impends, and the region where motion exists.

b) A block of weight w rests on a rough horizontal plank. The slope angle of the plank θ is gradually increased until the block starts to slip. Draw two graphs, both with θ along the x-axis. In one graph show the ratio of the normal force to the weight N/w as a function of θ. In the second graph, show the ratio of the friction force to the weight f/w. Indicate the region of no motion, the point where motion impends, and the region where motion exists.

2–16 A block weighing 20 N rests on a horizontal surface. The coefficient of static friction between block and surface is 0.40 and the coefficient of sliding friction is 0.20.

a) How large is the friction force exerted on the block?

b) How great will the friction force be if a horizontal force of 5 N is exerted on the block?

c) What is the minimum force which will start the block in motion?

d) What is the minimum force which will keep the block in motion once it has been started?

e) If the horizontal force is 10 N, how great is the friction force?

2–17 A block is pulled to the right at constant velocity by a 10-N force acting 30° above the horizontal. The coefficient of sliding friction between the block and the surface is 0.5. What is the weight of the block?

2–18 A block weighing 14 N is placed on an inclined plane and connected to a 10-N block by a cord passing over a small frictionless pulley, as in Fig. 2–9. The coefficient of sliding friction between the block and the plane is (1/7). For what two values of θ will the system move with constant velocity? [Hint: $\cos \theta = \sqrt{1 - \sin^2\theta}$.]

2–19 A block weighing 100 N is placed on an inclined plane of slope angle 30° and is connected to a second hanging block of weight w by a cord passing over a small frictionless pulley, as in Fig. 2–9. The coefficient of static friction is 0.40 and the coefficient of sliding friction is 0.30.

a) Find the weight w for which the 100-N block moves up the plane at constant speed.

b) Find the weight w for which it moves down the plane at constant speed.

c) For what range of values of w will the block remain at rest?

2–20 What force P at an angle ϕ above the horizontal is needed to drag a box of weight w at constant speed along a level floor if we are given that the coefficient of sliding friction between box and floor is μ?

2–21 A box of weight W is pushed at constant speed up a plane inclined at angle ϕ by a horizontal force of magnitude P. If the coefficient of kinetic friction is μ, derive an expression for P in terms of the other quantities. Show that, if ϕ exceeds a certain critical value, the box cannot be pushed up, no matter how large P is. Express the critical angle in terms of μ.

2–22 A block of wood of weight W is pushed at constant speed along the horizontal surface of a circular saw table by a stick which exerts a force P at an angle ϕ downward from the horizontal. If the coefficient of kinetic friction is μ, find the force P required, in terms of the other quantities. Show that if ϕ exceeds a certain critical value, the block cannot be pushed, no matter how great P is. Express the critical angle in terms of μ.

2–23 A rope is stretched across a glacial crevasse 20 m wide. At the center of the rope sits a mountaineer weighing 800 N (about 180 lb). The center of the rope is 4 m lower than the ends. What is the tension in the rope?

2–24 A concrete bucket weighing 2000 lb is suspended from a crane by a cable 50 ft long. What horizontal force is required to push the bucket to a position 5 ft to the side of the straight-down position?

2–25 A safe weighing 2000 N is to be lowered at constant speed down skids 4 m long, from a truck 2 m high.

a) If the coefficient of sliding friction between safe and skids is 0.30, will the safe need to be pulled down or held back?

b) How great a force parallel to the skids is needed?

2–26

a) If a force of 86 N parallel to the surface of a 20° inclined plane will push a 120-N block up the plane at constant speed, what force parallel to the plane will push it down at constant speed?

b) What is the coefficient of sliding friction?

2–27 Block A in Fig. 2–20 weighs 4 N and block B weighs 8 N. The coefficient of sliding friction between all surfaces is 0.25. Find the force P necessary to drag block B to the left at constant speed (a) if A rests on B and moves with it, (b) if A is held at rest, and (c) if A and B are connected by a light flexible cord passing around a fixed frictionless pulley.

(a)

(b)   (c)

Fig. 2–20

2–28 Block A, of weight w, slides down an inclined plane S of slope angle 37° at constant velocity while the plank B, also of weight w, rests on top of A. The plank is attached by a cord to the top of the plane (Fig. 2–21).

a) Draw a diagram of all the forces acting on block A.

b) If the coefficient of kinetic friction is the same between the surfaces A and B and between S and A, determine its value.

2–29 Two blocks, A and B, are placed as in Fig. 2–22 and connected by ropes to block C. Both A and B weigh 20 N and the coefficient of sliding friction between each block and the surface is 0.5. Block C descends with constant velocity.

a) Draw two separate force diagrams showing the forces acting on A and B.

b) Find the tension in the rope connecting blocks A and B.

c) What is the weight of block C?

Fig. 2–21

Fig. 2–22

Chapter 3

Equilibrium. Moment of a Force

3–1 MOMENT OF A FORCE

The effect produced on a body by a force of given magnitude and direction depends on the position of the *line of action* of the force. Thus in Fig. 3–1 the force F_1 would produce a counterclockwise rotation (together with a translation toward the right), while F_2 would produce clockwise rotation.

The line of action of a force can be specified by giving the perpendicular distance from some reference point to the line of action. In many instances, we shall be studying the motion of a body which is free to rotate about some axis, and which is acted on by a number of coplanar forces all lying in a plane perpendicular to the axis. It is then most convenient to select as the reference point the point at which the axis intersects the plane of the forces. The perpendicular distance from this point to the line of action of a force is called the *force arm* or the *moment arm* of

the force about the axis. The product of the magnitude of a force and its force arm is called the *moment* of the force about the axis, or the *torque*.

Thus Fig. 3–2 is a top view of a flat object, pivoted about an axis perpendicular to the plane of the diagram and passing through point O. The body is acted on by the forces F_1 and F_2, lying in the plane of the diagram. The moment arm of F_1 is the perpendicular distance OA, of length ℓ_1, and the moment arm of F_2 is the perpendicular distance OB of length ℓ_2.

Fig. 3–2 *The moment of a force about an axis is the product of the force and its moment arm.*

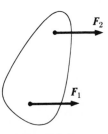

Fig. 3–1

35

The effect of the force F_1 is to produce counter-clockwise rotation about the axis, while that of F_2 is to produce clockwise rotation. To distinguish between these directions of rotation, we shall adopt the convention that counterclockwise moments are positive, and that clockwise moments are negative. Hence the moment Γ_1 (Greek "gamma") of the force F_1 about the axis through O is

$$\Gamma_1 = +F_1\ell_1,$$

and the moment Γ_2 of F_2 is

$$\Gamma_2 = -F_2\ell_2.$$

In the mks system, where the unit of force is the newton and the unit of length the meter, the unit of torque is the newton-meter. If forces are expressed in pounds and lengths in feet, torques are expressed in pound-feet.

3–2 THE SECOND CONDITION OF EQUILIBRIUM

We saw in Section 2–2 that when a body is in equilibrium under the action of several coplanar forces, the vector sum of the forces must be zero. In addition, if there are only two forces they must have the same line of action, and if there are three they must be *concurrent* (unless their lines of action are all parallel). The first of these requirements is satisfied by the *first condition for equilibrium*,

$$\boxed{\sum F_x = 0, \qquad \sum F_y = 0.}$$

The second requirement can be simply expressed in terms of the moments of the forces. Figure 3–3 again shows a flat object acted on by two forces F_1 and F_2. If the object is in equilibrium, the magnitudes of F_1 and F_2 are equal and both forces have the same line of action. Hence they have the same moment arm OA, of length ℓ, about an axis perpendicular to the plane of the body and passing through any arbitrary point O. Their moments about the axis are therefore equal in magnitude and opposite in sign, and the algebraic sum of their moments is zero.

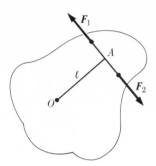

Fig. 3–3 *When a body is in equilibrium, the resultant moment of the forces about any axis is zero.*

In the case of three forces, we can replace any two by their vector sum, acting at the point of intersection of their lines of action. By the first condition, the third force must be equal and opposite, and we see that the second condition is satisfied only when it has the same line of action as the sum of the first two, in which case its moment must be the negative of that of the sum.

When a body is in equilibrium under the action of four or more forces, the lines of action need not all intersect at the same point, but the second condition of equilibrium can still be stated in terms of moments:

$$\boxed{\sum \Gamma = 0 \qquad \text{(about any arbitrary axis).}}$$

Although the choice of axis is arbitrary, one must, of course, use the *same* axis for *all* the moments.

Example 1 A rigid rod whose own weight is negligible (Fig. 3–4) is pivoted at point O and carries a body of weight w_1 at end A. Find the weight w_2 of a second body which must be attached at end B if the rod is to be in equilibrium, and find the force exerted on the rod by the pivot at O.

Figure 3–4(b) is the free-body diagram of the rod. The forces T_1 and T_2 are equal respectively to w_1 and w_2. The conditions of equilibrium, taking moments about an axis through O, perpendicular to the

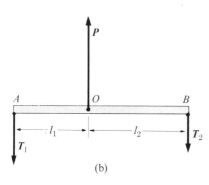

Fig. 3–4 *A rod in equilibrium under three parallel forces.*

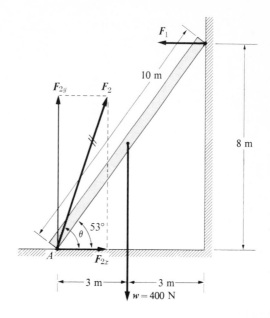

Fig. 3–5 *Forces on a ladder in equilibrium, leaning against a frictionless wall.*

diagram, give

$$\sum F_y = P - T_1 - T_2 = 0, \qquad \text{(First condition)}$$
$$\sum \Gamma_o = T_1\ell_1 - T_2\ell_2 = 0. \qquad \text{(Second condition)}$$

Let $\ell_1 = 1.2$ m, $\ell_2 = 1.6$ m, $w_1 = 20$ N. Then, from the equations above,

$$P = 35 \text{ N}, \qquad T_2 = w_2 = 15 \text{ N}.$$

To illustrate that the resultant moment about *any* axis is zero, let us compute moments about an axis through point A:

$$\sum \Gamma_A = P\ell_1 - T_2(\ell_1 + \ell_2)$$
$$= (35 \text{ N})(1.2 \text{ m}) - (15 \text{ N})(2.8 \text{ m}) = 0.$$

The point about which moments are computed need not lie on the rod. To verify this, let the reader calculate the resultant moment about a point 0.4 m to the left of A and 0.4 m above it.

Example 2 In **Fig.** 3–5 a ladder 10 m long, of weight 400 N, with its center of gravity at its center,

leans in equilibrium against a vertical frictionless wall and makes an angle of 53° with the horizontal (forming a 3–4–5 right triangle). We wish to find the magnitudes and directions of the forces F_1 and F_2.

If the wall is frictionless, F_1 is horizontal. The direction of F_2 is unknown (except in special cases, its direction does *not* lie along the ladder). Instead of considering its magnitude and direction as unknowns, it is simpler to resolve the force F_2 into (unknown) x- and y-components and solve for these. The magnitude and direction of F_2 may then be computed. The first condition of equilibrium therefore provides the equations

$$\left.\begin{array}{l} \sum F_x = F_2 \cos\theta - F_1 = 0, \\[2mm] \sum F_y = F_2 \sin\theta - 400 \text{ N} = 0. \end{array}\right\} \quad \text{(First condition)}$$

In writing the second condition, moments may be computed about an axis through any point. The resulting equation is simplest if one selects a point through which two or more forces pass, since these forces then do not appear in the equation. Let us therefore take moments about an axis through point A.

$$\sum \Gamma_A = F_1(8 \text{ m}) - (400 \text{ N})(3 \text{ m}) = 0. \qquad \text{(Second condition)}$$

From the second equation, $F_2 \sin \theta = 400 \text{ N}$, and from the third,

$$F_1 = \frac{1200 \text{ N-m}}{8 \text{ m}} = 150 \text{ N}.$$

Then from the first equation,

$$F_2 \cos \theta = 150 \text{ N}.$$

Hence

$$F_2 = \sqrt{(400 \text{ N})^2 + (150 \text{ N})^2} = 427.2 \text{ N}.$$

$$\theta = \tan^{-1} \frac{400 \text{ N}}{150 \text{ N}} = 69.5°.$$

(Note that this problem has already been solved by a different method in Example 4 in Section 2–5.)

Example 3 Figure 3–6(a) shows a human arm lifting a dumbbell, and Fig. 3–6(b) shows a free-body diagram for the forearm, showing the forces involved. The forearm is in equilibrium under the action of the weight w, the tension T in the tendon connected to the biceps muscle, and the forces exerted on it by the upper arm, through the elbow joint. We want to find the tendon tension and the components of force at the elbow, three unknown scalar quantities in all.

First we represent the tendon force in terms of its components T_x and T_y, using the given angle θ and the unknown magnitude T:

$$T_x = T \cos \theta, \qquad T_y = T \sin \theta.$$

Next we note that if moments about the elbow joint are taken, the resulting moment equation does not contain the unknowns E_x and E_y, since their lines of action pass through this point, as also does T_x. The moment equation is then simply

$$\ell w - dT_y = 0.$$

From this we immediately find

$$T_y = \frac{\ell w}{d} \quad \text{and} \quad T = \frac{\ell w}{d \sin \theta}.$$

To find E_x and E_y we could now use the first conditions for equilibrium, $\sum F_x = 0$ and $\sum F_y = 0$. Instead, for added practice in using moments, we take moments about the point A where the tendon is attached:

$$(\ell - d)w + dE_y = 0 \qquad \text{and} \qquad E_y = -\frac{(\ell - d)w}{d}.$$

The negative sign of this result shows that our initial guess for the direction of E_y was wrong; it is actually vertically *downward*.

Finally, we take moments about point B in the figure:

$$\ell w - hE_x = 0 \qquad \text{and} \qquad E_x = \frac{\ell w}{h}.$$

Fig. 3–6

(a)

(b)

We note that each stage of this calculation is simplified by choosing the point for calculating moments so as to eliminate one or more of the unknown quantities. In the last step, the force T has no moment about point B; thus when the moments of T_x and T_y are computed separately, they must add to zero. The reader is invited to verify this statement in detail.

As a specific example, suppose $w = 50$ N, $d = 0.1$ m, $\ell = 0.5$ m, and $\theta = 80°$. Since $\tan \theta = h/d$, we find

$$h = d \tan \theta = (0.1 \text{ m}) (5.67) = 0.567 \text{ m}.$$

From the previous general results, we find

$$T = \frac{(0.5 \text{ m})(50 \text{ N})}{(0.1 \text{ m})(0.985)} = 254 \text{ N},$$

$$E_y = \frac{(0.5 \text{ m} - 0.1 \text{ m})(50 \text{ N})}{0.1 \text{ m}} = 200 \text{ N},$$

$$E_x = \frac{(0.5 \text{ m})(50 \text{ N})}{0.567 \text{ m}} = 44 \text{ N}.$$

The magnitude of the force at the elbow is

$$E = \sqrt{E_x^2 + E_y^2} = 204 \text{ N}.$$

As mentioned above, we have not explicitly used the first condition for equilibrium, that the vector sum of the forces be zero. As a check, the reader should verify that this condition *is* satisfied by the above results.

Units of meters have been used in the above example; centimeters could have been used. In fact, *any* unit of distance may be used in a moment equation, provided the *same* unit is used throughout a given equation.

3–3 CENTER OF GRAVITY

Every particle of matter in a body is attracted by the earth, and the single force which we call the *weight* of the body is the resultant of all these forces of attraction. The direction of the force on each particle is toward the center of the earth, but the distance to the earth's center is so great that for all practical purposes the forces can be considered parallel to one

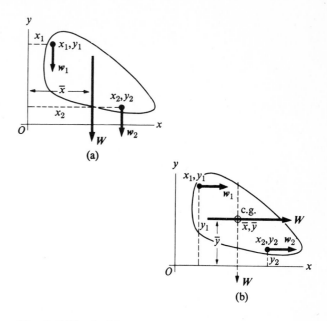

Fig. 3–7 *The body's weight W is the resultant of a large number of parallel forces. The line of action of $\mathbf{W}$ always passes through the center of gravity.*

another. Hence the weight of a body is the resultant of a large number of parallel forces.

Figure 3–7(a) shows a flat object of arbitrary shape in the xy-plane, the y-axis being vertical. Let the body be subdivided into a large number of small particles of weights w_1, w_2, etc., and let the coordinates of these particles be x_1 and y_1, x_2 and y_2, etc. The total weight W of the object is

$$W = w_1 + w_2 + \cdots = \sum w. \qquad (3\text{--}1)$$

Each particle's weight also contributes to the total torque acting on the body. Computing moments about point O, we see that the torque associated with particle 1 is $w_1 x_1$, that for particle 2 is $w_2 x_2$, and so on, and that the total torque is

$$w_1 x_1 + w_2 x_2 + \cdots = \sum wx.$$

For convenience in the present discussion we are taking clockwise torques to be positive, the opposite choice from that used in preceding sections.

Along what line must the total weight W act in order for the total torque to equal the above expres-

Fig. 3–8

sion? Suppose it acts along a line a distance $\bar{x}$ to the right of the origin. Then it must be true that $w_1 x_1 + w_2 x_2 + \cdots = W\bar{x}$, or

$$\bar{x} = \frac{w_1 x_1 + w_2 x_2 + \cdots}{W} = \frac{\sum wx}{W} = \frac{\sum wx}{\sum w}.$$

(3–2)

That is, the torque due to the weight of the object can always be obtained by assuming the total weight to be acting along a line determined by Eq. (3–2). Stated another way, in computing the torque due to the weight of an object, we must consider the weight as acting at a point whose x-coordinate $\bar{x}$ is given by Eq. (3–2). (See Fig. 3–7(a).)

Now let the object and the reference axes be rotated 90° clockwise or, which amounts to the same thing, let us consider the gravitational forces to be rotated 90° counterclockwise, as in Fig. 3–7(b). The total weight W is unaltered, but now it must act along a line a distance $\bar{y}$ above the origin, such that $W\bar{y} = \sum wy$, or

$$\bar{y} = \frac{w_1 y_1 + w_2 y_2 + \cdots}{w_1 + w_2 + \cdots} = \frac{\sum wy}{\sum w} = \frac{\sum wy}{W}.$$

(3–3)

The point of intersection of the lines of action of W in the two parts of Fig. 3–7 has the coordinates $\bar{x}$ and $\bar{y}$ and is called the *center of gravity* of the object. By considering some arbitrary orientation of the object, one can show that the line of action of W

always passes through the center of gravity, and thus that the torque of W can always be obtained correctly by taking W to act at the center of gravity.

If the centers of gravity of each of a number of bodies have been determined, the coordinates of the center of gravity of the combination can be computed from Eqs. (3–2) and (3–3), letting w_1, w_2, etc., be the weights of the bodies and x_1 and y_1, x_2 and y_2, etc., be the coordinates of the center of gravity of each.

Symmetry considerations are often useful in finding the position of the center of gravity. Thus the center of gravity of a homogeneous sphere, cube, circular disk or rectangular plate is at its center. That of a cylinder or right circular cone is on the axis of symmetry, and so on.

Example 1 Locate the center of gravity of the machine part in Fig. 3–8, consisting of a disk 4 cm in diameter and 2 cm long, and a rod 2 cm in diameter and 12 cm long, constructed of a homogeneous material.

By symmetry, the center of gravity lies on the axis and the center of gravity of each part is midway between its ends. The volume of the disk is 8π cm^3 and that of the rod is 12π m^3. Since the weights of the two parts are proportional to their volumes,

$$\frac{w(\text{disk})}{w(\text{rod})} = \frac{w_1}{w_2} = \frac{8\pi}{12\pi} = \frac{2}{3}.$$

Take the origin O at the left face of the disk, on the axis. Then

$$x_1 = 1 \text{ cm}, \qquad x_2 = 8 \text{ cm},$$

and

$$\bar{x} = \frac{w_1(1 \text{ cm}) + \frac{3}{2} w_1(8 \text{ cm})}{w_1 + \frac{3}{2} w_1} = 5.2 \text{ cm}.$$

The center of gravity is on the axis, 5.2 cm to the right of O.

The center of gravity of a flat object can be located experimentally as shown in Fig. 3–9. In part (a) the body is suspended from some arbitrary point

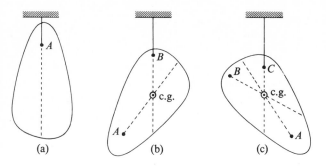

Fig. 3–9 Locating the center of gravity of a flat object.

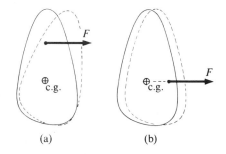

(a)   (b)

Fig. 3–10 A body is in rotational but not translational equilibrium when acted on by a force whose line of action passes through the center of gravity, as in (b).

Fig. 3–11 Traction arrangement for fractured femur. (a) Weights and pulleys support lower leg and provide tension to counteract muscle tension at fracture. There must be no sideways force at fracture. (b) Free-body diagram for lower leg, showing forces exerted by weights and pulleys.

(a)

(b)

A. When allowed to come to equilibrium, the center of gravity must lie on a vertical line through *A*. When the object is suspended from a second point *B*, as in part (b), the center of gravity lies on a vertical line through *B* and hence lies at the point of intersection of this line and the first. If the object is now suspended from a third point *C*, as in part (c), a vertical line through *C* will be found to pass through the point of intersection of the first two lines. The extent to which the three lines fail to meet at a single point gives an indication of the experimental error in this method of locating the center of gravity.

The center of gravity of a body has another important property. A force **F** whose line of action lies at one side or the other of the center of gravity, as in Fig. 3–10(a), will change both the translational and rotational motion of the body on which it acts. However, if the line of action passes through the center of gravity, as in part (b), only the translational motion is affected and the body remains in rotational equilibrium. Thus when a body is tossed in the air with a whirling motion, it continues to rotate at a constant rate, since the line of action of its weight passes through the center of gravity.

Example 2 Figure 3–11(a) shows an arrangement to maintain proper position of a factured femur. The thigh muscles tend to contract and pull the broken ends past each other; the tension T_3 prevents this, while T_1 and T_2 support the weight of the lower leg. It is essential that no vertical force be exerted at the position of the fracture; otherwise the broken ends would be pulled out of position.

Figure 3–11(b) shows a free-body diagram for the lower leg and the approximate location of its center of gravity. The force F is the muscle tension at the fracture; clearly, this must equal T_3, which in practice is determined by experience and trial. The other conditions for equilibrium are

$$T_1 + T_2 = W,$$

$$\frac{T_1\ell}{3} - \frac{2T_2\ell}{3} = 0,$$

where we have taken moments about the center of gravity. Simultaneous solution gives

$$T_1 = \tfrac{2}{3}W, \qquad T_2 = \tfrac{1}{3}W.$$

3–4 COUPLES

It sometimes happens that the forces on a body reduce to two forces of equal magnitude and opposite direction, having lines of action which are parallel but which do not coincide. Such a pair of forces is called a *couple*. A common example is afforded by the forces on a compass needle in the earth's magnetic field, as shown in Fig. 3–12. The north and south poles of the needle are acted on by equal forces, one toward the north and the other toward the south. Except when the needle points in the N–S direction, the two forces do not have the same line of action.

Figure 3–13 shows a couple consisting of two forces, each of magnitude F, separated by a perpendicular distance ℓ. The resultant R of the forces is

$$R = F - F = 0.$$

The fact that the resultant force is zero means that a couple has no effect in producing translation (as a whole) of the body on which it acts. The only effect of a couple is to produce rotation.

The resultant torque of the couple in Fig. 3–13 about an arbitrary point O is

$$\sum \Gamma_O = x_1 F - x_2 F$$
$$= x_1 F - (x_1 + \ell)F$$
$$= -\ell F.$$

Since the distances x_1 and x_2 do not appear in the result, we conclude that the torque of the couple is the same about all points in the plane of the forces forming the couple and is equal to the product of the magnitude of either force and the perpendicular distance between their lines of action.

A body acted on by a couple can be kept in equilibrium only by another couple of the same moment and in the opposite direction. As an example, the ladder in Fig. 3–5 can be considered as acted on by two couples, one formed by the forces $F_2 \sin \theta$ and w, the other by the forces $F_2 \cos \theta$ and F_1. The moment of the first is

$$\Gamma_1 = (3 \text{ m})(400 \text{ N}) = 1200 \text{ N-m}.$$

Fig. 3–12 *Forces on the poles of a compass needle.*

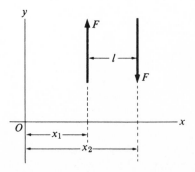

Fig. 3–13 *Two equal and opposite forces having different lines of action are called a couple. The moment of the couple is the same about all points, and is equal to ℓF.*

The moment of the second is

$$\Gamma_2 = (8 \text{ m})(150 \text{ N}) = 1200 \text{ N-m.}$$

The first moment is clockwise and the second is counterclockwise.

PROBLEMS

Section 2–5 and the note at the beginning of the problems in Chapter 2 are equally applicable to these problems. The only difference is that now we have a third "formula," $\sum \Gamma$ (about any axis) = 0.

3–1 A force F in the xy-plane has components of magnitudes F_x and F_y, respectively. The coordinates of its point of application are (x, y). The magnitude of the moment of force about the origin is $\Gamma_O = xF_y - yF_x$. Show that this equation is correct if the point (x, y) lies in any one of the four quadrants, and also that it is correct regardless of the direction of the force F.

3–2 You are given (a) a meter stick through which a number of holes have been bored so that its center of gravity is not at its midpoint, (b) a knife point, on which the meter stick can be pivoted, (c) a body whose weight is known to be w, and (d) a spool of thread. Using this equipment only, explain, with the aid of a diagram, how you would determine the weight of the meter stick.

3–3 A uniform plank 15 m long, weighing 400 N, rests symmetrically on two supports 8 m apart, as shown in Fig. 3–14. A boy weighing 640 N starts at point A and walks toward the right.

a) Construct in the same diagram two graphs showing the upward forces F_A and F_B exerted on the plank at points A and B, as functions of the coordinate x of the boy. Let 1 cm = 100 N vertically, and 1 cm = 1 m horizontally.

b) Find from your diagram how far beyond point B the boy can walk before the plank tips.

c) How far from the right end of the plank should support B be placed in order that the boy walk just to the end of the plank without causing it to tip?

3–4 The strut in Fig. 3–15 weighs 200 N and its center of gravity is at its center. Find (a) the tension in the cable and (b) the horizontal and vertical components of the force exerted on the strut at the wall.

Fig. 3–15

3–5 Find the tension in the cable BD in Fig. 3–16, and the horizontal and vertical components of the force exerted on the strut AB at pin A, using:

a) the first and second conditions of equilibrium ($\sum F_x = 0$, $\sum F_y = 0$, $\sum \Gamma = 0$), taking moments about an axis through point A perpendicular to the plane of the diagram;

b) the second condition of equilibrium only, taking moments first about an axis through A, then about an axis through B, and finally about an axis through D. The weight of the strut can be neglected.

c) Represent the computed forces by vectors in a diagram drawn to scale, and show that the lines of action of the forces exerted on the strut at points A, B, and C intersect at a common point.

Fig. 3–14

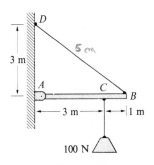

Fig. 3–16

3–6 End A of the bar AB in Fig. 3–17 rests on a friction-less horizontal surface, while end B is hinged. A horizontal force P of 60 N is exerted on end A. Neglect the weight of the bar. What are the horizontal and vertical components of the force exerted by the bar on the hinge at B?

Fig. 3–17

3–7 A single force is to be applied to the bar in Fig. 3–18 to maintain it in equilibrium in the position shown. The weight of the bar can be neglected.

a) What are the x- and y-components of the required force?

b) What is the tangent of the angle which the force must make with the bar?

c) What is the magnitude of the required force?

d) Where should the force be applied?

Fig. 3–18

3–8 A circular disk 0.5 m in diameter, pivoted about a horizontal axis through its center, has a cord wrapped around its rim. The cord passes over a frictionless pulley P and is attached to a body of weight 240 N. A uniform rod 2 m long is fastened to the disk, with one end at the center of the disk. The apparatus is in equilibrium, with the rod horizontal, as shown in Fig. 3–19.

a) What is the weight of the rod?

b) What is the new equilibrium direction of the rod when a second body weighing 20 N is suspended from the outer end of the rod, as shown by the dotted line?

Fig. 3–19

3–9 A roller whose diameter is 1.0 m weighs 360 N. What horizontal force is necessary to pull the roller over a brick 0.1 m high when the force is applied (a) at the center, (b) at the top?

3–10 The boom in Fig. 3–20 is uniform and weighs 2500 N. (a) Find the tension in the guy wire, and the horizontal and vertical components of the force exerted on the boom at its lower end. (b) Does the line of action of this force lie along the boom?

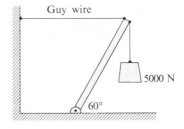

Fig. 3–20

3–11 Two ladders, 4 m and 3 m long, respectively, are hinged at point A and tied together by a horizontal rope 0.6 m above the floor, as in Fig. 3–21. The ladders weigh 400 N and 300 N respectively, and the center of gravity of each is at its center. If the floor is frictionless, find (a) the upward force at the bottom of each ladder, (b) the tension in the rope, and (c) the force which one ladder exerts on the other at point A. (d) If a load of 1000 N is now suspended from point A, find the tension in the rope.

Fig. 3–21

3–12 A uniform ladder 10 m long rests against a vertical frictionless wall with its lower end 6 m from the wall. The ladder weighs 400 N. The coefficient of static friction between the foot of the ladder and the ground is 0.40. A man weighing 800 N climbs slowly up the ladder.

a) What is the maximum frictional force which the ground can exert on the ladder at its lower end?

b) What is the actual frictional force when the man has climbed 5 m along the ladder?

c) How far along the ladder can the man climb before the ladder starts to slip?

3–13 One end of a meter stick is placed against a vertical wall, as in Fig. 3–22. The other end is held by a light cord making an angle θ with the stick. The coefficient of static friction between the end of the meter stick and the wall is 0.30.

a) What is the maximum value the angle θ can have if the stick is to remain in equilibrium?

b) Let the angle θ be 10°. A body of the same weight as the meter stick is suspended from the stick as shown by dotted lines, at a distance x from the wall. What is the minimum value of x for which the stick will remain in equilibrium?

c) When $\theta = 10°$, how large must the coefficient of static friction be so that the body can be attached at the left end of the stick without causing it to slip?

Fig. 3–22

3–14 One end of a post weighing 500 N rests on a rough horizontal surface with $\mu_s = 0.3$. The upper end is held by a rope fastened to the surface and making an angle of 37° with the post, as in Fig. 3–23. A horizontal force F is exerted on the post as shown.

a) If the force F is applied at the midpoint of the post, what is the largest value it can have without causing the post to slip?

Fig. 3–23

b) How large can the force be, without causing the post to slip, if its height above the surface equals 9/10 of the length of the post?

3–15 A uniform smooth rod of length ℓ and weight w rests at equilibrium in a smooth semispherical bowl of radius R, as shown in Fig. 3–24, where $R < \ell/2 < 2R$. If θ is the angle of equilibrium and P is the force exerted by the edge of the bowl on the rod, prove that (a) $P = (\ell/4R)w$, (b) $(\cos 2\theta)/(\cos \theta) = \ell/4R$.

Fig. 3–24

3–16 A door 3.5 m high and 1.5 m wide is hung from hinges 3 m apart and 0.25 m from the top and bottom of the door. The door weighs 300 N, its center of gravity is at its center, and each hinge carries half the weight of the door. Find the horizontal component of the force exerted on the door at each hinge.

3–17 A gate 4 m long and 2 m high weighs 400 N. Its center of gravity is at its center, and it is hinged at A and B. To relieve the strain on the top hinge, a wire CD is connected as shown in Fig. 3–25. The tension in CD is increased until the horizontal force at hinge A is zero.

Fig. 3–25

a) What is the tension in the wire *CD*?

b) What is the magnitude of the horizontal component of force at hinge *B*?

c) What is the combined vertical force exerted by hinges *A* and *B*?

3–18 A uniform rectangular block, 0.5 m high and 0.25 m wide, rests on a plank *AB*, as in Fig. 3–26. The coefficient of static friction between block and plank is 0.40.

a) In a diagram drawn to scale, show the line of action of the resultant normal force exerted on the block by the plank when the angle $\theta = 15°$.

b) If end *B* of the plank is slowly raised, will the block start to slide down the plank before it tips over? Find the angle θ at which it starts to slide, or at which it tips.

c) What would be the answer to part (b) if the coefficient of static friction were 0.60? If it were 0.50?

3–19 A rectangular block 0.25 m wide and 0.5 m high is dragged to the right along a level surface at constant speed by a horizontal force *P,* as shown in Fig. 3–27. The coefficient of sliding friction is 0.40, the block weighs 25 N, and its center of gravity is at its center.

a) Find the magnitude of the force *P.*

b) Find the position of the line of action of the normal force *N* exerted on the block by the surface, if the height $h = 0.125$ m.

c) Find the value of *h* at which the block just starts to tip.

3–20 A garage door is mounted on an overhead rail, as in Fig. 3–28. The wheels at *A* and *B* have rusted so that they do not roll, but slide along the track. The coefficient of sliding friction is 0.5. The distance between the wheels is 2 m, and each is 0.5 m in from the vertical sides of the door. The door is symmetrical and weighs 800 N. It is pushed to the left at constant velocity by a horizontal force *P.*

a) If the distance *h* is 1.5 m, what is the vertical component of the force exerted on each wheel by the track?

b) Find the maximum value *h* can have without causing one wheel to leave the track.

Fig. 3–28

3–21 The objects in Fig. 3–29 are constructed of wire bent into the shapes shown. Find the position of the center of gravity of each.

Fig. 3–26

Fig. 3–27

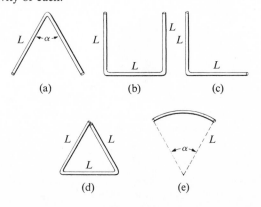

Fig. 3–29

3–22 A certain automobile has a wheelbase (distance between front and rear axles) of 120 inches. If 60 percent of the weight rests on the front wheels, how far behind the front wheels is the center of gravity?

3–23 Two men are carrying a ladder 20 ft long, weighing 200 lb. If one man can lift a maximum of 80 lb and lifts at one end, at what point should the other man lift?

3–24 A station wagon weighing 4000 lb has a wheelbase (distance between front and rear axles) of 120 inches. Ordinarily 2200 lb rests on the front wheels and 1800 lb on the rear wheels. A box weighing 500 lb is now placed on the tailgate, 4 ft behind the rear axle. How much total weight now rests on the front wheels? On the rear wheels?

3–25 A door 1.0 m wide and 2.5 m high is supported by two hinges, one 0.5 m from the top and the other 0.5 m from the bottom. Each hinge supports half the total weight of the door, which is 200 N. Assuming the door's center of gravity is at its center, find (a) the components of force exerted on the door by each hinge; (b) the magnitude and direction of the force exerted by each hinge.

3–26 Figure 3–30 is a top view of a triangular flat plate ABC resting on a horizontal frictionless surface. The length of side AB is 20 cm and that of CB is 15 cm. The plate is acted on by a couple consisting of the horizontal forces F_1 and F_1', each of magnitude 150 units.

a) In part (a) of the figure, the plate is kept in equilibrium by a second couple consisting of the forces F_2 and F_2'. Find the magnitude of each if the length $ab = 10$ cm.

b) Show that the plate is still in equilibrium (that is, the first and second conditions of equilibrium are satisfied) if the couple formed by F_2 and F_2' is displaced to the right, as in part (b).

c) Can equilibrum be maintained by the couple formed by the forces F_3 and F_3' in part (c) of the diagram? If so, find the magnitude of each.

d) Show in a diagram how the plate could be kept in equilibrium by a couple consisting of two forces perpendicular to side AC and applied at corners A and C. Find the magnitude of the required forces.

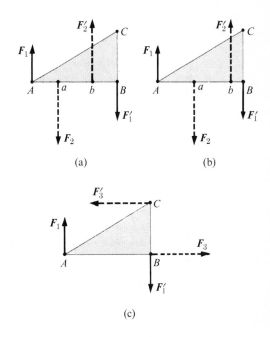

(a)   (b)

(c)

Fig. 3–30

3–27 Where is the center of gravity of the system made up of the earth and the moon?

Chapter 4

Rectilinear Motion

4–1 MOTION

Mechanics deals with the relations of force, matter, and motion. The preceding chapters have been concerned with forces, and we are now ready to discuss the mathematical methods of describing motion. This branch of mechanics is called *kinematics*.

Motion may be defined as a continuous change of position. In most actual motions, different points in a body move along different paths. The complete motion is known if we know how each point in the body moves, so to begin we consider only a moving point, or a very small body called a *particle*.

The position of a particle is conveniently specified by its projections onto the three axes of a rectangular coordinate system. As the particle moves along any path in space, its projections move in straight lines along the three axes. The actual motion can be reconstructed from the motions of these three projections, so we shall begin by discussing the motion of a single particle along a straight line, or *rectilinear motion*.

4–2 AVERAGE VELOCITY

Consider a particle moving along the x-axis, as in Fig. 4–1(a). The curve in Fig. 4–1(b) is a graph of its coordinate x plotted as a function of time t. At a time t_1 the particle is at point P in Fig. 4–1(a), where its coordinate is x_1, and at a later time t_2 it is at point Q, whose coordinate is x_2. The corresponding points on the coordinate-time graph in part (b) are lettered p and q.

The *displacement* of a particle as it moves from one point of its path to another is defined as the vector Δx drawn from the first point to the second.* Thus in Fig. 4–1(a) the vector PQ, of magnitude $x_2 - x_1 = \Delta x$, is the displacement. The *average velocity* of the particle is defined as the ratio of the displacement to the time interval $t_2 - t_1 = \Delta t$. We shall represent average velocity by the symbol $\bar{v}$ (the bar signifying an average value):

$$\bar{v} = \frac{\Delta x}{\Delta t}.$$

Average velocity is a vector, since the ratio of a vector to a scalar is itself a vector. Its direction is the same as that of the displacement vector. The magni-

---

* The Greek letter Δ is often used together with another symbol to represent a *change* in the quantity represented by that symbol. Thus Δt, read as a symbol, means "change in t." The elapsed time interval between time t_1 and t_2 is often written as $t_2 - t_1 = \Delta t$, and similarly Δx is the change in the vector x drawn from the origin to the position of the particle.

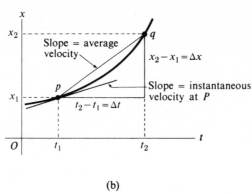

(a)   (b)

Fig. 4–1 (a) Particle moving on the x-axis. (b) Coordinate-time graph of the motion. The average velocity between t_1 and t_2 equals the slope of the chord pq. The instantaneous velocity at p equals the slope of the tangent at p.

tude of the average velocity is therefore

$$\bar{v} = \frac{x_2 - x_1}{t_2 - t_1} = \frac{\Delta x}{\Delta t}. \qquad (4\text{--}1)$$

In Fig. 4–1(b), the average velocity is represented by the slope of the chord pq; that is, the ratio of vertical to horizontal intervals on the triangle of which pq is the hypotenuse, evaluated using the scales to which x and t are plotted. The slope is the ratio of the "rise," $x_2 - x_1$ or Δx, to the "run," $t_2 - t_1$ or Δt.

Equation (4–1) can be written

$$x_2 - x_1 = \bar{v}(t_2 - t_1). \qquad (4\text{--}2)$$

Since our time-measuring device can be started at any instant, we can let $t_1 = 0$ and let t_2 be any arbitrary later time t. Then if x_0 is the coordinate when $t = 0$ (called the *initial position*), and x is the coordinate at time t, Eq. (4–2) becomes

$$x - x_0 = \bar{v}t. \qquad (4\text{--}3)$$

If the particle is at the origin when $t = 0$, then $x_0 = 0$ and Eq. (4–3) simplifies further to

$$x = \bar{v}t. \qquad (4\text{--}4)$$

4–3 INSTANTANEOUS VELOCITY

The velocity of a particle at some one instant of time, or at some one point of its path, is called its *instantaneous velocity*. This concept requires careful definition.

Suppose we wish to find the instantaneous velocity of the particle in Fig. 4–1 at the point P. The average velocity between points P and Q is associated with the entire displacement $\Delta \boldsymbol{x}$, and with the entire time interval Δt. Imagine the second point Q to be taken closer and closer to the first point P, and let the average velocity be computed over these shorter and shorter displacements and time intervals. The instantaneous velocity at the first point can then be defined as the *limiting value* of the average velocity when the second point is taken closer and closer to the first. Although the displacement then becomes extremely small, the time interval by which it must be divided becomes small also and the quotient therefore is not necessarily a small quantity.

In the notation of calculus, the limiting value of $\Delta x/\Delta t$, as Δt approaches zero, is written dx/dt and is called the *derivative* of x with respect to t.

Then if $\boldsymbol{v}$ represents the instantaneous velocity, its magnitude is

$$v = \lim_{\Delta t \to 0} \frac{\Delta x}{\Delta t} = \frac{dx}{dt}. \qquad (4\text{--}5)$$

Instantaneous velocity is also a vector, whose direction is the limiting direction of the displacement vector $\Delta \boldsymbol{x}$. Since Δt is necessarily positive, it follows that v has the same algebraic sign as Δx. Hence a positive velocity indicates motion toward the right along the x-axis, if we use the usual convention of signs.

As point Q approaches point P in Fig. 4–1(a), point q approaches point p in Fig. 4–1(b). In the limit, the slope of the chord pq equals the slope of the *tangent* to the curve at point p, due allowance being made for the scales to which x and t are plotted. *The instantaneous velocity at any point of a coordinate-time graph therefore equals the slope of the tangent to the graph at that point.* If the tangent slopes upward to the right, its slope is positive, the velocity is positive, and the motion is toward the right. If the tangent slopes downward to the right, the velocity is negative. At a point where the tangent is horizontal, its slope is zero and the velocity is zero.

If we express distance in meters and time in seconds, velocity is expressed in meters per second (m s^{-1}). Other common units of velocity are feet per second (ft s^{-1}), centimeters per second (cm s^{-1}), miles per hour (mi hr^{-1}), and knots (1 knot = 1 nautical mile per hour).

Example   Suppose the motion of the particle in Fig. 4–1 is described by the equation $x = a + bt^2$, where $a = 20$ cm and $b = 4$ cm s^{-2}.

a) Find the displacement of the particle in the time interval between $t_1 = 2$ s and $t_2 = 5$ s.

At time $t_1 = 2$ s, the position is

$$x_1 = 20 \text{ cm} + (4 \text{ cm s}^{-2})(2 \text{ s})^2 = 36 \text{ cm}.$$

At time $t_2 = 5$ s,

$$x_2 = 20 \text{ cm} + (4 \text{ cm s}^{-2})(5 \text{ s})^2 = 120 \text{ cm}.$$

The displacement is therefore

$$x_2 - x_1 = 120 \text{ cm} - 36 \text{ cm} = 84 \text{ cm}.$$

b) Find the average velocity in this time interval:

$$\bar{v} = \frac{x_2 - x_1}{t_2 - t_1} = \frac{84 \text{ cm}}{3 \text{ s}} = 28 \text{ cm s}^{-1}.$$

This corresponds to the slope of the chord pq in Fig. 4–1(b).

c) Find the instantaneous velocity at time $t_1 = 2$ s.

The position at time $t = 2 \text{ s} + \Delta t$ is

$$x = 20 \text{ cm} + (4 \text{ cm s}^{-2})(2 \text{ s} + \Delta t)^2$$

$$= 36 \text{ cm} + (16 \text{ cm s}^{-1})\Delta t + (4 \text{ cm s}^{-2})(\Delta t)^2 .$$

The displacement during the interval Δt is

$$\Delta x = 36 \text{ cm} + (16 \text{ cm s}^{-1})\Delta t + (4 \text{ cm s}^{-2})(\Delta t)^2$$

$$- 36 \text{ cm}$$

$$= (16 \text{ cm s}^{-1})\Delta t + (4 \text{ cm s}^{-2})(\Delta t)^2 .$$

The average velocity during Δt is

$$\bar{v} = \frac{\Delta x}{\Delta t} = 16 \text{ cm s}^{-1} + (4 \text{ cm s}^{-1})\Delta t.$$

For the instantaneous velocity at $t = 2$ s, we let Δt approach zero in the expression for $\bar{v}$: $v = 16 \text{ cm s}^{-1}$. This corresponds to the slope of the tangent at point p in Fig. 4–1(b). The instantaneous velocity may also be obtained directly by calculating the derivative of x with respect to t. Starting again with $x = a + bt^2$ and using Eq. (4–5), we find

$$v = \frac{dx}{dt} = \frac{d}{dt}(a + bt^2) = 2bt.$$

The instantaneous velocity at time $t_1 = 2$ s is

$$v = (2)(4 \text{ cm s}^{-2})(2 \text{ s}) = 16 \text{ cm s}^{-1} .$$

The term *speed* has two different meanings. It may mean the *magnitude* of the instantaneous velocity; two automobiles traveling at 50 mi hr^{-1}, one north and the other south, both have a speed of 50 mi hr^{-1}. In another sense, the *average speed* of a body means the total length of path covered, divided by the elapsed time. Thus, if an automobile travels 90 mi in 3hr, its average speed is 30 mi hr^{-1}, even if the trip starts and ends at the same point. The average *velocity*, in the latter case, would be zero, since the total displacement is zero.

4–4 AVERAGE AND INSTANTANEOUS ACCELERATION

When the velocity of a moving body changes continuously as the motion proceeds, the body is said to move with *accelerated motion*,

Fig. 4–2 (a) *Particle moving on the x-axis.* (b) *Velocity-time graph of the motion. The average accelera-tion between t_1 and t_2 equals the slope of the chord pq. The instan-taneous acceleration at p equals the slope of the tangent at p.*

Figure 4–2(a) shows a particle moving along the x-axis. The vector v_1 represents its instantaneous velocity at point P, and the vector v_2 represents its instantaneous velocity at point Q. Figure 4–2(b) is a graph of the magnitude of the instantaneous velocity v plotted as a function of time, points p and q corresponding to P and Q in part (a).

The *average acceleration* of the particle as it moves from P to Q is defined as the ratio of the change in velocity to the elapsed time.

$$\bar{a} = \frac{v_2 - v_1}{t_2 - t_1} = \frac{\Delta v}{\Delta t}, \qquad (4\text{–}6)$$

where t_1 and t_2 are the times corresponding to the velocities v_1 and v_2. Since v_1 and v_2 are vectors, the quantity $v_2 - v_1$ is a *vector difference* and must be found by the methods explained in Section 1–9. However, since in rectilinear motion both vectors lie in the same straight line, the magnitude of the vector difference in this special case equals the difference between the magnitudes of the vectors. The more general case, in which v_1 and v_2 are not in the same direction, will be considered in Chapter 6.

In Fig. 4–2(b), the magnitude of the average acceleration is represented by the slope of the chord pq.

The *instantaneous acceleration* of a body, that is, its acceleration at some one instant of time or at some one point of its path, is defined in the same way as instantaneous velocity. Let the second point Q in Fig. 4–2(a) be taken closer and closer to the first point P, and let the average acceleration be comput-ed over shorter and shorter intervals of time. The instantaneous acceleration at the first point is defined

as the limiting value of the average acceleration when the second point is taken closer and closer to the first:

$$a = \lim_{\Delta t \to 0} \frac{\Delta v}{\Delta t} = \frac{dv}{dt}. \qquad (4\text{–}7)$$

The direction of the instantaneous acceleration is the limiting direction of the vector change in ve-locity, Δv.

Instantaneous acceleration plays an important part in the laws of mechanics. Average acceleration is less frequently used. Hence from now on when the term "acceleration" is used we shall understand it to mean instantaneous acceleration.

The definition of acceleration just given applies to motion along any path, straight or curved. When a particle moves in a curved path, the *direction* of its velocity changes and this change in direction also gives rise to an acceleration, as will be explained in Chapter 6.

As point Q approaches point P in Fig. 4–2(a), point q approaches point p in Fig. 4–2(b) and the slope of the chord pq approaches the slope of the tangent to the velocity-time graph at point q. *The instantaneous acceleration at any point of the graph therefore equals the slope of the tangent to the graph at that point.*

The acceleration $a = dv/dt$ can be expressed in various ways. Since $v = dx/dt$, it follows that

$$a = \frac{dv}{dt} = \frac{d}{dt}\left(\frac{dx}{dt}\right) = \frac{d^2x}{dt^2}.$$

The acceleration is therefore the *second* derivative of the coordinate with respect to time.

We can also use the chain rule and write

$$a = \frac{dv}{dt} = \frac{dv}{dx}\frac{dx}{dt} = v\frac{dv}{dx}, \qquad (4\text{–}8)$$

which expresses the acceleration in terms of the *space rate of change in velocity*, dv/dx.

If we express velocity in feet per second and time in seconds, acceleration is expressed in feet per second, per second (ft s^{-1} s^{-1}). This is usually written as ft s^{-2}, and is read "feet per second squared." Other common units of acceleration are meters per second squared (m s^{-2}) and centimeters per second squared (cm s^{-2}).

When the absolute value of the magnitude of the velocity of a body is decreasing (in other words, when the body is slowing down), the body is said to be *decelerated* or to have a *deceleration*.

Example Suppose the velocity of the particle in Fig. 4–2 is given by the equation

$$v = m + nt^2,$$

where $m = 10$ cm s^{-1} and $n = 2$ cm s^{-3}.

a) Find the change in velocity of the particle in the time interval between $t_1 = 2$ s and $t_2 = 5$ s.

At time $t_1 = 2$ s,

$$v_1 = 10 \text{ cm s}^{-1} + (2 \text{ cm s}^{-3})(2 \text{ s})^2 = 18 \text{ cm s}^{-1}.$$

At time $t_2 = 5$ s,

$$v_2 = 10 \text{ cm s}^{-1} + (2 \text{ cm s}^{-3})(5 \text{ s})^2 = 60 \text{ cm s}^{-1}.$$

The change in velocity is therefore

$$v_2 - v_1 = 60 \text{ cm s}^{-1} - 18 \text{ cm s}^{-1} = 42 \text{ cm s}^{-1}.$$

b) Find the average acceleration in this time interval.

$$\bar{a} = \frac{v_2 - v_1}{t_2 - t_1} = \frac{42 \text{ cm s}^{-1}}{3 \text{ s}} = 14 \text{ cm s}^{-2}.$$

This corresponds to the slope of the chord pq in Fig. 4–2(b).

c) Find the instantaneous acceleration at time $t_1 = 2$ s.

At time $t = 2$ s $+ \Delta t$,

$$v = 10 \text{ cm s}^{-1} + (2 \text{ cm s}^{-3})(2 \text{ s} + \Delta t)^2$$

$$= 18 \text{ cm s}^{-1} + (8 \text{ cm s}^{-2})\Delta t + (2 \text{ cm s}^{-3})(\Delta t)^2.$$

The change in velocity during Δt is

$$\Delta v = 18 \text{ cm s}^{-1} + (8 \text{ cm s}^{-2})\Delta t + (2 \text{ cm s}^{-3})(\Delta t)^2$$

$$\quad - 18 \text{ cm s}^{-1}$$

$$= (8 \text{ cm s}^{-2})\Delta t + (2 \text{ cm s}^{-3})(\Delta t)^2.$$

The average acceleration during Δt is

$$\bar{a} = \frac{\Delta v}{\Delta t} = 8 \text{ cm s}^{-2} + (2 \text{ cm s}^{-3})\Delta t.$$

The instantaneous acceleration at $t = 2$ s, obtained by letting Δt approach zero, is $a = 8$ cm s^{-2}. This corresponds to the slope of the tangent at point p in Fig. 4–2(b).

The instantaneous acceleration can also be obtained directly by taking the time derivative of the instantaneous velocity: Starting with $v = m + nt^2$ and using Eq. (4–7), we find

$$a = \frac{dv}{dt} = \frac{d}{dt}(m + nt^2) = 2nt.$$

The instantaneous acceleration when $t = 2$ s is

$$a = (2)(2 \text{ cm s}^{-3})(2 \text{ s}) = 8 \text{ cm s}^{-2}.$$

4–5 RECTILINEAR MOTION WITH CONSTANT ACCELERATION

The simplest kind of accelerated motion is rectilinear motion in which the acceleration is constant, that is,

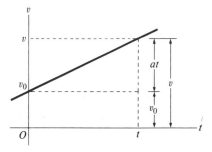

Fig. 4–3 *Velocity–time graph for rectilinear motion with constant acceleration.*

in which the velocity changes at the same rate throughout the motion. The velocity–time graph is then a straight line, as in Fig. 4–3, the velocity increasing by equal amounts in equal intervals of time. The slope of a chord between any two points on the line is the same as the slope of a tangent at any point, and the average and instantaneous accelerations are equal. Hence in Eq. (4–6) the average acceleration $\bar{a}$ can be replaced by the constant acceleration a and we have

$$a = \frac{v_2 - v_1}{t_2 - t_1}. \qquad (4\text{–}9)$$

Now let $t_1 = 0$ and let t_2 be any arbitrary time t. Let v_0 represent the velocity when $t = 0$ (called the *initial* velocity), and let v be the velocity at time t. Then the preceding equation becomes

$$a = \frac{v - v_0}{t - 0},$$

or

$$\boxed{v = v_0 + at.} \qquad (4\text{–}10)$$

The equation can be interpreted as follows: The acceleration a is the constant rate of change of velocity, or the change per unit time. The term at is the product of the change in velocity per unit time, a, and the duration of the time interval, t; therefore it equals the *total* change in velocity. The velocity v at the time t then equals the velocity v_0 at the time $t = 0$, plus the change in velocity at. Graphically, the ordinate v at time t, in Fig. 4–3, can be considered as the sum of two segments: one of length v_0 equal to the initial velocity, the other of length at equal to the change in velocity in time t.

To find the displacement of a particle moving with constant acceleration, we make use of the fact that when the acceleration is constant and the velocity–time graph is a straight line, as in Fig. 4–3, the average velocity in any time interval equals one-half the sum of the velocities at the beginning and the end of the interval. Hence the average velocity between zero and t is

$$\bar{v} = \frac{v_0 + v}{2}. \qquad (4\text{–}11)$$

(This is *not* true in general, when the acceleration is not constant and the velocity–time graph is curved as in Fig. 4–2.)

By definition, the average velocity is

$$\bar{v} = \frac{x_2 - x_1}{t_2 - t_1}.$$

Now let $t_1 = 0$ and let t_2 be any arbitrary time t. Let x_0 represent the position when $t = 0$ (the *initial* position) and let x be the position at time t. Then the preceding equation becomes

$$x - x_0 = \bar{v}t. \qquad (4\text{–}12)$$

Substituting the expression for $\bar{v}$ in Eq. (4–11), we obtain

$$x - x_0 = \left(\frac{v_0 + v}{2}\right)t. \qquad (4\text{–}13)$$

Two more very useful equations can be obtained from Eqs. (4–10) and (4–13), first by eliminating v and then by eliminating t. When we substitute in Eq. (4–13) the expression for v in Eq. (4–10), we get:

$$x - x_0 = \left(\frac{v_0 + v_0 + at}{2}\right)t,$$

or

$$\boxed{x - x_0 = v_0 t + \tfrac{1}{2}at^2.} \qquad (4\text{–}14)$$

When Eq. (4–10) is solved for t and the result substituted in Eq. (4–13), we have

$$x - x_0 = \left(\frac{v_0 + v}{2}\right)\left(\frac{v - v_0}{a}\right) = \frac{v^2 - v_0^2}{2a},$$

or finally,

$$\boxed{v^2 = v_0^2 + 2a(x - x_0).} \qquad (4\text{–}15)$$

Equations (4–10), (4–13), (4–14), and (4–15) are the *equations of motion with constant acceleration.*

The curve in Fig. 4–4 is the coordinate-time graph for motion with constant acceleration. That is, it is a graph of Eq. (4–14); the curve is a parabola. The slope of the tangent at $t = 0$ equals the initial

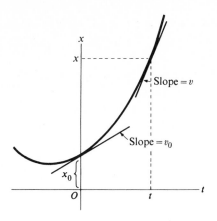

Fig. 4–4

velocity v_0, and the slope of the tangent at time t equals the velocity v at that time. It is evident that the slope continually increases, and measurements would show that the rate of increase is constant, that is, that the acceleration is constant.

A special case of motion with constant acceleration is that in which the acceleration is zero. The *velocity* is then constant and the equations of motion become simply

$$v = \text{constant}, \qquad x = x_0 + vt.$$

4–6 VELOCITY AND COORDINATE BY INTEGRATION

If the coordinate x of a particle moving on the x-axis is given as a function of time, the velocity can be found by differentiation, from the definition $v = dx/dt$. A second differentiation gives the acceleration, since $a = dv/dt$. We now consider the converse process: given the acceleration, to find the velocity and the cordinate. This can be done by the methods of *integral* calculus. We shall discuss first the indefinite and then the definite integral.

Suppose we are given the acceleration $a(t)$ as a function of time. Then, since

$$\frac{dv}{dt} = a(t),$$

we have

$$dv = a(t)dt, \qquad \int dv = \int a(t)dt,$$

and

$$v = \int a(t)dt + C_1, \qquad (4\text{–}16)$$

were C_1 is an integration constant whose value can be determined if the velocity is known at any time. It is customary to express C_1 in terms of the velocity v_0 when $t = 0$.

When the integral above has been evaluated, we have the velocity $v(t)$ as a function of time. Then, since

$$\frac{dx}{dt} = v(t),$$

we have

$$dx = v(t)dt, \qquad \int dx = \int v(t)dt,$$

$$x = \int v(t)dt + C_2, \qquad (4\text{–}17)$$

where C_2 is a second integration constant whose value can be determined if the coordinate is known at any time. It is customary to express C_2 in terms of the coordinate x_0 when $t = 0$.

If the acceleration is given as a function of x, we can use Eq. (4–8):

$$v\frac{dv}{dx} = a(x), \qquad \int v\,dv = \int a(x)dx,$$

$$\frac{v^2}{2} = \int a(x)dx + C_3. \qquad (4\text{–}18)$$

Example   Derive the equations of motion with constant acceleration, using the indefinite integral.

If a is constant, then from Eq. (4–16),

$$v = at + C_1.$$

But $v = v_0$ when $t = 0$, so

$$v_0 = 0 + C_1$$

and

$$v = v_0 + at,$$

which is Eq. (4–10).

From Eq. (4–17), when a is constant,

$$x = \int (v_0 + at)\,dt = v_0t + \tfrac{1}{2}at^2 + C_2.$$

If $x = x_0$ when $t = 0$, then $C_2 = x_0$ and

$$x = x_0 + v_0t + \tfrac{1}{2}at^2,$$

which is Eq. (4–14).

From Eq. (4–18), when a is constant,

$$\frac{v^2}{2} = ax + C_3.$$

If $v = v_0$ when $x = x_0$, then $C_3 = v_0^2/2 - ax_0$ and

$$v^2 = v_0^2 + 2a(x - x_0),$$

which is Eq. (4–15).

Consider next the definite integral. Let the area under the velocity–time graph in Fig. 4–5, between the vertical lines at t_1 and t_2, be subdivided into narrow rectangular strips of width Δt. The ordinate of the graph at any time t equals the instantaneous velocity v at that time. If the velocity remained constant at this value, the displacement Δx in the time interval between t and $t + \Delta t$ would equal $v\Delta t$. But this is the *area* of the shaded strip, since its height is v and its width is Δt. The *sum* of the areas of such rectangles, between t_1 and t_2, is approximately equal to the *total* displacement $x_2 - x_1$ in this time interval:

$$x_2 - x_1 \approx \sum v\Delta t.$$

The smaller the time intervals Δt, the more closely does $v\Delta t$ approach the actual displacement. In the limit, as Δt approaches zero, the sum of the areas becomes exactly equal to the total area under the curve, and to the total displacement $x_2 - x_1$. The limit of the sum of the areas is the definite integral from t_1 to t_2, so

$$x_2 - x_1 = \int_{t_1}^{t_2} v \, dt. \qquad (4–19)$$

The displacement in any time interval is therefore equal to the area between a velocity–time graph and the

time axis, bounded by vertical lines at the beginning and end of the interval.

In the same way, the area under an acceleration–time graph can be subdivided into vertical strips of height a and width Δt. If the acceleration remained constant, the change in velocity Δv in time Δt would equal $a\Delta t$, the area of a rectangular strip. The total change in velocity, $v_2 - v_1$, in the time interval from t_1 and t_2, is approximately equal to the sum of all such areas.

$$v_2 - v_1 \approx \sum a\Delta t.$$

In the limit, as Δt approaches zero,

$$v_2 - v_1 = \int_{t_1}^{t_2} a \, dt. \qquad (4–20)$$

The change in velocity in any time interval is therefore equal to the area between an acceleration–time graph and the time axis, bounded by vertical lines at the beginning and end of the interval.

Example Use the definite integral to find the velocity and coordinate, at any time t, of a body moving on the x-axis with constant acceleration. The initial velocity is v_0 and the initial coordinate is x_0.

As the limits of integration, we take $t_1 = 0$ and $t_2 = t$. Then, from Eq. (4–20),

$$v - v_0 = \int_0^t a \, dt = at,$$

which is Eq. (4–10).

From Eq. (4–19),

$$x - x_0 = \int_0^t (v_0 + at) \, dt = v_0 t + \tfrac{1}{2}at^2,$$

which is Eq. (4–14).

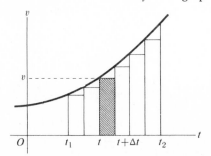

Fig. 4–5 *The area under a velocity–time graph equals the displacement.*

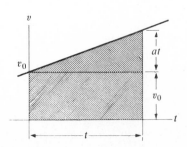

Fig. 4–6 *The area under a velocity–time graph equals the displacement $x - x_0$.*

It is not always necessary to use the methods of integral calculus to find the area under a graph. Figure 4–6 is the velocity–time graph for motion with constant acceleration. The area under the graph, between $t = 0$ and $t = t$, can be subdivided into a rectangle and a triangle. The area of the rectangle is $v_0 t$ and that of the triangle is $\frac{1}{2}(t)(at) = \frac{1}{2}at^2$. Since the displacement equals the total area,

$$x - x_0 = v_0 t + \tfrac{1}{2}at^2,$$

which is Eq. (4–14).

4–7 FREELY FALLING BODIES

The most common example of motion with (nearly) constant acceleration is that of a body falling toward the earth. In the absence of air resistance it is found that all bodies, regardless of their size or weight, fall with the same acceleration at the same point on the earth's surface, and if the distance covered is small compared to the radius of the earth, the acceleration remains constant throughout the fall. The effect of air resistance and the decrease in acceleration with altitude will be neglected. This idealized motion is spoken of as "free fall," although the term includes rising as well as falling motion.

The acceleration of a freely falling body is called acceleration due to gravity, or the acceleration of gravity, and is denoted by the letter g. At or near the earth's surface its magnitude is approximately 32 ft s^{-2}, 9.8 m s^{-2}, or 980 cm s^{-2}. More precise values, and small variations with latitude and elevation, will be considered later. On the surface of the moon, the acceleration of gravity is due to the attractive force exerted on a body by the moon rather than earth. On the moon, $g = 1.67$ m s^{-2} $= 5.47$ ft s^{-2}. Near the surface of the *sun*, $g = 274$ m s^{-2}!

Note. The quantity "g" is sometimes referred to simply as "gravity," or as "the force of gravity," both of which are incorrect. "Gravity" is a phenomenon, and the "force of gravity" means the force with which the earth attracts a body, otherwise known as the weight of the body. The letter "g" represents the *acceleration* caused by the force resulting from the phenomenon of gravity.

Example 1   A body is released from rest and falls freely. Compute its position and velocity after 1, 2, 3, and 4 s. Take the origin 0 at the elevation of the starting point, the y-axis vertical, and the upward direction as positive.

The initial coordinate y_0 and the initial velocity v_0 are both zero. The acceleration is downward, in the negative y-direction, so $a = -g = -32$ ft s^{-2}.

From Eqs. (4–14) and (4–10),

$$y = v_0 t + \tfrac{1}{2}at^2 = 0 - \tfrac{1}{2}gt^2 = (-16 \text{ ft s}^{-2})t^2,$$
$$v = v_0 + at = 0 - gt = (-32 \text{ ft s}^{-2})t.$$

When $t = 1$ s, $y = (-16 \text{ ft s}^{-2})(1 \text{ s})^2 = -16$ ft, and $v = (-32 \text{ ft s}^{-2})(1 \text{ s}) = 32$ ft s^{-1}. The body is therefore 16 ft below the origin (y is negative) and has a downward velocity (v is negative) of magnitude 32 ft s^{-1}.

The position and velocity at 2, 3, and 4 s are found in the same way. The results are illustrated in Fig. 4–7.

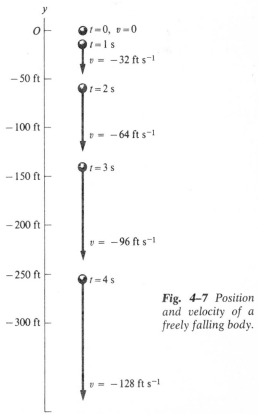

Fig. 4–7 *Position and velocity of a freely falling body.*

Example 2   A ball is thrown (nearly) vertically upward from the cornice of a tall building, leaving the thrower's hand with a speed of 48 ft s^{-1} and just missing the cornice on the way down, as in Fig. 4–8. The downward path is shown displaced to the right for clarity. Find (a) the position and velocity of the ball, 1 s and 4 s after leaving the thrower's hand; (b) the velocity when the ball is 20 ft above its starting point; (c) the maximum height reached and the time at which it is reached. Take the origin at the elevation at which the ball leaves the thrower's hand, the y-axis vertical and positive upward.

The initial position y_0 is zero. The initial velocity v_0 is +48 ft s^{-1}, and the acceleration is −32 ft s^{-2}. The velocity at any time is

$$v = v_0 + at = 48 \text{ ft s}^{-1} - (32 \text{ ft s}^{-2})t. \quad (4\text{–}21)$$

The coordinate at any time is

$$y = v_0 t + \tfrac{1}{2}at^2 = (48 \text{ ft s}^{-1})t - (16 \text{ ft s}^{-2})t^2.$$

The velocity at any coordinate is

$$v^2 = v_0{}^2 + 2ay = (48 \text{ ft s}^{-1})^2 - (64 \text{ ft s}^{-2})y. \quad (4\text{–}22)$$

a) When $t = 1$ s,

$$y = +32 \text{ ft}, \qquad v = +16 \text{ ft s}^{-1}.$$

The ball is 32 ft above the origin (y is positive) and it has an upward velocity (v is positive) of 16 ft s^{-1} (less than the initial velocity, as expected). When $t = 4$ s,

$$y = -64 \text{ ft}, \qquad v = -80 \text{ ft s}^{-1}.$$

The ball has passed its highest point and is 64 ft *below* the origin (y is negative). It has a *downward* velocity (v is negative) of magnitude 80 ft s^{-1}. Note that it is not necessary to find the highest point reached, or the time at which it was reached. The equations of motion give the position and velocity at *any* time, whether the ball is on the way up or the way down.

b) When the ball is 20 ft above the origin

$$y = +20 \text{ ft}$$

Fig. 4–8 *Position and velocity of a body thrown vertically upward.*

and

$$v^2 = 1024 \text{ ft}^2 \text{ s}^{-2}, \qquad v = \pm 32 \text{ ft s}^{-1}.$$

The ball passes this point *twice*, once on the way up and again on the way down. The velocity on the way up is +32 ft s^{-1}, and on the way down it is −32 ft s^{-1}.

c) At the highest point, $v = 0$. Hence

$$0 = (48 \text{ ft s}^{-1})^2 - (64 \text{ ft s}^{-2})y,$$

and

$$y = +36 \text{ ft}.$$

The time can now be found either from Eq. (4–21), setting $v = 0$, or from the equation following

Eq. (4–21), setting $y = 36$ ft. From either equation, we get

$$t = 1.5 \text{ s}.$$

Although at the highest point the velocity is instantaneously zero, the *acceleration* at this point is still -32 ft s^{-1}. The ball stops for an instant, but its velocity is continuously changing; the acceleration is constant throughout.

Figure 4–9 is a "multiflash" photograph of a freely falling golf ball. This photograph was taken with aid of a stroboscopic light source, developed originally by Dr. Harold E. Edgerton of the Massachusetts Institute of Technology. This source produces a series of intense flashes of light. The interval between successive flashes is controllable at will, and the duration of each flash is so short (a few millionths of a second) that there is no blur in the image of even a rapidly moving body. The camera shutter is left open during the entire motion, and as each flash occurs, the position of the ball at that instant is recorded on the photographic film.

The equally spaced light flashes subdivide the motion into equal time intervals Δt. Since the time intervals are all equal, the velocity of the ball between any two flashes is directly proportional to the separation of its corresponding images in the photograph. If the velocity were constant, the images would be equally spaced. The increasing separation of the images during the fall shows that the velocity is continually increasing or the motion is accelerated. By comparing two successive displacements of the ball, we can find the *change* in velocity in the corresponding time interval. Careful measurement, preferably on an enlarged print, shows that this change in velocity is the same in each time interval. In other words, the motion is one of *constant* acceleration.

4–8 RECTILINEAR MOTION WITH VARIABLE ACCELERATION

Motion with constant acceleration approximates the motion of some falling bodies, and of cars and of airplanes at the start of their journey. There are many

Fig. 4–9 *Multiflash photograph (retouched) of freely falling golf ball.*

important types of motion, however, in which the acceleration is variable, and it is worth while to develop the technique of dealing with such cases. Consider, for example, the motion of a body in the positive x-direction with an acceleration whose direction is opposite that of the velocity and whose magnitude is proportional to the speed. For such a motion

$$a = -kv,$$

where k is a constant. If the initial speed is v_0, how do the speed and the distance vary with the time?

Since $a = dv/dt$, we have

$$\frac{dv}{dt} = -kv, \quad \text{or} \quad \frac{dv}{v} = -k \, dt,$$

and since $v = v_0$ when $t = 0$,

$$\int_{v_0}^{v} \frac{dv}{v} = -k \int_{0}^{t} dt.$$

Integration yields the result

$$\ln \frac{v}{v_0} = -kt,$$

which may be written as

$$v = v_0 e^{-kt}, \tag{4–23}$$

showing that the velocity decays exponentially with the time. Exponential curves are found frequently in many diverse fields of physics. The decrease in activity of a radioactive substance, the discharge of a capacitor, and the dying out of mechanical, acoustical, and electrical oscillations are all described by exponential equations. Equation (4–23) shows that an infinite time would be required to reduce the speed to zero.

To find the distance x as a function of the time, we replace v by dx/dt. Thus

$$\frac{dx}{dt} = v_0 e^{-kt}.$$

Suppose that $x = 0$ when $t = 0$. Then

$$\int_{0}^{x} dx = v_0 \int_{0}^{t} e^{-kt} dt,$$

and

$$x = -\frac{v_0}{k} \left[e^{-kt} \right]_{0}^{t},$$

$$x = \frac{v_0}{k}(1 - e^{-kt}). \tag{4–24}$$

It follows from this equation that although an infinite time is necessary for the body to come to rest, in this infinite time the body will have gone only a finite distance v_0/k.

Other examples of variable acceleration will be found in the problems at the end of this chapter and also in Chapter 5.

4–9 VELOCITY COMPONENTS, RELATIVE VELOCITY

Velocity is a vector quantity involving both magnitude and direction. A velocity may therefore be resolved into components, or a number of velocity components combined into a resultant. As an example of the former process, suppose that a ship is steaming 30° E of N at 20 mi hr^{-1} in still water. Its velocity may be represented by the arrow in Fig. 4–10, and one finds by the usual method that its velocity component toward the east is 10 mi hr^{-1}, while toward the north it is 17.3 mi hr^{-1}.

The velocity of a body, like its position, can be specified only relative to some other body. The second body may be in motion relative to a third, and so on. Thus when we speak of "the velocity of an automobile," we usually mean its velocity relative to the earth. But the earth is in motion relative to the sun, the sun is in motion relative to some other star, and so on.

Fig. 4–10 *Resolution of a velocity vector into components.*

Fig. 4–11 *Vector v_{FE} is the velocity of the flatcars F relative to the earth E, and v_{AF} is the velocity of automobile A relative to the flatcars.*

Suppose a long train of flatcars is moving to the right along a straight level track, as in Fig. 4–11, and that a daring automobile driver is driving to the right along the flatcars. The vector v_{FE} in Fig. 4–11 represents the velocity of the flatcars F relative to the earth E, and the vector v_{AF} the velocity of the automobile A relative to the flatcars F. The velocity of the automobile relative to the earth, v_{AE}, is evidently equal to the sum of the relative velocities v_{AF} and v_{FE}:

$$v_{AE} = v_{AF} + v_{FE}. \qquad (4\text{–}25)$$

Thus if the flatcars are traveling relative to the earth at 30 mi hr^{-1} ($= v_{FE}$) and the automobile is traveling relative to the flatcars at 40 mi hr^{-1} ($= v_{AF}$), the velocity of the automobile relative to the earth (v_{AE}) is 70 mi hr^{-1}.

Imagine now that the flatcars are wide enough so that the automobile can drive on them in any direction. The velocity of the automobile relative to the earth is then the vector sum of its velocity relative to the flatcars and the velocity of the flatcars relative to the earth. The preceding equation is therefore a special case of the more general *vector* equation,

$$\boxed{v_{AE} = v_{AF} + v_{FE}} \qquad (4\text{–}26)$$

Thus if the automobile were driving transversely across the flatcars at 40 mi hr^{-1}, its velocity relative to the earth would be 50 mi hr^{-1} at an angle of 53° with the railway track.

In the special case in which v_{AF} and v_{FE} are along the same line, as in Fig. 4–11, the velocity v_{AE} is the *algebraic* sum of v_{AF} and v_{FE}. Thus if the automobile were traveling to the *left* with a velocity of 40 mi hr^{-1} relative to the flatcars, $v_{AF} = -40$ mi hr^{-1} and the velocity of the automobile relative to the earth would be -10 mi hr^{-1}; that is, it would be traveling to the left, relative to the earth.

Equation (4–26) can be extended to include any number of relative velocities. For example, if a bug B crawls along the floor of the automobile with a velocity relative to the automobile of v_{BA}, his velocity relative to the earth is the vector sum of his velocity relative to the automobile and that of the velocity of the automobile relative to the earth:

$$v_{BE} = v_{BA} + v_{AE}.$$

When this is combined with Eq. (4–26), we get

$$v_{BE} = v_{BA} + v_{AF} + v_{FE}. \qquad (4\text{–}27)$$

This equation illustrates the general rule for combining relative velocities.

1. Write each velocity with a double subscript in the *proper order*, meaning "velocity of (first subscript) relative to (second subscript)."

2. When adding relative velocities, the first letter of any subscript is to be the same as the last letter of the preceding subscript.

3. The first letter of the subscript of the first velocity in the sum, and the second letter of the subscript of the last velocity, are the subscripts, in that order, of the relative velocity represented by the sum. This somewhat lengthy statement should be clear when it is compared with Eq. (4–27).

Any of the relative velocities in an equation like Eq. (4–26) can be transferred from one side of the equation to the other, with sign reversed. Thus Eq. (4–26) can be written

$$v_{AF} = v_{AE} - v_{FE}.$$

The velocity of the automobile relative to the flatcar equals the *vector difference* between the velocities of automobile and flatcar, both relative to the earth.

One more point should be noted. The velocity of body A relative to body B, v_{AB}, is the negative of the velocity of B relative to A, v_{BA}:

$$v_{AB} = -v_{BA}.$$

That is, v_{AB} is equal in magnitude and opposite in direction to v_{BA}. If the automobile in Fig. 4–11 is traveling to the right at 40 mi hr^{-1}, relative to the flatcars, the flatcars are traveling to the left at 40 mi hr^{-1}, relative to the automobile.

Example 1   An automobile driver A, traveling relative to the earth at 65 km hr^{-1} on a straight level road, is ahead of motorcycle officer B traveling in the same direction at 80 km hr^{-1}. What is the velocity of B relative to A?

We have given

$$v_{AE} = 65 \text{ km hr}^{-1}, \qquad v_{BE} = 80 \text{ km hr}^{-1},$$

and we wish to find v_{BA}.

From the rule for combining velocities (along the same line), we get:

$$v_{BA} = v_{BE} + v_{EA}.$$

But

$$v_{EA} = -v_{AE},$$

so

$$v_{BA} = v_{BE} - v_{AE}$$
$$= 80 \text{ km hr}^{-1} - 65 \text{ km hr}^{-1}$$
$$= 15 \text{ km hr}^{-1},$$

and the officer is overtaking the driver at 15 km hr^{-1}.

Example 2   How would the relative velocity be altered if B were ahead of A?

Not at all. The relative *positions* of the bodies do not matter. The velocity of B relative to A is still + 15 km hr^{-1}, but he is now pulling ahead of A at this rate.

Example 3   The compass of an aircraft indicates that it is headed due north, and its airspeed indicator shows that it is moving through the air at 120 mi hr^{-1}. If there is a wind of 50 mi hr^{-1} from west to east, what is the velocity of the aircraft relative to the earth?

Let subscript A refer to the aircraft, and subscript F to the moving air (which now corresponds to the flatcar in Fig. 4–11). Subscript E refers to the earth. We have given

$$v_{AF} = 120 \text{ mi hr}^{-1}, \qquad \text{due north}$$
$$v_{FE} = 50 \text{ mi hr}^{-1}, \qquad \text{due east},$$

and we wish to find the magnitude and direction of v_{AE}:

$$v_{AE} = v_{AF} + v_{FE}.$$

The three relative velocities are shown in Fig. 4–12. It follows from this diagram that

$$v_{AE} = 130 \text{ mi hr}^{-1}, \qquad 22.5°\text{E of N}.$$

Example 4   In what direction should the pilot head in order to travel due north? What will then be his velocity relative to the earth? (The magnitude of his airspeed, and the wind velocity, are the same as in the preceding example.)

Fig. 4–12

Fig. 4–13

We now have given:

$$v_{AF} = 120 \text{ mi hr}^{-1}, \qquad \text{direction unknown,}$$

$$v_{FE} = 50 \text{ mi hr}^{-1}, \qquad \text{due east,}$$

and we wish to find v_{AE}, whose magnitude is unknown but whose direction is due north. (Note that both this and the preceding example require us to determine two unknown quantities. In the former example, these were the *magnitude and direction* of v_{AE}. In this example, the unknowns are the *direction* of v_{AF} and the *magnitude* of v_{AE}.)

The three relative velocities must still satisfy the *vector equation*

$$v_{AE} = v_{AF} + v_{FE}.$$

The problem can be solved graphically as follows. First, construct the vector v_{FE} (see Fig. 4–13), known in magnitude and direction. At the head of this vector draw a construction line of indefinite length due north, in the known direction of v_{AE}. With the tail of v_{FE} as a center, construct a circular arc of radius equal to the known magnitude of v_{AF}. Vectors v_{AF} and v_{AE} may then be drawn from the point of intersection of this arc and the construction line to the ends of the vector v_{FE}. We find on solving the right triangle that the magnitude of v_{AE} is 109 mi hr^{-1} and the direction of v_{AF} is 24.5° W of N. That is, the pilot should head 24.5° W of N, and his ground speed will be 109 mi hr^{-1}.

PROBLEMS

4–1 Suppose that a runner on a straight track covers a distance of 1 mile in exactly 4 minutes. What was his average velocity in (a) mi hr^{-1}? (b) ft s^{-1}? (c) cm s^{-1}?

4–2 A body moves along a straight line, its distance from the origin at any instant being given by the equation $x = 8t - 3t^2$, where x is in centimeters and t is in seconds. Find the average velocity of the body in the interval from $t = 0$ to $t = 1$ s, and in the interval from $t = 0$ to $t = 4$ s.

4–3 The motion of a certain body along the x-axis is described by the equation $x = (10 \text{ cm s}^{-2})t^2$. Compute the instantaneous velocity of the body at time $t = 3$ s, by letting Δt first equal 0.1 s, then 0.01 s, and finally 0.001 s. What limiting value do the results seem to be approaching?

4–4 An automobile is provided with a speedometer calibrated to read m s^{-1} rather than mi hr^{-1}. The following series of speedometer readings was obtained during a start.

Time (s)	0	2	4	6	8	10	12	14	16
Velocity (m s^{-1})	0	0	2	5	10	15	20	22	22

a) Compute the average acceleration during each 2-s interval. Is the acceleration constant? Is it constant during any part of the time?

b) Make a velocity-time graph of the data above, using scales of 1 in. = 2 s horizontally, and 1 in. = 5 m s^{-1} vertically. Draw a smooth curve through the plotted points. What distance is represented by 1 sq. in.? What is the displacement in the first 8 s? What is the acceleration when $t = 8$ s? When $t = 13$ s? When $t = 15$ s?

4–5 The graph in Fig. 4–14 shows the velocity of a body plotted as a function of time.

a) Find the instantaneous acceleration at $t = 3$ s, at $t = 7$ s, and at $t = 11$ s.

b) How far does the body go in the first 5 s? The first 9 s? The first 13 s?

Fig. 4–14

4–6 Figure 4–15 is a graph of the acceleration of a body moving on the x-axis. Sketch the graphs of its velocity and coordinate as functions of time, if $x = v = 0$ when $t = 0$.

Fig. 4–15

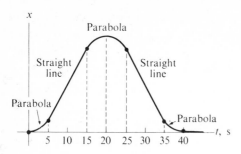

Fig. 4–16

4–7 Figure 4–16 is a graph of the coordinate of a body moving on the x-axis. Sketch the graphs of its velocity and acceleration as functions of the time.

4–8 Each of the following changes in velocity takes place in a 10-s interval. What is the magnitude, the algebraic sign, and the direction of the average acceleration in each interval?

a) At the beginning of the interval a body is moving toward the right along the x-axis at 5 m s^{-1}, and at the end of the interval it is moving toward the right at 20 m s^{-1}.

b) At the beginning it is moving toward the right at 20 m s^{-1}, and at the end it is moving toward the right at 5 m s^{-1}.

c) At the beginning it is moving toward the left at 5 m s^{-1}, and at the end it is moving toward the left at 20 m s^{-1}.

d) At the beginning it is moving toward the left at 20 m s^{-1}, and at the end it is moving toward the left at 5 m s^{-1}.

e) At the beginning it is moving toward the right at 20 m s^{-1}, and at the end it is moving toward the left at 20 m s^{-1}.

f) At the beginning it is moving toward the left at 20 m s^{-1}, and at the end it is moving toward the right at 20 m s^{-1}.

g) In which of the above instances is the body decelerated?

4–9 The makers of a certain automobile advertise that it will accelerate from 15 to 50 mi hr^{-1} in 13 s. Compute (a) the acceleration in ft s^{-2}, and (b) the distance the car travels in this time, assuming the acceleration to be constant.

4–10 An airplane taking off from a landing field has a run of 500 m. If it starts from rest, moves with constant

acceleration, and makes the run in 30 s, with what velocity in m s^{-1} did it take off?

4–11 An automobile starts from rest and acquires a velocity of 40 km hr^{-1} in 15 s.

a) Compute the acceleration in kilometers per hour per second, assuming it to be constant.

b) If the automobile continues to gain velocity at the same rate, how many more seconds are needed for it to acquire a velocity of 60 km hr^{-1}?

c) Find the distances covered by the automobile in parts (a) and (b).

4–12 A body moving with constant acceleration covers the distance between two points 60 m apart in 6 s. Its velocity as it passes the second point is 15 m s^{-1}.

a) What is the acceleration?

b) What is its velocity at the first point?

4–13 A ball is released from rest and rolls down an inclined plane, requiring 4 s to cover a distance of 100 cm. (a) What was its acceleration in cm s^{-2}? (b) How many centimeters would it have fallen vertically in the same time?

4–14 The "reaction time" of the average automobile driver is about 0.7 s. (The reaction time is the interval between the perception of a signal to stop and the application of the brakes.) If an automobile can decelerate at 16 ft s^{-2}, compute the total distance covered in coming to a stop after a signal is observed: (a) from an initial velocity of 30 mi hr^{-1}, (b) from an initial velocity of 60 mi hr^{-1}.

4–15 At the instant the traffic lights turn green, an automobile that has been waiting at an intersection starts ahead with a constant acceleration of 2 m s^{-2}. At the same instant a truck, traveling with a constant velocity of 10 m s^{-1}, overtakes and passes the automobile. (a) How far beyond its starting point will the automobile overtake the truck? (b) How fast will it be traveling?

4–16 The engineer of a passenger train traveling at 30 m s^{-1} sights a freight train whose caboose is 200 m ahead on the same track. The freight train is traveling in the same direction as the passenger train with a velocity of 10 m s^{-1}. The engineer of the passenger train immediately applies the brakes, causing a constant deceleration of 1 m s^{-2}, while the freight train continues with constant speed. (a) Will there be a collision? (b) If so, where will it take place?

4–17 A sled starts from rest at the top of a hill and slides down with a constant acceleration. The sled is 140 ft from the top of the hill 2 s after passing a point which is 92 ft from the top. Four seconds after passing the 92-ft point it is 198 ft from the top, and 6 s after passing the point it is 266 ft from the top. (a) What is the average velocity of the sled during each of the 2-s intervals after passing the 92-ft point? (b) What is the acceleration of the sled? (c) What was the velocity of the sled when it passed the 92-ft point? (d) How long did it take to go from the top to the 92-ft point? (e) How far did the sled go during the first second after passing the 92-ft point? (f) How long does it take the sled to go from the 92-ft point to the midpoint between the 92-ft and the 140-ft mark? (g) What is the velocity of the sled as it passes the midpoint in part (f)?

4–18 A subway train starts from rest at a station and accelerates at a rate of 2 m s^{-2} for 10 s. It then runs at constant speed for 30 s, and decelerates at 4 m s^{-2} until it stops at the next station. Find the *total* distance covered.

4–19 A body starts from rest, moves in a straight line with constant acceleration, and covers a distance of 64 ft in 4 s. (a) What was the final velocity? (b) How long a time was required to cover half the total distance? (c) What was the distance covered in one-half the total time? (d) What was the velocity when half the total distance had been covered? (e) What was the velocity after one-half the total time?

4–20 The speed of an automobile going north is reduced from 30 to 20 m s^{-1} in a distance of 125 m. Find (a) the magnitude and direction of the acceleration, assuming it to be constant, (b) the elapsed time, (c) the distance in which the car can be brought to rest from 20 m s^{-1}, assuming the acceleration of part (a).

4–21 An automobile and a truck start from rest at the same instant, with the automobile initially at some distance behind the truck. The truck has a constant acceleration of 2 m s^{-2} and the automobile an acceleration of 3 m s^{-2}. The automobile overtakes the truck after the truck has moved 75 m.

a)   How long does it take the auto to overtake the truck?

b)   How far was the auto behind the truck initially?

c)   What is the velocity of each when they are abreast?

4–22

a)   With what velocity must a ball be thrown vertically upward in order to rise to a height of 20 m?

b)   How long will it be in the air?

4–23 A ball is thrown vertically downward from the top of a building, leaving the thrower's hand with a velocity of 30 ft s^{-1}.

a)   What will be its velocity after falling for 2 s?

b)   How far will it fall in 2 s?

c)   What will be its velocity after falling 30 ft?

d)   If it moved a distance of 3 ft while in the thrower's hand, find its acceleration while in his hand.

e)   If the ball was released at a point 120 ft above the ground, in how many seconds will it strike the ground?

f)   What will the velocity of the ball be when it strikes the ground?

4–24 A balloon, rising vertically with a velocity of 5 m s^{-1}, releases a sandbag at an instant when the balloon is 20 m above the ground.

a)   Compute the position and velocity of the sandbag at the following times after its release: $\frac{1}{4}$ s, $\frac{1}{2}$ s, 1 s, 2 s.

b)   How many seconds after its release will the bag strike the ground?

c)   With what velocity will it strike?

4–25 A stone is dropped from the top of a tall cliff, and 1 s later a second stone is thrown vertically downward with a velocity of 20 m s^{-1}. How far below the top of the cliff will the second stone overtake the first?

4–26 A ball dropped from the cornice of a building takes 0.25 s to pass a window 3 m high. How far is the top of the window below the cornice?

4–27 A ball is thrown nearly vertically upward from a point near the cornice of a tall building. It just misses the cornice on the way down, and passes a point 160 ft below its starting point 5 s after it leaves the thrower's hand.

a)   What was the initial velocity of the ball?

b)   How high did it rise above its starting point?

c)   What were the magnitude and direction of its velocity at the highest point?

d)   What were the magnitude and direction of its acceleration at the highest point?

e)   What was the magnitude of its velocity as it passed a point 64 ft below the starting point?

4–28 A juggler performs in a room whose ceiling is 3 m above the level of his hands. He throws a ball vertically upward so that it just reaches the ceiling.

a)   With what initial velocity does he throw the ball?

b)   What time is required for the ball to reach the ceiling?

He throws a second ball upward with the same initial

velocity, at the instant that the first ball is at the ceiling.

c) How long after the second ball is thrown do the two balls pass each other?

d) When the balls pass each other, how far are they above the juggler's hands?

4–29 An object is thrown vertically upward. It has a speed of 32 ft s^{-1} when it has reached one-half its maximum height.

a) How high does it rise?

b) What are its velocity and acceleration 1 s after it is thrown?

c) 3 s after?

d) What is the average velocity during the first half second?

4–30 A student determined to test the law of gravity for himself walks off a skyscraper 300 m high, stopwatch in hand, and starts his free fall (zero initial velocity). Five seconds later, Superman arrives at the scene and dives off the roof to save the student.

a) What must Superman's initial velocity be in order that he catch the student just before the ground is reached?

b) What must be the maximum height of the skyscraper so that even Superman can't save him? (Assume that Superman's acceleration is that of any freely falling body.)

4–31 A ball is thrown vertically upward from the ground and a student gazing out of the window sees it moving upward past him at 5 m s^{-1}. The window is 10 m above the ground.

a) How high does the ball go above the ground?

b) How long does it take to go from a height of 10 m to its highest point?

c) Find its velocity and acceleration $\frac{1}{2}$ s after it left the ground.

4–32 A ball is thrown vertically upward from the ground with a velocity of 30 m s^{-1}.

a) How long will it take to rise to its highest point?

b) How high does the ball rise?

c) How long after projection will the ball have a velocity of 10 ms^{-1} upward?

d) Of 10 m s^{-1} downward?

e) When is the displacement of the ball zero?

f) When is the magnitude of the ball's velocity equal to half its velocity of projection?

g) When is the magnitude of the ball's displacement equal to half the greatest height to which it rises?

h) What are the magnitude and direction of the acceleration while the ball is moving upward?

i) While moving downward?

j) When at the highest point?

4–33 A ball rolling on an inclined plane moves with a constant acceleration. One ball is released from rest at the top of an inclined plane 18 m long and reaches the bottom 3 s later. At the same instant that the first ball is released, a second ball is projected upward along the plane from its bottom with a certain initial velocity. The second ball is to travel part way up the plane, stop, and return to the bottom so that it arrives simultaneously with the first ball.

a) Find the acceleration.

b) What must be the initial velocity of the second ball?

c) How far up the plane will it travel?

4–34 The rocket-driven sled Sonic Wind No. 2, used for investigating the physiological effects of large accelerations and decelerations, runs on a straight, level track 3500 ft long. Starting from rest, it can reach a speed of 1000 mi hr^{-1} in 1.8 s.

a) Compute the acceleration, assuming it to be constant.

b) What is the ratio of this acceleration to that of a freely-falling body, g?

c) What is the distance covered?

d) A magazine article states that at the end of a certain run the speed of the sled was decreased from 632 mi hr^{-1} to zero in 1.4 s, and that as the sled decelerated its passenger was subjected to more than 40 times the pull of gravity (that is, the deceleration was greater than 40 g). Are these figures consistent?

4–35 The first stage of a rocket to launch an earth satellite will, if fired vertically upward, attain a speed of 4000 mi hr^{-1} at a height of 36 mi above the earth's surface, at which point its fuel supply will be exhausted.

a) Assuming constant acceleration, find the time to reach a height of 36 mi.

b) How much higher would the rocket rise if it continued to "coast" vertically upward?

4–36 Suppose the acceleration of gravity were only 1.0 m s^{-2}, instead of 10 m s^{-2}.

a) Estimate the height to which you could jump vertically from a standing start.

b) How high could you throw a baseball?

c) Estimate the maximum height of a window from which you would care to jump to a concrete sidewalk below. (Each story of an average building is about 4 m high.)

d) With what speed, in meters per second, would you strike the sidewalk?

e) How many seconds would be required?

4–37 After the engine of a moving motorboat is cut off, the boat has an acceleration in the opposite direction to its velocity and directly proportional to the square of its velocity. That is, $dv/dt = -kv^2$, where k is a constant.

a) Show that the magnitude v of the velocity at a time t after the engine is cut off is given by

$$\frac{1}{v} = \frac{1}{v_0} + kt.$$

b) Show that the distance x traveled in a time t is

$$x = \frac{1}{k}\ln(v_0 kt + 1).$$

c) Show that the velocity after traveling a distance x is

$$v = v_0 e^{-kx}.$$

As a numerical example, suppose the engine is cut off when the velocity $v_0 = 20$ ft s^{-1}, and that the velocity decreases to 10 ft s^{-1} in a time of 15 s.

d) Find the numerical value of the constant k, and the unit in which it is expressed.

e) Find the acceleration at the instant the engine is cut off.

f) Construct graphs of x, v, and a for a time of 20 s. Let 1 in. = 5 s horizontally, and 1 in. = 100 ft, 5 ft s^{-1}, and 0.5 ft s^{-2} vertically.

4–38 The equation of motion of a body suspended from a spring and oscillating vertically is $y = A\sin\omega t$, where A and ω are constants.

a) Find the velocity of the body as a function of time.

b) Find its acceleration as a function of time.

c) Find its velocity as a function of its coordinate.

d) Find its acceleration as a function of its coordinate.

e) What is the maximum distance of the body from the origin?

f) What is its maximum velocity?

g) What is its maximum acceleration?

h) Sketch graphs of y, v, and a as functions of time.

4–39 The acceleration of a body suspended from a spring and oscillating vertically is $a = -Ky$, where K is a constant and y is the coordinate measured from the equilibrium position. Suppose that a body moving in this way is given an initial velocity v_0 at the coordinate y_0. Find the expression for the velocity v of the body as a function of its coordinate y. [*Hint:* Use the expression $a = v\,dv/dy$.]

4–40 The motion of a body falling from rest in a resisting medium is described by the equation

$$\frac{dv}{dt} = A - Bv,$$

where A and B are constants. In terms of A and B, find

a) the initial acceleration, and

b) the velocity at which the acceleration becomes zero (the terminal velocity).

c) Show that the velocity at any time t is given by

$$v = \frac{A}{B}(1 - e^{-Bt}).$$

4–41 Two cars, A and B, travel in a straight line. The distance of A from the starting point is given as a function of time by $x_A = 4t + 4t^2$, and the distance of B from the starting point is $x_B = 2t^2 + 2t^3$.

a) Which car is ahead just after they leave the starting point?

b) At what times are the cars at the same point?

c) At what times is the velocity of B relative to A zero?

d) At what times is the distance from A to B neither increasing nor decreasing?

4–42 A hypothetical spaceship takes a straight-line path from the earth to the moon, a distance of about 400,000 km. Suppose it accelerates at 10 m s^{-2} for the first 10 min of the trip, then travels at constant speed until the last 10 min, when it decelerates at 10 m s^{-2}, just coming to rest as it reaches the moon.

a) What is the maximum speed attained?

b) What total time is required for the trip?

c) What fraction of the total distance is traveled at constant speed?

4–43 A "moving sidewalk" in an airport terminal building moves 1 m s^{-1} and is 150 m long. If a man steps on at one end and walks 2 m s^{-1} relative to the moving sidewalk, how much time does he require to reach the opposite end if he walks (a) in the same direction the sidewalk is moving; (b) in the opposite direction?

4–44 The driver of a car wishes to pass a truck which is traveling at a constant speed of 20 m s^{-1} (about 50 mi hr^{-1}). The car's maximum acceleration at this speed is 0.5 m s^{-2}. Initially the vehicles are separated by 25 m, and the car pulls back into the truck's lane after it is 25 m ahead of the truck. The car is 5 m long and the truck 20 m.

a) How much time is required for the car to pass the truck?

b) What distance does the car travel during this time?

c) What is the final speed of the car, assuming its acceleration while passing the truck is constant?

4–45 Two piers A and B are located on a river, one mile apart. Two men must make round trips from pier A to pier B and return. One man is to row a boat at a velocity of 4 mi hr^{-1} relative to the water, and the other man is to walk on the shore at a velocity of 4 mi hr^{-1}. The velocity of the river is 2 mi hr^{-1} in the direction from A to B. How long does it take each man to make the round trip?

4–46 A passenger on a ship traveling due east with a speed of 18 knots observes that the stream of smoke from the ships' funnels makes an angle of 20° with the ship's wake. The wind is blowing from south to north. Assume that the smoke acquires a velocity (with respect to the earth) equal to the velocity of the wind, as soon as it leaves the funnels. Find the velocity of the wind.

4–47 An airplane pilot wishes to fly due north. A wind of 60 mi hr^{-1} is blowing toward the west. If the flying speed of the plane (its speed in still air) is 180 mi hr^{-1}, in what direction should the pilot head? What is the speed of the plane over the ground? Illustrate with a vector diagram.

4–48 An airplane pilot sets a compass course due west and maintains an air speed of 240 km hr^{-1}. After flying for $\frac{1}{2}$ hr, he finds himself over a town which is 150 km west and 40 km south of his starting point.

a) Find the wind velocity, in magnitude and direction.

b) If the wind velocity were 120 km hr^{-1} due south, in what direction should the pilot set his course in order to travel due west? Take the same air speed of 240 km hr^{-1}.

4–49 When a train has a speed of 10 m s^{-1} eastward, raindrops which are falling vertically with respect to the earth make traces which are inclined 30° to the vertical on the windows of the train.

a) What is the horizontal component of a drop's velocity with respect to the earth? With respect to the train?

b) What is the velocity of the raindrop with respect to the earth? With respect to the train?

4–50 A river flows due north with a velocity of 2 m s^{-1}. A man rows a boat across the river, his velocity relative to the water being 3 m s^{-1} due east.

a) What is his velocity relative to the earth?

b) If the river is 1000 m wide, how far north of his starting point will he reach the opposite bank?

c) How long a time is required to cross the river?

4–51

a) In what direction should the rowboat in Problem 4–50 be headed in order to reach a point on the opposite bank directly east from the start?

b) What will be the velocity of the boat relative to the earth?

c) How long a time is required to cross the river?

4–52 A motorboat is observed to travel 10 mi hr^{-1} relative to the earth in the direction 37° north of east. If the velocity of the boat due to the wind is 2 mi hr^{-1} eastward and that due to the current is 4 mi hr^{-1} southward, what is the magnitude and direction of the velocity of the boat due to its own power?

Chapter 5

Newton's Second Law. Gravitation

5-1 INTRODUCTION

In the preceding chapters we have discussed separately the concepts of force and acceleration. We have made use, in problems in equilibrium, of Newton's first law, which states that when the resultant force on a body is zero, the acceleration of the body is also zero. The next logical step is to ask how a body behaves when the resultant force on it is *not* zero. The answer to this question is contained in Newton's second law, which states that when the resultant force is not zero the body moves with accelerated motion, and the acceleration, with a given force, depends on a property of the body known as its *mass*.

This part of mechanics, which includes the study both of motion and of the forces that bring about the motion, is called *dynamics*. In its broadest sense, dynamics includes nearly the whole of mechanics. Statics treats of special cases in which the acceleration is zero, and kinematics deals with motion only.

It will be assumed in this chapter that all velocities are small compared with the velocity of light, so that relativistic considerations do not arise. We shall also assume that unless otherwise stated all velocities and accelerations are measured relative to an inertial reference system (Section 2–3), and shall consider rectilinear motion only. Motion in a curved path will be discussed in the next chapter.

5-2 NEWTON'S SECOND LAW. MASS

We know from experience that an object at rest never starts to move by itself; a push or pull must be exerted on it by some other body. Similarly, a force is required to slow down or stop a body already in motion, and to make a moving body deviate from straight-line motion requires a sideways force. All these processes (speeding up, slowing down, or changing direction) involve a change in either the magnitude or direction of the velocity. Thus in each case the body is *accelerated*, and an external force must act on it to produce the acceleration.

Let us consider some experiments. A small body (a particle) rests on a level frictionless surface, and moves to the right along the x-axis of an inertial reference system, as in Fig. 5–1(a). A horizontal force F, measured by a spring balance calibrated as described in Section 1–4, is exerted on the body. We find that the velocity of the body increases as long as the force acts. In other words, the body has an acceleration a toward the right. If the magnitude of the force F is kept constant, the velocity is found to increase at a constant rate. If the force is altered, the rate of change of velocity alters in the same proportion. Doubling the force doubles the rate of change of velocity, halving the force halves the rate of change, etc. If the force is reduced to zero, the rate

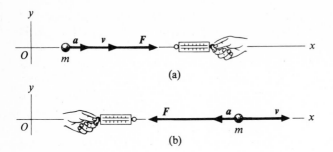

Fig. 5-1 *The acceleration **a** is proportional to the force **F** and is in the same direction as the force.*

of change of velocity is zero and the body continues to move with constant velocity.

Before the time of Galileo and Newton, it was generally believed that a force was necessary just to keep a body moving, even on a level frictionless surface or in "outer space." Galileo and Newton realized that *no* force is necessary to keep a body moving, once it has been set in motion, and that the effect of a force is not to *maintain* the velocity of a body, but to *change* its velocity. The *rate of change* of velocity, for a given body, is directly proportional to the force exerted on it.

In Fig. 5-1(b), the velocity of the body is also toward the right, but the force is toward the left. Under these conditions the body moves more slowly. If the force continues to act, it will ultimately reverse the body's direction of motion. The acceleration will then be toward the *left*, in the same direction as the force **F**. Hence we conclude not only that the *magnitude* of the acceleration is proportional to that of the force, but that the *direction* of the acceleration is the *same* as that of the force, regardless of the direction of the velocity.

To say that the rate of change of the velocity of a body is directly proportional to the force exerted on it is to say that the ratio of the force to the rate of change of velocity is a constant, regardless of the magnitude of the force. This constant ratio of force to rate of change of velocity is called the *mass m* of the body. Thus

$$m = \frac{F}{dv/dt} = \frac{F}{a},$$

or

$$\boldsymbol{F} = m\frac{d\boldsymbol{v}}{dt} = m\boldsymbol{a}. \qquad (5\text{-}1)$$

By writing this as a vector equation, we automatically include the experimental fact that the direction of the acceleration **a** is the same as that of the force **F**. Mass is, however, a *scalar* quantity.

The mass of a body can be thought of as the *force per unit of acceleration*. For example, if the acceleration of a certain body is found to be 5 ft s^{-2} when the force is 20 lb, the mass of the body is

$$m = \frac{20 \text{ lb}}{5 \text{ ft s}^{-2}} = 4 \text{ lb ft}^{-1} \text{ s}^2,$$

and a force of 4 lb must be exerted on the body for every ft s^{-2} of acceleration.

When a large force is needed to give a body a certain acceleration (i.e., speed it up, slow it down, or deviate it if it is in motion), the mass of the body is large; if only a small force is needed for the same acceleration, the mass is small. Thus the mass of a body is a quantitative measure of the property described in everyday language as *inertia*.

In *rectilinear* motion, the force **F** acting on a body and its velocity **v** always have the same action line, as in Fig. 5-1. If the direction of the force is *not* the same as that of the velocity, the body is deflected sideways and moves in a curved path. We shall see in the next chapter, however, that Eq. (5-1) applies in this case also, except that the change in velocity $\Delta \boldsymbol{v}$, or the acceleration **a**, includes the change in *direction* as well as the change in *magnitude* of the velocity. In every case, then, *the vector force equals the product of the mass and the vector acceleration.*

When two vectors are equal, their rectangular components are equal also. Hence the vector equation (5-1) is equivalent (for forces and accelerations in the *xy*-plane) to the pair of scalar equations

$$F_x = m\frac{dv_x}{dt} = ma_x, \quad F_y = m\frac{dv_y}{dt} = ma_y. \quad (5\text{-}2)$$

This means that each component of the force can be considered to produce its own component of acceleration. It follows that if any number of forces act on a body simultaneously, as is often the case in problems of practical interest, the forces may be resolved into x- and y-components, the algebraic sums $\sum F_x$ and $\sum F_y$ may be computed, and the components of acceleration are given by

$$\sum F_x = m\frac{dv_x}{dt} = ma_x, \qquad \sum F_y = m\frac{dv_y}{dt} = ma_y.$$

This pair of equations is equivalent to the single vector equation

$$\boxed{\sum F = m\frac{dv}{dt} = ma,} \qquad (5–3)$$

where we write the lefthand side explicitly as $\sum F$ to emphasize that the acceleration is determined by the *resultant* of *all* of the *external* forces acting *on* the body.

Equation (5–3) is the mathematical statement of Newton's *second law of motion*. If we think of the equation as solved for a or dv/dt, the law can be stated: *The rate of change of the velocity of a particle, or its acceleration, is equal to the resultant of all external forces exerted on the particle divided by the mass of the particle, and is in the same direction as the resultant force.* The acceleration is to be measured relative to an inertial system.

It is necessary to state the law for a *particle*, because when a resultant force acts on an extended body, the body may be set in rotation and not all particles in it have the same acceleration. We shall discuss this more fully later on, but it may be said at this point that the acceleration of the center of gravity of a body is the same as that of a particle whose mass equals that of the body.

5–3 SYSTEMS OF UNITS

Nothing was said in the preceding discussion regarding the *units* in which force, mass, and acceleration are to be expressed. It is evident, however, from the equation $F = ma$, that these must be such that *unit force inparts unit acceleration to unit mass.*

Several systems of units are in common use in the United States, including the meter-kilogram-second (mks) system, the centimeter-gram-second (cgs) system, and the British system. This book will use principally the mks system, not only because it is

convenient and widely used in scientific work, but also because there seems little doubt that the metric system (mks, cgs, or a combination) will eventually be adopted worldwide for commerce and industry as well as scientific work. The United States and Britain are the only two major countries that do not use the metric system, and Britain is beginning to convert.

In the mks system, the mass of the standard kilogram is the unit of mass, and the unit of acceleration is 1 m s^{-2}. The unit force in this system is then *that force which gives a standard kilogram an acceleration of* 1 m s^{-2}. This force is called *one newton* (1 N). One newton is approximately equal to one-quarter of a pound-force (more precisely, $1 \text{ N} = 0.22481 \text{ lb.}$) Thus in the mks system,

$$F(\text{N}) = m(\text{kg}) \times a(\text{m s}^{-2}).$$

The mass unit in the *centimeter-gram-second* (cgs) system is one gram, equal to $1/1000$ kg, and the unit of acceleration is 1 cm s^{-2}. The unit force in this system is then *that force which gives a body whose mass is one gram, an acceleration of* 1 cm s^{-2}. This force is called *one dyne* (1 dyn). Since $1 \text{ kg} = 10^3$ g and $1 \text{ m s}^{-2} = 10^2 \text{ cm s}^{-2}$, it follows that $1 \text{ N} = 10^5 \text{ dyn}$. In the cgs system,

$$F(\text{dyn}) = m(\text{g}) \times a(\text{cm s}^{-2}).$$

In the mks and cgs systems, we first selected units of mass and acceleration, and defined the unit of force in terms of these. In the *British engineering system*, we first select a unit of *force* (1 lb) and a unit of acceleration (1 ft s^{-2}), and then define the unit of mass as *the mass of a body whose acceleration is* 1 ft s^{-2} *when the resultant force on the body is* 1 lb. This unit of mass is called *one slug*. (The origin of the name is obscure; it may have arisen from the concept of mass as inertia or *sluggishness*.) Then in the engineering system,

$$F(\text{lb}) = m(\text{slugs}) \times a(\text{ft s}^{-2}).$$

The pound is used in everyday life as a unit of quantity of matter (e.g., a pound of butter) but properly speaking it is a unit of *force* or *weight*. Thus a pound of butter is that quantity which has a weight

Table 5–1

System of units	Force	Mass	Acceleration
mks	newton (N)	kilogram (kg)	$m\ s^{-2}$
cgs	dyne (dyn)	gram (g)	$cm\ s^{-2}$
engineering	pound (lb)	slug	$ft\ s^{-2}$

of 1 lb. A useful fact in converting between mks and British units is that an object with a *mass* of 1 kg has a *weight* of about 2.205 lb.

The units of force, mass, and acceleration in the three systems are summarized in Table 5–1.

Example 1 A constant horizontal force of 2 N is applied to a body of mass 4 kg, resting on a level frictionless surface. The acceleration of the body is

$$a = \frac{F}{m} = \frac{2\,N}{4\,kg} = 0.5\ m\ s^{-2}.$$

Since the force is constant, the acceleration is constant also. Hence if the initial position and velocity of the body are known, the velocity and position at any later time can be found from the equations of motion with constant acceleration.

Example 2 A body of mass 0.2 kg is given an initial velocity of 0.4 m s^{-1} toward the right along a level laboratory tabletop. The body is observed to slide a distance of 1.0 m along the table before coming to rest. What was the magnitude and direction of the friction force f acting on it?

In the absence of further information, let us assume that the friction force is constant. The acceleration is then constant also; from the equations of motion with constant acceleration, we have

$$v^2 = v_0{}^2 + 2ax, \qquad 0 = (0.4\ m\ s^{-1})^2 + (2a)(1.0\ m),$$
$$a = -0.08\ m\ s^{-2}.$$

The negative sign means that the acceleration is toward the *left* (although the velocity is toward the right). The friction force on the body is

$$f = ma = (0.2\ kg)(-0.08\ m\ s^{-2}) = -0.016\ N,$$

and is toward the left also. (A force of equal magnitude, but directed toward the right, is exerted on the table by the sliding body.)

Example 3 The acceleration of a certain body is found to be 5 ft s^{-2} when the resultant force on the body is 20 lb. The mass of the body is

$$m = \frac{F}{a} = \frac{20\ lb}{5\ ft\ s^{-2}} = 4\ lb\ ft^{-1}\ s^2 = 4\ slugs.$$

5–4 NEWTON'S LAW OF UNIVERSAL GRAVITATION

Throughout our study of mechanics we have been continually encountering the force of gravitational attraction between a body and the earth. We now wish to study this phenomenon of gravitation in more detail.

The law of universal gravitation was discovered by Newton, and was published by him in 1686. There seems to be some evidence that Newton was led to deduce the law from speculations concerning the fall of an apple toward the earth, but his first published calculations to justify its correctness had to do with the motion of the moon around the earth.

Newton's law of gravitation may be stated: *Every particle of matter in the universe attracts every other particle with a force which is directly proportional to the product of the masses of the particles and inversely proportional to the square of the distance between them.* Thus

$$F_g = G\frac{mm'}{r^2}, \qquad (5\text{–}4)$$

where F_g is the gravitational force on either particle, m and m' are their masses, r is the distance between them and G is a universal constant called the *gravitational constant*, whose numerical value depends on the units in which force, mass, and length are expressed.

The gravitational forces acting on the particles form an action–reaction pair. Although the masses of the particles may be different, forces of *equal* magni-

Fig. 5–2 *Principle of the Cavendish balance.*

tude act on each, and the action line of both forces lies along the line joining the particles.

Newton's law of gravitation refers to the force between two *particles*. It can also be shown, however, that the force of attraction exerted on or by a homogeneous sphere is the same as if the mass of the sphere were concentrated at its center. The proof is not difficult but is too long to give here, and we shall simply state as a fact that *the gravitational force exerted on or by a homogeneous sphere is the same as if the entire mass of the sphere were concentrated in a point at its center.* Thus if the earth were a homogeneous sphere, of mass m_E, the force exerted by it on a small body of mass m, at a distance r from its center, would be

$$F_g = G \frac{m m_E}{r^2}.$$

A force of the same magnitude would be exerted *on* the earth by the body.

The magnitude of the gravitational constant G can be found experimentally by measuring the force of gravitational attraction between two bodies of known masses m and m', at a known separation. For bodies of moderate size the force is extremely small, but it can be measured with an instrument invented by the Rev. John Michell and first used for this purpose by Sir Henry Cavendish in 1798. The same type of instrument was also used by Coulomb for studying forces of electrical and magnetic attraction and repulsion.

The Cavendish balance consists of a light, rigid T-shaped member (Fig. 5–2) supported by a fine vertical fiber such as a quartz thread or a thin metallic ribbon. Two small spheres of mass m are mounted at the ends of the horizontal portion of the T, and a small mirror M, fastened to the vertical portion, reflects a beam of light onto a scale. To use the balance, two large spheres of mass m' are brought up to the positions shown. The forces of gravitational attraction between the large and small spheres result in a *couple* which twists the system through a small angle, thereby moving the reflected light beam along the scale.

By using an extremely fine fiber, the deflection of the mirror may be made sufficiently large so that the gravitational force can be measured quite accurately. The gravitational constant, measured in this way, is found to be

$$G = 6.670 \times 10^{-11} \, \text{N m}^2 \, \text{kg}^{-2}.$$

Example 1   The mass m of one of the small spheres of a Cavendish balance is 0.001 kg, the mass m' of one of the large spheres is 0.5 kg, and the center-to-center distance between the spheres is 0.05 m. The gravitational force on each sphere is

$$F_g = 6.67 \times 10^{-11} \, \text{N m}^2 \, \text{kg}^{-2} \frac{(0.001 \, \text{kg})(0.5 \, \text{kg})}{(0.05 \, \text{m})^2}$$

$$= 1.33 \times 10^{-11} \, \text{N},$$

or about one hundred-billionth of a newton!

Example 2   Suppose the spheres in Example 1 are placed 0.05 m from each other at a point in space far removed from all other bodies. What is the acceleration of each, relative to an inertial system?

The acceleration a of the smaller sphere is

$$a = \frac{F_g}{m} = \frac{1.33 \times 10^{-11} \, \text{N}}{1.0 \times 10^{-3} \, \text{kg}} = 1.33 \times 10^{-8} \, \text{m s}^{-2}.$$

The acceleration a' of the larger sphere is

$$a' = \frac{F_g}{m'} = \frac{1.33 \times 10^{-11} \, \text{N}}{0.5 \, \text{kg}} = 2.67 \times 10^{-11} \, \text{m s}^{-2}.$$

In this case, the accelerations are *not* constant because the gravitational force increases as the spheres approach each other.

5–5 MASS AND WEIGHT

The *weight* of a body can now be defined more generally than in the preceding chapters as *the resultant gravitational force exerted on the body by all other bodies in the universe.* At or near the surface of the earth, the force of the earth's attraction is so much greater than that of any other body that for practical purposes all other gravitational forces can be neglected and the weight can be considered as arising solely from the gravitational attraction of the earth. Similarly, at the surface of the moon, or of another planet, the weight of a body results almost entirely from the gravitational attraction of the moon or the planet. Thus if the earth were a homogeneous sphere of radius R and mass m_E, the weight w of a small body of mass m at or near its surface would be

$$w = F_g = \frac{Gmm_E}{R^2}. \tag{5–5}$$

The weight of a given body varies by a few tenths of a percent from point to point on the earth's surface, partly because of local deposits of ore, oil, or other substances whose density differs from the average, and partly because the earth is not a perfect sphere but is flattened somewhat at the poles. Also, the weight of a given body decreases inversely with the square of its distance from the earth's center, and at a radial distance of two earth radii, for example, it has decreased to one-quarter of its value at the earth's surface.

The *apparent* weight of a body at the surface of the earth differs slightly in magnitude and direction from the earth's force of gravitational attraction because of the rotation of the earth about its axis. For the present, we shall ignore the small difference between the apparent weight of a body and the force of the earth's gravitational attraction, and shall assume that the earth is an inertial reference system. Then when a body is allowed to fall freely, the force accelerating it is its weight $\mathbf{w}$ and the acceleration produced by this force is the acceleration due to gravity, $\mathbf{g}$. The general relation

$$\mathbf{F} = m\mathbf{a}$$

therefore becomes, for the special case of a freely falling body,

$$\boxed{w = mg.} \tag{5–6}$$

Since

$$w = mg = G\frac{mm_E}{R^2},$$

it follows that

$$g = \frac{Gm_E}{R^2}, \tag{5–7}$$

showing that the acceleration due to gravity is the same for *all* bodies (since m cancelled out) and is very nearly constant (since G and m_E are constants and R varies only slightly from point to point on the earth).

The weight of a body is a force, and must be expressed in terms of the unit of force in the particular system of units one is using. Thus in the mks system the unit of weight is 1 N, in the cgs system it is 1 dyn, and in the engineering system it is 1 lb. Equation (5–6) indicates the relation between the mass and weight of a body in any consistent set of units.

For example, the weight of a standard kilogram, at a point where $g = 9.80 \text{ m s}^{-2}$, is

$$w = mg = 1 \text{ kg} \times 9.80 \text{ m s}^{-2} = 9.80 \text{ N}.$$

At a second point where $g = 9.78 \text{ m s}^{-2}$, the weight is

$$w = 9.78 \text{ N}.$$

Thus, unlike the mass of a body, which is a constant, the weight varies from one point to another. On the moon, where $g = 1.67 \text{ m s}^{-2}$, the weight is 1.67 N, but the mass is still 1 kg.

The weight of a body whose mass is 1 slug, at a point where $g = 32.0 \text{ ft s}^{-2}$, is

$$w = mg = 1 \text{ slug} \times 32.0 \text{ ft s}^{-2} = 32.0 \text{ lb},$$

and the mass of a man who weighs 160 lb at this point is

$$m = \frac{w}{g} = \frac{160 \text{ lb}}{32.0 \text{ ft s}^{-2}} = 5 \text{ slugs}.$$

If we insert for the weight w in the equation $w = mg$, the gravitational force F_g as given by Newton's law of gravitation, we obtain, after cancelling the mass m,

$$m_E = \frac{R^2 g}{G},$$

where R is the earth's radius. All of the quantities on the right are known, so the mass of the earth, m_E, can be calculated. Taking $R = 6370$ km $= 6.37 \times 10^6$ m, and $g = 9.80$ m s^{-2}, we find

$$m_E = 5.98 \times 10^{24} \text{ kg}.$$

The volume of the earth is

$$V = \frac{4}{3} \pi R^3 = 1.08 \times 10^{21} \text{ m}^3.$$

The mass of a body divided by its volume is known as its average *density*. (The density of water is 1 g cm^{-3} = 1000 kg m^{-3}.) The average density of the earth is therefore

$$\frac{m_E}{V} = 5.5 \text{ g cm}^{-3} = 5500 \text{ kg m}^{-3}.$$

The density of most rock near the earth's surface, such as granites and gneisses, is about 3 g cm^{-3} = 3000 kg m^{-3}, so the interior of the earth must have much higher density. Some rock found on the surface, notably basaltic rock, has a density of about 5 g cm^{-3}.

As with many other physical quantities, the mass of a body can be measured in several different ways. One is to use the relation by which the quantity is defined, which in this case is the ratio of the force on the body to its acceleration. A measured force is applied to the body, its acceleration is measured, and the unknown mass is obtained by dividing the force by the acceleration. This method is used exclusively to measure masses of atomic particles.

The second method consists of finding by trial some other body whose mass (a) is equal to that of the given body, and (b) is already known. Consider first a method of determining when two masses are equal. At the same point on the earth's surface all bodies fall freely with the same acceleration g. Since the weight w of a body equals the product of its mass m and the acceleration g, it follows that if, at the same point, the weights of the two bodies are equal, their *masses* are equal also. The *equal-arm balance* is

an instrument by means of which one can determine very precisely when the weights of two bodies are equal, and hence when their masses are equal.

It will be seen from the preceding discussion that the property of matter called *mass* makes itself evident in two very different ways. The force of gravitational attraction between two particles is said to be proportional to the product of their masses, and in this sense mass can be considered as *that property of matter by virtue of which every particle exerts a force of attraction on every other particle.* We may call this property *gravitational mass.* On the other hand, Newton's second law is concerned with an entirely different property of matter, namely, the fact that a force (not necessarily gravitational) must be exerted on a particle in order to accelerate it, i.e., to change its velocity, either in magnitude or direction. This property can be called *inertial mass.* It is not at all obvious that the gravitational mass of a particle should be the same as its inertial mass, but experiment shows that the two are in fact the same. That is, if we have to push twice as hard on body A as we do on body B to produce a given acceleration, then the force of gravitational attraction between body A and some third body C is twice as great as the gravitational attraction between body B and body C, the distance between them being the same. Thus when an equal-arm balance is used to compare masses, it is actually *gravitational* mass that is being measured. Therefore the property represented by m in Newton's second law can be operationally defined as the result obtained by the prescribed methods of using an equal-arm balance.

5–6 APPLICATIONS OF NEWTON'S SECOND LAW

We now give a number of applications of Newton's second law to specific problems. In all these examples, and in the problems at the end of the chapter, it will be assumed that the acceleration due to gravity is 9.80 m s^{-2} or 32.0 ft s^{-2}, unless otherwise specified. In later chapters the approximate value $g = 10$ m s^{-2}, which is in error by only about 2%, will often be used for convenience.

Example 1 A block whose mass is 10 kg rests on a horizontal surface. What constant horizontal force T is required to give it a velocity of 4 m s^{-1} in 2 s,

Fig. 5–3

Fig. 5–4 *The resultant force is* $T - w$.

starting from rest, if the friction force between the block and the surface is constant and is equal to 5 N? Assume that all forces act at the center of the block. (See Fig. 5–3.)

The mass of the block is given. The y-component of its acceleration is zero. The x-component of acceleration can be found from the data on the velocity acquired in a given time. Since the forces are constant, the x-acceleration is constant and, from the equations of motion with constant acceleration,

$$a_x = \frac{v - v_0}{t} = \frac{4\ \mathrm{m\,s^{-1}} - 0}{2\ \mathrm{s}} = 2\ \mathrm{m\,s^{-2}}.$$

The resultant of the x-forces is

$$\sum F_x = T - f,$$

and that of the y-forces is

$$\sum F_y = N - w.$$

Hence, from Newton's second law in component form,

$$T - f = ma_x, \qquad N - w = ma_y = 0.$$

From the second equation, we find that

$$N = w = mg = (10\ \mathrm{kg})(9.80\ \mathrm{m\,s^{-2}}) = 98.0\ \mathrm{N},$$

and from the first,

$$T = f + ma_x = 5\ \mathrm{N} + (10\ \mathrm{kg})(2\ \mathrm{m\,s^{-2}}) = 25\ \mathrm{N}.$$

Example 2 An elevator and its load have a total mass of 800 kg. Find the tension T in the supporting cable when the elevator, originally moving down-

ward at 10 m s^{-1}, is brought to rest with constant acceleration in a distance of 25 m. (See Fig. 5–4).

The weight of the elevator is

$$w = mg = (800\ \mathrm{kg})(9.8\ \mathrm{m\,s^{-2}}) = 7840\ \mathrm{N}.$$

From the equations of motion with constant acceleration,

$$v^2 = v_0^2 + 2ay, \qquad a = \frac{v^2 - v_0^2}{2y}.$$

The initial velocity v_0 is -10 m s^{-1}; the final velocity v is zero. If we take the origin at the point where the deceleration begins, then $y = -25$ m. Hence

$$a = \frac{0 - (-10\ \mathrm{m\,s^{-1}})^2}{2(-25\ \mathrm{m})} = 2\ \mathrm{m\,s^{-2}}.$$

The acceleration is therefore positive (upward). From the free-body diagram (Fig. 5–4) the resultant force is

$$\sum F = T - w = T - 7840\ \mathrm{N}.$$

Since $F = ma$,

$$T - 7840\ \mathrm{N} = (800\ \mathrm{kg})(2\ \mathrm{m\,s^{-2}}) = 1600\ \mathrm{N},$$

$$T = 9440\ \mathrm{N}.$$

The tension must be *greater* than the weight by 1600 N to cause the upward acceleration when the elevator stops.

Example 3 In the above example, with what force do the feet of a passenger press downward on the floor, if the passenger's mass is 80 kg?

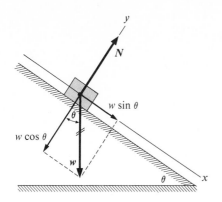

Fig. 5–5 *A block on a frictionless inclined plane.*

We first find the force the floor exerts *on* the passenger's feet, which is the reaction force to the force required. The forces on the passenger are his weight $(80 \text{ kg})(9.8 \text{ m s}^{-2}) = 784 \text{ N}$ and the upward force F caused by the floor. Thus Newton's second law applied to the man, whose acceleration is the same as that of the elevator, gives

$$F - 784 \text{ N} = (80 \text{ kg})(2 \text{ m s}^{-2}) = 160 \text{ N},$$

and

$$F = 944 \text{ N}.$$

Thus the man pushes down on the floor with a force of 944 N while the elevator is stopping, and he *feels* a greater strain in his legs and feet than while standing at rest. If he stands on a bathroom scale calibrated in newtons, what will the scale read?

Example 4   What is the acceleration of a block on a frictionless plane inclined at an angle θ with the horizontal?

The only forces acting on the block are its weight w and the normal force N exerted by the plane (Fig. 5–5).

Take axes parallel and perpendicular to the surface of the plane and resolve the weight into x- and y- components. Then

$$\sum F_y = N - w \cos \theta,$$
$$\sum F_x = w \sin \theta.$$

But we know that $a_y = 0$, so, from the equation $\sum F_y = ma_y$, we find that $N = w \cos \theta$. From the equation $\sum F_x = ma_x$, we have

$$w \sin \theta = ma_x,$$

and since $w = mg$,

$$a_x = g \sin \theta.$$

The mass does not appear in the final result, which means that any block, regardless of its mass, will slide on a frictionless inclined plane with an acceleration down the plane of $g \sin \theta$. In particular, when $\theta = 0$, $a_x = 0$, and when the plane is vertical, $\theta = 90°$ and $a_x = g$, as we should expect.

Example 5   Refer to Fig. 2–3 (Chapter 2). Let the mass of the block be 4 kg and that of the rope be 0.5 kg. If the force F_1 is 9 N, what are the forces F_1', F_2, and F_2'? The surface on which the block moves is level and frictionless.

We know from Newton's third law that $F_1 = F_1'$ and that $F_2 = F_2'$. Hence $F_1' = 9 \text{ N}$. The force F_2 could be computed by applying Newton's second law to the block, if its acceleration were known, or the force F_2' could be computed by applying this law to the rope if its acceleration were known. The acceleration is not given, but it can be found by considering the block and rope together as a single system. The vertical forces on this system need not be considered. Since there is no friction, the resultant *external* force acting *on* the system is the force F_1. (The forces F_2 and F_2' are *internal* forces when we consider block and rope as a single system, and the force F_1' does not act on the system, but *on the man.*) Then, from Newton's second law,

$$\sum F = ma,$$
$$9 \text{ N} = (4 \text{ kg} + 0.5 \text{ kg})a,$$
$$a = 2 \text{ m s}^{-2}.$$

We can now apply Newton's second law to the block:

$$\sum F = ma,$$
$$F_2 = (4 \text{ kg})(2 \text{ m s}^{-2}) = 8 \text{ N}.$$

If we consider the rope alone, the resultant force on it is

$$\sum F = F_1 - F_2' = 9\,\text{N} - F_2',$$

and from the second law,

$$9\,\text{N} - F_2' = (0.5\,\text{kg})(2\,\text{m s}^{-2}) = 1\,\text{N},$$

$$F_2' = 8\,\text{N}.$$

In agreement with Newton's third law, which was tacitly used when the forces F_2 and F_2' were omitted in considering the system as a whole, we find that F_2 and F_2' are equal in magnitude. Note, however, that the forces F_1 and F_2 are *not* equal and opposite (the rope is not in equilibrium) and that these forces are *not* an action–reaction pair.

Example 6 In Fig. 5–6, a block of weight w_1 (mass $= m_1$) moves on a level frictionless surface, connected by a light flexible cord passing over a small frictionless pulley to a second hanging block of weight w_2(mass $= m_2$). What is the acceleration of the system, and what is the tension in the cord connecting the two blocks?

The diagram shows the forces acting on each block. The forces exerted on the blocks by the cord can be considered an action–reaction pair, so we have used the same symbol T for each. For the block on the surface,

$$\sum F_x = T = m_1 a,$$

$$\sum F_y = N - w_1 = 0.$$

Since the cord connecting the two blocks is inextensible, the accelerations are the same. Applying Newton's second law to the hanging block, we obtain

$$\sum F_y = w_2 - T = m_2 a.$$

Addition of the first and third equations gives

$$w_2 = (m_1 + m_2)a,$$

or

$$a = \frac{w_2}{m_1 + m_2},$$

which says that the acceleration of the *entire system*

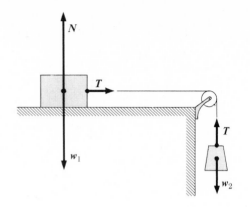

Fig. 5–6

equals the *resultant external force* (w_2) divided by the *total mass* ($m_1 + m_2$). Since $w_2 = m_2 g$,

$$a = g\,\frac{m_2}{m_1 + m_2}.$$

Eliminating a from the first and third equations, we get

$$T = w_2\,\frac{m_1}{m_1 + m_2},$$

so that T is *only a fraction of* w_2. Although the earth pulls on the *hanging* block with a force w_2, the force exerted on the *sliding* block is only a fraction of w_2. It is not the earth that pulls on the sliding block, but the cord, whose tension *must be less than* w_2 if the hanging block is to accelerate downward.

Example 7 A body of mass m is suspended from a spring balance attached to the roof of an elevator, as shown in Fig. 5–7. What is the reading of the balance if the elevator has an acceleration a, relative to the earth? Consider the earth's surface to be an inertial reference system.

The forces on the body are its weight w (the gravitational force f_g exerted on it by the earth) and the upward force T exerted on it by the balance. The body is at rest relative to the elevator, and hence has an acceleration a relative to the earth. (We take the downward direction as positive.) The resultant force on the body is $w - T$, so, from Newton's second law,

$$w - T = ma, \qquad T = w - ma.$$

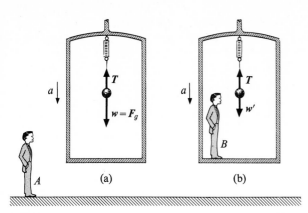

Fig. 5–7 *(a) To observer A, the body has a downward acceleration a, and he writes* **w − T = ma.** *(b) To observer B, the acceleration of the body is zero. He writes* **w′ = T.**

By Newton's third law, the body pulls down on the balance with a force equal and opposite to **T**, or equal to $(w - ma)$, so the balance reading equals $(w - ma)$.

If the same body were suspended in equilibrium from a balance attached to the earth, the balance reading would equal the weight w. To an observer riding in the elevator, the body *appears* to be in equilibrium and hence *appears* to be acted on by a downward force **w′** equal in magnitude to the balance reading, as in Fig. 5–7(b). This *apparent* force **w′** can be called the *apparent weight* of the body. The gravitational force **w** will be called the *true weight*. Then

$$w' = w - ma. \qquad (5\text{–}8)$$

If the elevator is at rest, or moving vertically (either up or down) with constant velocity, $a = 0$ and the apparent weight equals the true weight. If the acceleration is downward, as in Fig. 5–7, so that a is positive, the apparent weight is less than the true weight; the body appears "lighter." If the acceleration is *upward*, a is negative, the apparent weight is greater than the true weight, and the body appears "heavier." If the elevator falls freely, $a - g$, and since the true weight w also equals mg, the apparent weight is zero and the body *appears* "weightless." It is in this sense that an astronaut orbiting the earth in a space capsule is said to be "weightless."

The physiological effect of this condition is exactly the same as though the body were in outer space with no gravitational force at all. The physiological effect of prolonged weightlessness is an interesting medical problem which is just beginning to be explored, with the advent of manned spacecraft. Gravity plays some role in blood distribution in the body, and one reaction to weightlessness is a decrease in volume of blood through increased excretion of water. In some cases astronauts returning to earth have experienced a temporary impairment of sense of balance and a greater tendency toward motion sickness.

Example 8   Figure 5–8(a) represents a simple *accelerometer*. A small body is fastened at one end of a light rod pivoted freely at the point *P*. When the system has an acceleration a toward the right, the rod makes an angle θ with the vertical. (In a practical

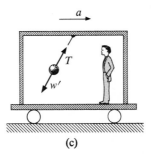

Fig. 5–8 *(a) A simple accelerometer. (b) The forces on the body are* **w** *and* **T.** *(c) The apparent weight* **w′** *is equal and opposite to* **T.**

instrument, some form of damping must be provided to keep the rod from swinging when the acceleration changes. The rod might hang in a tank of mineral oil.)

As shown in the free-body diagram, Fig. 5–8(b), two forces are exerted on the body; its weight w and the tension T in the rod. (We neglect the weight of the rod.) The resultant horizontal force is

$$\sum F_x = T \sin \theta,$$

and the resultant vertical force is

$$\sum F_y = T \cos \theta - w.$$

The x-acceleration is the acceleration a of the system, and the y-acceleration is zero. Hence

$$T \sin \theta = ma, \qquad T \cos \theta = w.$$

When the first equation is divided by the second, and w replaced by mg, we get

$$a = g \tan \theta,$$

and the acceleration a is proportional to the tangent of the angle θ.

Let us consider this problem from the standpoint of an observer riding with the accelerometer, as in Fig. 5–8(c). To him, the body *appears* in equilibrium and hence appears to be acted on by a force w', equal and opposite to T, and which he calls the *apparent weight* of the body. To obtain the expression for w', let us write Newton's second law in general vector form. The resultant force on the body is the vector sum of w and T, and hence

$$w + T = ma, \qquad T = -w + ma.$$

The apparent weight w' is equal and opposite to T, so $w' = -T$ and

$$w' = w - ma. \tag{5–9}$$

This is the general form of Eq. (5–8) for the apparent weight w' of a body, in a system which has an acceleration a relative to an inertial system. In this example, *the apparent weight differs in both magnitude and direction from the true weight.*

Up to this point, applications of Newton's second law have been limited to cases where the resultant force acting on a body was constant, thereby imparting to the body a constant acceleration. Such cases are very important, but make only slight demands on one's mathematical knowledge. When the resultant force is variable, however, the acceleration is not constant and the simple equations of motion with constant acceleration do not apply. We conclude this section with two examples of motion under the action of a variable force.

Example 9   Assume the earth to be a nonrotating, homogeneous sphere. Discuss the motion of a freely falling body (or a body projected vertically upward), taking into account the variation of the gravitational force on the body with its distance from the earth's center. Neglect air resistance.

The gravitational force on the body at a distance r from the earth's center is Gmm_E/r^2, and from Newton's second law its acceleration is

$$g = \frac{w}{m} = -\frac{Gm_E}{r^2},$$

where the positive direction is upward (or better, radially outward).

It was shown in Eq. (4–8) that the acceleration can be expressed as

$$g = v \frac{dv}{dr}.$$

Then

$$v \frac{dv}{dr} = -\frac{Gm_E}{r^2}, \quad \int_{v_1}^{v_2} v \, dv = -Gm_E \int_{r_1}^{r_2} \frac{dr}{r^2},$$

where v_1 and v_2 are the velocities at the radial distances r_1 and r_2. It follows that

$$v_2{}^2 - v_1{}^2 = 2Gm_E \left(\frac{1}{r_2} - \frac{1}{r_1} \right). \tag{5–10}$$

As an illustration, let us find the initial velocity v_1 required to project a body vertically upward so that it rises to a height above the earth's surface equal to the earth's radius R. Then $v_2 = 0$, $r_1 = R$, $r_2 = 2R$,

and

$$v_1{}^2 = \frac{Gm_E}{R}. \tag{5–11}$$

Let g_0 represent the acceleration of gravity at the earth's surface, where $r = R$. Then

$$Gm_E = g_0R^2,$$

and Eq. (5–11) can be written

$$v_1{}^2 = g_0R. \tag{5–12}$$

How does this compare with the velocity that would be required if the acceleration had the *constant* value g_0?

Example 10 A resisting force that varies directly with the speed is found frequently in nature. Any small spherical body of radius r, like a raindrop, an oil droplet, or a steel sphere, moving with a small velocity v through a viscous fluid (liquid or gas) is subjected to a force R, where

$$\boldsymbol{R} = -6\pi\eta r\boldsymbol{v},$$

and η is the viscosity. This relation is known as *Stokes' law*. Letting

$$k = 6\pi\eta r,$$

we may write Stokes' law simply as

$$\boldsymbol{R} = -k\boldsymbol{v}.$$

A small sphere falling through a viscous fluid is subjected to three vertical forces, as shown in Fig. 5–9: the weight $\boldsymbol{w}$, the buoyant force $\boldsymbol{B}$, and the resisting force $\boldsymbol{R}$.

Let us suppose that the sphere starts from rest and that the positive y-direction is downward. Under

these conditions

$$\sum F_y = w - B - kv = ma.$$

At first, when $v = 0$, the resisting force is zero and the initial acceleration a_0 is positive:

$$a_0 = \frac{w - B}{m}. \tag{5–13}$$

The sphere speeds up and, after a while, when v becomes large enough, the resisting force equals $w - B$ and there is no resultant force acting on the sphere. At this moment the acceleration is zero and the speed undergoes no further increase. The maximum or *terminal speed* v_T may therefore be calculated by setting $a = 0$; thus

$$w - B - kv_T = 0$$

or

$$v_T = \frac{w - B}{k}. \tag{5–14}$$

Figure 5–10 shows how the acceleration, velocity, and position vary with time.

To find the relation between the speed and the time during the interval before the terminal speed is reached, we go back to Newton's second law,

$$m\frac{dv}{dt} = w - B - kv.$$

After rearranging terms and replacing $(w - B)/k$ by v_T, we get

$$\frac{dv}{v - v_T} = -\frac{k}{m}\,dt.$$

Since $v = 0$ when $t = 0$,

$$\int_0^v \frac{dv}{v - v_T} = -\frac{k}{m}\int_0^t dt,$$

whence

$$\ln\frac{v_T - v}{v_T} = -\frac{k}{m}t,$$

or

$$1 - \frac{v}{v_T} = e^{-(k/m)t},$$

Fig. 5–9 *Forces acting on a small sphere falling through a viscous fluid.*

and finally

$$v = v_T(1 - e^{-(k/m)t}). \qquad (5-15)$$

An important concept related to an exponentially varying quantity is the *relaxation time*, t_R, whose meaning can be seen from Fig. 5–10(b). Suppose the acceleration remained constant at the initial value a_0, as indicated by the dotted line. The relaxation time can be defined as the time that *would* be required to reach the terminal velocity with this constant acceleration, and evidently

$$t_R = \frac{v_T}{a_0} = \frac{(w - B)/k}{(w - B)/m} = \frac{m}{k}.$$

Equation (5–15) can now be written more simply as

$$v = v_T(1 - e^{-t/t_R}). \qquad (5-16)$$

(a)

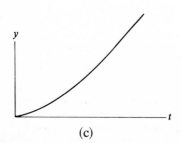

(b)

(c)

Fig. 5–10 *Graphs of acceleration, velocity, and position versus time for a body falling in a viscous medium.*

At the instant when t equals the relaxation time, $t/t_R = 1$ and

$$v = v_T(1 - e^{-1}) = 0.63v_T.$$

Thus about 63% of the final velocity is actually acquired in a time equal to the relaxation time.

PROBLEMS

For problem work, use the approximate value $g = 9.80$ m s^{-2}. A force diagram should be constructed for each problem.

5–1

a) What is the mass of a body which weighs 1 N at a point where $g = 9.80$ m s^{-2}?

b) What is the mass of a body which weighs 1 dyn at a point where $g = 980$ cm s^{-2}?

c) What is the mass of a standard pound body?

5–2

a) At what distance from the earth's center would a standard kilogram weigh 1 N?

b) At what distance would a 1-g body weigh 1 dyn?

c) At what distance would a 1-slug body weigh 1 lb?

5–3 If action and reaction are always equal in magnitude and opposite in direction, why don't they always cancel each other and leave no net force for acceleration of the body?

5–4 The mass of a certain object is 10 kg.

a) What would its mass be if taken to the planet Mars?

b) Is the expression $F = ma$ valid on Mars?

c) Newton's second law is sometimes written in the form $F = wa/g$ instead of $F = ma$. Would this expression be valid on Mars?

5–5 A constant horizontal force of 10 lb acts on a body on a *smooth horizontal* plane. The body starts from rest and is observed to move 250 ft in 5 s.

a) What is the mass of the body?

b) If the force ceases to act at the end of 5 s, how far will the body move in the next 5 s?

5–6 A .22 rifle bullet, traveling at 360 m s^{-1}, strikes a block of soft wood, which it penetrates to a depth of 0.1 m. The mass of the bullet is 1.8 g. Assume a constant retarding force.

a) How long a time was required for the bullet to stop?

b) What was the decelerating force, in newtons? in pounds?

5-7 A body of mass 15 kg rests on a frictionless horizontal plane and is acted on by a horizontal force of 30 N.

a) What acceleration is produced?

b) How far will the body travel in 10 s?

c) What will be its velocity at the end of 10 s?

5-8 A body of mass 0.5 kg is at rest at the origin, $x = 0$, on a horizontal frictionless surface. At time $t = 0$ a force of 0.001 N is applied to the body parallel to the x-axis, and 5 s later this force is removed.

a) What are the position and velocity of the body at $t = 5$ s?

b) If the same force is again applied at $t = 15$ s, what are the position and velocity of the body at $t = 20$ s?

5-9 A body of mass 10 kg is moving with a constant velocity of 5 m s^{-1} on a horizontal surface. The coefficient of sliding friction between body and surface is 0.20.

a) What horizontal force is required to maintain the motion?

b) If the force is removed, how soon will the body come to rest?

5-10 A hockey puck leaves a player's stick with a velocity of 10 m s^{-1} and slides 40 m before coming to rest. Find the coefficient of friction between the puck and the ice.

5-11 An electron (mass $= 9 \times 10^{-31}$ kg) leaves the cathode of a radio tube with zero initial velocity and travels in a straight line to the anode, which is 1 cm away. It reaches the anode with a velocity of 6×10^{6} m s^{-1}. If the acceleration force was constant, compute (a) the accelerating force, in newtons, (b) the time to reach the anode, (c) the acceleration. (The gravitational force on the electron may be neglected.)

5-12 Give arguments either for or against the statement that "the only reason an apple falls downward to meet the earth instead of the earth falling upward to meet the apple is that the earth, being so much more massive, exerts the greater pull."

5-13 In an experiment using the Cavendish balance to measure the gravitational constant G, it is found that a sphere of mass 0.8 kg attracts another sphere of mass 0.004 kg with a force of 13×10^{-11} N when the distance between the centers of the spheres is 0.04 m. The acceleration of gravity at the earth's surface is 9.80 m s^{-2}, and the radius of the earth is 6400 km. Compute the mass of the earth from these data.

5-14 Two spheres, each of mass 6.4 kg, are fixed at points A and B (Fig. 5-11). Find the magnitude and direction of the initial acceleration of a sphere of mass 0.010 kg if released from rest at point P and acted on only by forces of gravitational attraction of the spheres at A and B.

Fig. 5-11

5-15 The mass of the moon is about one eighty-first, and its radius one-fourth, that of the earth. What is the acceleration due to gravity on the surface of the moon?

5-16 In round numbers, the distance from the earth to the moon is 250,000 mi, the distance from the earth to the sun is 93 million mi, the mass of the earth is 6×10^{27} g, and the mass of the sun is 2×10^{33} g. Approximately what is the ratio of the gravitational pull of the sun on the moon to that of the earth on the moon?

5-17 An elevator with mass 2000 kg rises with an acceleration of 1 m s^{-2}. What is the tension in the supporting cable?

5-18 A 4-kg block is accelerated upward by a cord whose breaking strength is 60 N. Find the maximum acceleration which can be given the block without breaking the cord.

5-19 A 5-kg block is supported by a cord and pulled upward with an acceleration of 2 m s^{-2}.

a) What is the tension in the cord?

b) After the block has been set in motion, the tension in the cord is reduced to 49 N. What sort of motion will the block perform?

c) If the cord is now slackened completely, the block is observed to move up 2 m farther before coming to rest. With what velocity was it traveling?

5-20 A block weighing 10 lb is held up by a string which can be moved up or down. What conclusions can you draw regarding magnitude and direction of the acceleration and velocity of the upper end of the string when the tension in the string is (a) 5 lb? (b) 10 lb? (c) 15 lb?

5–21 A body hangs from a spring balance supported from the roof of an elevator.

a) If the elevator has an upward acceleration of 4 ft s^{-2} and the balance reads 45 lb, what is the true weight of the body?

b) Under what circumstances will the balance read 35 lb?

c) What will the balance read if the elevator cable breaks?

5–22 A transport plane is to take off from a level landing field with two gliders in tow, one behind the other. Each glider has a mass of 1200 kg, and the friction force or drag on each may be assumed constant and equal to 2000 N. The tension in the towrope between the transport plane and the first glider is not to exceed 10,000 N.

a) If a velocity of 40 m s^{-1} is required for the take-off, how long a runway is needed?

b) What is the tension in the towrope between the two gliders while the planes are accelerating for the take-off?

5–23 If the coefficient of friction between tires and road is 0.5, what is the shortest distance in which an automobile can be stopped when traveling at 60 mi hr^{-1}?

5–24 A 40-kg packing case is on the floor of a truck. The coefficient of static friction between the case and the truck floor is 0.40, and the coefficient of sliding friction is 0.25. Find the magnitude and direction of the friction force acting on the case (a) when the truck is accelerating at 2 m s^{-2}, (b) when it is decelerating at 3 m s^{-2}.

5–25 A balloon is descending with a constant acceleration a, less than the acceleration due to gravity g. The weight of the balloon, with its basket and contents, is w. What weight, W, of ballast should be released so that the balloon will begin to be accelerated upward with constant acceleration a? Neglect air resistance.

5–26 A 64-lb block is pushed up a 37° inclined plane by a horizontal force of 100 lb. The coefficient of sliding friction is 0.25. Find (a) the acceleration, (b) the velocity of the block after it has moved a distance of 20 ft along the plane, and (c) the normal force exerted by the plane. Assume that all forces act at the center of the block.

5–27 A block rests on an inclined plane which makes an angle θ with the horizontal. The coefficient of sliding

friction is 0.50, and the coefficient of static friction is 0.75.

a) As the angle θ is increased, find the minimum angle at which the block starts to slip.

b) At this angle, find the acceleration once the block has begun to move.

c) How long a time is required for the block to slip 20 ft along the inclined plane?

5–28

a) What constant horizontal force is required to drag a 16-lb block along a horizontal surface with an acceleration of 4 ft s^{-2} if the coefficient of sliding friction between block and surface is 0.5?

b) What weight, hanging from a cord attached to the 16-lb block and passing over a small frictionless pulley, will produce this acceleration?

5–29 A 20-kg box rests on the flat floor of a truck. The coefficient of friction between box and floor is 0.1. The truck stops at a stop sign and then starts, with an acceleration of 2 m s^{-2}. If the box is 5 m from the rear of the truck when it starts, how much time elapses before it falls off the rear of the truck? How far does the truck travel in this time?

5–30 A 160-lb man stands on a bathroom scale in an elevator. As the elevator starts moving, the scale reads 200 lb.

a) Find the acceleration of the elevator (magnitude and direction).

b) What is the acceleration if the scale reads 120 lb?

c) If the scale reads zero, should the man worry? Explain.

5–31 A short commuter train consists of a locomotive and two cars. The mass of the locomotive is 6000 kg and that of each car 2000 kg. The train pulls away from a station with an acceleration of 0.5 m s^{-2}.

a) Find the tension in the coupler joining the locomotive to the first car, and in the coupler joining the two cars.

b) What total horizontal force must the locomotive wheels exert on the track?

5–32 A loaded elevator with very worn cables has a total mass of 2000 kg, and the cables can withstand a maximum tension of 24,000 N.

a) What is the maximum acceleration for the elevator if the cables are not to break?

b) What is the answer for part (a) if the elevator is taken to the moon?

5–33 A block of mass 5 kg resting on a horizontal surface is connected by a cord passing over a light frictionless pulley to a hanging block of mass 5 kg. The coefficient of friction between the block and the horizontal surface is 0.5. Find (a) the tension in the cord, and (b) the acceleration of each block.

5–34 A block having a mass of 2 kg is projected up a long 30° incline with an initial velocity of 22 m s^{-1}. The coefficient of friction between the block and the plane is 0.3.

a) Find the friction force acting on the block as it moves up the plane.

b) How long does the block move up the plane?

c) How *far* does the block move up the plane?

d) How long does it take the block to slide from its position in part (c) to its starting point?

e) With what velocity does it arrive at this point?

f) If the mass of the block had been 5 kg instead of 2 kg, would the answers in the preceding parts be changed?

5–35 A 30-lb block on a level frictionless surface is attached by a cord passing over a small frictionless pulley to a hanging block originally at rest 4 ft above the floor. The hanging block strikes the floor in 2 s. Find (a) the weight of the hanging block, (b) the tension in the string while both blocks were in motion.

5–36 Two blocks, each having mass 20 kg, rest on frictionless surfaces, as shown in Fig. 5–12. Assuming the pulleys to be light and frictionless, compute (a) the time required for block A to move 1 m down the plane, starting from rest, (b) the tension in the cord connecting the blocks.

Fig. 5–12

5–37 A block of mass 0.2 kg rests on top of a block of mass 0.8 kg. The combination is dragged along a level surface at constant velocity by a hanging block of mass 0.2 kg as in Fig. 5–13(a).

a) The first 0.2-kg block is removed from the 0.8-kg block and attached to the hanging block, as in Fig. 5–13(b). What is now the acceleration of the system?

b) What is the tension in the cord attached to the 0.8-kg block in part (b)?

(a)

Fig. 5–13

(b)

5–38 Block A in Fig. 5–14 has a mass of 2 kg and block B 20 kg. The coefficient of friction between B and the horizontal surface is 0.1.

a) What is the mass of block C if the acceleration of B is 2 m s^{-2} toward the right?

b) What is the tension in each cord when B has the acceleration stated above?

Fig. 5–14

5–39 Two blocks connected by a cord passing over a small frictionless pulley rest on frictionless planes, as shown in Fig. 5–15.

a) Which way will the system move?

b) What is the acceleration of the blocks?

c) What is the tension in the cord?

Fig. 5–15

5–40 Two 0.2-kg blocks hang at the ends of a light flexible cord passing over a small frictionless pulley, as in Fig. 5–16. A 0.1-kg block is placed on the right, and removed after 2 s.

a) How far will each block move in the first second after the 0.1-kg block is removed?

b) What was the tension in the cord before the 0.1-kg block was removed? After it was removed?

c) What was the tension in the cord supporting the pulley before the 0.1-kg block was removed? Neglect the weight of the pulley.

Fig. 5–16

5–41 In terms of m_1, m_2, and g, find the acceleration of both blocks in Fig. 5–17. Neglect all friction and the masses of the pulleys.

Fig. 5–17

5–42 The masses of bodies A and B in Fig. 5–18 are 20 kg and 10 kg, respectively. They are initially at rest on the

Fig. 5–18

floor and are connected by a weightless string passing over a weightless and frictionless pulley. An upward force F is applied to the pulley. Find the accelerations a_1 of body A and a_2 of body B when F is (a) 98 N, (b) 196 N, (c) 394 N, (d) 788 N.

5–43 The two blocks in Fig. 5–19 are connected by a heavy uniform rope of mass 4 kg. An upward force of 200 N is applied as shown.

a) What is the acceleration of the system?

b) What is the tension at the top of the heavy rope?

c) What is the tension at the midpoint of the rope?

Fig. 5–19

5–44 Two blocks with masses of 4 kg and 8 kg, respectively, are connected by a string and slide down a 30° inclined plane, as in Fig. 5–20. The coefficient of sliding friction between the 4-kg block and the plane is 0.25, and between the 8-kg block and the plane it is 0.50.

a) Calculate the acceleration of each block.

b) Calculate the tension in the string.

Fig. 5–20

5–45 Two bodies with masses of 5 kg and 2 kg, respectively, hang 1 m above the floor from the ends of a cord 3 m long passing over a frictionless pulley. Both bodies start from rest. Find the maximum height reached by the 2-kg body.

5–46 A man who weighs 160 lb stands on a platform which weighs 80 lb. He pulls a rope which is fastened to the platform and runs over a pulley on the ceiling. With what force does he have to pull in order to give himself and the platform an upward acceleration of 2 ft s^{-2}?

5–47 What acceleration must the cart in Fig. 5–21 have in order that the block A will not fall? The coefficient of friction between the block and the cart is μ. How would the behavior of the block be described by an observer on the cart?

Fig. 5–21

5–48 The left end of the weightless rod shown in Fig. 5–22 is hinged to a cart. A heavy particle is attached to the right end. If the cart has an acceleration a to the right, find the angle θ. How would this situation be described by an observer on the cart?

Fig. 5–22

5–49 Which way will the accelerometer in Fig. 5–8 deflect under the following conditions?

a) The cart is moving toward the right and traveling faster.

b) The cart is moving toward the right and traveling slower.

c) The cart is moving toward the left and traveling faster.

d) The cart is moving toward the left and traveling slower.

e) The cart is at rest on a sloping surface.

f) The cart is given an upward velocity on a frictionless inclined plane. It first moves up, then stops, and then moves down. What is the deflection in each stage of the motion?

5–50

a) In terms of the acceleration of gravity at the surface of the earth, g_0, and the earth's radius R, find the velocity with which a body must be projected vertically upward, in the absence of air resistance, to rise to an infinite distance above the earth's surface. This is called the *escape velocity*.

b) In terms of the same quantities, find the velocity with which a body will strike the earth's surface if it falls from rest toward the earth from an infinitely distant point.

c) Compute both of these velocities in miles per hour.

d) Explain why the velocities are not infinitely large.

5–51 The mass of the motorboat in Problem 4–37 is 100 slugs. Find the force decelerating the boat when its speed is (a) 20 ft per sec, and (b) 10 ft per sec. (c) If the boat is being towed at 10 ft per sec, what is the tension in the towline?

5–52 A body of mass $m = 5$ kg falls from rest in a viscous medium. The body is acted on by a net constant downward force of 20 N, and by a viscous retarding force proportional to its speed and equal to $5v$, where v is the speed in meters per second.

a) Find the initial acceleration, a_0.

b) Find the acceleration when the speed is 3 m s^{-1}.

c) Find the speed when the acceleration equals $0.1a_0$.

d) Find the terminal velocity, v_T.

e) Find the relaxation time, t_R.

f) Find the coordinate, velocity, and acceleration 2 s after the start of the motion.

g) Find the time required to reach a speed $0.9v_T$.

h) Construct a graph of v versus t, for a time of 3 s.

5–53 A body falls from rest through a medium which exerts a resisting force that varies directly with the square of the velocity $(R = -kv^2)$.

a) Draw a diagram showing the direction of motion and indicate with the aid of vectors all of the forces acting on the body.

b) Apply Newton's second law and infer from the resulting equation the general properties of the motion.

c) Show that the body acquires a terminal velocity and calculate it.

d) Find the relaxation time.

e) Derive the equation for the velocity at any time.

5–54 A particle of mass m, originally at rest, is subjected to a force whose direction is constant but whose magnitude varies with the time according to the relation

$$F = F_0 \left[1 - \left(\frac{t - T}{T} \right)^2 \right],$$

where F_0 and T are constants. The force acts only for the time interval $2T$.

a) Make a rough graph of F versus t.

b) Prove that the speed v of the particle after a time $2T$ has elapsed is equal to $4F_0T/3m$.

c) Choose numbers for v, T, and m that might be appropriate to a batted baseball, and calculate the force F_0. Judge whether the answer is sensible.

5–55 A fifty-lb monkey with downcast eyes has a firm hold on a light rope that passes over a frictionless pulley and is attached to a fifty-lb. bunch of bananas, as shown in Fig. 5–23. The monkey happens to glance upward, sees the bananas, and starts to climb the rope to get at the bananas.

a) As he climbs, do the bananas move up, down, or remain at rest?

b) As he climbs, does the distance between him and the bananas decrease, increase, or remain constant?

c) The monkey releases his hold on the rope. What about the distance between him and the bananas while he is falling?

d) Before he reaches the ground, he grabs the rope to stay his fall. What do the bananas do?

Fig. 5–23

Chapter 6

Motion in a Plane

6–1 MOTION IN A PLANE

Thus far we have discussed only motion along a straight line, or *rectilinear motion*. In this chapter we shall consider *plane motion*, that is, motion in a curved path which lies in a fixed plane. Examples of such motion are the flight of a thrown or batted baseball, a projectile shot from a gun, a ball whirled at the end of a cord, the motion of the moon or of a satellite around the earth, and the motion of the planets around the sun.

If the motion is referred to a set of rectangular coordinate axes x and y, the equation of the path expresses y as a function of x, $y = f(x)$. Very often one is interested in the position of the moving body as a function of time. If s is the distance along the path from some fixed point to the position of the body, its position at any time is given by an equation of the form $s = f(t)$. It is usually simpler, however, to deal with the x- and y-coordinates separately and to describe the motion by the two equations

$$x = f_1(t), \qquad y = f_2(t).$$

These can be considered as *parametric equations* of the path, expressing the coordinates x and y in terms of the parameter t.

We shall consider several kinds of problems in plane motion. In some the motion of the particle is

known and we wish to determine its velocity and acceleration and the resultant force acting on it. An example is that of a ball attached to a cord and whirled in a circle at constant speed. What is the tension in the cord? In others, the force acting on a particle may be known at every point of space and we may wish to find the equation of motion of the particle. Examples are the orbit of a planet around the sun, or the path followed by a rocket.

6–2 AVERAGE AND INSTANTANEOUS VELOCITY

Consider a particle moving along the curved path in Fig. 6–1(a). Points P and Q represent two positions of the particle. Its displacement as it moves from P to Q is the vector $\Delta \mathbf{s}$. Just as in the case of rectilinear motion, the average velocity $\bar{\mathbf{v}}$ of the particle is defined as the vector displacement $\Delta \mathbf{s}$ divided by the elapsed time Δt:

$$\boxed{\text{Average velocity } \bar{\mathbf{v}} = \Delta \mathbf{s}/\Delta t.} \qquad (6\text{–}1)$$

The average velocity would be the same for any path that would take the particle from P to Q in the time interval Δt.

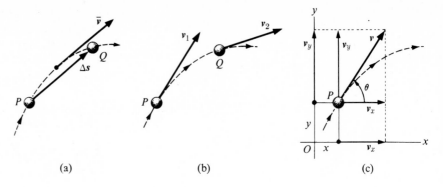

Fig. 6–1 (a) The vector $v = \Delta s/\Delta t$ represents the average velocity between P and Q. (b) Vectors v_1 and v_2 represent the instantaneous velocities at P and Q. (c) The velocities v_x and v_y of the projections of P are the rectangular components of v.

(a)   (b)   (c)

Average velocity is a vector quantity, in the same direction as the vector Δs. Because it is associated with the entire displacement Δs, the vector $\bar{v}$ has been constructed in Fig. 6–1(a) at a point midway between P and Q.

The *instantaneous* velocity v at the point P is defined in magnitude and direction as the *limit approached by the average velocity* when point Q is taken closer and closer to point P:

$$\text{Instantaneous velocity } v = \lim_{\Delta t \to 0} \frac{\Delta s}{\Delta t} = \frac{ds}{dt}. \quad (6\text{–}2)$$

As point Q approaches point P, the direction of the vector Δs approaches that of the tangent to the path at P, so that the instantaneous velocity vector at any point is *tangent* to the path at that point. The instantaneous velocities at point P and Q are shown in Fig. 6–1(b).

In Fig. 6–1(c), the motion of a particle is referred to a rectangular coordinate system. As the particle moves along its path, its projections onto the x- and y-axes move along these axes in rectilinear motion. If in a time interval Δt these projections move by amounts Δx and Δy, respectively, then the average velocities of the projections which we may call $\bar{v}_x$ and $\bar{v}_y$, are given by

$$\bar{v}_x = \frac{\Delta x}{\Delta t}, \qquad \bar{v}_y = \frac{\Delta y}{\Delta t}.$$

We see that these are equal to the x- and y-components of the average velocity vector. Similarly, the instantaneous velocities of these projections, denoted by v_x and v_y, are equal respectively to the x- and y-

components of the instantaneous velocity v, and we may write

$$v_x = \lim_{\Delta t \to 0} \frac{\Delta x}{\Delta t} = \frac{dx}{dt}, \qquad v_y = \lim_{\Delta t \to 0} \frac{\Delta y}{\Delta t} = \frac{dy}{dt}.$$

The magnitude of the instantaneous velocity is given by

$$|v| = \sqrt{v_x{}^2 + v_y{}^2},$$

and the angle θ in Fig. 6–1(c) by

$$\tan \theta = \frac{v_y}{v_x}.$$

Thus we may represent velocity, a vector quantity, in terms of its components or in terms of its magnitude and direction, just as with force, another vector quantity.

It is well to remember that the direction of the instantaneous velocity of a particle at any point in its path is *always* in a direction tangent to the path at that point, no matter how complicated the motion.

6–3 AVERAGE AND INSTANTANEOUS ACCELERATION

In Fig. 6–2(a), the vectors v_1 and v_2 represent the instantaneous velocities, at points P and Q, of a particle moving in a curved path. The velocity v_2 necessarily differs in *direction* from the velocity v_1. The diagram has been constructed for a case in which it also differs in *magnitude*, although in special cases the magnitude of the velocity may remain constant.

The *average acceleration* $\bar{a}$ of the particle as it moves from P to Q is defined, just as in the case of

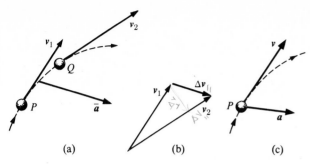

(a)   (b)   (c)

Fig. 6–2 (a) The vector **a** = Δ**v**/Δt represents the average acceleration between P and Q. (b) Construction for obtaining Δ**v** = **v**₂ − **v**₁. (c) Instantaneous velocity **v** and instantaneous acceleration **a** at point P. Vector **v** is tangent to path; vector **a** points toward concave side of path.

rectilinear motion, as the *vector change in velocity*, Δ**v**, divided by the time interval Δt:

$$\text{Average acceleration } \bar{\boldsymbol{a}} = \frac{\Delta \boldsymbol{v}}{\Delta t}. \qquad (6\text{–}3)$$

Average acceleration is a vector quantity, in the same direction as the vector Δ**v**.

The vector change in velocity, Δ**v**, means the vector difference **v**₂ − **v**₁:

$$\Delta \boldsymbol{v} = \boldsymbol{v}_2 - \boldsymbol{v}_1,$$

or

$$\boldsymbol{v}_2 = \boldsymbol{v}_1 + \Delta \boldsymbol{v}.$$

As explained in Section 1–9, the vector difference Δ**v** can be found by drawing the vectors **v**₁ and

v₂ from a common point, as in Fig. 6–2(b), and constructing the vector from the tip of **v**₁ to the tip of **v**₂. Then **v**₂ is the vector sum of **v**₁ and Δ**v**.

The average acceleration vector, $\bar{\boldsymbol{a}} = \Delta \boldsymbol{v}/\Delta t$, is shown in Fig. 6–2(a) at a point midway between P and Q.

The *instantaneous acceleration* **a** at point P is defined in magnitude and direction as the limit approached by the average acceleration when point Q approaches point P and Δ**v** and Δt both approach zero:

$$\text{Instantaneous acceleration } \boldsymbol{a} = \lim_{\Delta t \to 0} \frac{\Delta \boldsymbol{v}}{\Delta t} = \frac{d\boldsymbol{v}}{dt}. \quad (6\text{–}4)$$

The instantaneous acceleration vector at point P is shown in Fig. 6–2(c). Note that it does *not* have the same direction as the velocity vector. Reference to the construction of Fig. 6–2(b) will show that the acceleration vector must always lie on the *concave* side of the curved path.

6–4 COMPONENTS OF ACCELERATION

Figure 6–3(a) again shows the motion of a particle referred to a rectangular coordinate system. The accelerations of the projections of the particle onto the x- and y-axes are

$$a_x = \frac{dv_x}{dt} = \frac{d^2 x}{dt^2}, \qquad a_y = \frac{dv_y}{dt} = \frac{d^2 y}{dt^2}.$$

These accelerations, however, are also the rectangular components of the acceleration **a** of the

Fig. 6–3 In (a) the acceleration **a** is resolved into rectangular components **a**ₓ and **a**ᵧ. In (b) it is resolved into a normal component **a**⊥ and a tangential component **a**∥.

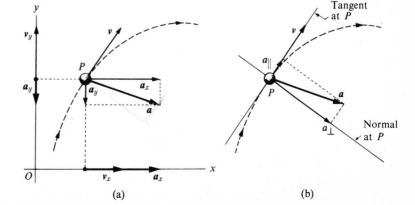

(a)   (b)

particle. Thus, if the components a_x and a_y are known, the magnitude and direction of the acceleration a can be obtained, just as with velocity. When the acceleration is known, the force on the particle can be found, in magnitude and direction, from Newton's second law,

$$F = ma.$$

Conversely, if the force F is known at every point, the acceleration a and its components a_x and a_y can be found from Newton's second law.

The acceleration of a particle moving in a curved path can also be resolved into rectangular components $a_\perp$ and $a_\parallel$, in directions *normal* and *parallel* (or tangential) to the path, as shown in Fig. 6–3(b). Unlike the rectangular components referred to a set of fixed axes, the normal and tangential components do not have fixed directions in space. They do, however, have a direct physical significance. The parallel component $a_\parallel$ arises from a change in the *magnitude* of the velocity vector v, while the normal component $a_\perp$ arises from a change in the *direction* of the velocity.

This is illustrated in Fig. 6–4, which corresponds to Fig. 6–2(b). The vector from O to A represents the velocity v_1 of a particle at point P in Fig. 6–2(a), and the vector from O to B represents the velocity v_2 at

point Q. The change in velocity, Δv, is given by the vector from A to B.

Note that the vectors Δv in Figs. 6–4 and 6–2(b) have the same direction and (apart from the difference in scale of the two diagrams) the same length.

The vector from O to C in Fig. 6–4 has the same length as v_1. The vector Δv can be resolved into components represented by the vectors from A to C, and from C to B. The length of the vector from C to B equals the difference in *length* between the vectors v_2 and v_1. That is, this vector represents the change in *magnitude* of the velocity, and when divided by Δt, gives the component of average acceleration resulting from this change in magnitude.

If the magnitude of the velocity did *not* change between points P and Q, the velocity v_2 at point Q would be represented by the vector from O to C in Fig. 6–4. In this case, however, *there would still be a change in the vector velocity*, represented by the vector from A to C. This change would result from the change in *direction* of the velocity vector, and when divided by Δt would give the component of average acceleration resulting from this change in direction. That is, *motion in a curved path with constant speed is accelerated motion*, because velocity is a vector quantity which can change in magnitude, in direction, or, as in Fig. 6–4, in both.

Now suppose that point Q in Fig. 6–2(a) approaches point P. The vector from O to B in Fig. 6–4 then swings upward toward the vector v_1. The angle ϕ becomes smaller and smaller and the angle θ approaches 90°. The vector from A to C becomes more and more nearly perpendicular to v_1 and the vector from C to B becomes more and more nearly parallel to v_1. In the limit, the vector from A to C becomes normal to v_1 (and hence normal to the path) and the vector from C to B becomes parallel to v_1 (and hence tangent to the path). Thus although the vectors labeled $\Delta v_\perp$ and $\Delta v_\parallel$ in Fig. 6–4 are not normal and parallel to v_1 in this figure, they become so in the limit as point Q approaches point P. The limiting value of $\Delta v_\perp/\Delta t$ equals the normal component of acceleration $a_\perp$, and the limiting value of $\Delta v_\parallel/\Delta t$ equals the parallel component of acceleration $a_\parallel$.

Is the acceleration resulting from a change in the *direction* of a velocity as "real" as that arising from a

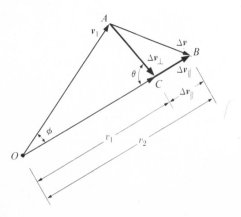

Fig. 6–4 *The vector Δv is resolved into normal and tangential components $\Delta v_\perp$ and $\Delta v_\parallel$. The normal component is the change in velocity resulting from the change in direction of v; the tangential component is the change resulting from a change in the magnitude of v.*

change in its *magnitude?* From a purely kinematical viewpoint the answer is of course "Yes," since by definition acceleration equals the vector rate of change of velocity. A more satisfying answer is that a force must be exerted on a body to change its direction of motion, as well as to increase or decrease its speed. In the absence of an external force, a body continues to move not only with constant speed but also *in a straight line.* When a body moves in a *curved* path, a transverse force must be exerted on it to deviate it sidewise. The ratio of transverse (or normal) force to normal acceleration is the mass of the body, and is equal to the ratio of tangential force to tangential acceleration. That is, if $F_\perp$ and $F_\parallel$ are the normal and tangential components of the force on a body moving in a curved path, Newton's second law (with the same value of m) applies to both of these components:

$$F_\perp = ma_\perp, \qquad F_\parallel = ma_\parallel.$$

These considerations are useful in atomic physics. The usual experimental method of measuring the mass of an individual ion is to project it into a magnetic field. The magnetic field exerts a transverse force on the ion and the mass of the ion is obtained by dividing this force by the measured transverse acceleration.

It follows from the discussion above that if the force on a particle is always normal to the path, then $F_\parallel = 0$, $a_\parallel = 0$, and the particle has no tangential component of acceleration. The *magnitude* of the velocity then remains constant and the only effect of the force is to change the direction of motion, that is, to deviate the particle sidewise.

If the force has no normal component, then $F_\perp = 0$, $a_\perp = 0$, and there is no change in the *direction* of the velocity; that is, the particle moves in a straight line.

6–5 MOTION OF A PROJECTILE

Any object that is given an initial velocity and which subsequently follows a path determined by the gravitational force acting on it and by the frictional resistance of the atmosphere is called a *projectile.* Thus the term applies to a missile shot from a gun, a rocket after its fuel is exhausted, a bomb released

from an airplane, or a thrown or batted baseball. The motion of a freely falling body discussed in Chapter 4 is a special case of projectile motion. The path followed by a projectile is called its *trajectory.*

The gravitational force on a projectile is directed toward the center of the earth and is inversely proportional to the square of the distance from the earth's center. Here we shall consider only trajectories which are of sufficiently short range so that the gravitational force can be considered constant in magnitude and direction. The motion will be referred to axes fixed with respect to the earth. Since this is not an inertial system, it is not strictly correct to use Newton's second law to relate the force on the projectile to its acceleration. However, for trajectories of short range, the error is very small. Finally, all effects of air resistance will be ignored, so that our results apply only to motion in a vacuum on a flat, nonrotating earth. These simplifying assumptions form the basis of an idealized *model* of the physical problem, in which we neglect unimportant details and focus attention on the most important aspects of the phenomenon.

Since the only force on a projectile in this idealized case is its weight, considered constant in magnitude and direction, the motion is best referred to a set of rectangular coordinate axes. We shall take the x-axis horizontal, the y-axis vertical, and the origin at the point where the projectile starts its free flight, for example, at the muzzle of a gun or the point where it leaves the thrower's hand. The x-component of the force on the projectile is then zero and the y-component is the weight of the projectile, $-mg$. Then from Newton's second law,

$$a_x = \frac{F_x}{m} = 0, \qquad a_y = \frac{F_y}{m} = \frac{-mg}{m} = -g.$$

That is, the horizontal component of acceleration is zero and the vertical component is downward and equal to that of a freely falling body. Since zero acceleration means constant velocity, the motion can be described as a combination of *horizontal motion with constant velocity and vertical motion with constant acceleration.*

Consider next the velocity of the projectile. In Fig. 6–5, x- and y-axes have been constructed with the origin at the point where the projectile begins its free flight. We shall set $t = 0$ at this point. The velocity at the origin is represented by the vector $\boldsymbol{v}_0$, called the *initial velocity* or the *muzzle velocity* if the projectile is shot from a gun. The angle θ_0 is the *angle of departure*. The initial velocity has been resolved into a horizontal component v_{0x}, of magnitude $v_0 \cos \theta_0$, and a vertical component v_{0y}, of magnitude $v_0 \sin \theta_0$.

Since the horizontal velocity component v_x is constant, we have at any later time t,

$$v_x = v_{0x} = v_0 \cos \theta_0.$$

The vertical acceleration is $-g$, so the vertical velocity component at time t is

$$v_y = v_{0y} - gt = v_0 \sin \theta_0 - gt.$$

These components can be added vectorially to find the resultant velocity v. Its magnitude is

$$v = \sqrt{v_x{}^2 + v_y{}^2},$$

and the angle θ it makes with the horizontal is given by

$$\tan \theta = \frac{v_y}{v_x}.$$

The velocity vector $\boldsymbol{v}$ is tangent to the trajectory, so its direction is the same as that of the trajectory.

The coordinates of the projectile at any time can now be found from the equations of motion with constant velocity, and with constant acceleration, developed in Sec. 4–5. The x-coordinate is

$$x = v_{0x} t = (v_0 \cos \theta_0)t, \tag{6–5}$$

and the y-coordinate is

$$y = v_{0y} t - \tfrac{1}{2}gt^2 = (v_0 \sin \theta_0)t - \tfrac{1}{2}gt^2. \tag{6–6}$$

The two preceding equations give the equation of the trajectory in terms of the parameter t. The equation in terms of x and y can be obtained by eliminating t. We find $t = x/v_0 \cos \theta_0$ and

$$y = (\tan \theta_0)x - \frac{g}{2v_0{}^2 \cos^2 \theta_0} x^2. \tag{6–7}$$

The quantities v_0, $\tan \theta_0$, $\cos \theta_0$, and g are constants, so the equation has the form

$$y = ax - bx^2,$$

which will be recognized as the equation of a *parabola*.

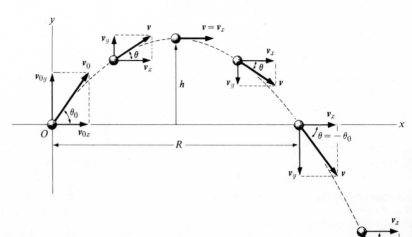

Fig. 6–5 *Trajectory of a body projected with an initial velocity v_0 at an angle of departure θ_0. The distance R is the horizontal range.*

Fig. 6-6 Trajectory of a body projected horizontally.

Example 1   A ball is projected horizontally with a velocity v_0 of magnitude 5 m s^{-1}. Find its position and velocity after $\frac{1}{2}$ s (see Fig. 6-6).

In this case, the departure angle is zero. The initial vertical velocity component is therefore zero. The horizontal velocity component equals the initial velocity and is constant.

The x- and y-coordinates, when $t = \frac{1}{2}$ s, are

$$x = v_x t = (5 \text{ m s}^{-1})(\tfrac{1}{2} \text{ s}) = 2.5 \text{ m},$$

$$y = -\tfrac{1}{2}gt^2 = -\tfrac{1}{2}(10 \text{ m s}^{-2})(\tfrac{1}{2} \text{ s})^2 = -1.25 \text{ m}.$$

The components of velocity are

$$v_x = v_0 = 5 \text{ m s}^{-1},$$

$$v_y = -gt = (-10 \text{ m s}^{-2})(\tfrac{1}{2} \text{ s}) = -5 \text{ m s}^{-1}.$$

The resultant velocity is

$$v = \sqrt{v_x^2 + v_y^2} = 5\sqrt{2} \text{ m s}^{-1}.$$

The angle θ is

$$\theta = \tan^{-1}\frac{v_y}{v_x} = \tan^{-1}\frac{-5 \text{ m s}^{-1}}{5 \text{ m s}^{-1}} = -45°.$$

That is, the velocity at $t = \frac{1}{2}$ s is 45° *below* the horizontal.

Example 2   In Fig. 6-5, let $v_0 = 160$ ft s^{-1}, $\theta_0 = 53°$. Then

$$v_{0x} = v_0 \cos\theta_0 = (160 \text{ ft s}^{-1})(0.60) = 96 \text{ ft s}^{-1},$$

$$v_{0y} = v_0 \sin\theta_0 = (160 \text{ ft s}^{-1})(0.80) = 128 \text{ ft s}^{-1}.$$

a) Find the position of the projectile, and the magnitude and direction of its velocity, when $t = 2.0$ s.

$$x = (96 \text{ ft s}^{-1})(2.0 \text{ s}) = 192 \text{ ft},$$

$$y = (128 \text{ ft s}^{-1})(2.0 \text{ s}) - \tfrac{1}{2}(32 \text{ ft s}^{-2})(2.0 \text{ s})^2$$

$$= 192 \text{ ft}.$$

$$v_x = 96 \text{ ft s}^{-1},$$

$$v_y = 128 \text{ ft s}^{-1} - (32 \text{ ft s}^{-2})(2.0 \text{ s}) = 64 \text{ ft s}^{-1},$$

$$\theta = \tan^{-1}\frac{64 \text{ ft s}^{-1}}{96 \text{ ft s}^{-1}} = \tan^{-1}0.667 = 33.5°.$$

b) Find the time at which the projectile reaches the highest point of its flight, and find the elevation of this point.

At the highest point, the vertical velocity v_y is zero. If t_1 is the time at which this point is reached,

$$v_y = 0 = 128 \text{ ft s}^{-1} - (32 \text{ ft s}^{-2})t_1,$$

$$t_1 = 4 \text{ s}.$$

The elevation h of the point is the value of y when $t = 4$ s.

$$h = (128 \text{ ft s}^{-1})(4 \text{ s}) - \tfrac{1}{2}(32 \text{ ft s}^{-2})(4 \text{ s})^2 = 256 \text{ ft}.$$

c) Find the *horizontal range R*, that is, the horizontal distance from the starting point to the point at which the projectile returns to its original elevation and at which, therefore, $y = 0$. Let t_2 be the time at which this point is reached. Then

$$y = 0 = (128 \text{ ft s}^{-1})t_2 - \tfrac{1}{2}(32 \text{ ft s}^{-2})t_2^2.$$

This quadratic equation has two roots,

$$t_2 = 0 \quad \text{and} \quad t_2 = 8 \text{ s},$$

corresponding to the two points at which $y = 0$. Evidently the time desired is the second root, $t_2 = 8$ s, which is just twice the time to reach the highest point. The time of descent therefore equals the time of rise.

The horizontal range R is the value of x when $t = 8$ s:

$$R = v_x t_2 = (96 \text{ ft s}^{-1})(8 \text{ s}) = 768 \text{ ft}.$$

The vertical velocity at this point is

$$v_y = 128 \text{ ft s}^{-1} - (32 \text{ ft s}^{-2})(8 \text{ s}) = -128 \text{ ft s}^{-1}.$$

That is, the vertical velocity has the same magnitude as the initial vertical velocity, but the opposite direction. Since v_x is constant, the angle below the horizontal at this point equals the angle of departure.

d) If unimpeded, the projectile continues to travel beyond its horizontal range. For example, the projectile might have been fired from the edge of a high cliff, so that negative values of y are possible. It is left as an exercise to compute the position and velocity at a time 10 s after the start, corresponding to the last position shown in Fig. 6–5. The results are:

$$x = 960 \text{ ft}, \quad y = -320 \text{ ft},$$

$$v_x = 96 \text{ ft s}^{-1}, \quad v_y = -192 \text{ ft s}^{-1}.$$

Example 3 Figure 6–7 illustrates an interesting experimental demonstration of the properties of projectile motion. A ball (shown by the open circle) is projected directly toward a second ball (the solid circle). The second ball is released from rest at the instant the first is projected, and it is found that the balls collide as shown regardless of the value of the initial velocity. To show that this must happen, we note that the initial elevation of the second ball is $x \tan \theta_0$, and that in time t it falls a distance $\frac{1}{2}gt^2$. Its elevation at the instant of collision is therefore

$$y = x \tan \theta_0 - \tfrac{1}{2}gt^2.$$

But t is also the time required for the first ball to traverse the horizontal distance x with constant horizontal velocity $v_0 \cos \theta_0$, so

$$x = v_0 \cos \theta_0 t.$$

Solving this for t and substituting in the above equation, we find

$$y = x \tan \theta_0 - \tfrac{1}{2}g\left(\frac{x}{v_0 \cos \theta}\right)^2,$$

which is the same as the elevation of the first ball as given by Eq. (6–7).

For any given initial velocity, there is one particular angle of departure for which the horizontal range is a maximum. To find this angle, we first work out a general algebraic expression for the range. From Eq. (6–6), the projectile returns to earth ($y = 0$) at a time t_2 given by $2v_0 \sin \theta_0/g$. The range R is the value of x at this time, and is

$$R = v_x t_2 = (v_0 \cos \theta_0)\left(\frac{2v_0 \sin \theta_0}{g}\right)$$

$$= \frac{2v_0{}^2 \sin \theta_0 \cos \theta_0}{g} = \frac{v_0{}^2 \sin 2\theta_0}{g}. \qquad (6\text{–}8)$$

Since the sine of an angle can never be greater than unity, we see that the *maximum* value of R is $v_0{}^2/g$, corresponding to the case $2\theta_0 = 90°$ or $\theta_0 = 45°$. Thus the maximum horizontal range is attained when the departure angle is 45°.

From the standpoint of gunnery, what we usually want to know is what the departure angle should be for a given muzzle velocity v_0 in order to hit a target whose position R is known. Let us assume target and gun are at the same elevation. Then, from Eq. (6–8),

$$\theta_0 = \tfrac{1}{2} \sin^{-1}\frac{Rg}{v_0{}^2}.$$

Fig. 6–7

Fig. 6–8 *Trajectory of a body projected at an angle with the horizontal.*

Fig. 6–9 *An angle of departure of 45° gives the maximum horizontal range.*

Provided R is less than the maximum range, this equation has two solutions for values of θ_0 between 0° and 90°.

Thus if $R = 800$ ft, $g = 32$ ft s^{-2}, and $v_0 = 200$ ft s^{-1},

$$\theta_0 = \tfrac{1}{2}\sin^{-1}\frac{(800 \text{ ft})(32 \text{ ft s}^{-2})}{(200 \text{ ft s}^{-2})^2}$$

$$= \tfrac{1}{2}\sin^{-1}0.64.$$

But $\sin^{-1}0.64 = 40°$, or $180° - 40°$. Therefore

$$\theta_0 = 20° \quad \text{or} \quad 70°.$$

Either of these angles gives the same range. Of course both the time of flight and the maximum height reached are greater for the high-angle trajectory.

Figure 6–8 is a multiflash photograph of the trajectory of a ball; x- and y-axes and the initial velocity vector have been added. The horizontal distances between consecutive positions are all equal, showing that the horizontal velocity component is constant. The vertical distances first decrease and then increase, showing that the vertical motion is accelerated.

Figure 6–9 is a composite photograph of the three trajectories of a ball projected from a spring gun with departure angles of 30°, 45°, and 60°. It will be seen that the horizontal ranges are (nearly) the same for the 30° and 60° angles and that both are less than the range when the angle is 45°. As the departure angle is altered, there is a variation in the initial velocity which is imparted to the ball.

If the departure angle is *below* the horizontal, as for instance in the motion of a ball after it rolls off a sloping roof, or the trajectory of a bomb released from a dive bomber, exactly the same principles apply. The horizontal velocity component remains constant and equal to $v_0 \cos \theta_0$. The vertical motion is the same as that of a body projected *downward* with an initial velocity $v_0 \sin \theta_0$, which is now a negative quantity.

6–6 CIRCULAR MOTION

The acceleration of a particle moving in a curved path can be resolved into components normal and tangential to the path. There is a simple relation between the normal component of acceleration, the speed of the particle, and the radius of curvature of the path. We now derive this for the special case of motion in a circle.

Figure 6–10(a) represents a particle moving in a circular path of radius R with center at O. Vectors v_1 and v_2 represent its velocities at points P and Q. The vector change in velocity, Δv, is obtained in Fig. 6–10(b), which is the same as Fig. 6–4. Vectors $\Delta v_\perp$ and $\Delta v_\parallel$ are the normal and tangential components of Δv.

The triangles OPQ and opq in Fig. 6–10(a) and (b) are similar, since both are isosceles triangles and their long sides are mutually perpendicular. Hence

$$\frac{\Delta v_\perp}{v_1} = \frac{\Delta s}{R} \quad \text{or} \quad \Delta v_\perp = \frac{v_1}{R}\Delta s.$$

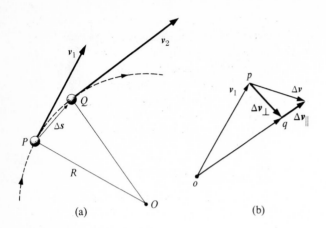

(a)

(b)

Fig. 6–10 *Construction for finding change in velocity, $\Delta \boldsymbol{v}$, of particle moving in a circle.*

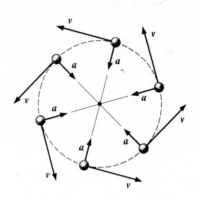

Fig. 6–11 *Velocity and acceleration vectors of a particle in uniform circular motion.*

The magnitude of the average normal acceleration $\bar{a}_\perp$ is therefore

$$\bar{a}_\perp = \frac{\Delta v_\perp}{\Delta t} = \frac{v_1}{R} \frac{\Delta s}{\Delta t}.$$

The *instantaneous* acceleration $a_\perp$ at point P is the limiting value of this expression, as point Q is taken closer and closer to point P:

$$a_\perp = \lim_{\Delta t \to 0} \frac{v_1}{R} \frac{\Delta s}{\Delta t} = \frac{v_1}{R} \lim_{\Delta t \to 0} \frac{\Delta s}{\Delta t}.$$

But the limiting value of $\Delta s / \Delta t$ is the speed v_1 at point P, and since P can be any point of the path, we can drop the subscript from v_1 and let v represent the speed at any point. Then

$$\boxed{a_\perp = \frac{v^2}{R}.} \qquad (6\text{–}9)$$

The magnitude of the instantaneous normal acceleration is therefore equal to the square of the speed divided by the radius. The direction is inward along the radius, toward the center of the circle. Because of this it is called a *central*, a *centripetal*, or a *radial* acceleration. (The term "centripetal" means "seeking a center.")

The unit of radial acceleration is the same as that of an acceleration resulting from a change in the *magnitude* of a velocity. Thus if a particle travels with

a speed of 4 m s^{-1} in a circle of radius 2 m, its radial acceleration is

$$a = \frac{(4 \text{ m s}^{-1})^2}{2 \text{ m}} = 8 \text{ m s}^{-2}.$$

If the *speed* of the particle changes, it will also have a *tangential* component of acceleration, defined as

$$a_\parallel = \lim_{\Delta t \to 0} \frac{\Delta v_\parallel}{\Delta t}.$$

If the speed is constant, there is no tangential component of acceleration and the acceleration is purely normal, resulting from the continuous change in *direction* of the velocity. In general, there will be both tangential and normal components of acceleration.

Figure 6–11 shows the directions of the velocity and acceleration vectors at a number of points, for a particle revolving in a circle with a velocity of constant magnitude.

A *centrifuge* is a device for whirling an object with a high velocity. The consequent large radial acceleration is equivalent to increasing the value of g, and such processes as sedimentation (settling of particles or precipitates out of a solution) which would otherwise take place only slowly, can be greatly accelerated in this way. Very high speed centrifuges, called ultracentrifuges, have been operated at speeds as high as 180,000 rev min^{-1}, and some small experimental units have been driven as fast as 1,300,000 rev min^{-1}.

Centrifuges producing accelerations up to the order of 100,000 g are used routinely in medical laboratories. Applications include increasing of sedimentation rates and actual measurements of molecular weights of large molecules such as proteins and viruses by measurement of sedimentation rate.

Example The moon revolves about the earth in a circle (very nearly) of radius $R = 239,000$ mi or 12.6×10^8 ft, and requires 27.3 days, or 23.4×10^5 s, to make a complete revolution.

a) What is the acceleration of the moon toward the earth?

The velocity of the moon is

$$v = \frac{2\pi R}{T} = \frac{2\pi(12.6 \times 10^8 \text{ ft})}{23.4 \times 10^5 \text{ s}} = 3383 \text{ ft s}^{-1}.$$

Its radial acceleration is therefore

$$a = \frac{v^2}{R} = \frac{(3383 \text{ ft s}^{-1})^2}{12.6 \times 10^8 \text{ ft}} = 0.00908 \text{ ft s}^{-2}.$$

b) If the gravitational force exerted on a body by the earth is inversely proportional to the square of the distance from the earth's center, the acceleration produced by this force should vary in the same way. Therefore, if the acceleration of the moon is caused by the gravitational attraction of the earth, the ratio of the moon's acceleration to that of a falling body at the earth's surface should equal the ratio of the square of the earth's radius (3950 mi or 2.09×10^7 ft) to the square of the radius of the moon's orbit. Is this true?

The ratio of the two accelerations is

$$\frac{8.96 \times 10^{-3} \text{ ft s}^{-2}}{32.2 \text{ ft s}^{-2}} = 2.82 \times 10^{-4}.$$

The ratio of the squares of the distances is

$$\frac{(2.09 \times 10^7 \text{ ft})^2}{(12.6 \times 10^8 \text{ ft})^2} = 2.75 \times 10^{-4}.$$

The agreement is very close, although not exact because we have used average values.

Newton used the above calculation to justify his hypothesis that gravitation was truly *universal* and that the earth's pull extended out indefinitely into space. The numerical values available in Newton's time were not highly precise. While he did not obtain as close an agreement as that above, he states that he found his results to "answer pretty nearly," and he concluded that his hypothesis was verified.

6–7 CENTRIPETAL FORCE

Having obtained an expression for the normal or *radial* acceleration of a particle moving in a circle, we can now use Newton's second law to find the radial force on the particle. Since the magnitude of the radial acceleration equals v^2/R, and its direction is toward the center, the magnitude of the radial force on a particle of mass m is

$$F = m\frac{v^2}{R}. \qquad (6\text{–}10)$$

The direction of this force is toward the center also, and it is called a *centripetal force*. It is unfortunate that it has become common practice to characterize the force by the adjective "centripetal," since this seems to imply that there is some difference in nature between centripetal forces and other forces. This is not the case. Centripetal forces, like other forces, are pushes and pulls exerted by sticks and strings, or arise from the action of gravitational or other causes. The term "centripetal" refers to the *effect* of the force, that is, to the fact that it results in a change in the *direction* of the velocity of the body on which it acts, rather than a change in the *magnitude* of this velocity.

Anyone who has ever tied an object to a cord and whirled it in a circle will realize the necessity of exerting this inward, centripetal force. If the cord breaks, the direction of the velocity ceases to change (unless other forces are acting) and the object flies off along a tangent to the circle.

Example 1 A small body of mass 0.2 kg revolves uniformly in a circle on a horizontal frictionless surface, attached by a cord 0.2 m long to a pin set in the surface. If the body makes two complete revolu-

Fig. 6–12

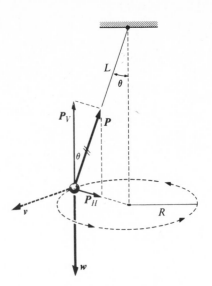

Fig. 6–13 *The conical pendulum.*

tions per second, find the force P exerted on it by the cord. (See Fig. 6–12.)

The circumference of the circle is

$$2\pi(0.2 \text{ m}) = 0.4\pi \text{ m}$$

so the velocity is 0.8π m s^{-1}. The magnitude of the centripetal acceleration is

$$a = \frac{v^2}{R} = \frac{(0.8\pi \text{ m s}^{-1})^2}{0.2 \text{ m}} = 31.6 \text{ m s}^{-2}.$$

Since the body has no vertical acceleration, the forces N and w are equal and opposite and the force P is the resultant force. Therefore

$$P = ma = (0.2 \text{ kg})(31.6 \text{ m s}^{-2})$$

$$= 6.3 \text{ N}.$$

Example 2   Figure 6–13 represents a small body of mass m revolving in a horizontal circle with velocity v of constant magnitude at the end of a cord of length L. As the body swings around its path, the cord sweeps over the surface of a cone. The cord makes an angle θ with the vertical, so the radius of the circle in which the body moves is $R = L \sin \theta$ and the magnitude of the velocity v equals

$2\pi L \sin \theta/T$, where T is the time for one complete revolution.

The forces exerted on the body when it is in the position shown are its weight w and the tension P in the cord. (Note that the force diagram in Fig. 6–13 is exactly like that in Fig. 5–8(b). The only difference is that in this case the acceleration a is the *radial* acceleration, v^2/R.) Let P be resolved into a horizontal component P_H and a vertical component P_V, of magnitudes $P \sin \theta$ and $P \cos \theta$, respectively. The body has no vertical acceleration, so the vertical forces $P \cos \theta$ and w are equal in magnitude, and the resultant inward, radial, or centripetal force is the horizontal component $P \sin \theta$, which is equal to the mass m times the radial (or centripetal) acceleration:

$$P \sin \theta = m\frac{v^2}{R}, \qquad P \cos \theta = w.$$

When the first of these equations is divided by the second, and w is replaced by mg, the result is

$$\tan \theta = \frac{v^2}{Rg}. \tag{6–11}$$

When we make use of the relations $R = L \sin \theta$ and $v = 2\pi L \sin \theta/T$, Eq. (6–11) becomes

$$\cos \theta = \frac{gT^2}{4\pi^2 L}, \tag{6–12}$$

or

$$T = 2\pi\sqrt{L(\cos\theta)/g}. \qquad (6\text{–}13)$$

Equation (6–12) indicates how the angle θ depends on the time of revolution T and the length L of the cord. For a given length L, $\cos\theta$ decreases as the time is made shorter, and the angle θ increases. The angle never becomes 90°, however, since this requires that $T = 0$ or $v = \infty$.

Equation (6–13) is similar in form to the expression for the time of swing of a simple pendulum, which will be derived in Chapter 11. Because of this similarity, the present device is called a *conical pendulum.*

Some readers may wish to add to the forces shown in Fig. 6–13 an outward, "centrifugal" force, to "keep the body out there," or to "keep it in equilibrium." ("Centrifugal" means "fleeing a center.") Let us examine this point of view. In the first place, to look for a force to "*keep* the body out there" is an example of faulty observation, because the body doesn't stay there! A moment later it will be at a different position on its circular path. At the instant shown it is moving in the direction of the velocity vector v, and unless a resultant force acts on it, it will, according to Newton's first law, continue to move in this direction. If an outward force *were* acting on it, equal and opposite to the inward component of the force P, there would be no resultant inward force to deviate it sidewise from its present direction of motion.

Those who wish to add a force to "keep the body in equilibrium" forget that the term *equilibrium* refers to a state of rest, or of motion *in a straight line* with constant speed. Here, the body is *not* moving in a straight line, but in a circle. It is *not* in equilibrium, but has an acceleration toward the center of the circle and must be acted on by a resultant or *un*balanced force to produce this acceleration. In this example there is *no outward force* on the body!

Example 3 Figure 6–14(a) represents an automobile or a railway car rounding a flat curve of radius R, on a level road or track. The forces acting on it are its weight w, the normal force N, and the centripetal force P. The force P must be provided by friction, in the case of an automobile, or by a force exerted by the rails against the flanges on the wheels of a railway car.

In order not to have to rely on friction, or to reduce wear on the rails and flanges, the road or the track may be banked, as shown in Fig. 6–14(b). The normal force N then has a vertical component of magnitude $N\cos\theta$, and a horizontal component of magnitude $N\sin\theta$ toward the center, which provides the centripetal force. The banking angle θ can be computed as follows. If v is the velocity and R the radius, then

$$N\sin\theta = \frac{mv^2}{R}.$$

Since there is no vertical acceleration, $N\cos\theta = w$.

Dividing the first equation by the second, and replacing w by mg, we obtain

$$\tan\theta = \frac{v^2}{Rg}.$$

The tangent of the angle of banking is proportional to the square of the speed and inversely

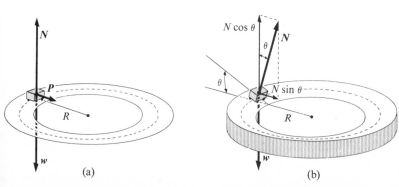

Fig. 6–14 (a) *Forces on a vehicle rounding a curve on a level track.* (b) *Forces when the track is banked.*

proportional to the radius. For a given radius no one angle is correct for all speeds. Hence in the design of highways and railroads, curves are banked for the *average speed* of the traffic over them. The same considerations apply to the correct banking angle of a plane when it makes a turn in level flight.

Note that the banking angle is given by the same expression as that for the angle with the vertical made by the cord of a conical pendulum. In fact, the force diagrams in Figs. 6–13 and 6–14 are identical.

6–8 MOTION IN A VERTICAL CIRCLE

Figure 6–15 represents a small body attached to a cord of length R and whirling in a *vertical* circle about a fixed point O to which the other end of the cord is attached. The motion, while circular, is not *uniform*, since the speed increases on the way down and decreases on the way up.

The forces on the body at any point are its weight $w = mg$ and the tension T in the cord. Let the weight be resolved into a normal component, of magnitude $w \cos \theta$, and a tangential component of magnitude $w \sin \theta$, as in Fig. 6–15. The resultant tangential and normal forces are then

$$F_\parallel = w \sin \theta \quad \text{and} \quad F_\perp = T - w \cos \theta.$$

The tangential acceleration, from Newton's second law, is

$$a_\parallel = \frac{F_\parallel}{m} = g \sin \theta,$$

Fig. 6–15 *Forces on a body whirling in a vertical circle with center at O.*

and is the same as that of a body sliding on a frictionless inclined plane of slope angle θ. The normal (radial) acceleration $a_\perp = v^2/R$, is

$$a_\perp = \frac{F_\perp}{m} = \frac{T - w \cos \theta}{m} = \frac{v^2}{R},$$

and the tension in the cord is therefore

$$T = m\left(\frac{v^2}{R} + g \cos \theta\right). \qquad (6\text{–}14)$$

At the lowest point of the path, $\theta = 0$, $\sin \theta = 0$, $\cos \theta = 1$. Hence at this point $F_\parallel = 0$, $a_\parallel = 0$, and the acceleration is purely radial (upward). The magnitude of the tension, from Eq. (6–14), is

$$T = m\left(\frac{v_{\text{bot}}^2}{R} + g\right).$$

At the highest point, $\theta = 180°$, $\sin \theta = 0$, $\cos \theta = -1$, and the acceleration is again purely radial (downward). The tension is

$$T = m\left(\frac{v_{\text{top}}^2}{R} - g\right). \qquad (6\text{–}15)$$

With motion of this sort, it is a familiar fact that there is a certain critical speed v_c at the highest point, below which the cord becomes slack and the path is no longer circular. To find this speed, set $T = 0$ in Eq. (6–15):

$$0 = m\left(\frac{v_c^2}{R} - g\right), \qquad v_c = \sqrt{Rg}.$$

The multiflash photographs of Fig. 6–16 illustrate another case of motion in a vertical circle, a small ball "looping-the-loop" on the inside of a vertical circular track. The inward normal force exerted on the ball by the track takes the place of the tension T shown in Fig. 6–15.

In Fig. 6–16(a), the ball is released from an elevation such that its speed at the top of the track is greater than the critical speed, $\sqrt{Rg}$. In Fig. 6–16(b), the ball starts from a lower elevation and reaches the top of the circle with a speed such that its own weight is slightly larger than the requisite centripetal force. In other words, the track would have to pull *outward*

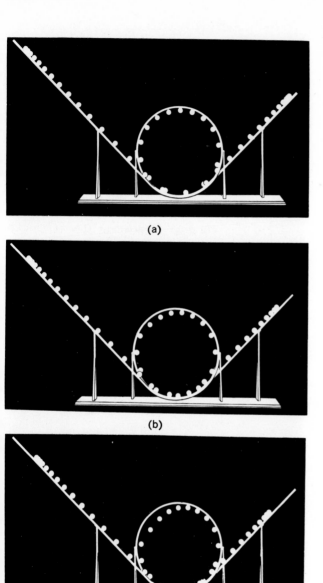

Fig. 6-16 *Multiflash photographs of a ball looping-the-loop in a vertical circle.*

Fig. 6-17 (a) *Forces on a body swinging in a vertical circle.* (b) *The radial and tangential components of acceleration are combined to obtain the resultant acceleration a.*

made from a still lower elevation, the ball leaves the track sooner, and the parabolic path is clearly evident.

Example In Fig. 6-17, a small body of mass m = 0.10 kg swings in a vertical circle at the end of a cord of length R = 1.0 m. If its speed v = 2.0 m s^{-1} when the cord makes an angle θ = 30° with the vertical, find (a) the radial and tangential components of its acceleration at this instant, (b) the magnitude and direction of the resultant acceleration, and (c) the tension T in the cord.

a) The radial component of acceleration is

$$a_\perp = \frac{v^2}{R} = \frac{(2.0 \text{ m s}^{-1})^2}{1.0 \text{ m}} = 4.0 \text{ m s}^{-2}.$$

The tangential component of acceleration, due to the tangential force $mg \sin \theta$, is

$$a_\| = g \sin \theta = (9.8 \text{ m s}^{-2})(0.50) = 4.9 \text{ m s}^{-2}.$$

b) The magnitude of the resultant acceleration (see Fig. 6-17b) is

$$a = \sqrt{a_\perp{}^2 + a_\|{}^2} = 6.3 \text{ m s}^{-2}.$$

to maintain the circular motion. Since this is impossible, the ball leaves the track and moves for a short distance in a parabola. This parabola soon intersects the circle, however, and the remainder of the trip is completed successfully. In Fig. 6-16(c), the start is

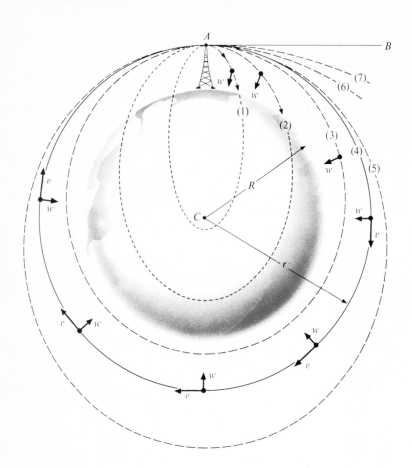

Fig. 6–18 *Trajectories of a body projected from point A in the direction AB with different initial velocities.*

The angle ϕ is

$$\phi = \tan^{-1} \frac{a_\parallel}{a_\perp} = 51°.$$

c) The tension in the cord is given by $F_\perp = ma_\perp$:
$T - mg \cos \theta = mv^2/R$, so

$$T = m\left(\frac{v^2}{R} + g \cos \theta\right) = 1.3 \text{ N}.$$

Note that the magnitude of the tangential acceleration is not constant but is proportional to the sine of the angle θ. Hence the equations of motion with constant acceleration *cannot* be used to find the speed at other points of the path. We shall show in the next chapter, however, how the speed at any point can be found from *energy* considerations.

6-9 MOTION OF A SATELLITE

In discussing the trajectory of a projectile in Section 6–5, we assumed that the gravitational force on the projectile (its weight **w**) had the same direction and magnitude at all points of the trajectory. These conditions are approximately satisfied if the projectile remains near the earth's surface and if its trajectory is small compared to the earth's radius. Under these conditions the trajectory is a parabola.

In reality, the gravitational force is directed toward the center of the earth and is inversely proportional to the square of the distance from the earth's center, so that it is not constant in either magnitude or direction. It can be shown that under an inverse square force directed toward a fixed point the trajectory must always be a *conic section* (ellipse, circle, parabola, or hyperbola).

Suppose that a very tall tower could be constructed as in Fig. 6–18, and that a projectile were launched from point A at the top of the tower in the "horizontal" direction AB. If the initial velocity is not too great, the trajectory will be like that numbered (1), which is a portion of an ellipse with the earth's center C at one focus. (If the trajectory is so short that changes in the magnitude and direction of w can be neglected, the ellipse *approximates* a parabola.)

The trajectories numbered from (2) to (7) illustrate the effect of increasing the initial velocity. (Any effect of the earth's atmosphere is neglected.) Trajectory (2) is again a portion of an ellipse. Trajectory (3), which just misses the earth, is a *complete* ellipse and the projectile has become an earth satellite. Its velocity when it returns to point A is the same as its initial velocity, and in the absence of retarding forces it will repeat its motion indefinitely. (The earth's rotation will have moved the tower to a different point by the time the satellite returns to point A, but does not affect the orbit.)

Trajectory (4) is a special case in which the orbit is a circle. Trajectory (5) is again an ellipse, (6) is a parabola, and (7) is a hyperbola. Trajectories (6) and (7) are not closed orbits.

All manmade earth satellites have orbits like (3), (4), or (5). Many are very nearly circles; for simplicity we shall consider circular orbits only. Let us calculate the velocity required for such an orbit, and the time of one revolution. The centripetal acceleration of the satellite in its circular orbit is produced by the gravitational force on the satellite, which is equal to the product of the mass and the centripetal (or radial) acceleration. The acceleration may in turn be obtained from the velocity of the satellite and the radius of the orbit.

Thus

$$w = F_g = G\left(\frac{mm_E}{r^2}\right) = m\left(\frac{v^2}{r}\right),$$

and from the last two terms,

$$v^2 = \frac{Gm_E}{r}, \qquad v = \sqrt{\frac{Gm_E}{r}}. \qquad (6\text{–}16)$$

The larger the radius r, the smaller the orbital velocity.

The speed of the satellite may also be expressed in terms of the acceleration of gravity g at the earth's surface, which according to Eq. (5–7) is given by $g = Gm_E/R^2$. Combining this relation with Eq. (6–16), we obtain

$$v = R\sqrt{\frac{g}{r}}. \qquad (6\text{–}17)$$

The acceleration, given by $a_\perp = v^2/r$ (*not* v^2/R!) can also be expressed in terms of g :

$$a_\perp = \frac{R^2}{r^2} g. \qquad (6\text{–}18)$$

This is, of course, the acceleration of gravity at radius r; the satellite, like any projectile, is a freely falling body. As expected, the acceleration is less than g at the earth's surface, in the ratio of the square of the radii.

The period T, or the time required for one complete revolution, equals the circumference of the orbit, $2\pi r$, divided by the velocity v:

$$T = \frac{2\pi r}{v} = \frac{2\pi r}{R\sqrt{g/r}} = \frac{2\pi}{R\sqrt{g}} r^{3/2}, \qquad (6\text{–}19)$$

and the larger the radius, the longer the period.

Example   An earth satellite revolves in a circular orbit at a height of 300 km (about 200 miles) above the earth's surface.
a) What is the velocity of the satellite, assuming the earth's radius to be 6400 km and g to be 9.80 m s^{-2}?

From Eq. (6–17),

$$v = R\sqrt{\frac{g}{r}}$$

$$= (6.40 \times 10^6 \text{ m})\left(\frac{9.80 \text{ m s}^{-2}}{6.70 \times 10^6 \text{ m}}\right)^{1/2}$$

$$= 7740 \text{ m s}^{-1} = 25,400 \text{ ft s}^{-1}$$

$$= 17,300 \text{ mi hr}^{-1}.$$

b) What is the period T?

$$T = \frac{2\pi r}{v} = 90.6 \text{ min} = 1.51 \text{ hr.}$$

c) What is the radial acceleration of the satellite?

$$a_\perp = \frac{v^2}{r} = 8.94 \text{ m s}^{-2}.$$

This is equal to the free-fall acceleration at a height of 300 km above the earth.

6–10 EFFECT OF THE EARTH'S ROTATION ON g

Because of the earth's rotation it is not precisely an inertial frame of reference, and the acceleration of a falling body is not precisely equal to the ratio of its true weight (the earth's gravitational attraction) to its mass. The small correction associated with earth's rotation will be discussed next.

Figure 6–19 is a cutaway view of our rotating earth, with three observers each holding a body of mass m, hanging from a string. Each body is attracted toward the earth's center with a force $F_g = Gmm_E/R^2$, which we now designate by w_0. Let us consider the earth's center as the origin of an inertial reference system. The string exerts a force T on the body.

Except at the pole, each body is carried along by the earth's rotation and moves in a circle with center on the earth's axis. It therefore has a radial acceleration $a_\perp$ equal to v^2/R, toward the axis. The resultant or vector sum of the forces T and w_0 must therefore be such as to provide the requisite radial acceleration. That is, in general vector form,

$$T + w_0 = ma_\perp.$$

At some arbitrary latitude θ, T and w_0 have directions as shown in the diagram, such that their resultant F points toward the point O' (the center of the circle in which the body moves) and equals $ma_\perp$. Thus we see that, except at the pole and the equator, a plumb line does not point toward the earth's center.

At the pole, where $a_\perp = 0$, the force T is equal and opposite to the true weight w_0.

Fig. 6–19 *The resultant of the forces T and w_0 is the centripetal force, equal to mv^2/r.*

At the equator, where the direction of the radial acceleration is toward the earth's center O, T and w_0 have the same action line but the magnitude of w_0 is greater than that of T.

Since each body appears in equilibrium to its respective observer, the *apparent* weight of each body, which we now represent by w, is a force equal and opposite to T, as shown in the inset diagram for the body at latitude θ. That is,

$$w = -T = w_0 - ma_\perp.$$

At an intermediate latitude, the apparent weight w differs in both magnitude and direction from the true weight w_0.

At the pole, the apparent and true weights are equal. At the equator,

$$w = w_0 - ma_\perp,$$

Table 6–1 VARIATIONS OF g WITH LATITUDE AND ELEVATION

Station	North latitude	Elevation, m	g, m s^{-2}	g, ft s^{-2}
Canal Zone	9°	0	9.78243	32.0944
Jamaica	18°	0	9.78591	32.1059
Bermuda	32°	0	9.79806	32.1548
Denver	40°	1638	9.79609	32.1393
Cambridge, Mass.	42°	0	9.80398	32.1652
Standard station			9.80665	32.1740
Greenland	70°	0	9.82534	32.2353

where $a_\perp$ is the radial acceleration at the equator. Dividing by m, we get

$$\frac{w}{m} = \frac{w_0}{m} - a_\perp,$$

where w/m is the observed acceleration g relative to the surface of the earth of a body falling at the equator, and w_0/m is the acceleration g_0 relative to the center of the earth. Thus, at the equator, $g = g_0 - a_\perp$.

Let us calculate the magnitude of the radial acceleration $a_\perp$ at the equator. The equatorial velocity v, equal to the earth's circumference divided by the time of one rotation, is

$$v = \frac{2\pi(6.4 \times 10^6 \text{ m})}{8.04 \times 10^4 \text{ s}} = 465 \text{ m s}^{-1},$$

and hence

$$a_R = \frac{v^2}{R} = 0.034 \text{ m s}^{-2} = 3.4 \text{ cm s}^{-2}.$$

Thus if the free-fall acceleration g_0 at the equator is 9.880 m s^{-2}, the observed acceleration g is 9.766 m s^{-2}. Table 6–1 lists the measured free-fall acceleration g at a number of points. A part of the variation results from the fact that the earth is not spherical, that the observation points are at different elevations, or that there are local variations in earth density. It will be seen, however, that the change in the radial acceleration between equator and poles has the right magnitude to account for most of the differences.

A similar discussion can be applied to the phenomenon of "weightlessness" in satellites. A space vehicle in orbit is a freely falling body, with an acceleration $a_\perp$ toward the earth's center equal to the value of the acceleration of gravity g at its orbit radius. The apparent weight w is given as before by

$$w = w_0 - ma_\perp = mg - ma_\perp.$$

But, in this case,

$$g = a_\perp,$$

so

$$w = 0.$$

It is in this sense that an astronaut in the vehicle is said to be "weightless" or in a state of "zero g." The vehicle is like the freely falling elevator discussed in Section 5–6, Example 7, except that it has a large and constant tangential speed along its orbit.

PROBLEMS

6–1 A ball rolls off the edge of a tabletop 4 ft above the floor, and strikes the floor at a point 6 ft horizontally from the edge of the table.

a) Find the time of flight.

b) Find the initial velocity.

c) Find the magnitude and direction of the velocity of the ball just before it strikes the floor. Draw a diagram to scale.

6–2 A block slides off a horizontal tabletop 1 m high with a velocity of 3 m s^{-1}. Find (a) the horizontal distance from the table at which the block strikes the floor, and (b) the horizontal and vertical components of its velocity when it reaches the floor.

6–3 A level-flight bomber, flying at 300 ft s^{-1}, releases a bomb at an elevation of 6400 ft.

a) How much time is required for the bomb to reach the earth?

b) How far does it travel horizontally?

c) Find the horizontal and vertical components of its velocity when it strikes.

6–4 A block passes a point 3 m from the edge of a table with a velocity of 4 m s^{-1}. It slides off the edge of the table, which is 1 m high, and strikes the floor 1 m from the edge of the table. What was the coefficient of sliding friction between block and table?

6–5 A golf ball is driven horizontally from an elevated tee with a velocity of 25 m s^{-1}. It strikes the fairway 2.5 s later.

a)   How far has it fallen vertically?

b)   How far has it traveled horizontally?

c)   Find the horizontal and vertical components of its velocity, and the magnitude and direction of its resultant velocity, just before it strikes.

6–6 A level-flight bombing plane, flying at an altitude of 1024 ft with a velocity of 240 ft s^{-1}, is overtaking a motor torpedo boat traveling at 80 ft s^{-1} in the same direction as the plane. At what distance astern of the boat should a bomb be released in order to hit the boat?

6–7 A .22 calibre rifle bullet is fired in a horizontal direction with a muzzle velocity of 300 m s^{-1}. In the absence of air resistance, how far will it have dropped in traveling a horizontal distance of (a) 20 m? (b) 40 m? (c) 60 m? (d) How far will it drop in one second?

6–8 A ball is projected with an initial upward velocity component of 20 m s^{-1} and a horizontal velocity component of 25 m s^{-1}.

a)   Find the position and velocity of the ball after 2 s, 3 s, 6 s.

b)   How long a time is required to reach the highest point of the trajectory?

c)   How high is this point?

d)   How long a time is required for the ball to return to its original level?

e)   How far has it traveled horizontally during this time? Show your results in a neat sketch, large enough to show all features clearly.

6–9 A batted baseball leaves the bat at an angle of 30° above the horizontal, and is caught by an outfielder 400 ft from the plate.

a)   What was the initial velocity of the ball?

b)   How high did it rise?

c)   How long was it in the air?

6–10 A spring gun projects a golf ball at an angle of 45° above the horizontal. The horizontal range is 10 m.

a)   What is the maximum height to which the ball rises?

b)   For the same initial speed, what are the two angles of departure for which the range is 6 m?

c)   Sketch all three trajectories to scale in the same diagram.

6–11 Suppose the departure angle θ_0 in Fig. 6–7 is 15° and the distance x is 5 m. Where will the balls collide if the muzzle velocity of the first is (a) 20 m s^{-1}, (b) 5 m s^{-1}? Sketch both trajectories. (c) Will a collision take place if the departure angle is below horizontal?

6–12 If a baseball player can throw a ball a maximum distance of 60 m over the ground, what is the maximum vertical height to which he can throw it? Assume the ball to have the same initial speed in each case.

6–13 A player kicks a football at an angle of 37° with the horizontal and with an initial velocity of 48 ft s^{-1}. A second player standing at a distance of 100 ft from the first in the direction of the kick starts running to meet the ball at the instant it is kicked. How fast must he run in order to catch the ball before it hits the ground?

6–14 A baseball leaves the bat at a height of 4 ft above the ground, traveling at an angle of 45° with the horizontal, and with a velocity such that the horizontal range would be 400 ft. At a distance of 360 ft from home plate is a fence 30 ft high. Will the ball be a home run?

6–15 A projectile shot at an angle of 60° above the horizontal strikes a building 30 m away from a point 15 m above the point of projection.

a)   Find the velocity of projection.

b)   Find the magnitude and direction of the velocity of the projectile when it strikes the building.

6–16

a)   What must be the velocity of a projectile fired vertically upward to reach an altitude of 60 km?

b)   What velocity is required to reach the same height if the gun makes an angle of 45° with the vertical?

c)   Compute the time required to reach the highest point in both trajectories.

d)   How far would a plane traveling at 500 km hr^{-1} move in this time?

6–17 The angle of elevation of an antiaircraft gun is 70° and the muzzle velocity is 2700 ft s^{-1}. For what time after firing should the fuse be set if the shell is to explode at an altitude of 5000 ft?

6–18 A trench mortar fires a projectile at an angle of 53° above the horizontal with a muzzle velocity of 200 ft s^{-1}. A tank is advancing directly toward the mortar on level ground at a speed of 10 ft s^{-1}. What should be the distance from mortar to tank at the instant the mortar is fired in order to score a hit?

6–19 A projectile is fired with an initial speed v_0 at a departure angle θ_0, from the foot of an inclined plane of slope angle α (Fig. 6–20).

a) What is the range R measured along the plane?

b) Show that the expression for R reduces to that for the horizontal range when $\alpha = 0$.

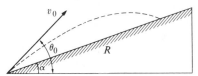

Fig. 6–20

6–20 A bomber, diving at an angle of 53° with the vertical, releases a bomb at an altitude of 800 m. The bomb is observed to strike the ground 5 s after its release.

a) What was the velocity of the bomber?

b) How far did the bomb travel horizontally during its flight?

c) What were the horizontal and vertical components of its velocity just before striking?

6–21 A 10-kg stone is dropped from a cliff in a high wind. The wind exerts a steady horizontal 50-N force on the stone as it falls. Is the path of the stone a straight line, a parabola, or some more complicated path? Explain.

6–22 A lump of ice slides 10 m down a smooth, sloping roof making an angle of 30° with the horizontal. The edge of the roof is 10 m above a sidewalk which extends 5 m out from the side of the building. Will the ice land on the sidewalk or in the street?

6–23 The second stage of a three-stage rocket designed to launch an earth satellite will exhaust its fuel at a height of about 130 mi above the earth's surface, when its speed will be approximately 11,000 mi hr^{-1}. The rocket will then "coast" to a height of about 300 mi, at which point its velocity will be horizontal. Assume a flat earth and a constant value of g.

a) Find the direction of the trajectory at the 130-mi height.

b) Find the horizontal distance traveled while the rocket rises from 130 mi to 300 mi.

c) Find the speed at the 300-mi height.

d) At a height of 300 mi, the third stage will separate from the second and the latter will fall to the earth. How far will it travel horizontally while falling?

e) How long a time will be required?

6–24 A man is riding on a flatcar traveling with a constant velocity of 30 ft s^{-1} (Fig. 6–21). He wishes to throw a ball through a stationary hoop 16 ft above the height of his hands in such a manner that the ball will move horizontally as it passes through the hoop. He throws the ball with a velocity of 40 ft s^{-1} with respect to himself.

a) What must be the vertical component of the initial velocity of the ball?

b) How many seconds after he releases the ball will it pass through the hoop?

c) At what horizontal distance in front of the hoop must he release the ball?

Fig. 6–21

6–25 A particle moves in the xy-plane with coordinates given as functions of time by the equations $x = 2t + t^3$, $y = 6 - t^2$, where x and y are measured in meters and t in seconds.

a) Find the x and y components of velocity as functions of time.

b) Find the magnitude and direction of the velocity at time $t = 2$ s.

c) If the particle has a mass of 2 kg, find the components of force on the particle at time $t = 2$ s.

d) In part (c), find the magnitude and direction of force at time $t = 2$ s.

6–26 A particle moves in the xy-plane with coordinates given as functions of time by the equations $x = R \cos \omega t$, $y = R \sin \omega t$, where R and ω are constants.

a) Show that the particle moves in a circle centered at the origin, with radius R.

b) Determine the components of velocity of the particle. From these find the magnitude of velocity. Show that it is constant and equal to $R\omega$.

c) Determine the components of acceleration of the particle. From these find the magnitude of acceleration. Show that it is constant and equal to $R\omega^2$. Use the result of part (b) to show that this is also equal to v^2/R.

d) Comparing the acceleration components to the expressions for the x and y coordinates, show that the acceleration is always in a direction opposite to that from the origin to the particle.

6–27 At time $t = 0$ a body is moving east at 0.1 m s^{-1}. At time $t = 2$ s it is moving $25°$ north of east at 0.14 m s^{-1}. Find graphically the magnitude and direction of its change in velocity and the magnitude and direction of its average acceleration.

6–28 An automobile travels around a circular track 2000 m in circumference at a constant speed of 30 m s^{-1}.

a) Show in a scale diagram the velocity vectors of the automobile at the beginning and end of a time interval $\Delta t = 5$ s. Let 1 cm = 5 m s^{-1}.

b) Find graphically the change in velocity, Δv, in this time interval.

c) Find the magnitude of the average acceleration during this interval, $\Delta v/\Delta t$.

d) What is the radial acceleration, v^2/R ?

e) If the track is 10 m wide, what should be the elevation of the outer circumference above the inner circumference, for the speed above?

6–29 The radius of the earth's orbit around the sun (assumed circular) is 93×10^6 mi, and the earth travels around this orbit in 365 days.

a) What is the magnitude of the orbital velocity of the earth, in miles per hour?

b) What is the radial acceleration of the earth toward the sun, in feet per second squared?

6–30 A model of a helicopter rotor has four blades, each 2 m long, and is rotated in a wind tunnel at 1500 rev min^{-1}.

a) What is the linear speed of the blade tip, in meters per second?

b) What is the radial acceleration of the blade tip, in terms of the acceleration of gravity, g?

c) A pressure-measuring device with a mass of 0.1 kg is mounted at the blade tip. Find the centripetal force on it, and compare with its weight.

6–31 A highway curve of radius 1600 ft is to be banked so that a car traveling 50 mi hr^{-1} will have no tendency to skid sideways. At what angle should it be banked?

6–32 A flat (unbanked) curve on a highway has a radius of 800 ft, and a car rounds the curve at a speed of 50 mi hr^{-1}. What must be the minimum coefficient of friction to prevent sliding?

6–33 A stone of mass 1 kg is attached to one end of a string 1 m long, of breaking strength 500 N, and is whirled in a horizontal circle on a frictionless tabletop. The other end of the string is kept fixed. Find the maximum velocity the stone can attain without breaking the string.

6–34 An unbanked circular highway curve on level ground makes a turn of $90°$. The highway carries traffic at 60 mi hr^{-1}, and the centripetal force on a vehicle is not to exceed $\frac{1}{10}$ of its weight. What is the minimum length of the curve, in miles?

6–35 A coin placed on a 12-in. record will revolve with the record when it is brought up to a speed of $33\frac{1}{3}$ rev min^{-1}, provided the coin is not more than 4 in. from the axis.

a) What is the coefficient of static friction between coin and record?

b) How far from the axis can the coin be placed, without slipping, if the turntable rotates at 45 rev min^{-1}?

6–36

a) At how many revolutions per second must the apparatus of Fig. 6–22 rotate about the vertical axis in order that the cord shall make an angle of $45°$ with the vertical?

$a = 0.1$ m

$L = 0.2$ m

$m = 0.2$ kg   **Fig. 6–22**

b) What is then the tension in the cord?

c) Find the angle θ which the cord makes with the vertical if the system is rotating at 1.5 rev s^{-1}. (Set up the general equation relating the angle θ to the number of revolutions per second, n, the lengths a and L, and the acceleration of gravity, g. Then find by trial the angle θ which satisfies this equation.)

6–37 The "Giant Swing" at a country fair consists of a vertical central shaft with a number of horizontal arms attached at its upper end. Each arm supports a seat suspended from a cable 5 m long, the upper end of the cable being fastened to the arm at a point 4 m from the central shaft.

a) Find the time of one revolution of the swing if the cable supporting a seat makes an angle of 30° with the vertical.

b) Does the angle depend on the weight of the passenger for a given rate of revolution?

6–38 The 4-kg block in Fig. 6–23 is attached to a vertical rod by means of two strings. When the system rotates about the axis of the rod, the strings are extended as shown in the diagram.

a) How many revolutions per minute must the system make in order that the tension in the upper cord shall be 60 N?

b) What is then the tension in the lower cord?

6–39 A bead can slide without friction on a circular hoop of radius 0.1 m in a vertical plane. The hoop rotates at a constant rate of 2 rev s^{-1} about a vertical diameter, as in Fig. 6–24.

a) Find the angle θ at which the bead is in vertical equilibrium. (Of course it has a radial acceleration toward the axis.)

b) Is it possible for the beads to "ride" at the same elevation as the center of the hoop?

c) What will happen if the hoop rotates at 1 rev s^{-1}?

6–40 A curve of 200-m radius on a level road is banked at the correct angle for a velocity of 15 m s^{-1}. If an automobile rounds this curve at 30 m s^{-1}, what is the minimum coefficient of friction between tires and road so that the automobile will not skid? Assume all forces to act at the center of gravity.

6–41 An airplane in level flight is said to make a *standard turn* when it makes a complete circular turn in 2 min.

a) What is the banking angle of a standard turn if the speed of the airplane is 100 m s^{-1}?

b) What is the radius of the circle in which it turns?

c) What is the centripetal force on the airplane, expressed as a fraction (or multiple) of its weight?

6–42 The radius of the circular track in Fig. 6–16 is 0.4 m and the mass of the ball is 0.1 kg.

a) Find the critical velocity at the highest point of the track.

b) If the actual velocity at the highest point of the track is twice the critical velocity, find the force exerted by the ball against the track.

6–43 The pilot of a divebomber who has been diving at a velocity of 400 mi hr^{-1} pulls out of the dive by changing his course to a circle in a vertical plane.

a) What is the minimum radius of the circle in order that the acceleration at the lowest point shall not exceed "$7g$"?

b) How much does a 180-lb pilot apparently weigh at the lowest point of the pullout?

6–44 A cord is tied to a pail of water and the pail is swung in a vertical circle of radius 1 m. What must be the minimum velocity of the pail at the highest point of the circle if no water is to spill from the pail?

6–45 The radius of a Ferris wheel is 5 m and it makes one revolution in 10 s.

a) Find the difference between the apparent weight of a passenger at the highest and lowest points, expressed

Fig. 6–23

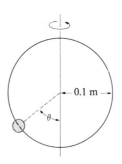

Fig. 6–24

as a fraction of his weight. (That is, find the difference between the upward force exerted on the passenger by the seat at these two points.)

b) What would the time of one revolution be if his apparent weight at the highest point were zero?

c) What would then be his apparent weight at the lowest point?

d) What would happen, at this rate of revolution, if his seat belt broke at the highest point, and if he did not hang onto his seat?

6–46 A ball is held at rest in position A in Fig. 6–25 by two light cords. The horizontal cord is cut and the ball starts swinging as a pendulum. What is the ratio of the tension in the supporting cord, in position B, to that in position A?

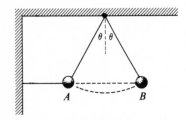

Fig. 6–25

6–47 The questions are often asked: "What keeps an earth satellite moving in its orbit?" and "What keeps the satellite up?"

a) What are your answers to these questions?

b) Are your answers applicable to the moon?

6–48 What is the period of revolution of a manmade satellite of mass m which is orbiting the earth in a circular path of radius 8000 km? (Mass of earth $= 5.98 \times 10^{24}$ kg.)

6–49 There exist several satellites moving in a circle in the earth's equatorial plane and at such a height above the earth's surface that they remain always above the same point. Find the height of such a satellite.

6–50 It is desired to launch a satellite 600 km above the earth in a circular orbit. If suitable rockets are used to reach this elevation, what horizontal orbital velocity must be imparted to the satellite? The radius of the earth is 6380 km.

6–51 An earth satellite rotates in a circular orbit of radius 4400 mi (about 400 mi above the earth's surface) with an orbital speed of 17,000 mi hr^{-1}.

a) Find the time of one revolution.

b) Find the acceleration of gravity at the orbit.

6–52 What would be the length of a day if the rate of revolution of the earth were such that $g = 0$ at the equator?

6–53 The weight of a man as determined by a spring balance at the equator is 180 lb. By how many ounces does this differ from the true force of gravitational attraction at the same point?

Chapter 7

Work and Energy

7-1 WORK

In everyday life, the word *work* is applied to any form of activity that requires the exertion of muscular or mental effort. In physics, however, the term is used in a very specific sense. Figure 7–1 represents a body moving in a horizontal direction which we shall take as the *x*-axis. A constant force, *F*, at an angle θ with the direction of motion, is exerted on the body by some outside agent. The work *W* of this force (or the work done by the force), when its point of application undergoes a displacement *s*, is defined as the product of the magnitudes of the displacement and of the component of the force in the direction of the displacement.

The magnitude of the component of *F* in the direction of *s* is $F \cos \theta$. Then

$$W = F \cos \theta \cdot s. \qquad (7\text{–}1)$$

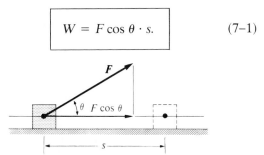

Fig. 7–1 *The work done by the force F in a displacement s is (F cos θ) · s.*

Work is a *scalar* quantity, equal to the product of $F \cos \theta$ and *s*.

Work is done only when a force is exerted on a body while the body at the same time moves in such a way that the force has a component along the line of motion of its point of application. If the component of the force is in the *same direction* as the displacement, the work *W* is *positive*. If it is *opposite* to the displacement, the work is *negative*. If the force is at *right angles* to the displacement, it has no component in the direction of the displacement and the work is *zero*.

Thus, when a body is lifted, the work done by the lifting force is positive; when a spring is stretched, the work done by the stretching force is positive; when a gas is compressed in a cylinder, again the work of the compressing force is positive. On the other hand, the work of the gravitational force on a body being lifted is negative, since the (downward) gravitational force is opposite to the (upward) displacement. When a body slides on a fixed surface, the work of the frictional force exerted *on the body* is negative, since this force is always opposite to the displacement of the body. No work is done by the frictional force acting *on the fixed surface* because there is no motion of this surface. Also, although it would be considered "hard work" to hold a heavy object stationary at arm's length, no work

113

Fig. 7-2 *An object on a rough horizontal surface moving to the right under the action of a force **P** inclined at an angle θ.*

would be done in the technical sense because there is no motion. Even if one were to walk along a level floor while carrying the object, no work would be done, since the (vertical) supporting force has no component in the direction of the (horizontal) motion. Similarly, the work of the normal force exerted on a body by a surface on which it moves is zero, as is the work of the centripetal force on a body moving in a circle.

The unit of work in any system is the unit of force multiplied by the unit of distance. In the mks system the unit of force is the newton and the unit of distance is the meter; thus in this system the unit of work is one *newton meter* (1 N m) which is also called one *joule* (1 J). The unit of work in the cgs system is one *dyne centimeter* (1 dyn cm), called one *erg*. Since 1 m = 100 cm and 1 N = 10^5 dyn, it follows that

$$1 \, \text{N m} = 10^7 \, \text{dyn cm}, \quad \text{or} \quad 1 \, \text{J} = 10^7 \, \text{erg}.$$

In the engineering system, the unit of work is one *foot pound* (1 ft lb):

$$1 \, \text{J} = 0.7376 \, \text{ft lb}, \quad 1 \, \text{ft lb} = 1.356 \, \text{J}.$$

When several external forces act on a body, we may wish to consider the works of the separate forces. Each of these may be computed from the definition of work in Eq. (7-1). Then, since work is a scalar quantity, the total work is the algebraic sum of the individual works.

Example Figure 7-2 shows a box being dragged along a horizontal surface by a constant force **P** making a constant angle θ with the direction of

motion. The other forces on the box are its weight **w**, the normal upward force **N** exerted by the surface, and the friction force **f**. What is the work of each force when the box moves a distance *s* along the surface to the right?

The component of **P** in the direction of motion is $P \cos \theta$. The work of the force **P** is therefore

$$W_P = (P \cos \theta)s.$$

The forces **w** and **N** are both at right angles to the displacement. Hence,

$$W_w = 0, \quad W_N = 0.$$

The friction force **f** is opposite to the displacement, so the work of the friction force is

$$W_f = -fs.$$

Since work is a scalar quantity, the total work W of all forces on the body is the algebraic (not the vector) sum of the individual works:

$$W = W_P + W_w + W_N + W_f$$
$$= (P \cos \theta)s + 0 + 0 - fs$$
$$= (P \cos \theta - f)s.$$

But ($P \cos \theta - f$) is the *resultant* force on the body. Hence *the total work of all forces is equal to the work of the resultant force.*

Suppose that $w = 100$ N, $P = 50$ N, $f = 15$ N, $\theta = 37°$, and $s = 20$ m. Then

$$W_P = (P \cos \theta)s$$
$$= (50 \, \text{N})(0.8)(20 \, \text{m}) = 800 \, \text{N m},$$
$$W_f = -fs = (-15 \, \text{N})(20 \, \text{m}) = -300 \, \text{N m},$$
$$W = W_P + W_f = 500 \, \text{N m}.$$

As a check, the total work may be expressed as

$$W = (P \cos \theta - f)s$$
$$= (40 \, \text{N} - 15 \, \text{N})(20 \, \text{m}) = 500 \, \text{N m}.$$

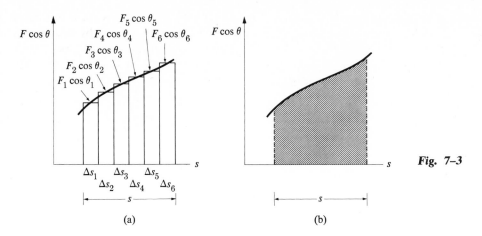

Fig. 7–3

(a)   (b)

7–2 WORK DONE BY A VARYING FORCE

In the preceding section we defined the work of a *constant* force. In many important cases, however, the work is done by a force which varies in magnitude or direction during the displacement of the body on which it acts. Thus when a spring is stretched slowly, the force required to stretch it increases steadily as the spring elongates; when a body is projected vertically upward, the gravitational force exerted on it by the earth decreases inversely with the square of its distance from the earth's center.

The work of a varying force can be found as follows. In Fig. 7–3(a), the curved line is a graph of the component $F \cos \theta$, for a force which varies in some arbitrary way (in magnitude and direction) with the distance s. Let the distance be subdivided into short segments Δs_1, Δs_2, etc., and approximate the varying force component by one that remains constant at the value $F_1 \cos \theta_1$ over the distance Δs_1, then increases to the constant value $F_2 \cos \theta_2$ over the distance Δs_2, etc., as indicated by the zigzag line. The work that would be done by the constant force $F_1 \cos \theta_1$ in the displacement Δs_1 would be $F_1 \cos \theta_1 \Delta s_1$; the work done in the displacement Δs_2 would be $F_2 \cos \theta_2 \Delta s_2$, etc. The work total W would be

$$W = F_1 \cos \theta_1 \Delta s_1 + F_2 \cos \theta_2 \Delta s_2 + \cdots .$$

We may take these intervals smaller and smaller and, in the limit of very small intervals, the sum becomes an integral:

$$W = \int_{s_1}^{s_2} F \cos \theta \, ds. \qquad (7\text{–}2)$$

We also see that the total work during the displacement from s_1 to s_2 is represented graphically by the *area* under the graph of $F \cos \theta$ between these limits, as shown in Fig. 7–3(b).

Example 1 Compute the work required to stretch a spring a distance X from its unstretched length, if the force required increases in direct proportion to the amount of elongation.

To keep a spring stretched an amount x beyond its unstretched length requires a force F at one end and a force of equal magnitude and opposite direction at the other end. If the elongation x is not too great, F is directly proportional to x, and

$$F = kx, \qquad (7\text{–}3)$$

where k is a constant called the *force constant* or the *stiffness* of the spring. This direct proportion between force and elongation, for elongations that are not too great, was discovered by Robert Hooke in 1678 and is known as *Hooke's law*. It will be discussed more fully in a later chapter.

Suppose that forces equal in magnitude and opposite in direction are exerted on the ends of a spring, and that the forces are gradually increased from zero. Let one end of the spring be kept fixed. The work of the force at the fixed end will be zero, but work will be done by the varying force $\boldsymbol{F}$ at the moving end. This force is in the same direction as the displacement. In Fig. 7–4, the force F is plotted vertically and the displacement x of the moving end representing the elongation of the spring, is plotted horizontally.

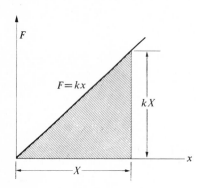

Fig. 7–4 *The work done in stretching a spring is equal to the area of the shaded triangle.*

The work required to stretch the spring from $x = 0$ (no elongation) to $x = X$ is

$$W = \int_0^X F\,dx = \int_0^X kx\,dx = \tfrac{1}{2}kX^2. \quad (7\text{–}4)$$

This result can also be obtained graphically; the area of the shaded triangle in Fig. 7–4, representing the total work, is equal to half the product of base and altitude, or

$$W = \tfrac{1}{2}(X)(kX) = \tfrac{1}{2}kX^2$$

in agreement with the above result.

The work of a force can be expressed in several ways. If θ is the angle between the force vector $\boldsymbol{F}$ and the infinitesimal displacement vector $d\boldsymbol{s}$, the component $F_\parallel$ in the direction of $d\boldsymbol{s}$ is equal to $F \cos \theta$, so

$$W = \int_{s_1}^{s} F_\parallel\,ds = \int_{s_1}^{s_2} F \cos \theta\,ds.$$

In Chapter 1 we discussed the *addition* and *subtraction* of vectors. The definition of work suggests a third process of vector algebra, namely, *scalar multiplication* of two vectors. In computing the work of a force, we multiply the magnitude of the vector $d\boldsymbol{s}$ by the magnitude of the component of another vector $\boldsymbol{F}$

in the direction of $d\boldsymbol{s}$. The product is called the *scalar product* or the *dot product* of the vectors. Thus if $\boldsymbol{A}$ and $\boldsymbol{B}$ are any two vectors, their scalar or dot product is defined as a scalar, equal to the product of the magnitudes of the vectors and the cosine of the angle θ between their positive directions:

$$\boldsymbol{A} \cdot \boldsymbol{B} = AB \cos \theta.$$

Evidently, $\boldsymbol{A} \cdot \boldsymbol{B} = \boldsymbol{B} \cdot \boldsymbol{A}$. In this new notation, the work of a force can be written as

$$W = \int_{s_1}^{s_2} \boldsymbol{F} \cdot d\boldsymbol{s}.$$

In the special case in which the force $\boldsymbol{F}$ is constant in magnitude and makes a constant angle with the direction of motion of its point of application, $F \cos \theta$ is constant and may be taken outside the integral sign. If in addition we measure diplacements from the starting point of the motion and let $s_1 = 0$ and $s_2 = s$, the work of the force is

$$W = \int_0^s F \cos \theta\,ds = F \cos \theta \int_0^s ds = (F \cos \theta)s.$$

If the force is constant and in the same direction as the motion or opposite to the motion, the angle θ equals zero or $180°$, $\cos \theta = \pm 1$, and

$$W = \pm Fs.$$

That is, in this *very special case only*, we can say, "Work equals force times distance." It is important to remember, however, that the *general* definition of the work of a force is

$$\boxed{W = \int_{s_1}^{s_2} \boldsymbol{F} \cdot d\boldsymbol{s} = \int_{s_1}^{s_2} F_\parallel\,ds = \int_{s_1}^{s_2} F \cos \theta\,ds.} \quad (7\text{–}5)$$

Example 2   A small object of weight $\boldsymbol{w}$ hangs from a string of length ℓ, as shown in Fig. 7–5. A *variable* horizontal force $\boldsymbol{P}$, which starts at zero and gradually increases, is used to pull the object very slowly (so that equilibrium exists at all times) until the string makes an angle θ with the vertical. Calculate the work of the force $\boldsymbol{P}$.

Since the object is in equilibrium, the sum of the horizontal forces equals zero, whence

$$P = T \sin \theta.$$

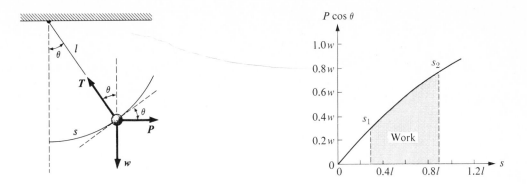

Fig. 7-5 *A variable horizontal force **P** acts on a small object while the displacement varies from zero to s.*

Fig. 7-6 *The work of a force **P** when its point of application moves from $s_1 = 0.3\ell$ to $s_2 = 0.9\ell$ is the shaded area under the curve of P cos θ, plotted against s.*

Equating the sum of the vertical forces to zero, we find

$$w = T \cos \theta.$$

Dividing these two equations, we get

$$P = w \tan \theta.$$

The point of application of **P** swings through the arc *s*. Since $s = \ell\theta$, $ds = \ell\, d\theta$ and

$$W = \int \mathbf{P} \cdot d\mathbf{s} = \int P \cos \theta\, ds$$

$$= \int_0^\theta w \tan \theta \cos \theta\, \ell\, d\theta$$

$$= w\ell \int_0^\theta \sin \theta\, d\theta = w\ell(1 - \cos \theta). \quad (7\text{-}6)$$

A graph of $P \cos \theta\ (= w \sin \theta)$ plotted against *s* $(= \ell\theta)$ is shown in Fig. 7-6. The work during any displacement is represented by the area under this curve. The shaded area is the work when the point of application of **P** moves from a position where $s_1 = 0.3\ell$ to a place where $s_2 = 0.9\ell$.

7-3 WORK AND KINETIC ENERGY

Consider again the body illustrated in Fig. 7-2. The constant resultant force *F* on the body is $F = P \cos \theta - f$, and from Newton's second law,

$$F = ma.$$

Suppose the speed of the body increases from v_1 to v_2 while the body undergoes a displacement *s*. Then, since *a* is constant,

$$v_2^2 = v_1^2 + 2as,$$

$$a = \frac{v_2^2 - v_1^2}{2s}.$$

Hence

$$F = m\frac{v_2^2 - v_1^2}{2s},$$

and

$$Fs = \tfrac{1}{2}mv_2^2 - \tfrac{1}{2}mv_1^2. \quad (7\text{-}7)$$

The product *Fs* is the work *W* of the resultant force *F*. The quantity $\tfrac{1}{2}mv^2$, one-half the product of the mass of the body and the square of its speed, is called its *kinetic energy*, E_k.

$$\boxed{E_k = \tfrac{1}{2}mv^2.} \quad (7\text{-}8)$$

The first term on the right side of Eq. (7-7) containing the final speed v_2, is the final kinetic energy of the body, E_{k2}, and the second term is the initial kinetic energy, E_{k1}. The difference between these terms is the *change* in kinetic energy, and we have the important result that *the work of the result-ant external force on a body is equal to the change in kinetic energy of the body.* As we have shown, the work of the resultant force is equal to the algebraic sum of the works of the individual forces

$$W = E_{k2} - E_{k1} = \Delta E_k. \qquad (7\text{–}9)$$

Kinetic energy, like work, is a *scalar* quantity. The kinetic energy of a moving body depends only on its speed, or the *magnitude* of its velocity, but not on the *direction* in which it is moving. The *change* in kinetic energy depends only on the work $W = Fs$ and not on the individual values of F and s. That is, the force F could have been large and the displacement s small, or the reverse might have been true. If the mass m and the speeds v_1 and v_2 are known, the *work* of the resultant force can be found without any knowledge of the force F and the displacement s.

If the work W is *positive*, the final kinetic energy is greater than the initial kinetic energy and the kinetic energy *increases*. If the work is *negative*, the kinetic energy *decreases*. In the special case in which the work is *zero*, the kinetic energy remains *constant*.

In computing the kinetic energy of a body, consistent units must be used for m and v. In the mks system, m must be in kilograms and v in meters per second. In the cgs system m must be expressed in grams and v in centimeters per second. In the engineering system, m must be in slugs and v in feet per second. The corresponding units of kinetic energy are $1 \text{ kg m}^2 \text{ s}^{-2}$, $1 \text{ g cm}^2 \text{ s}^{-2}$, and $1 \text{ slug ft}^2 \text{ s}^{-2}$. However, the unit of kinetic energy in any system is equal to the unit of work in that system, and kinetic energy is customarily expressed in joules, ergs, or foot pounds. That is,

$$1 \text{ kg m}^2 \text{ s}^{-2} = 1 \text{ N m}^{-1} \text{ s}^2 \cdot \text{m}^2 \text{ s}^{-2} = 1 \text{ N m} = 1 \text{ J}.$$

In the same way,

$$1 \text{ g cm}^2 \text{ s}^{-2} = 1 \text{ erg}, \qquad 1 \text{ slug ft}^2 \text{ s}^{-2} = 1 \text{ ft lb}.$$

Example   Refer again to the body in Fig. 7–2 and the numerical values given in the example of Section 7–1. The total work of the external forces was shown to be 500 N m. Hence the kinetic energy of the body increases by 500 N m. To verify this, suppose the initial speed v_1 is 4 m s^{-1}. The mass of the body is

$m = w/g = (100 \text{ N})/(10 \text{ m s}^{-2}) = 10 \text{ kg}$. The initial kinetic energy is

$$E_{k1} = \tfrac{1}{2}mv_1^2 = \tfrac{1}{2}(10 \text{ kg})(4 \text{ m s}^{-1})^2 = 80 \text{ J}.$$

To find the final kinetic energy we must first find the acceleration:

$$a = \frac{F}{m} = \frac{40 \text{ N} - 15 \text{ N}}{10 \text{ kg}} = 2.5 \text{ m s}^{-2}.$$

Then

$$v_2^2 = v_1^2 + 2as = (4 \text{ m s}^{-1})^2 + 2(2.5 \text{ m s}^{-2})(20 \text{ m})$$
$$= 116 \text{ m}^2 \text{ s}^{-2},$$

and

$$E_{k2} = \tfrac{1}{2}(10 \text{ kg})(116 \text{ m}^2 \text{ s}^{-2}) = 580 \text{ J}.$$

The increase in kinetic energy is therefore 500 J.

Equation (7–9) was derived for the special case of straight-line motion with a constant resultant force, but it is also valid for a curved path and a variable force. To show this, we consider the curve in Fig. 7–7, representing the path of a particle of mass m moving in the xy-plane and acted on by a *resultant* force $\mathbf{F}$ which may vary in magnitude and direction from point to point of the path. Let us resolve the force into a component $\mathbf{F}_{\parallel}$ along the path and a component $\mathbf{F}_{\perp}$ normal to the path.

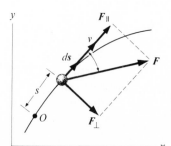

Fig. 7–7 *Path of a particle in the xy-plane.*

The component $\mathbf{F}_{\perp}$, at right angles to the velocity $\mathbf{v}$, is a *centripetal* force, and its only effect is to change the *direction* of the velocity. The effect of the component $\mathbf{F}_{\parallel}$ is to change the *magnitude* of the velocity.

Fig. 7–8 *Work of the gravitational force* **w** *during the motion of an object from one point in a gravitational field to another.*

(a)   (b)   (c)

Let s be the distance of the particle from some fixed point O, measured along the path. In general, the magnitude of $F_\parallel$ will be a function of s. From Newton's second law,

$$F_\parallel = m\frac{dv}{dt}.$$

Since $F_\parallel$ is a function of s, we use the chain rule and write

$$\frac{dv}{dt} = \frac{dv}{ds}\frac{ds}{dt} = v\frac{dv}{ds}.$$

Then we have

$$F_\parallel = mv\frac{dv}{ds},$$

$$F_\parallel\, ds = mv\, dv.$$

If v_1 is the velocity when $s = s_1$, and v_2 the velocity when $s = s_2$, it follows that

$$\int_{s_1}^{s_2} F_\parallel\, ds = \int_{v_1}^{v_2} mv\, dv. \qquad (7\text{–}10)$$

The integral on the left is the *work* W of the force $\mathbf{F}$, between the points s_1 and s_2. This integral can be evaluated only when the component $F_\parallel$ is known as a function of s, or when both F and s are known as functions of another variable. The integral on the right side of Eq. (7–10), however, can always

be evaluated:

$$\int_{v_1}^{v_2} mv\, dv = \tfrac{1}{2}mv_2{}^2 - \tfrac{1}{2}mv_1{}^2$$

$$= E_{k2} - E_{k1}.$$

Equation (7–10) can therefore be written

$$W = E_{k2} - E_{k1},$$

which is Eq. (7–9).

7–4 GRAVITATIONAL POTENTIAL ENERGY

Suppose a body of mass m (and of weight $\mathbf{w} = m\mathbf{g}$) moves vertically, as in Fig. 7–8(a), from a point where its center of gravity is at a height y_1 above an arbitrarily chosen plane (the reference level) to a point at a height y_2. For the present, we shall consider only displacements near the earth's surface, so that variations of gravitational force with distance from the earth's center can be neglected. The downward gravitational force on the body is then constant and equal to $\mathbf{w}$. Let $\mathbf{P}$ represent the resultant of all other forces acting on the body and let W' be the work of these forces. The direction of the gravitational force $\mathbf{w}$ is opposite to the upward displacement, and the work of this force is

$$W_{\text{grav}} = -w(y_2 - y_1) = -(mgy_2 - mgy_1). \quad (7\text{–}11)$$

The reader should convince himself that the work of the gravitational force is given by $-mg(y_2 - y_1)$ whether the body moves up or down.

Now suppose that the body starts at the same elevation y_1 but is moved up to the elevation y_2 along some arbitrary path, as in Fig. 7–8(b). Part (c) of the figure is an enlarged view of a small portion of the path. The work of the gravitational force is

$$W_{\text{grav}} = \int_{s_1}^{s_2} w \cos \theta \, ds.$$

Let ϕ represent the angle between ds and its vertical component dy. Then $dy = ds \cos \phi$, and since $\phi = 180° - \theta$,

$$\cos \phi = -\cos \theta, \qquad \cos \theta \, ds = -dy,$$

and

$$W_{\text{grav}} = -\int_{y_1}^{y_2} w \, dy = -w(y_2 - y_1)$$
$$= -(mgy_2 - mgy_1). \qquad (7\text{–}12)$$

The work of the gravitational force, therefore, depends only on the initial and final *elevations* and not on the path. If these points are at the same elevation the work is zero.

Let W' be the work done by forces other than the gravitational force. Since the *total* work equals the change in kinetic energy,

$$W' + W_{\text{grav}} = E_{k2} - E_{k1},$$
$$W' - (mgy_2 - mgy_1) = (\tfrac{1}{2}mv_2^2 - \tfrac{1}{2}mv_1^2). \quad (7\text{–}13\text{a})$$

The quantities $\tfrac{1}{2}mv_2^2$ and $\tfrac{1}{2}mv_1^2$ depend only on the final and initial *speeds*; the quantities mgy_2 and mgy_1 depend only on the final and initial *elevations*. Let us therefore rearrange this equation, transferring the quantities mgy_2 and mgy_1 from the "work" side of the equation to the "energy" side:

$$W' = (\tfrac{1}{2}mv_2^2 - \tfrac{1}{2}mv_1^2) + (mgy_2 - mgy_1). \quad (7\text{–}13\text{b})$$

The left side of Eq. (7–13b) contains only the work of the force $\boldsymbol{P}$. The terms on the right depend only on the final and initial states of the body (its speed and elevation) and not specifically on the way in which it moved. The quantity mgy, the product of the weight mg of the body and the height y of its center of gravity above the reference level, is called its *gravitational potential energy*, E_p.

$$\boxed{E_p(\text{gravitational}) = mgy.} \qquad (7\text{–}14)$$

The first expression in parentheses on the right of Eq. (7–13b) is the change in kinetic energy of the body, and the second is the change in its gravitational potential energy. Equation (7–13b) can also be written

$$W' = (\tfrac{1}{2}mv_2^2 + mgy_2) - (\tfrac{1}{2}mv_1^2 + mgy_1). \quad (7\text{–}13\text{c})$$

The sum of the kinetic and potential energy of the body is called its *total mechanical energy*. The first expression in parentheses on the right of Eq. (7–13c) is the final value of the total mechanical energy, and the second is the initial value. Hence, *the work of all forces acting on the body,* **with the exception of the gravitational force,** *equals the change in the total mechanical energy of the body.* If the work W' is positive, the mechanical energy increases. If W' is negative, the energy decreases.

In the special case in which the *only* force on the body is the gravitational force, the work W' is zero. Equation (7–13c) can then be written

$$\tfrac{1}{2}mv_2^2 + mgy_2 = \tfrac{1}{2}mv_1^2 + mgy_1.$$

Under these conditions, then, the *total mechanical energy remains constant*, or is *conserved*. This is a special case of the principle of *the conservation of mechanical energy.*

Example 1   A man holds a ball of mass $m = 0.2$ kg at rest in his hand. He then throws the ball vertically upward. In this process, his hand moves up 0.5 m and the ball leaves his hand with an upward velocity of 20 m s^{-1}. Discuss the motion of the ball from the work–energy standpoint, assuming $g = 10$ m s^{-2}.

First, consider the throwing process. Take the reference level at the initial position of the ball. Then $E_{k1} = 0$, $E_{p1} = 0$. Take point 2 at the point where the ball leaves the thrower's hand. Then

$$E_{p2} = mgy_2 = (0.2\ \text{kg})(10\ \text{m s}^{-2})(0.5\ \text{m}) = 1.0\ \text{J},$$

$$E_{k2} = \tfrac{1}{2}mv_2{}^2 = \tfrac{1}{2}(0.2\ \text{kg})(20\ \text{m s}^{-1})^2 = 40\ \text{J}.$$

Let P represent the upward force exerted on the ball by the man in the throwing process. The work W' is then the work of this force and is equal to the sum of changes in kinetic and potential energy of the ball.

The kinetic energy of the ball increases by 40 J and its potential energy by 1 J. The work W' of the upward force P is therefore 41 J.

If the force P is constant, the work of this force is given by

$$W' = P(y_2 - y_1),$$

and the force P is then

$$P = \frac{W'}{y_2 - y_1} = \frac{41\ \text{J}}{0.5\ \text{m}} = 82\ \text{N}.$$

However, the *work* of the force P is 41 J, whether or not the force is constant.

Now consider the flight of the ball after it leaves the thrower's hand. In the absence of air resistance, the only force on the ball is then its weight $\boldsymbol{w} = m\boldsymbol{g}$. Hence the total mechanical energy of the ball remains constant. The calculations will be simplified if we take a new reference level at the point where the ball leaves the thrower's hand. Calling this point 1, we have

$$E_{k1} = 40\ \text{J}, \qquad E_{p1} = 0,$$

$$E_k + E_p = 40\ \text{J},$$

and the total mechanical energy at any point of the path equals 40 J.

Suppose we wish to find the speed of the ball at a height of 15 m above the reference level. Its potential energy at this elevation is 30 J. (Why?) Its kinetic energy is therefore 10 J. To find its speed, we have

$$\tfrac{1}{2}mv^2 = E_k, \qquad v = \pm\sqrt{2E_k/m} = \pm 10\ \text{m s}^{-1}.$$

The significance of the $\pm$ sign is that the ball passes this point *twice*, once on the way up and again on the way down. Its *potential* energy at this point is the same whether it is moving up or down. Hence its kinetic energy is the same and its *speed* is the same. The algebraic sign of the speed is $+$ when the ball is moving up and $-$ when it is moving down.

Next, let us find the height of the highest point reached. At this point $v = 0$ and $E_k = 0$. Therefore $E_p = 40\ \text{J}$, and the ball rises to a height h above the point where it leaves the thrower's hand, given by $mgh = (2\ \text{N})h = 40\ \text{J}$. Thus $h = 20\ \text{m}$.

Finally, suppose we were asked to find the speed at a point 30 m above the reference level. The potential energy at this point would be 60 J. But the *total* energy is only 40 J, so the ball never reaches a height of 30 m.

Example 2 A body slides down a curved track which is one quadrant of a circle of radius R, as in Fig. 7–9. If it starts from rest and there is no friction, find its speed at the bottom of the track. The motion of this body is exactly the same as that of a body attached to one end of a string of length R, the other end of which is held at point O.

The equations of motion with constant acceleration cannot be used, since the acceleration decreases during the motion. (The slope angle of the track becomes smaller and smaller as the body descends.) However, if there is no friction, the only force on the body in addition to its weight is the normal force N exerted on it by the track. The work of this force is zero because at each point it is perpendicular to the small element of displacement near that point. Thus

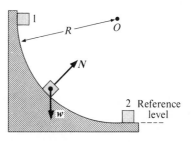

Fig. 7–9 *An object sliding down a frictionless curved track.*

$W' = 0$ and mechanical energy is conserved. Take point 1 at the starting point and point 2 at the bottom of the track. Take the reference level at point 2. Then $y_1 = R$, $y_2 = 0$, and

$$E_{k2} + E_{p2} = E_{k1} + E_{p1},$$
$$\tfrac{1}{2}mv_2{}^2 + 0 = 0 + mgR,$$
$$v_2 = \pm\sqrt{2gR}.$$

The speed is therefore the same as if the body had fallen *vertically* through a height R. (What is now the significance of the $\pm$ sign?)

As a numerical example, let $R = 1$ m. Then

$$v = \pm\sqrt{2(9.8 \text{ m s}^{-2})(1 \text{ m})} = \pm 4.43 \text{ m s}^{-1}.$$

Example 3 Suppose a body of mass 0.5 kg slides down a track of radius $R = 1$ m, like that in Fig. 7–9, but its speed at the bottom is only 3 m s^{-1}. What was the work of the frictional force acting on the body?

In this case, $W' = W_f$, and

$$W_f = (\tfrac{1}{2}mv_2{}^2 - \tfrac{1}{2}mv_1{}^2) + (mgy_2 - mgy_1)$$
$$= \tfrac{1}{2}(0.5 \text{ kg})(9 \text{ m}^2 \text{ s}^{-2}) - 0$$
$$+ 0 - (0.5 \text{ kg})(9.8 \text{ m s}^{-2})(1 \text{ m})$$
$$= 2.25 \text{ J} - 4.9 \text{ J} = -2.65 \text{ J}.$$

The frictional work was therefore -2.65 J, and the total mechanical energy *decreased* by 2.65 J. The mechanical energy of a body is *not* conserved when friction forces act on it.

Example 4 In the absence of air resistance, the only force on a projectile is its weight, and the mechanical energy of the projectile remains constant. Figure 7–10 shows two trajectories of a projectile with the same initial speed (and hence the same total energy) but with different angles of departure. At all points at the same elevation the potential energy is the same; hence the kinetic energy is the same and the speed is the same.

Example 5 A small object of weight w hangs from a string of length ℓ, as shown in Fig. 7–11. A *variable* horizontal force P which starts at zero and gradually increases is used to pull the object very slowly until the string makes an angle θ with the vertical. Calculate the work of the force P.

The sum of the works W' of all the forces other than the gravitational force must equal the change of kinetic energy plus the change of gravitational potential energy. Hence

$$W' = W_p + W_T = \Delta E_k + \Delta E_p.$$

Since T is perpendicular to the path of its point of application, $W_T = 0$; and since the body was pulled very slowly at all times, the change of kinetic

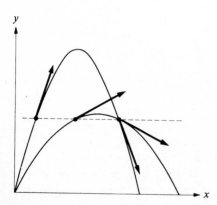

Fig. 7–10 *For the same initial speed, the speed is the same at all points at the same elevation.*

Fig. 7–11 $\Delta y = \ell(1 - \cos \theta)$.

energy is also zero. Hence

$$W_P = \Delta E_p = w\,\Delta y,$$

where Δy is the distance that the object has been raised. From Fig. 7–1, Δy is seen to be $\ell(1 - \cos\theta)$. Therefore

$$W_P = w\ell(1 - \cos\theta).$$

The reader is referred to Example 2 in Section 7–2, where this same problem was solved by performing the integration in the equation $W_p = \int P \cos\theta\, ds$. The two answers, of course, are identical, but how much more simply the energy principle leads to the final result!

Thus far in this section it has been assumed that the changes in elevation are small enough so that the gravitational force on a body can be considered constant. But as discussed in Section 5–5, the gravitational force is given in general by

$$F_g = \frac{Gmm_E}{r^2},$$

where m_E is the mass of the earth and r is the distance from the earth's center. When r increases from r_1 to r_2, the work of the gravitational force is

$$W_{grav} = -Gmm_E \int_{r_1}^{r_2} \frac{dr}{r^2} = \frac{Gmm_E}{r_2} - \frac{Gmm_E}{r_1}.$$

Setting the total work equal to the change in kinetic energy and rearranging terms, we get, instead of Eq. (7–13),

$$W' = \left(\tfrac{1}{2}mv_2{}^2 - \frac{Gmm_E}{r_2}\right) - \left(\tfrac{1}{2}mv_1{}^2 - \frac{Gmm_E}{r_1}\right).$$

The quantity $-G(mm_E/r)$ is therefore the general expression for the gravitational potential energy of a body attracted by the earth:

$$E_p(\text{gravitational}) = -G\frac{mm_E}{r}. \qquad (7\text{–}14)$$

The total mechanical energy of the body, the sum of its kinetic energy and potential energy, is

$$E = E_k + E_p = \tfrac{1}{2}mv^2 - G\frac{mm_E}{r}.$$

If the only force on the body is the gravitational force, then $W' = 0$ and the total mechanical energy remains constant, or is *conserved*.

Example 6   Use the principle of conservation of mechanical energy to find the velocity with which a body must be projected vertically upward, in the absence of air resistance, (a) to rise to a height above the earth's surface equal to the earth's radius, R, and (b) to escape from the earth.

a) Let v_1 be the initial velocity. Then

$$r_1 = R, \qquad r_2 = 2R, \qquad v_2 = 0,$$

and

$$\tfrac{1}{2}mv_1{}^2 - G\frac{mm_E}{R} = 0 - G\frac{mm_E}{2R},$$

or

$$v_1{}^2 = \frac{Gm_E}{R},$$

in agreement with Eq. (5–11) obtained in Example 9, Section 5–6.

b) When v_1 is the escape velocity, $r_1 = R$, $r_2 = \infty$, $v_2 = 0$. Then

$$\tfrac{1}{2}mv_1{}^2 - G\frac{mm_E}{R} = 0 \qquad \text{or} \qquad v_1{}^2 = \frac{2Gm_E}{R}.$$

It may be puzzling at first sight that the general expression for gravitational potential energy should contain a minus sign. The reason for this lies in the choice of a reference state or reference level in which the potential energy is considered zero. If we set $E_p = 0$ in Eq. (7–14) and solve for r, we get

$$r = \infty.$$

That is, *the gravitational potential energy of a body is now considered to be zero when the body is at an infinite distance from the earth*. Since the potential energy *decreases* as the body approaches the earth, it must be *negative* at any finite distance from the earth. The *change* in potential energy of a body as it moves from one point to another is the same, whatever the choice of reference level, and it is only changes in potential energy that are significant.

Finally, we show that the general expression for the change in potential energy reduces to Eq. (7–11) for small changes in elevation near the earth's surface. The increase in potential energy of a body of mass m, when its distance from the earth's center increases from r_1 to r_2, is

$$E_{p2} - E_{p1} = -G\frac{mm_E}{r_2} - \left(-G\frac{mm_E}{r_1}\right)$$

$$= Gmm_E\left(\frac{1}{r_1} - \frac{1}{r_2}\right)$$

$$= Gmm_E\left(\frac{r_2 - r_1}{r_1 r_2}\right)$$

If points 1 and 2 are at elevations y_1 and y_2 above the earth's surface, then

$$r_2 - r_1 = y_2 - y_1.$$

Furthermore, if y_1 and y_2 are both small compared with the earth's radius R, the product $r_1 r_2$ is very nearly equal to R^2. The preceding equation then reduces to

$$E_{p2} - E_{p1} = \frac{Gmm_E}{R^2}(y_2 - y_1).$$

But

$$\frac{Gm_E}{R^2} = g,$$

where g is the acceleration of gravity at the earth's surface, so

$$E_{p2} - E_{p1} = mg(y_2 - y_1),$$

which is the same as the expression derived earlier when variations in g were neglected.

7–5 ELASTIC POTENTIAL ENERGY

Figure 7–12 shows a body of mass m on a level surface. One end of a spring is attached to the body and the other end of the spring is fixed. Take the origin of coordinates at the position of the body when the spring is unstretched (Fig. 7–12a). An outside agent exerts a force P of sufficient magnitude to cause the spring to stretch. As soon as the slightest extension takes place, a force F is created within the spring which acts in a direction opposite that of P. The force F is called an *elastic force*. If the force P is reduced or made zero, the elastic force will restore

the spring to its original unstretched condition. It may therefore be referred to as a *restoring force*.

As shown in Section 7–2, for a force obeying Hooke's law, stretching the string to a final elongation x requires an amount of work $\frac{1}{2}kx^2$. Thus the work done *by the spring* (the restoring force) during this elongation is $-\frac{1}{2}kx^2$. Similarly, in a displacement from an initial elongation x_1 to a final elongation x_2, the elastic restoring force of the spring does an amount of work W_{el} given by

$$W_{el} = -\tfrac{1}{2}kx_2{}^2 - (-\tfrac{1}{2}kx_1{}^2).$$

Let W' stand for the work of the applied force P. Setting the total work equal to the change in kinetic energy of the body, we have

$$W' + W_{el} = \Delta E_k,$$
$$W' - (\tfrac{1}{2}kx_2{}^2 - \tfrac{1}{2}kx_1{}^2) = (\tfrac{1}{2}mv_2{}^2 - \tfrac{1}{2}mv_1{}^2).$$

The quantities $\frac{1}{2}kx_2{}^2$ and $\frac{1}{2}kx_1{}^2$ depend only on the initial and final positions of the body and not specifically on the way in which it moved. Let us therefore transfer them from the "work" side of the equation to the "energy" side. Then

$$\boxed{W' = (\tfrac{1}{2}mv_2{}^2 - \tfrac{1}{2}mv_1{}^2) + (\tfrac{1}{2}kx_2{}^2 - \tfrac{1}{2}kx_1{}^2).}\quad(7\text{–}15)$$

The quantity $\frac{1}{2}kx^2$, one-half the product of the force constant and the square of the coordinate of the body, is called the *elastic potential energy* of the body, E_p. (The symbol E_p is used for any form of potential energy.)

$$\boxed{E_p \text{ (elastic)} = \tfrac{1}{2}kx^2.}\quad(7\text{–}16)$$

Hence the work W' of the force P equals the sum of the change in the kinetic energy of the body and the change in its elastic potential energy.

Equation (7–15) can also be written

$$W' = (\tfrac{1}{2}mv_2{}^2 + \tfrac{1}{2}kx_2{}^2) - (\tfrac{1}{2}mv_1{}^2 + \tfrac{1}{2}kx_1{}^2).$$

The sum of the kinetic and potential energies of the body is its total mechanical energy; and *the work*

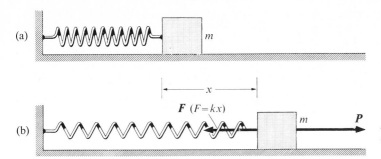

(a)

(b)

x

F $(F = kx)$

m

P

Fig. 7-12 *When an applied force **P** produces an extension x of a spring, an elastic restoring force **F** is created within the spring, where F = kx.*

of all forces acting on the body, **with the exception of the elastic force,** *equals the change in the total mechanical energy of the body.*

If the work W' is positive, the mechanical energy increases. If W' is negative, it decreases. In the special case in which W' is zero, the mechanical energy remains constant or is *conserved.*

Example 1 Let the force constant k of the spring in Fig. 7-12 be 24 N m^{-1}, and let the mass of the body be 4 kg. The body is initially at rest, and the spring is initially unstretched. Suppose that a constant force P of 10 N is exerted on the body, and that there is no friction. What will be the speed of the body when it has moved 0.5 m?

The equations of motion with constant acceleration cannot be used, since the resultant force on the body varies as the spring is stretched. However, the speed can be found from energy considerations:

$$W' = \Delta E_k + \Delta E_p,$$
$$(10\text{ N})(0.5\text{ m}) = \tfrac{1}{2}(4\text{ kg})v_2{}^2 - 0$$
$$+ \tfrac{1}{2}(24\text{ N m}^{-1})(0.25\text{ m}^2) - 0,$$
$$v_2 = 1\text{ m s}^{-1}.$$

Example 2 Suppose the force P ceases to act when the body has moved 0.5 m. *How much farther* will the body move before coming to rest?

The elastic force is now the only force, and mechanical energy is conserved. The kinetic energy is

$\tfrac{1}{2}mv^2 = 2$ J, and the potential energy is $\tfrac{1}{2}kx^2 = 3$ J. The total energy is therefore 5 J (equal to the work of the force P). When the body comes to rest, its kinetic energy is zero and its potential energy is therefore 5 J. Hence

$$\tfrac{1}{2}kx_{\text{max}}{}^2 = 5\text{ J}, \qquad x_{\text{max}} = 0.645\text{ m}.$$

7-6 CONSERVATIVE AND DISSIPATIVE FORCES

We have seen that when an object acted on by gravity is moved from any position above a zero reference level to any other position, the work of the gravitational force is independent of the path and equal to the difference between the final and the initial values of a function called the *gravitational potential energy.* If the gravitational force alone acts on the object, the total mechanical energy (the sum of the kinetic and gravitational potential energies) is constant or conserved; the gravitational force is therefore called a *conservative force.* Thus, if the object is ascending, the work of the gravitational force is accomplished at the expense of the kinetic energy. When the object descends to its original level, the work of the gravitational force serves to increase the kinetic energy back to its original value. Thus the work done by gravity on ascent is completely recovered on descent. Complete recoverability is an important aspect of the work of a conservative force.

When an object attached to a spring is moved from one value of the spring extension to any other value, the work of the elastic force is also independent of the path and equal to the difference between the final and initial values of a function called the *elastic potential energy*. If the elastic force alone acts on the object, the sum of the kinetic and elastic potential energies is conserved; therefore the elastic force is also a conservative force. If the object moves so as to increase the extension of the spring, the work of the elastic force is accomplished at the expense of the kinetic energy. If, however, the extension is decreasing, then the work of the elastic force serves to increase the kinetic energy so that this work is completely recovered also.

To summarize, we see that the work of a conservative force has the following properties:

1.   It is independent of the path.

2.   It is equal to the difference between the final and initial values of an energy function.

3.   It is completely recoverable.

Contrast a conservative force with a friction force exerted on a moving object by a fixed surface. The work of the friction force *does depend* on the path; the longer the path between two given points, the greater the work. There is *no* function the difference of two values of which equals the work of the friction force. When we slide an object on a rough fixed surface back to its original position, the friction force reverses, and instead of recovering the work done in the first displacement, we must again do work on the return trip. In other words, frictional work is *not* completely recoverable. When the friction force acts alone, the total mechanical energy is *not* conserved. The friction force is therefore called a *nonconservative* or a *dissipative* force. *The mechanical energy of a body is conserved only when no dissipative forces act on it.*

We find that when friction forces act on a moving body, another form of energy is involved. The more general principle of conservation of energy includes this other form of energy, along with kinetic and potential energy; and, when it is included the *total* energy of any system remains constant. We shall study this general conservation principle more fully in a later chapter.

Example   Example 3 in Section 7–4 illustrates the motion of a body acted on by a dissipative friction force. The initial mechanical energy of the body is its initial potential energy of 4.9 J. Its final mechanical energy is its final kinetic energy of 2.25 J. The frictional work W_f is −2.65 J. A quantity of thermal energy equivalent to 2.65 J is developed as the body slides down the track. The sum of this energy and the final mechanical energy equals the initial mechanical energy, and the total energy of the system is conserved.

7–7 INTERNAL WORK

Figure 7–13(a) shows a man standing on frictionless roller skates on a level surface, facing a rigid wall. Suppose that he sets himself in motion backward by pushing against the wall. The external forces *on the man* are his weight w, the upward forces N_1 and N_2 exerted by the ground, and the horizontal force P exerted by the wall. (The latter is the reaction to the force with which the man pushes against the wall.) The works of w and of N are zero because they are perpendicular to the motion. The force P is the unbalanced horizontal force which imparts to the system a horizontal acceleration. *The work of P, however, is zero because there is no motion of its point of application.* We are therefore confronted with a curious situation in which a force is responsible for acceleration, but its work, being zero, is not equal to the increase in kinetic energy of the system!

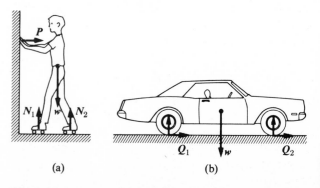

Fig. 7–13 (a) *External forces acting on a man who is pushing against a wall. The work of these forces is zero.* (b) *External forces on an automobile. The work of these forces is zero. In both cases, the work of the internal force is responsible for the increase in kinetic energy.*

At this point the concept of *internal work* is useful. Although internal forces play no role in accelerating the system, their points of application may move in such a way that work is done. In this case, an *internal* muscular force is exerted within the man away from the wall. (Think of his lengthening arm as an expanding spring.) Since the point of application of this force moves in the same direction, the work W_1 of the internal force is not zero. This is the work responsible for the increase in kinetic energy.

When both external and internal forces act on the particles of a system, the total work W of *all* forces, external and internal, is the sum of the works W_e and W_i and is equal to the change in the total kinetic energy of the system:

$$W = W_e + W_i = \Delta E_k .$$

The same principle holds in the case of an accelerated automobile. The portions of the rubber tires that are in contact with the rough roadway push back on the ground; the reactions to these forces, designated by $\boldsymbol{Q}_1$ and $\boldsymbol{Q}_2$ in Fig. 7–13(b), are the external horizontal forces acting *on* the automobile which are responsible for imparting the horizontal acceleration to the system. Since, however, the portions of the tires momentarily in contact with the road are at rest with respect to the road, the works of the forces $\boldsymbol{Q}_1$ and $\boldsymbol{Q}_2$ are zero. As a result of the expanding gases in the cylinders of the automobile engine there are many internal forces, some of which do work; these internal forces are responsible for the increase in kinetic energy.

7–8 INTERNAL POTENTIAL ENERGY

If any of the *external* forces on the particles of a system are conservative (gravitational or elastic), the work of these forces can be transferred to the energy side of the work–energy equation and called the change in *external potential energy of the system*. In many instances, the *internal* forces depend only on the distances between *pairs* of particles. The internal work then depends only on the initial and final distances between the particles, and not on the particular way in which they moved. The work of these internal forces can then also be transferred to the energy side of the work–energy equation and called the change in *internal potential energy of the system*. Internal potential energy is a property of the system as a whole and cannot be assigned to any specific particle.

Let W' represent the works of all external and internal forces that have *not* been transferred to the energy side of the work–energy equation and called changes in external and internal potential energy. Let $E_p{}^e$ and $E_p{}^i$ represent the external and internal potential energies. The work–energy equation then takes the form

$$W' = \Delta E_k + \Delta E_p{}^e + \Delta E_p{}^i, \qquad (7\text{–}17)$$

where the total mechanical energy now includes the kinetic energy of the system and both its external and internal potential energy. If the work W' is zero, the total mechanical energy is conserved.

Example 1   Consider a system consisting of two particles in "outer space," very far from all other matter. No external forces act on the system, no external work is done when the bodies move, and the system has no external potential energy. A gravitational force of attraction acts between the bodies, which depends only on the distance between them. We can therefore say that the system has an internal potential energy. If the particles start from rest and accelerate toward each other, the total kinetic energy of the system increases and its internal potential energy decreases. The sum of its kinetic energy and internal potential energy remains constant.

Energy considerations *alone* do not suffice to tell us how much kinetic energy is gained by each body separately. This can be determined, however, from *momentum* considerations, as will be explained in the next chapter.

Example 2   In Section 7–5 we computed the elastic potential energy of a body acted on by a force exerted by a spring. This force was an *external* force acting on the body, and the elastic potential energy $\frac{1}{2}kx^2$ should properly be called the *external* elastic potential energy *of the body*.

Let us now consider the spring itself. The spring is a *system* composed of an enormous number of molecules which exert internal forces on one another. When the spring is stretched, the distances between its molecules change. The intermolecular forces depend only on the distances between pairs of molecules, so that a stretched spring has internal elastic potential energy. To calculate this, suppose the spring is slowly stretched from its no-load length by equal and opposite forces applied at its ends. The work of these forces was shown to equal $\frac{1}{2}kx^2$, where x is the elongation of the spring above its no-load length. Let us retain this work on the left side of the work–energy equation so that it becomes the work W' in Eq. (7–17). There is no change in the kinetic energy of the spring and no change in its external potential energy. If we call the internal potential energy zero when $x = 0$, the *change* in internal potential energy in the stretching process equals the final potential energy $E_p{}^i$. Then

$$W' = \Delta E_p{}^i = E_p{}^i, \qquad E_p{}^i = \tfrac{1}{2}kx^2.$$

Hence the *internal elastic potential energy of a stretched spring* is equal to $\frac{1}{2}kx^2$. Note that the internal potential energy can be calculated without any detailed information regarding the intermolecular forces.

Example 3   A block of mass m, initially at rest, is dropped from a height h onto a spring whose force constant is k. Find the maximum distance y that the spring will be compressed. (See Fig. 7–14).

This is a process for which the principle of the conservation of mechanical energy holds. At the moment of release, the kinetic energy is zero. At the moment when maximum compression occurs, there is also no kinetic energy. Hence, the loss of gravitational potential energy of the block equals the gain of elastic potential energy of the spring. As shown in Fig. 7–14, the total fall of the block is $h + y$, whence

$$mg(h + y) = \tfrac{1}{2}ky^2,$$

or

$$y^2 - \frac{2mg}{k}\,y - \frac{2mgh}{k} = 0.$$

Fig. 7–14 *The total fall of the block is h + y.*

Therefore,

$$y = \frac{1}{2}\left[\frac{2mg}{k} \pm \sqrt{\left(\frac{2mg}{k}\right)^2 + \frac{8mgh}{k}}\,\right].$$

The positive root is the desired result; the negative root corresponds to the height to which the block and spring would rebound if they were fastened together after contact.

7–9 POWER

Time considerations are not involved in the definition of work. The same amount of work is done in raising a given weight through a given height whether the work is done in 1 s, or 1 hr, or 1 yr. In many instances, however, it is necessary to consider the *rate* at which work is done as well as the total amount of work accomplished. The rate at which work is done by a working agent is called the *power* developed by that agent.

If a quantity of work ΔW is done in a time interval Δt, the average power $\bar{P}$ is defined as

$$\text{Average power} = \frac{\text{work done}}{\text{time interval}},$$

$$\bar{P} = \frac{\Delta W}{\Delta t}.$$

If the rate at which work is done is not constant, this ratio may vary; in this case we may define an *instantaneous* power as the limiting value of this quotient as Δt approaches zero:

$$P = \lim_{\Delta t \to 0} \frac{\Delta W}{\Delta t} = \frac{dW}{dt}. \qquad (7\text{--}18)$$

The mks unit of power is one joule per second $(1\ \text{J s}^{-1})$, which is called one *watt* (1 W). Since this is a rather small unit, the kilowatt $(1\ \text{kW} = 10^3\ \text{W})$ and the megawatt $(1\ \text{MW} = 10^6\ \text{W})$ are commonly used. The cgs power unit is one erg per second $(1\ \text{erg s}^{-1})$. No single term is assigned to this unit.

In the engineering system, where work is expressed in foot-pounds and time in seconds, the unit of power is one foot-pound per second. Since this unit is inconveniently small, a larger unit called the *horsepower* (hp) is in common use: $1\ \text{hp} = 550\ \text{ft lb s}^{-1} = 33{,}000\ \text{ft lb min}^{-1}$. That is, a 1-hp motor running at full load is doing 33,000 ft-lb of work every minute it runs.

A common misconception is that there is something inherently *electrical* about a watt or a kilowatt. This is not the case. It is true that electrical power is usually expressed in watts or kilowatts, but the power consumption of an incandescent lamp could equally well be expressed in horsepower, or an automobile engine rated in kilowatts.

From the relations between the newton, pound, meter, and foot, we can show that $1\ \text{hp} = 746\ \text{W} = 0.746\ \text{kW}$, or about $\frac{3}{4}$ of a kilowatt, a useful figure to remember.

Since power is energy or work per unit time, the units of power may be used to define new units of work or energy. One such unit in common use is the kilowatt-hour, a unit of energy commonly used for electrical energy.

One kilowatt-hour is the work done in one hour by an agent working at the constant rate of one kilowatt.

Since such an agent does 1000 J of work each second, the work done in 1 hr is $3600 \times 1000 =$ 3,600,000 J:

$$1\ \text{kwh} = 3.6 \times 10^6\ \text{J} = 3.6\ \text{MJ}.$$

Note that the horsepower-hour and the kilowatt-hour are units of *work* or *energy*, not power.

Although energy is an abstract physical quantity, it nevertheless has a monetary value. A newton of force or a meter per second of velocity are not things which are bought and sold as such, but a kilowatt-hour of energy is a quantity offered for sale at a definite market rate. In the form of electrical energy, a kilowatt-hour can be purchased at a price varying from a few tenths of a cent to a few cents, depending on the locality and the quantity purchased.

7-10 POWER AND VELOCITY

Suppose that a force $\mathbf{F}$ is exerted on a particle while the particle moves a distance Δs along its path. If $F_{\parallel}$ is the magnitude of the component of $\mathbf{F}$ tangent to the path, then the work of $\mathbf{F}$ is given by $\Delta W = F_{\parallel}\,\Delta s$, and the average power is

$$\bar{P} = \frac{\Delta W}{\Delta t} = F_{\parallel}\frac{\Delta s}{\Delta t} = F_{\parallel}\bar{v}.$$

The instantaneous power is therefore

$$P = F_{\parallel}v, \qquad (7\text{--}19)$$

where v is the instantaneous velocity. Equation (7–19) may also be written in terms of the scalar product:

$$P = \mathbf{F} \cdot \mathbf{v}. \qquad (7\text{--}20)$$

Example   A jet airplane engine develops a thrust of 15,000 N (roughly 3000 lb). When the plane is flying at 300 m s^{-1} (roughly 600 mph), what horsepower is developed?

$$P = Fv = (1.5 \times 10^4\ \text{N})(300\ \text{m s}^{-1}) = 4.5 \times 10^6\ \text{W}$$

$$= \frac{4.5 \times 10^6}{746}\ \text{hp} = 6030\ \text{hp}.$$

7–11 MASS AND ENERGY

At sufficiently high speeds (comparable to the speed of light) the laws of newtonian mechanics are no longer precisely correct but must be replaced by the more general relations predicted by the special relativity, to be studied in Chapter 14. One aspect of this generalization is that kinetic energy is no longer precisely $\frac{1}{2}mv^2$, but is given instead by

$$E_k = \frac{mc^2}{\sqrt{1 - v^2/c^2}} - mc^2, \qquad (7\text{–}21)$$

where c is the speed of light, $c = 3.00 \times 10^8 \text{ m s}^{-1}$.

This appears quite different from $\frac{1}{2}mv^2$, but when v is small it reduces to $\frac{1}{2}mv^2$. To see this, we apply the binomial theorem to the radical in Eq. (7–21):

$$(1 - a)^{-\frac{1}{2}} = 1 + \tfrac{1}{2}a + \tfrac{3}{8}a^2 + \cdots,$$

where, in this case, $a = v^2/c^2$. Equation (7–21) becomes

$$E_k = mc^2\left(1 + \frac{v^2}{2c^2} + \frac{3v^4}{8c^4} + \cdots\right) - mc^2$$

$$= \tfrac{1}{2}mv^2 + \tfrac{3}{8}m\left(\frac{v^4}{c^2}\right) + \cdots$$

When v is much smaller than c, powers of v^2/c^2 become very small and may be neglected, and what remains is simply $\frac{1}{2}mv^2$.

The relativistic kinetic energy expression was predicted theoretically by Lorentz and Einstein, and has been verified experimentally with a variety of observations on rapidly moving particles such as electrons. One interesting feature is that the expression becomes very large as v becomes nearly equal to c, and predicts an infinitely large energy at $v = c$. This suggests physically that no particle may have a speed as large as c, since this would require an infinite quantity of energy. The relation of Eq. (7–21) to the nonrelativistic expression $\frac{1}{2}mv^2$ is shown in Fig. 7–15.

Equation (7–21) may be reinterpreted in an interesting way. Suppose a particle with mass m has

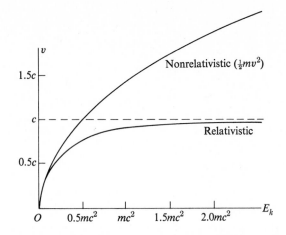

Fig. 7–15 *Relation between the speed of a particle and its kinetic energy. The speed never exceeds c, no matter how great the kinetic energy. At speeds small compared to c, the kinetic energy is approximately $(1/2)mv^2$, but at larger speeds this expression deviates more and more from the relativistically correct equation.*

an intrinsic energy equal to mc^2, not related directly to its speed. Then the total energy of a moving particle is the sum

$$E = E_k + mc^2 = \frac{mc^2}{\sqrt{1 - v^2/c^2}}. \qquad (7\text{–}22)$$

We may then call mc^2 the "rest energy" of the particle. The validity of this concept may be tested by considering situations in which the mass of a particle changes; we may investigate whether, when mass changes, there is a corresponding change in kinetic energy, so that the *total* energy is conserved.

Here is an example from the field of nuclear physics. When the nucleus of a lithium atom is struck by a rapidly moving proton (the nucleus of a hydrogen atom), a momentary union of the two nuclei takes place, after which the compound nucleus breaks up into two alpha particles. (Alpha particles are the nuclei of helium atoms.) The alpha particles recoil in almost opposite directions and move initially with very high velocities. Their combined kinetic energy is much *greater* than the kinetic energy of the original proton. The source of this kinetic energy is the so-called "binding energy" of the nuclear particles, which is a form of internal potential energy. That is, the internal potential energy of the assemblage of protons and neutrons that makes up the unstable composite nucleus is larger than the potential energy when the same number of particles is combined in the form of two helium nuclei. A crude analogy is that of two masses forced apart by a compressed spring, but tied together by a cord. If the cord is cut, the internal potential energy of the spring is transformed into kinetic energy of the recoiling masses.

The mass of a proton is 1.6715×10^{-27} kg. The mass of a lithium nucleus is 11.6399×10^{-27} kg, and that of an alpha particle is 6.6404×10^{-27} kg. Hence the mass of the original system is

$$(1.6715 + 11.6399) \times 10^{-27}\, \text{kg} = 13.3114 \times 10^{-27}\, \text{kg}.$$

The mass of the two alpha particles is

$$2(6.6404 \times 10^{-27}\, \text{kg}) = 13.2808 \times 10^{-27}\, \text{kg}.$$

Thus, the final mass is less than the initial by 0.0306×10^{-27} kg, corresponding to an energy difference $(0.0306 \times 10^{-27}\, \text{kg})(3.00 \times 10^8\, \text{ms}^{-1})^2 = 0.275 \times 10^{-11}$ J. In order for energy to be conserved, the final total kinetic energy must be *greater* than the initial by this amount. Experiments have confirmed this prediction and hence also confirmed the validity of the "rest energy" concept.

Mass changes occur also in ordinary chemical reactions, but they are so small as not to be directly observable. For example, when two molecules of hydrogen H_2, and one of oxygen O_2, combine to form two molecules of water (H_2O), an energy of about 4×10^{-19} J is released for each water molecule produced. This corresponds to a mass change of $4 \times 10^{-19}\, \text{J}/c^2 = 0.44 \times 10^{-37}$ kg. Since the mass of a water molecule is about 3×10^{-26} kg, this is a mass change of less than 1 part in 10^{11}; such small mass changes in chemical reactions have never been observed, but in nuclear reactions the mass changes may be as large as 10^{-4}, roughly ten million times as great, and are readily observable.

Thus the concepts of *mass* and *energy* emerge as two aspects of a single conserved entity that may be called *mass–energy*. The relation between mass

changes and energy changes can also be described by reinterpreting Eq. (7–21) to mean that the mass of a body actually depends on its speed. In this case, mass increase with speed is another way of describing kinetic energy. Although the concept of "relativistic mass increase" is widely used in the literature, it can be misleading. In any event, it is not necessary, and will not be used in this book.

PROBLEMS

7–1 The locomotive of a freight train exerts a constant force of 60,000 N on the train while drawing it at 50 km hr^{-1} on a level track. How much work does it do in a distance of 1 km?

7–2 A 40-kg block is pushed a distance of 5 m along a level floor at constant speed by a force at an angle of 30° below the horizontal. The coefficient of friction between block and floor is 0.25. How much work is done by the force?

7–3 A horse is towing a canal boat, the towrope making an angle of 10° with the towpath. If the tension in the rope is 100 lb, how much work is done while moving 100 ft along the towpath?

7–4 A block is pushed 2 m along a fixed horizontal surface by a horizontal force of 2N. The opposing force of friction is 0.4 N.

a) How much work is done by the 2-N force?

b) What is the work of the friction force?

7–5 A body is attracted toward the origin with a force given by $F = -6x^3$ where F is in pounds and x in feet. (a) What force is required to hold the body at point a, 1 ft from the origin? (b) At point b, 2 ft from the origin? (c) How much work must be done to move the body from point a to point b?

7–6 The force exerted by a gas in a cylinder on a piston whose area is A is given by $F = pA$, where p is the force per unit area, or *pressure*. The work W in a displacement of the piston from x_1 to x_2 is

$$W = \int_{x_1}^{x_2} F\, dx = \int_{x_1}^{x_2} pA\, dx = \int_{V_1}^{V_2} p\, dV,$$

where dV is the accompanying infinitesimal change of volume of the gas.

a) During an expansion of a gas at constant temperature (isothermal) the pressure depends on the volume according to the relation

$$p = \frac{nRT}{V},$$

where n and R are constants and T is the constant temperature. Calculate the work in expanding isothermally from volume V_1 to volume V_2.

b) During an expansion of a gas at constant entropy (adiabatic) the pressure depends on the volume according to the relation

$$p = \frac{K}{V^{\gamma}},$$

where K and γ are constants. Calculate the work in expanding adiabatically from V_1 to V_2.

7–7

a) Compute the kinetic energy of an 1800-lb automobile traveling at 30 mi hr^{-1}.

b) How many times as great is the kinetic energy if the velocity is doubled?

7–8 Compute the kinetic energy, in joules, of a 2-g rifle bullet traveling at 500 m s^{-1}.

7–9 An electron strikes the screen of a cathode-ray tube with a velocity of 10^7 m s^{-1}. Compute its kinetic energy in joules. The mass of an electron is 9×10^{-31} kg.

7–10 What is the potential energy of an 800-kg elevator at the top of the Empire State Building 380 m above street level? Assume the potential energy at street level to be zero.

7–11 What is the increase in potential energy of a 1-kg body when lifted from the floor to a table 1 m high?

7–12 A meter stick whose mass is 0.3 kg is pivoted at one end, as in Fig. 7–16, and displaced through an angle of 60°. What is the increase in its potential energy?

60°

Fig. 7–16

7–13 The force in newtons required to stretch a certain spring a distance of x m beyond its unstretched length is given by $F = 100x$.

a) What force will stretch the spring 0.1 m? 0.2 m? 0.4 m?

b) How much work is required to stretch the spring 0.1 m? 0.2 m? 0.4 m?

7–14 The scale of a certain spring balance reads from zero to 1200 N and is 0.1 m long.

a) What is the potential energy of the spring when it is stretched 0.1 m? 0.05 m?

b) When a 60-kg mass hangs from the spring?

7–15 A body moves a distance of 10 m under the action of a force which has the constant value of 5.5 N for the first 6 m and then decreases to a value of 2 N, as shown by the graph in Fig. 7–17.

a) How much work is done in the first 6 m of the motion?

b) How much work is done in the last 4 m?

Fig. 7–17

7–16 A block weighing 16 lb is pushed 20 ft along a horizontal frictionless surface by a horizontal force of 8 lb. The block starts from rest.

a) How much work is done? What becomes of this work?

b) Check your answer by computing the acceleration of the block, its final velocity, and its kinetic energy.

7–17 In the preceding problem, suppose the block had an initial velocity of 10 ft s^{-1}, other quantities remaining the same.

a) How much work is done?

b) Check by computing the final velocity and the increase in kinetic energy.

7–18 A 5-kg block is lifted vertically at a constant velocity of 4 m s^{-1} through a height of 12 m.

a) How great a force is required?

b) How much work is done? What becomes of this work?

7–19 A 12-kg block is pushed 20 m up the sloping surface of a plane inclined at an angle of 37° to the horizontal, by a constant force F of 120 N acting parallel to the plane. The coefficient of friction between the block and plane is 0.25.

a) What is the work of the force F?

b) Compute the increase in kinetic energy of the block.

c) Compute the increase in potential energy of the block.

d) Compute the work done against friction. What becomes of this work?

e) What can you say about the sum of (b), (c), and (d)?

7–20 A man of mass 80 kg sits on a platform suspended from a movable pulley and raises himself by a rope passing over a fixed pulley (Fig. 7–18). Assuming no friction losses, find

a) the force he must exert,

b) the increase in his energy when he raises himself 1 m. Answer part (b) by calculating his increase in potential energy, and also by computing the product of the force on the rope and the length of rope passing through his hands.

Fig. 7–18

7–21 A barrel of mass 120 kg is suspended by a rope 10 m long.

a) What horizontal force is necessary to hold the barrel sideways 2 m from the vertical?

b) How much work is done in moving it to this position?

7–22 The system in Fig. 7–19 is released from rest with the 12-kg block 3 m above the floor. Use the principle of conservation of energy to find the velocity with which the block strikes the floor. Neglect friction and inertia of the pulley.

Fig. 7–19

7–23 The spring of a spring gun has a force constant of 500 N m^{-1}. It is compressed 0.05 m and a ball of mass 0.01 kg is placed in the barrel against the compressed spring.

a) Compute the maximum velocity with which the ball leaves the gun when released.

b) Determine the maximum velocity if a constant resisting force of 10 N acts on the ball.

7–24 A block of mass 1 kg is forced against a horizontal spring of negligible mass, compressing the spring an amount $x_1 = 0.2$ m. When released, the block moves on a horizontal tabletop a distance $x_2 = 1.0$ m before coming to rest. The spring constant k is 100 N m^{-1} (Fig. 7–20). What is the coefficient of friction, μ, between the block and the table?

$x_1 = 0.2$ m

$x_2 = 1.0$ m

Fig. 7–20

7–25 A 2-kg block is dropped from a height of 0.4 m onto a spring whose force constant k is 1960 N m^{-1}. Find the maximum distance the spring will be compressed.

7-26 A 16-lb projectile is fired from a gun with a muzzle velocity of 800 ft s^{-1} at an angle of departure of 45°. The angle is then increased to 90° and a similar projectile is fired with the same muzzle velocity.

a) Find the maximum height attained by each projectile.

b) Show that the total energy at the top of the trajectory is the same in the two cases.

c) Using the energy principle, find the height attained by a similar projectile if fired at an angle of 30°.

7–27 A block of mass 2 kg is released from rest at point A on a track which is one quadrant of a circle of radius 1 m (Fig. 7–21). It slides down the track and reaches point B with a velocity of 4 m s^{-1}. From point B it slides on a level surface a distance of 3 m to point C, where it comes to rest.

a) What was the coefficient of sliding friction on the horizontal surface?

b) How much work was done against friction as the body slid down the circular arc from A to B?

A

B C

Fig. 7–21

7–28 A small sphere of mass m is fastened to a weightless string of length 0.5 m to form a pendulum. The pendulum is swinging so as to make a maximum angle of 60° with the vertical.

a) What is the velocity of the sphere when it passes through the vertical position?

b) What is the instantaneous acceleration when the pendulum is at its maximum deflection?

7–29 A ball is tied to a cord and set in rotation in a vertical circle. Prove that the tension in the cord at the lowest point exceeds that at the highest point by six times the weight of the ball.

7–30 A small body of mass m slides without friction around the loop-the-loop apparatus shown in Fig. 7–22. It starts from rest at point A at a height $3R$ above the bottom of the loop. When it reaches point B at the end of a horizontal diameter of the loop, compute

a) its radial acceleration,

b) its tangential acceleration, and

c) its resultant acceleration. Show these accelerations in a diagram, approximately to scale.

Fig. 7–22

7–31 A meter stick, pivoted about a horizontal axis through its center, has a body of mass 2 kg attached to one end and a body of mass 1 kg attached to the other. The mass of the meter stick can be neglected. The system is released from rest with the stick horizontal. What is the velocity of each body as the stick swings through a vertical position?

7-32 A variable force P is maintained tangent to a frictionless cylindrical surface of radius a, as shown in Fig. 7–23. By slowly varying this force, a block of weight w is moved and the spring to which it is attached is stretched from position 1 to position 2. The spring, of force constant k, is unstretched in position 1. Calculate the work of the force P, (a) by integration, (b) by use of the energy principle.

7–33 A 5-kg block is pushed up a frictionless plane inclined at 30° to the horizontal. It is pushed 2.0 m along the plane by a constant force parallel to the plane, of magnitude

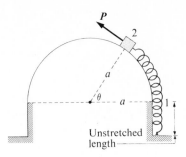

Unstretched length

Fig. 7–23

100 N. If its speed at the bottom is 2 m s^{-1}, what is its speed at the top?

7–34 In Problem 7–33, suppose the plane is not frictionless but has $\mu = 0.1$.

a) What is the speed of the block at the top?

b) What fraction of the work done by the force is dissipated by friction?

7–35 An 80-kg man jumps from a height of 2 m onto a platform mounted on springs. As the springs compress, the platform is pushed down a maximum distance of 0.2 m below its initial position, and it then rebounds.

a) What is the man's speed at the instant the platform is depressed 0.1 m?

b) If the man had just stepped gently onto the platform how much would it have been pushed down?

7–36 In Problem 7–30, if point A is at a height h (not equal to $3R$) above the bottom of the loop, what is the minimum value of h (in terms of R) so that the body moves around the loop without falling off at the top?

7–37 A ball of mass 0.5 kg is tied to a string of length 1.0 m, and the other end of the string is tied to a rigid support. The ball is held straight out from the point of support, with the string pulled taut, and is then released.

a) What is the speed of the ball at the lowest point of its motion?

b) What is the tension in the string at this point?

7–38 A skier starts at the top of a large frictionless spherical snowball, with very small initial velocity, and skis straight down the side. At what point does he lose contact with the snowball and fly off at a tangent? That is, at the instant he loses contact with the snowball, what angle does a radial line from the center to the skier make with the vertical?

7–39 A small 4-lb body is fastened to a weightless rod 5 ft long to form a pendulum, as shown in Fig. 7–24. The body is pulled aside until the rod makes an angle of 53° with the vertical.

a) With what tangential speed v_A must the body be started from point A so that it will reach C, the highest point, with a tangential speed of 10 ft s^{-1}?

b) With what speed does it pass through the lowest point B when started from A with the speed v_A?

c) What is the tension in the rod as the body passes through B?

d) If the body is started from A with the tangential speed v_A of part (a) in the direction opposite to that shown, with what speed will it then arrive at C?

Imagine that the rod is replaced by a weightless string and that the body is given the same speed v_A as in part (a) in the same direction as that shown in Fig. 7–24.

e) At what point D does the string become slack?

f) What is the speed at D?

g) What happens after the body passes D?

h) What is the minimum speed that the body can have at the top point C when tied to a string?

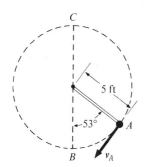

Fig. 7–24

7–40 The potential energy of a diatomic molecule, as given by Lennard-Jones, is the following function of the distance r between the atoms:

$$E_p(r) = \epsilon_0\left[\left(\frac{r_0}{r}\right)^{12} - 2\left(\frac{r_0}{r}\right)^{6}\right].$$

Prove that (a) r_0 is the intermolecular separation when the potential energy is a minimum, (b) the minimum potential energy is $-\epsilon_0$, and (c) the intermolecular separation when $E_p(r) = 0$ is equal to $r_0/\sqrt[6]{2}$. (d) Sketch the graph of $E_p(r)$.

7–41 A particle originally at rest at the origin is constrained to move along the x-axis. Its potential energy is a function of x, $E_p(x)$, and its total energy is a constant E. Prove that

the time t to go to a point where the coordinate is x is

$$t = \int_0^x \frac{dx}{\sqrt{(2/m)[E - E_p(x)]}}.$$

7–42 An earth satellite of mass m revolves in a circle at a height above the earth equal to twice the earth's radius, $2R$. In terms of m, R, the gravitational constant G, and the mass of the earth m_E, what are the values of (a) the kinetic energy of the satellite, (b) its gravitational potential energy, and (c) its total mechanical energy?

7–43 What average horsepower is developed by an 80-kg man when climbing in 10 s a flight of stairs which rises 6 m vertically? Express this power in watts and kilowatts.

7–44 The hammer of a pile driver has a mass of 500 kg and must be lifted a vertical distance of 2 m in 3 s. What horsepower engine is required?

7–45 A ski tow is to be operated on a 37° slope 800 ft long. The rope is to move at 8 mi hr^{-1} and power must be provided for 80 riders at one time, with average weight 150 lb. Estimate the horsepower required to operate the tow.

7–46

a) If energy costs 5 cents per kwh, how much is one horsepower-hour worth?

b) How many ft-lb can be purchased for one cent?

7–47 Compute the monetary value of the kinetic energy of the projectile of a 14-in. naval gun, at the rate of 2 cents per kwh. The projectile has a mass of 600 kg and its muzzle velocity is 800 m s^{-1}.

7–48 At 5 cents per kwh, what does it cost to operate a 10-hp motor for 8 hr?

7–49 The engine of an automobile develops 20 hp when the automobile is traveling at 50 km hr^{-1}.

a) What is the resisting force in newtons?

b) If the resisting force is proportional to the velocity, what horsepower will drive the car at 25 km hr^{-1}? At 100 km hr^{-1}?

7–50 The engine of a motorboat delivers 40 hp to the propeller while the boat is moving at 20 mi hr^{-1}. What would be the tension in the towline if the boat were being towed at the same speed?

7–51 A man whose mass is 70 kg walks up to the third floor of a building. This is a vertical height of 12 m above the street level.

a) How many joules of work has he done?

b) By how much has he increased his potential energy?

c) If he climbs the stairs in 20 s, what was his rate of working in horsepower?

7–52 A pump is required to lift 800 kg (about 200 gallons) of water per minute from a well 10 m deep and eject it with a speed of 20 m s^{-1}.

a) How much work is done per minute in lifting the water?

b) How much in giving it kinetic energy?

c) What horsepower engine is needed?

7–53 A 2000-kg elevator starts from rest and is pulled upward with a constant acceleration of 4 m s^{-2}.

a) Find the tension in the supporting cable.

b) What is the velocity of the elevator after it has risen 15 m?

c) Find the kinetic energy of the elevator 3 s after it starts.

d) How much is its potential energy increased in the first 3 s?

e) What horsepower is required when the elevator is traveling 8 m s^{-1}?

7–54 An automobile of mass 1200 kg has a speed of 30 m s^{-1} on a horizontal road when the engine is developing 50 hp. What is its speed, with the same horsepower, if the road rises 1 m in 20 m? Assume all friction forces to be constant.

7–55

a) If 20 hp are required to drive a 1200 kg automobile at 50 km hr^{-1} on a level road, what is the retarding force of friction, windage, etc?

b) What power is necessary to drive the car at 50 km hr^{-1} up a 10 percent grade (ie., one rising 10 m vertically in 100 m horizontally)?

c) What power is necessary to drive the car at 50 km hr^{-1} *down* a 2 percent grade?

d) Down what grade would the car coast at 50 km hr^{-1}?

7–56 The total consumption of electrical energy in the United States is of the order of 10^{19} joules per year.

a) What is the average rate of energy consumption, in watts?

b) If the population of the United States is 200 million, what is the average rate of energy consumption per person?

7–57 The sun transfers energy to the earth by radiation, at a rate of approximately 1.4 kW per square meter of surface. If this energy could be collected and converted to electrical energy with 100 percent efficiency, how great an area would be required to collect the energy cited in the previous problem?

7–58

a) Calculate the distance from the surface of the moon to the point P where the gravitational fields of the moon and of the earth cancel.

b) If the potential energy of a 1-kg object is zero at an infinite distance from the moon and the earth, what is its potential energy at point P?

c) Assume that the Apollo 8 space capsule acquired a speed of 12,000 m s^{-1} while still close to the earth and then coasted on an orbit designed to bring it close to the moon. What would be its speed at point P?

d) What would be its speed almost at the surface of the moon?

e) At what speed would a space vehicle execute a circular orbit a short distance from the surface of the moon?

f) What would be the period of the motion in part (e)? [*Note*: The distance between the centers of the moon and of the earth is 3.84×10^8 m; the radius of the moon is 1.74×10^6 m; the radius of the earth is 6.37×10^6 m; the mass of the moon is 7.34×10^{22} kg; the mass of the earth is 5.98×10^{24} kg; the gravitational constant is 6.67×10^{-11} N m^2 kg^{-2}.]

7–59 A nuclear bomb containing 20 kg of plutonium explodes. The rest mass of the products of the explosion is less than the original rest mass by one ten-thousandth of the original rest mass.

a) How much energy is released in the explosion?

b) If the explosion takes place in 1 μs, what is the average power developed by the bomb?

c) How much water could the released energy lift to a height of 1 km?

7–60 In a hypothetical nuclear-fusion reactor two deuterium nuclei combine or "fuse" to form one helium nucleus. The mass of a deuterium nucleus, expressed in atomic mass units (u) is 2.0147 u; that of a helium nucleus is 4.0039 u

$(1 \text{ u} = 1.66 \times 10^{-27} \text{ kg})$.

a) How much energy is released when 1 kg of deuterium undergoes fusion?

b) The annual consumption of electrical energy in the United States is of the order of 10^{19} J. How much deuterium must react to produce this much energy?

7–61 Compute the kinetic energy of an electron (mass 9.11×10^{-31} kg) using both the nonrelativistic and relativistic expressions, and compare the two results, for speeds of

a) $1.0 \times 10^7 \text{ m s}^{-1}$;

b) $2.0 \times 10^8 \text{ m s}^{-1}$.

Chapter 8

Impulse and Momentum

8–1 IMPULSE AND MOMENTUM

In the preceding chapter, the concepts of work and energy were developed from Newton's laws of motion. We shall see next how two similar concepts, those of *impulse* and *momentum*, also arise from these laws.

Let us consider again a particle of mass m moving along a straight line. For the moment we assume that the force F on this particle is constant and directed along the line of motion. If the particle's velocity at some initial time $t = 0$ is v_0, the velocity v at a later time t is given by

$$v = v_0 + at,$$

where the constant acceleration a is given by F/m. Making this substitution in the above equation, multiplying through by m, and rearranging, we obtain

$$Ft = mv - mv_0. \qquad (8\text{--}1)$$

The lefthand side, the product of the force and the time during which it acts, is called the *impulse* of the force. More generally, if a constant force F acts from time t_1 to time t_2, the impulse of the force is defined to be

$$\boxed{\text{Impulse} = F(t_2 - t_1).}$$

The right side of Eq. (8–1) contains the product of mass and velocity of the particle at two different times. This product is also given a special name, *momentum*. It is also sometimes called *linear momentum* to distinguish it from a similar quantity called *angular* momentum, to be discussed later. The symbol p is often used for momentum.

$$\boxed{\text{Momentum} = p = mv.}$$

In terms of these newly-defined quantities, the content of Eq. (8–1) is that the impulse of the force from time zero to time t is equal to the change of momentum during that interval, that is, the momentum at the end of the interval minus that at the beginning. We see that there is nothing special about the two times zero and t, so we may also say that if the particle's velocity at time t_1 is v_1, and its velocity at time t_2 is v_2, then

$$\boxed{F(t_2 - t_1) = mv_2 - mv_1.} \qquad (8\text{--}2)$$

The above relation between impulse and momentum change has the same form as the relation between work and kinetic-energy change developed in Chapter 7. There are important differences, however. First, impulse is a product of a force and a *time* interval, while work is a product of a force and a distance, and depends on the angle between force and displacement. Furthermore, both force and velocity are *vector* quantities, so impulse and momentum are also vector quantities, unlike work and kinetic energy, which are scalars. In motion along a straight line, where only one component of a vector is involved, the force and velocity may have components along this line which are either positive or negative.

Example 1   A particle of mass 2 kg moves along the x-axis with an initial velocity of 3 m s^{-1}. A force $F = -6$ N (i.e., in the negative x-direction) is applied for a period of 3 s. Find the final velocity.

From Eq. (8–2),

$$(-6 \text{ N})(3 \text{ s}) = (2 \text{ kg})v_2 - (2 \text{ kg})(3 \text{ m s}^{-1})$$

or

$$v_2 = -6 \text{ m s}^{-1}.$$

The particle's final velocity is in the negative x-direction.

The unit of impulse, in any system, equals the product of the units of force and time in that system. Thus, in the mks system, the unit is one newton second (1 N · s), in the cgs system it is one dyne second (1 dyn · s), and in the British system it is one pound second (1 lb · s).

The unit of momentum in the mks system is one kilogram meter per second (1 kg m s^{-1}), in the cgs system it is one gram centimeter per second (1 g · cm · s^{-1}), and in the British system it is one slug foot per second (1 slug · ft · s^{-1}). Since

$$1 \text{ kg} \cdot \text{m} \cdot \text{s}^{-1} = (1 \text{ kg} \cdot \text{m} \cdot \text{s}^{-2})(1 \text{ s}) = 1 \text{ N} \cdot \text{s},$$

it follows that the unit of momentum in any system equals the unit of impulse in that system.

The concept of impulse can be generalized to include forces that vary with time; both the magnitude and the direction of the force may vary. We consider a particle of mass m moving in the xy-plane, as in Fig. 8–1, acted on by a varying resultant force F. Newton's second law states that

$$F = m \frac{dv}{dt},$$

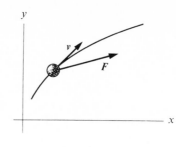

Fig. 8–1 *Particle moving in the xy-plane.*

or

$$F \, dt = m \, dv.$$

If v_1 is the velocity when $t = t_1$, and v_2 the velocity when $t = t_2$, it follows that

$$\int_{t_1}^{t_2} F \, dt = \int_{v_1}^{v_2} m \, dv. \qquad (8\text{–}3)$$

The integral on the left is the *impulse of the force F* in the time interval $t_2 - t_1$, and is a *vector quantity:*

$$\boxed{\text{Impulse} = \int_{t_1}^{t_2} F \, dt.}$$

This integral can be evaluated, of course, only when the force is known as a function of the time. The integral on the right, however, always yields the result

$$\int_{v_1}^{v_2} m \, dv = mv_2 - mv_1.$$

Thus Eq. (8–3) may be written as

$$\int_{t_1}^{t_2} F \, dt = mv_2 - mv_1, \qquad (8\text{–}4)$$

In the special case when F is constant, this reduces to Eq. (8–2).

Equation (8–4) is also equivalent (for forces and velocities in the xy-plane) to the two scalar equations

$$\int_{t_1}^{t_2} F_x \, dt = mv_{x_2} - mv_{x1},$$

$$\int_{t_1}^{t_2} F_y \, dt = mv_{y_2} - mv_{y1}, \qquad (8\text{–}5)$$

The impulse of any force component, or any force whose direction is constant, can be represented graphically by plotting the force vertically and the time horizontally, as in Fig. 8–2. The *area* under the curve, between vertical lines at t_1 and t_2, is equal to the impulse of the force in this time interval.

If the impulse of a force is *positive* the momentum of the body on which it acts *increases* algebraically. If the impulse is *negative*, the momentum *decreases*. If the impulse is *zero*, there is no change in momentum.

Example 2 Consider the changes in momentum produced by the following forces:

a) A body moving on the *x*-axis is acted on for 2 s by a constant force of 10 N toward the right.

b) The body is acted on for 2 s by a constant force of 10 N toward the right and then for 2 s by a constant force of 20 N toward the left.

c) The body is acted on for 2 s by a constant force of 10 N toward the right, then for 1 s by a constant force of 20 N toward the left. The three forces are shown graphically in Fig. 8–3.

a) The impulse of the force is $(+10 \text{ N})(2 \text{ s})$ $= +20 \text{ N s}$. Hence the momentum of *any* body on which the force acts increases by 20 kg m s^{-1}. This change is the same whatever the mass of the body and whatever the magnitude and direction of its initial velocity.

Suppose the mass of the body is 2 kg and that it is initially at rest. Its *final* momentum then equals its *change* in momentum and its final *velocity* is 10 m s^{-1} toward the right. (The reader should verify by computing the acceleration.)

Had the body been initially moving toward the right at 5 m s^{-1}, its initial momentum would have been 10 kg m s^{-1} toward the right.

Had the body been moving initially toward the *left* at 5 m s^{-1}, its initial momentum would have been -10 kg m s^{-1}, its final momentum $+10$ kg m s^{-1}, and its final velocity 5 m s^{-1} toward the right. That is, the

constant force of 10 N toward the right would first have brought the body to rest and then given it a velocity in the direction opposite to its initial velocity.

b) The impulse of this force is

$$(+10 \text{ N})(2 \text{ s}) - (20 \text{ N})(2 \text{ s}) = -20 \text{ N s}.$$

The momentum of *any* body on which it acts is decreased by 20 kg m s^{-1}. The reader should examine various possibilities, as in the preceding example.

c) The impulse of this force is

$$(+10 \text{ N})(2 \text{ s}) - (20 \text{ N})(1 \text{ s}) = 0.$$

Hence the momentum of any body on which it acts is not changed. Of course, the momentum of the body is increased during the first 2 s but it is *decreased* by an equal amount in the next second. As an exercise, describe the motion of a body of mass 2 kg, moving initially to the left at 5 m s^{-1}, and acted on by this force. It will help to construct a graph of velocity versus time.

Example 3 A ball of mass 0.4 kg is thrown against a brick wall. When it strikes the wall it is moving horizontally to the left at 30 m s^{-1}, and it rebounds horizontally to the right at 20 m s^{-1}. Find the impulse of the force exerted on the ball by the wall.

The initial momentum of the ball is (0.4 kg) $(-30 \text{ m s}^{-1}) = -12 \text{ kg m s}^{-1}$. The final momentum is $+8$ kg m s^{-1}. The *change* in momentum is

$$mv_2 - mv_1 = 8 \text{ kg ms}^{-1} - (-12 \text{ kg m s}^{-1})$$

$$= 20 \text{ kg ms}^{-1}.$$

Fig. 8–2

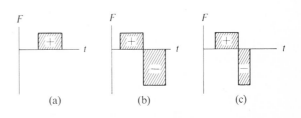

Fig. 8–3

Hence, the impulse of the force exerted on the ball was 20 N s. Since the impulse is *positive*, the force must be toward the right.

Note that the *force* exerted on the ball cannot be found without further information regarding the collision. The general nature of the force–time graph is shown by one of the curves in Fig. 8–4. The force is zero before impact, rises to a maximum, and decreases to zero when the ball leaves the wall. If the ball is relatively rigid, like a baseball, the time of collision is small and the maximum force is large, as in curve (a). If the ball is more yielding, like a tennis ball, the collision time is larger and the maximum force is less, as in curve (b). In any event, the *area* under the force–time graph must equal 20 N s.

For an idealized case in which the force is constant and the collision time is 1 m s (10^{-3} s), as represented by the horizontal straight line, the force is 20,000 N.

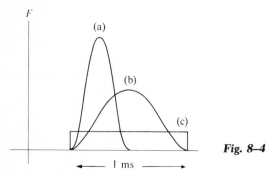

Fig. 8–4

8–2 CONSERVATION OF MOMENTUM

Whenever there is a force of interaction between two particles, the momentum of each particle is changed as a result of the force exerted on it by the other. (The force may be gravitational, electric, magnetic, or of any other origin.) Furthermore, since, by Newton's *third* law, the force on one particle is always equal in magnitude and opposite in direction to that on the other, the impulses of the forces are equal in magnitude and opposite in direction. It follows that *the vector change in momentum of either particle*, in any time interval, *is equal in magnitude and opposite in direction to the vector change in momentum of the other*. The *net change in momentum of the* **system** (the two particles together) *is therefore zero*.

The pair of action–reaction forces are *internal* forces of the system, and we conclude that *the total momentum of a system of bodies cannot be changed by internal forces between the bodies*. Hence, if the *only* forces acting on the particles of a system are *internal* forces (that is, if there are no *external* forces), the total momentum of the system remains constant in magnitude and direction. This is the *principle of conservation of linear momentum*: *When no resultant external force acts on a system, the total momentum of the system remains constant in magnitude and direction.*

The principle of conservation of momentum is one of the most fundamental and important principles of mechanics. Note that it is more general than the principle of conservation of mechanical energy; mechanical energy is conserved *only* when the internal forces are *conservative*. The principle of conservation of momentum holds whatever the nature of the internal forces, conservative or not.

Example 1   Figure 8–5 shows a body A of mass m_A moving toward the right on a level frictionless surface (or a level air track) with a velocity v_{A1}. It collides with a second body B of mass m_B moving toward the left with a velocity v_{B1}. Since there is no friction and the resultant vertical force on the system is zero, the only forces on the bodies are the internal action–reaction forces which they exert on each other in the collision process, and the momentum of the system remains constant in magnitude and direction.

Let v_{A2} and v_{B2} represent the velocities of A and B after the collision. Then

$$m_A v_{A1} + m_B v_{B1} = m_A v_{A2} + m_B v_{B2}. \quad (8–6)$$

Fig. 8–5

Example 2   In Fig. 8–6, body A of mass m_A is initially moving toward the right with a velocity v_{A1}. It collides with body B, initially at rest, after which the bodies separate and move with velocities v_{A2} and

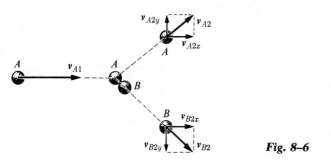

Fig. 8–6

v_{B2}. No forces act on the system except those in the collision process.

This example illustrates the *vector* nature of momentum, that is the x- and y-components of momentum are *both* conserved. Let us take the x-axis in the direction of v_{A1}. The initial x-momentum is $m_A v_{A1}$ and the initial y-momentum is zero.

The final x-momentum of the system is

$$m_A v_{A2x} + m_B v_{B2x},$$

and the final y-momentum is

$$m_A v_{A2y} - m_B v_{B2y}.$$

Therefore

$$m_A v_{A2x} + m_B v_{B2x} = m_A v_{A1}, \qquad (8\text{–}7)$$

$$m_A v_{A2y} - m_B v_{B2y} = 0. \qquad (8\text{–}8)$$

8–3 COLLISIONS

Suppose we are given the masses and initial velocities of two colliding bodies, and we wish to compute their velocities after the collision. If the collision is "head on," as in Fig. 8–5, Eq. (8–6) provides one equation for the velocities v_{A2} and v_{B2}. If it is of the type shown in Fig. 8–6, Eqs. (8–7) and (8–8) provide two equations for the four velocity components v_{A2x}, v_{A2y}, v_{B2x}, and v_{B2y}. Hence, momentum considerations *alone* do not suffice to completely determine the final velocities; we must have more information about the collision process.

If the forces of interaction between the bodies are *conservative*, the total kinetic energy is the same

before and after the collision and the collision is said to be *completely elastic*. Such a collision is closely approximated if one end of an inverted U-shaped steel spring is attached to one of the bodies, as in Fig. 8–7. When the bodies collide the spring is momentarily compressed and some of the original kinetic energy is momentarily converted to elastic potential energy. The spring then expands, and when the bodies separate, this potential energy is reconverted to kinetic energy.

At the opposite extreme from a completely elastic collision is one in which the colliding bodies stick together and move *as a unit* after the collision. Such a collision is called a *completely inelastic* collision, and will result if the bodies in Fig. 8–5 are provided with a coupling mechanism like that between two freight cars, or if the bodies in Fig. 8–7 are locked together at the instant when their velocities become equal and the spring is compressed.

Fig. 8–7

8–4 INELASTIC COLLISIONS

For the special case of a completely inelastic collision between two bodies A and B, we have, from the definition of such a collision,

$$v_{A2} = v_{B2} = v_2.$$

When this is combined with the principle of conservation of momentum, we obtain

$$m_A v_{A1} + m_B v_{B1} = (m_A + m_B) v_2, \qquad (8\text{–}9)$$

and the final velocity can be computed if the initial velocities and the masses are known.

The kinetic energy of the system before collision is

$$E_{k1} = \tfrac{1}{2} m_A v_{A1}^2 + \tfrac{1}{2} m_B v_{B1}^2.$$

The final kinetic energy is

$$E_{k2} = \tfrac{1}{2} (m_A + m_B) v_2^2.$$

For the special case in which body B is *initially* at rest, $v_{B1} = 0$, and the ratio of the final to the initial kinetic energy is

$$\frac{E_{k2}}{E_{k1}} = \frac{(m_A + m_B)v_2^2}{m_A v_{A1}^2}.$$

Inserting the expression for v_2 from Eq. (8–9), we find that this reduces to

$$\frac{E_{k2}}{E_{k1}} = \frac{m_A}{m_A + m_B}.$$

The right side is necessarily less than unity, so *the total kinetic energy always decreases in an inelastic collision.*

Example 1 Suppose the collision in Fig. 8–5 is completely inelastic and that the masses and velocities have the values shown. The velocity after the collision is then

$$v_2 = \frac{m_A v_{A1} + m_B v_{B1}}{m_A + m_B} = 0.5 \text{ m s}^{-1}.$$

Since v_2 is positive, the system moves to the right after the collision.

The kinetic energy of body A before the collision is

$$\tfrac{1}{2}m_A v_{A1}^2 = 10 \text{ J},$$

and that of body B is

$$\tfrac{1}{2}m_B v_{B1}^2 = 6 \text{ J}.$$

The total kinetic energy before collision is therefore 16 J.

Note that the kinetic energy of body B is positive, although its velocity v_{B1} and its momentum $m v_{B1}$ are both negative. The kinetic energy after the collision is $\tfrac{1}{2}(m_A + m_B)v_2^2 = 1$ J.

Hence, far from remaining constant, the final kinetic energy is only $\tfrac{1}{16}$ of the original, and $\tfrac{15}{16}$ is "lost" in the collision. If the bodies couple together like two freight cars, most of this energy is converted to elastic waves which are eventually absorbed.

If there is a spring between the bodies, as in Fig. 8–7, and the bodies are locked together when their velocities become equal, the energy is trapped as potential energy in the compressed spring. If all these forms of energy are taken into account, the *total* energy of the system is conserved although its *kinetic* energy is not. However, *momentum is always conserved* in a collision, whether or not the collision is elastic.

Example 2 The *ballistic pendulum* is a device for measuring the velocity of a bullet. The bullet is allowed to make a completely inelastic collision with a body of much greater mass. The momentum of the system immediately after the collision equals the original momentum of the bullet, but since the velocity is very much smaller, it can be determined more easily. Although the ballistic pendulum has now been superseded by other devices, it is still an important laboratory experiment for illustrating the concepts of momentum and energy.

In Fig. 8–8, the pendulum, consisting perhaps of a large wooden block of mass m', hangs vertically by two cords. A bullet of mass m, traveling with a velocity v, strikes the pendulum and remains embedded in it. If the collision time is very small compared with the time of swing of the pendulum, the supporting cords remain practically vertical during this time. Hence, no external horizontal forces act on the system during the collision, and the horizontal momentum is conserved. Then if V represents the velocity of bullet and block immediately after the collision,

$$mv = (m + m')V, \qquad v = \frac{m + m'}{m} V.$$

The kinetic energy of the system, immediately after the collision, is $E_k = \tfrac{1}{2}(m + m')V^2.$

Fig. 8–8 *The ballistic pendulum.*

The pendulum now swings to the right and upward until its kinetic energy is converted to gravitational potential energy. (Small frictional effects can be neglected.) Hence,

$$\tfrac{1}{2}(m + m')V^2 = (m + m')gy,$$

$$V = \sqrt{2gy},$$

and

$$v = \frac{m + m'}{m}(\sqrt{2gy}).$$

By measuring m, m', and y, we can compute the original velocity v of the bullet.

It is important to remember that kinetic energy is not conserved *in the collision*. The ratio of the kinetic energy of bullet and pendulum, after the collision, to the original kinetic energy of the bullet, is

$$\frac{\tfrac{1}{2}(m + m')V^2}{\tfrac{1}{2}mv^2} = \frac{m}{m + m'}.$$

Thus, if $m' = 1000$ g and $m = 1$ g, only about 0.1 percent of the original energy remains as kinetic energy; 99.9 percent is converted to internal energy.

8–5 ELASTIC COLLISIONS

Consider next a perfectly elastic "head-on" or *central* collision between two bodies A and B. The bodies separate after the collision and have different velocities, v_{A2} and v_{B2}. Since kinetic energy and momentum are *both* conserved, we have:

Conservation of kinetic energy

$$\tfrac{1}{2}m_Av_{A1}{}^2 + \tfrac{1}{2}m_Bv_{B1}{}^2 = \tfrac{1}{2}m_Av_{A2}{}^2 + \tfrac{1}{2}m_Bv_{B2}{}^2.$$

Conservation of momentum

$$m_Av_{A1} + m_Bv_{B1} = m_Av_{A2} + m_Bv_{B2}.$$

Hence if the masses and initial velocities are known, we have two independent equations, from which the final velocities can be found. Simultaneous solution of these equations gives

$$(v_{B2} - v_{A2}) = -(v_{B1} - v_{A1}), \qquad (8\text{--}10)$$

$$v_{A2} = \frac{2m_Bv_{B1} + v_{A1}(m_A - m_B)}{m_A + m_B}, \qquad (8\text{--}11)$$

$$v_{B2} = \frac{2m_Av_{A1} - v_{B1}(m_A - m_B)}{m_A + m_B}. \qquad (8\text{--}12)$$

The difference $(v_{B2} - v_{A2})$ is the velocity of B relative to A after the collision, while $(v_{B1} - v_{A1})$ is its relative velocity before the collision. Thus, from Eq. (8–10), *the relative velocity of two particles in a central, completely elastic collision is unchanged in magnitude but reversed in direction.*

For the special case in which body B is at rest before the collision, $v_{B1} = 0$ and Eqs. (8–11) and (8–12) simplify to

$$v_{A2} = \frac{m_A - m_B}{m_A + m_B}v_{A1}, \qquad v_{B2} = \frac{2m_A}{m_A + m_B}v_{A1}.$$

If the masses of A and B are equal, $v_{A2} = 0$ and $v_{B2} = v_{A1}$. That is, the first body comes to rest and the second moves off with a velocity equal to the original velocity of the first. Both the momentum and kinetic energy of the first are completely transferred to the second.

When the masses are unequal, the kinetic energies after the collision are

$$(E_{k2})_A = \tfrac{1}{2}m_Av_{A2}{}^2 = \left(\frac{m_A - m_B}{m_A + m_B}\right)^2(E_{k1})_A,$$

$$(E_{k2})_B = \tfrac{1}{2}m_Bv_{B2}{}^2 = \left(\frac{4m_Am_B}{(m_A + m_B)^2}\right)(E_{k1})_A.$$

It is of interest to derive the expression for the *fractional decrease* in kinetic energy of body a, that is, the ratio of its decrease in kinetic energy to its original kinetic energy. Since, in an elastic collision, the energy lost by A equals the energy gained by B, this ratio is

$$\frac{(E_{k2})_B}{(E_{k1})_A} = \frac{4m_Am_B}{(m_A + m_B)^2} = 4\left(\frac{m_A}{m_B}\right)\frac{1}{[1 + (m_A/m_B)]^2}.$$

The fractional decrease in kinetic energy of body A is plotted in Fig. 8–9 as a function of the ratio m_A/m_B (note the logarithmic scale). The energy loss approaches zero as m_A/m_B approaches zero (a very small body colliding with a very large one) and also as m_A/m_B approaches infinity (a very large body colliding with a very small one). In the former case, the first body rebounds with practically its original velocity. In the latter, the first body continues to

Fig. 8–9 *Fractional loss in energy in a head-on elastic collision, plotted as a function of the ratio of the masses of the colliding bodies.*

move with very nearly its original velocity. The maximum fractional energy decrease occurs when $m_A = m_B$, and for this ratio of masses the fractional energy loss equals unity, as shown above.

This fact has an important bearing on the problem of slowing down the rapidly moving neutrons in the moderator of a nuclear reactor. As neutrons pass through matter, they make occasional elastic collisions with nuclei. In a head-on collision, the greatest loss of kinetic energy results when the mass of the nucleus equals that of the neutron, that is, when the nucleus is that of ordinary hydrogen. The greater the nuclear mass, the less of the neutron's energy is transferred to the nucleus.

Example Suppose the collision illustrated in Fig. 8–5 is completely elastic. What are the velocities of A and B after the collision?

From the principle of conservation of momentum,

$$(5 \text{ kg})(2 \text{ m s}^{-1}) + (3 \text{ kg})(-2 \text{ m s}^{-1})$$
$$= (5 \text{ kg})v_{A2} + (3 \text{ kg})v_{B2},$$
$$5v_{A2} + 3v_{B2} = 4 \text{ m s}^{-1}.$$

Since the collision is completely elastic,

$$v_{B2} - v_{A2} = -(v_{B1} - v_{A1}) = 4 \text{ m s}^{-1}.$$

Solving these equations simultaneously, we obtain

$$v_{A2} = -1 \text{ m s}^{-1}, \qquad v_{B2} = 3 \text{ m s}^{-1}.$$

Both bodies therefore reverse their directions of motion, A traveling to the left at 1 m s^{-1} and B to the right at 3 m s^{-1}.

The total kinetic energy after the collision is

$$\tfrac{1}{2}(5 \text{ kg})(-1 \text{ m s}^{-1})^2 + \tfrac{1}{2}(3 \text{ kg})(3 \text{ m s}^{-1})^2 = 16 \text{ J},$$

which equals the total kinetic energy before the collision.

8–6 RECOIL

Figure 8–10 shows two blocks A and B, between which there is a compressed spring. When the system is released from rest, the spring exerts equal and opposite forces on the blocks until it has expanded to its natural unstressed length. It then drops to the surface, while the blocks continue to move. The original momentum of the system is zero, and if frictional forces can be neglected, the resultant external force on the system is zero. The momentum of the system therefore remains constant and equal to zero. Then if v_A and v_B are the velocities acquired by A and B, we have

$$m_A v_A + m_B v_B = 0, \qquad \frac{v_A}{v_B} = -\frac{m_B}{m_A}.$$

The velocities are of opposite sign and their magnitudes are inversely proportional to the corresponding masses.

The original kinetic energy of the system is also zero. The final kinetic energy is

$$E_k = \tfrac{1}{2}m_A v_A^2 + \tfrac{1}{2}m_B v_B^2.$$

The source of this energy is the original elastic potential energy of the system. The ratio of kinetic energies is

$$\frac{\tfrac{1}{2}m_A v_A^2}{\tfrac{1}{2}m_B v_B^2} = \frac{m_A}{m_B}\left(\frac{v_A}{v_B}\right)^2 = \frac{m_B}{m_A}.$$

Fig. 8–10 *Conservation of momentum in recoil.*

Thus, although the momenta are equal in magnitude, the kinetic energies are inversely proportional to the corresponding masses, the body of smaller mass receiving the larger share of the original potential energy. The reason is that the change in *momentum* of a body equals the *impulse* of the force acting on it, while the change in *kinetic energy* equals the *work* of the force. The forces on the two bodies are equal in magnitude and act for equal times, so they produce equal and opposite changes in momentum. The points of application of the forces, however, do not move through equal distances (except when $m_A = m_B$), since the acceleration, velocity, and displacement of the smaller body are greater than those of the larger. Hence more *work* is done on the body of smaller mass.

Considerations like those above also apply to the firing of a rifle. The initial momentum of the system is zero. When the rifle is fired, the bullet and the powder gases acquire a forward momentum, and the rifle (together with any system to which it is attached) acquires a rearward momentum of the same magnitude. The ratio of velocities cannot be expressed as simply as in the case discussed above. The bullet travels with a definite velocity, but different portions of the powder gases have different velocities. Because of the relatively large mass of the rifle compared with that of the bullet and powder charge, the velocity and kinetic energy of the rifle are much smaller than those of the bullet and powder gases.

In the processes of radioactive decay and nuclear fission, a nucleus splits into two or more parts, which fly off in different directions. Although the nature of the forces which act in these processes is not yet completely understood, it has been verified by many experiments that the principle of conservation of momentum applies to them also. The total (vector) momentum of the products equals the original momentum of the system. The source of the kinetic energy of the products is the so-called *binding energy* of the original nucleus, analogous to the chemical energy of the powder charge in a rifle.

Figure 8–11 shows the explosive break-up (fission) of one uranium atom into two fragments of approximately equal mass. The atom was at rest in

Fig. 8–11 *Cloud-chamber photograph of the fission of an atom of uranium. [From Boggild, Brostøm, and Lauritzen, Phys. Rev.* **59,** *275, 1941.]*

the thin foil that stretches horizontally in the picture, then it was "triggered off" by a passing neutron, and the two fission products recoiled from each other in opposite directions. While they cannot, of course, be directly observed, the two fragments give themselves away indirectly by the thin streaks of fog or clouds which can be made to condense along their paths.* It is evident that the fragments are slowed down to a stop after a few centimeters in the medium through which they have to travel, but, from such data as the length and density of the tracks, we can deduce the speeds of separation. From the law of conservation of momentum, applied to the fission fragments, we can then obtain directly the ratio of their masses, an important aid in the identification of the nuclear reaction.

In this simplified account we are neglecting the contribution to momentum of minor fission products not apparent in this picture, and of the incident neutron. In the "background" can be seen the result of collisions between other incident neutrons and the atmosphere in the cloud chamber.

---

* The equipment for this purpose, a most important tool in contemporary research, is called a *cloud chamber.* It originated in a design devised in 1895 by the British Nobel Prize physicist C. T. R. Wilson.

(a)   (b)

Fig. 8–12

8–7 ROCKET PROPULSION

A rocket is propelled by the ejection of a portion of
its mass to the rear. The forward force on the rocket
is the reaction to the backward force on the ejected
material, and as more material is ejected, the mass of
the rocket decreases. The problem is best handled by
impulse–momentum considerations. In order not to
bring in too many complicating forces, we shall
consider a rocket fired vertically upward, and neglect
air resistance and variations in g.

Figure 8–12(a) represents the rocket at a time t
after take-off, when its mass is m and its upward
velocity is v. The total momentum at this instant is
thus mv. In a short time interval dt, a mass dm of gas
is ejected from the rocket. Let v_r represent the
downward velocity of this gas *relative to the rocket*.
The velocity v' of the gas relative to the earth is then

$$v' = v - v_r,$$

and its momentum is

$$dm\, v' = dm(v - v_r).$$

At the end of the time interval dt, the mass of rocket
and unburned fuel has decreased to $m - dm$, and its
velocity has increased to $v + dv$. Its momentum is
therefore

$$(m - dm)(v + dv).$$

Thus the *total* momentum at time $t + dt$ is

$$(m - dm)(v + dv) + dm(v - v_r).$$

Figure 8–12(b) represents rocket and ejected gas at
this time.

We now make use of the impulse–momentum
relation; the product of the resultant external force F
on a system, and the time interval dt during which it
acts, is equal to the change in momentum of the
system. If air resistance is neglected, the external
force F on the rocket is its weight, $-mg$. (We take the
upward direction as positive.) The change in momentum, in time dt, is the difference between the momentum of the system at the end and at the beginning of
the time interval. Hence

$$-mg\, dt = [(m - dm)(v + dv) + dm(v - v_r)] - mv$$

$$= m\, dv - dm\, v_r + dm\, dv.$$

The term $dm\, dv$ may be dropped because it is a
product of two small quantities and thus is much
smaller than the other terms. Dropping this term,
dividing by dt, and rearranging, we obtain

$$m\left(\frac{dv}{dt}\right) = v_r\left(\frac{dm}{dt}\right) - mg. \qquad (8\text{--}13)$$

The ratio dv/dt is the acceleration of the rocket,
so the left side of this equation (mass times acceleration) equals the resultant force on the rocket. The
first term on the right equals the upward thrust on
the rocket, and the resultant force equals the difference between this thrust and the weight of the rocket,
mg. It will be seen that the upward thrust is proportional both to the relative velocity v_r of the ejected
gas and to the mass of gas ejected per unit time,
dm/dt.

The acceleration is

$$\frac{dv}{dt} = \frac{v_r}{m}\left(\frac{dm}{dt}\right) - g. \qquad (8\text{--}14)$$

As the rocket rises, the value of g decreases, according to Newton's law of gravitation. (In "outer space,"
far from all other bodies, g becomes negligibly
small.) If the values of v_r and dm/dt remain approximately constant while the fuel is being consumed and
m continuously decreases, the acceleration *increases*
until all the fuel is burned.

Equation (8–13) can be integrated to find a relation between the velocity at any time and the remaining mass. From Eq. (8–13),

$$dv = v_r \frac{dm}{m} - g\,dt.$$

Now dm is a positive quantity, representing the mass ejected in time dt, so the change in mass of the rocket in that time is $-dm$. Thus, in computing the total mass change *in the rocket* we must change the sign of the term containing dm.

Let the mass and velocity at time $t = 0$ be m_0 and v_0, respectively; then

$$\int_{v_0}^{v} dv = -\int_{m_0}^{m} v_r \frac{dm}{m} - \int_0^t g\,dt,$$

and

$$v - v_0 = -v_r \ln \frac{m}{m_0} - gt,$$

$$v = v_0 + v_r \ln \frac{m_0}{m} - gt. \qquad (8\text{–}15)$$

Example 1 In the first second of its flight, a rocket ejects 1/60 of its mass with a relative velocity of 2400 m s^{-1}. What is the acceleration of the rocket?

We have: $dm = m/60$, $dt = 1$ s. From Eq. (8–14),

$$\frac{dv}{dt} = \frac{2400 \text{ m s}^{-1}}{(60)(1 \text{ s})} - 10 \text{ m s}^{-2}$$

$$= 30 \text{ m s}^{-2}.$$

Example 2 Suppose the ratio of initial mass to final mass for the rocket above is 4 and that the fuel is consumed in a time $t = 60$ s. The velocity at the end of this time, from Eq. (8–15), is

$$v = (2400 \text{ m s}^{-1})(\ln 4) - (10 \text{ m s}^{-2})(60 \text{ s})$$

$$= 2727 \text{ m s}^{-1}.$$

At the start of the flight, when the velocity of the rocket is zero, the ejected gases are moving downward, relative to the earth, with a velocity equal to the relative velocity v_r. When the velocity of the rocket has increased to v_r, the ejected gases have a velocity zero relative to the earth. When the rocket velocity becomes greater than v_r, the velocity of the ejected gases is in the same direction as that of the rocket. Thus the velocity acquired by the rocket can be greater (and is often much greater) than the relative velocity v_r. In the example above, where the final velocity of the rocket was 2727 m s^{-1} and the relative velocity was 2400 m s^{-1}, the last portion of the ejected fuel had an upward velocity of $(2727 - 2400)\text{m s}^{-1} = 327 \text{ m s}^{-1}$.

8–8 GENERALIZATIONS

The momentum–impulse relations and the principle of conservation of momentum have been developed from Newton's laws of motion, which are strictly valid only for bodies moving with speeds small compared to the speed of light. Thus it should not be surprising that when v is *not* small compared to c, modifications are needed.

The correct relativistic generalization, as predicted theoretically by Lorentz and Einstein and also verified experimentally, is to define momentum not as mv, but as

$$\boxed{\text{Momentum} = p = \frac{mv}{\sqrt{1 - v^2/c^2}}.} \qquad (8\text{–}16)$$

When v is much smaller than c, this reduces to mv, but in general the momentum of a particle is larger than mv. Equation (8–16) can be interpreted as saying that the mass of the particle increases by the factor $1/(1 - v^2/c^2)^{1/2}$ when it is moving, but an equally valid and, in some ways, less misleading view is simply that the definition of momentum is no longer mv, but instead is given by Eq. (8–16). In either case, it is found that in collisions where there is no resultant external force, *the principle of conservation of momentum is still obeyed*, provided the generalized definition of momentum is used.

A further generalization of the concept of momentum is needed when we deal with particles that have no mass in the usual sense. One such particle is the *photon*, or *quantum* of electromagnetic wave energy, which will be discussed more fully in later chapters. An electromagnetic wave, such as light or a radio wave, transfers energy in packages of size proportional to the frequency of the radiation. For a wave of frequency f, the energy E in each package or *photon* is given by

$$E = hf, \qquad (8\text{--}17)$$

where h is a fundamental physical constant called *Planck's constant*. In mks units, $h = 6.62 \times 10^{-34}$ J s.

The theory of relativity predicts that any particle with energy must also have momentum, whether it has mass or not. For a massless particle the momentum turns out to be the energy divided by the speed of light. Thus, from Eq. (8–17) we obtain, for the momentum of a massless particle,

$$\boxed{p = \frac{hf}{c} = \frac{h}{\lambda},} \qquad (8\text{--}18)$$

where $\lambda = c/f$ is the wavelength of the radiation.

Example   A sodium atom initially at rest emits a photon of yellow light, recoiling in the opposite direction to the direction of emission. What is the atom's recoil velocity?

The wavelength of the light is about $\lambda = 589 \times 10^{-9}$ m, and the momentum of a photon is

$$p = \frac{h}{\lambda} = \frac{6.62 \times 10^{-34}\,\text{J s}}{589 \times 10^{-9}\,\text{m}} = 1.12 \times 10^{-27}\,\text{kg m s}^{-1}.$$

For conservation of momentum, the atom must recoil with a momentum of this magnitude in the opposite direction. The mass of a sodium atom is about 3.8×10^{-26} kg, so its speed after emission is

$$v = \frac{1.12 \times 10^{-27}\,\text{kg m s}^{-1}}{3.8 \times 10^{-26}\,\text{kg}} = 0.029\,\text{m s}^{-1}.$$

In conclusion, we emphasize again that the principle of conservation of momentum, like that of conservation of energy, has much greater generality than the newtonian context in which it was originally developed. As indicated briefly above, the definition of momentum can be generalized to relativistic mechanics and to quantum mechanics. With these

generalizations, conservation of momentum is believed to be a *universal* conservation law; *no exception has ever been found.*

PROBLEMS

8–1

a)   What is the momentum of a 10,000-kg truck whose velocity is 20 m s^{-1}? What velocity must a 5,000-kg truck attain in order to have (b) the same momentum, (c) the same kinetic energy?

8–2   A baseball has a mass of about 0.2 kg.

a)   If the velocity of a pitched ball is 30 m s^{-1}, and after being batted it is 50 m s^{-1} in the opposite direction, find the change in momentum of the ball and the impulse of the blow.

b)   If the ball remains in contact with the bat for 0.002 s, find the average force of the blow.

8–3   A bullet having a mass of 0.05 kg, moving with a velocity of 400 m s^{-1}, penetrates a distance of 0.1 m into a wooden block firmly attached to the earth. Assume the decelerating force constant. Compute (a) the deceleration of the bullet, (b) the decelerating force, (c) the time of deceleration, (d) the impulse of the collision. Compare the answer to part (d) with the initial momentum of the bullet.

8–4   A bullet emerges from the muzzle of a gun with a velocity of 300 m s^{-1}. The resultant force on the bullet, while it is in the gun barrel, is given by

$$F = 400 - \frac{4 \times 10^5}{3}\,t,$$

where F is in newtons and t in seconds.

a)   Construct a graph of F versus t.

b)   Compute the time required for the bullet to travel the length of the barrel, assuming the force becomes zero just at the end of the barrel.

c)   Find the impulse of the force.

d)   Find the mass of the bullet.

8–5   A box, initially sliding on the floor of a room, is eventually brought to rest by friction. Is the momentum of the box conserved? If not, does this process contradict the principle of conservation of momentum? What becomes of the original momentum of the box?

8–6   Compare the damage to an automobile (and its occupants) in the following circumstances:

a) The automobile makes a completely inelastic head-on collision with an identical automobile traveling with the same speed in the opposite direction, and (b) it makes a completely inelastic head-on collision with a vertical rock cliff.

c) Which would be worse (for the occupants of a light car), to collide head-on with a truck traveling in the opposite direction with a momentum of equal magnitude, or to collide head-on with a truck having the same kinetic energy?

8–7

a) An empty freight car of mass 10,000 kg rolls at 2 m s^{-1} along a level track and collides with a loaded car of mass 20,000 kg, standing at rest with brakes released. If the cars couple together, find their speed after the collision.

b) Find the decrease in kinetic energy as a result of the collision.

c) With what speed should the loaded car be rolling toward the empty car, in order that both shall be brought to rest by the collision?

8–8 When a bullet of mass 20 g strikes a ballistic pendulum of mass 10 kg, the center of gravity of the pendulum is observed to rise a vertical distance of 7 cm. The bullet remains embedded in the pendulum.

a) Calculate the original velocity of the bullet.

b) What fraction of the original kinetic energy of the bullet remains as kinetic energy of the system immediately after the collision?

c) What fraction of the original momentum remains as momentum of the system?

8–9 A bullet weighing 0.01 lb is shot through a 2-lb wood block suspended on a string 5 ft long. The center of gravity of the block is observed to rise a distance of 0.0192 ft. Find the speed of the bullet as it emerges from the block if the initial speed is 1000 ft s^{-1}.

8–10 When a bullet of mass 10 g strikes a ballistic pendulum of mass 2 kg, the center of gravity of the pendulum is observed to rise a vertical distance of 10 cm. The bullet remains embedded in the pendulum. Calculate the velocity of the bullet.

8–11 A frame of mass 200 g, when suspended from a certain coil spring, is found to stretch the spring 10 cm. A lump of putty of mass 200 g is dropped from rest onto the frame from a height of 30 cm (Fig. 8–13). Find the maximum distance the frame moves downward.

Fig. 8–13

30 cm

8–12 A bullet of mass 2 g, traveling in a horizontal direction with a velocity of 500 m s^{-1}, is fired into a wooden block of mass 1 kg, initially at rest on a level surface. The bullet passes through the block and emerges with its velocity reduced to 100 m s^{-1}. The block slides a distance of 20 cm along the surface from its initial position.

a) What was the coefficient of sliding friction between block and surface?

b) What was the decrease in kinetic energy of the bullet?

c) What was the kinetic energy of the block at the instant after the bullet passed through it?

8–13 A rifle bullet weighing 0.02 lb is fired with a velocity of 2500 ft s^{-1} into a ballistic pendulum weighing 10 lb, suspended from a cord 3 ft long. Compute (a) the vertical height through which the pendulum rises, (b) the initial kinetic energy of the bullet, (c) the kinetic energy of bullet and pendulum after the bullet is embedded in the pendulum.

8–14 A 5-g bullet is fired horizontally into a 3-kg wooden block resting on a horizontal surface. The coefficient of sliding friction between block and surface is 0.20. The bullet remains embedded in the block, which is observed to slide 25 cm along the surface. What was the velocity of the bullet?

8–15 A bullet of mass 2 g, traveling at 500 m s^{-1}, is fired into a ballistic pendulum of mass 1 kg suspended from a cord 1 m long. The bullet penetrates the pendulum and emerges with a velocity of 100 m s^{-1}. Through what vertical height will the pendulum rise?

8–16 A rifle bullet of mass 0.01 kg strikes and embeds itself in a block of mass 0.99 kg which rests on a horizontal frictionless surface and is attached to a coil spring, as shown in Fig. 8–14. The impact compresses the spring 10 cm. Calibration of the spring shows that a force of 1.0 N is required to compress the spring 1 cm.

a) Find the maximum potential energy of the spring.

b) Find the velocity of the block just after the impact.

c) What was the initial velocity of the bullet?

Fig. 8–14

|←10 cm→|

8–17 A 4000-lb automobile going eastward on Chestnut Street at 40 mi hr^{-1} collides with a 4-ton truck which is going southward across Chestnut Street at 15 mi hr^{-1}. If they become coupled on collision, what is the magnitude and direction of their velocity immediately after colliding?

8–18 On a frictionless table, a 3-kg block moving 4 m s^{-1} to the right collides with an 8-kg block moving 1.5 m s^{-1} to the left.

a) If the two blocks stick together, what is the final velocity?

b) If the two blocks make a completely elastic head-on collision, what are their final velocities?

c) How much mechanical energy is dissipated in the collision of part (a)?

8–19 Two blocks of mass 300 g and 200 g are moving toward each other along a horizontal frictionless surface with velocities of 50 cm s^{-1} and 100 cm s^{-1}, respectively.

a) If the blocks collide and stick together, find their final velocity.

b) Find the loss of kinetic energy during the collision.

c) Find the final velocity of each block if the collision is completely elastic.

8–20

a) Prove that when a moving body makes a perfectly inelastic collision with a second body of equal mass, initially at rest, one-half of the original kinetic energy is "lost."

b) Prove that when a very heavy particle makes a perfectly elastic collision with a very light particle that is at rest, the light one goes off with twice the velocity of the heavy one.

8–21 A 10-g block slides with a velocity of 20 cm s^{-1} on a smooth level surface and makes a head-on collision with a 30-g block moving in the opposite direction with a velocity of 10 cm s^{-1}. If the collision is perfectly elastic, find the velocity of each block after the collision.

8–22 A block of mass 200 g, sliding with a velocity of 12 cm s^{-1} on a smooth, level surface, makes a perfectly elastic head-on collision with a block of mass m, initially at rest. After the collision, the velocity of the 200-g block is 4 cm s^{-1} in the same direction as its initial velocity. Find (a) the mass m, and (b) its velocity after the collision.

8–23 A body of mass 600 g is initially at rest. It is struck by a second body of mass 400 g initially moving with a velocity of 125 cm s^{-1} toward the right along the x-axis. After the collision, the 400-g body has a velocity of 100 cm s^{-1} at an angle of 37° above the x-axis in the first quadrant. Both bodies move on a horizontal frictionless plane.

a) What is the magnitude and direction of the velocity of the 600-g body after the collision?

b) What is the loss of kinetic energy during the collision?

8–24 A small steel ball moving with speed v_0 in the positive x-direction makes a perfectly elastic, noncentral collision with an identical ball originally at rest. After impact, the first ball moves with speed v_1 in the first quadrant at an angle θ_1 with the x-axis and the second with speed v_2 in the fourth quadrant at an angle θ_2 with the x-axis.

a) Write the equations expressing conservation of linear momentum in the x-direction, and in the y-direction.

b) Square these equations and add them.

c) At this point, introduce the fact that the collision is perfectly elastic.

d) Prove that $\theta_1 + \theta_2 = \pi/2$.

8–25 A jet of liquid of cross-sectional area A and density ρ moves with speed v_J in the positive x-direction and impinges against a perfectly smooth blade B which deflects the stream at right angles but does not slow it down, as shown in Fig. 8–15.

a) If the blade is *stationary*, prove that the rate of arrival of mass at the blade is $\Delta m/\Delta t = \rho A v_J$.

b) If the impulse–momentum theorem is applied to a small mass Δm, prove that the x-component of the force acting on this mass for the time interval Δt is given by

$$F_x = -\frac{\Delta m}{\Delta t} v_J.$$

c) Prove that the *steady* force exerted *on* the blade in the x-direction is

$$F_x = \rho A v_J{}^2.$$

If the blade moves to the right with a speed v_B ($v_B < v_J$), derive the equations for:

d)   the rate of arrival of mass at the moving blade,

e)   the force F_x on the blade, and

f)   the power delivered to the blade.

Fig. 8-15

8-26 A stone whose mass is 100 g rests on a horizontal frictionless surface. A bullet of mass 2.5 g, traveling horizontally at 400 m s^{-1}, strikes the stone and rebounds horizontally at right angles to its original direction with a speed of 300 m s^{-1}.

a)   Compute the magnitude and direction of the velocity of the stone after it is struck.

b)   Is the collision perfectly elastic?

8-27 A hockey puck B rests on a smooth ice surface and is struck by a second puck, A, which was originally traveling at 30 m s^{-1} and which is deflected 30° from its original direction (Fig. 8-16). Puck B acquires a velocity at 45° with the original velocity of A.

a)   Compute the speed of each puck after the collision.

b)   Is the collision perfectly elastic? If not, what fraction of the original kinetic energy of puck A is "lost"?

Fig. 8-16

8-28 Imagine that a ball of mass 200 g rolls back and forth between opposite sides of a billiard table 1 m wide, with a velocity that remains constant in magnitude but reverses in direction at each collision with the cushions. The magnitude of the velocity is 4 m s^{-1}.

a)   What is the change in momentum of the ball at each collision?

b)   How many collisions per unit time are made by the ball with one of the cushions?

c)   What is the average time rate of change of momentum of the ball as a result of these collisions?

d)   What is the average force exerted by the ball on a cushion?

e)   Sketch a graph of the force exerted on the ball as a function of time, for a time interval of 5 s. [*Note:* This problem illustrates how one computes the average force exerted by the molecules of a gas on the walls of the containing vessel.]

8-29 A projectile is fired at an angle of departure of 60° and with a muzzle velocity of 400 m s^{-1}. At the highest point of its trajectory the projectile explodes into two fragments of equal mass, one of which falls vertically.

a)   How far from the point of firing does the other fragment strike if the terrain is level?

b)   How much energy was released during the explosion?

8-30 A railroad handcar is moving along straight frictionless tracks. In each of the following cases, the car initially has a total mass (car and contents) of 200 kg and is traveling with a velocity of 4 m s^{-1}. Find the *final velocity* of the car in each of the three cases.

a)   A 20-kg mass is thrown sideways out of the car with a velocity of 2 m s^{-1} relative to the car.

b)   A 20-kg mass is thrown backwards out of the car with a velocity of 4 m s^{-1} relative to the car.

c)   A 20-kg mass is thrown into the car with a velocity of 6 m s^{-1} relative to the ground and opposite in direction to the velocity of the car.

8-31 A bullet weighing 0.02 lb is fired with a muzzle velocity of 2700 ft s^{-1} from a rifle weighing 7.5 lb.

a)   Compute the recoil velocity of the rifle, assuming it free to recoil.

b)   Find the ratio of the kinetic energy of the bullet to that of the rifle.

8-32 Block A in Fig. 8-17 has a mass of 1 kg, and block B has a mass of 2 kg. The blocks are forced together, compressing a spring S between them, and the system is released from rest on a level frictionless surface. The spring is not fastened to either block and drops to the surface after it has expanded. Block B acquires a speed of 0.5 m s^{-1}. How much potential energy was stored in the compressed spring?

Fig. 8–17

8–33 A steel ball of mass 0.5 kg is dropped from a height of 4 m onto a horizontal steel slab. The collision is elastic, and the ball rebounds to its original height.

a) Calculate the impulse delivered to the ball during impact.

b) If the ball is in contact with the slab for 0.002 s, find the average force on the ball during impact.

8–34 A baseball of mass 0.25 kg is struck by a bat. Just before impact, the ball is traveling horizontally at 40 m s^{-1}, and it leaves the bat at an angle of 30° above horizontal with a speed of 60 m s^{-1}. If the ball and bat were in contact for 0.05 s, find the horizontal and vertical components of the average force on the ball.

8–35 In Problem 8–27, suppose the collision is perfectly elastic, and A is deflected 30° from its initial direction. Find the final speed of each puck and the direction of B's velocity.

8–36 A neutron of mass m collides elastically with a nucleus of mass M. Show that if the neutron's initial kinetic energy is E_0, the maximum kinetic energy it can *lose* during the collision is $4mME_0/(M + m)^2$.

8–37 An open-topped freight car of mass 10,000 kg is coasting without friction along a level track. It is raining very hard, with the rain falling vertically downward. The car is originally empty and moving with a velocity of 1 m s^{-1}. What is the velocity of the car after it has traveled long enough to collect 1000 kg of rain water?

8–38 A neutron of mass 1.67×10^{-27} kg, moving with a velocity of 2×10^4 m s^{-1}, makes a head-on collision with a boron nucleus of mass 17.0×10^{-27} kg, originally at rest.

a) If the collision is completely inelastic, what is the final kinetic energy of the system, expressed as a fraction of the original kinetic energy?

b) If the collision is perfectly elastic, what fraction of its original kinetic energy does the neutron transfer to the boron nucleus?

8–39 A nucleus, originally at rest, decays radioactively by emitting an electron of momentum 9.22×10^{-21} kg m s^{-1}, and at right angles to the direction of the electron a neutrino with momentum 5.33×10^{-21} kg m s^{-1}.

a) In what direction does the residual nucleus recoil?

b) What is its momentum?

c) If the mass of the residual nucleus is 3.90×10^{-25} kg, what is its kinetic energy?

8–40 An 80-kg man standing on ice throws a 0.2-kg ball horizontally with a speed of 30 m s^{-1}.

a) With what speed and in what direction will the man begin to move?

b) If the man throws 4 such balls every 3 s, what is the average force acting on him? [*Hint:* Average force equals average rate of change of momentum.]

8–41 Find the average recoil force on a machine gun firing 120 shots per minute. The mass of each bullet is 10 g, and the muzzle velocity is 800 m s^{-1}.

8–42 A rifleman, who together with his rifle has a mass of 100 kg, stands on roller skates and fires 10 shots horizontally from an automatic rifle. Each bullet has a mass of 10 g and a muzzle velocity of 800 m s^{-1}.

a) If the rifleman moves back without friction, what is his velocity at the end of the 10 shots?

b) If the shots were fired in 10 s, what was the average force exerted on him?

c) Compare his kinetic energy with that of the 10 bullets.

8–43 A rocket burns 0.05 kg of fuel per second, ejecting it as a gas with a velocity of 5,000 m s^{-1}.

a) What force does this gas exert on the rocket? Give the result in dynes and newtons.

b) Would the rocket operate in free space?

c) If it would operate in free space, how would you steer it? Could you brake it?

8–44 This problem illustrates the advantage of using a multi-stage rather than a single-stage rocket. Suppose that the first stage of a two-stage rocket has a total mass of 12,000 kg, of which 9000 kg is fuel. The total mass of the second stage is 1000 kg, of which 750 kg is fuel. Assume that the relative velocity v_r of ejected material is constant, and neglect any effect of gravity. (The latter effect is small during the firing period if the rate of fuel consumption is large.)

a) Suppose that the entire fuel supply carried by the two-stage rocket were utilized in a single-stage rocket of the same total mass of 13,000 kg. What would be the velocity of the rocket, starting from rest, when its fuel was exhausted?

b) What is the velocity when the fuel of the first stage is exhausted, if the first stage carries the second stage with it to this point? This velocity then becomes the initial velocity of the second stage.

c) What is the final velocity of the second stage?

d) Sputnik I's velocity was about 8 km s^{-1}. What value of v_r would be required to give the second stage of the above rocket a velocity of this magnitude?

8–45 If a single-stage rocket, fired vertically from rest at the earth's surface, burns its fuel in a time of 30 s, and the relative velocity $v_r = 3000$ m s^{-1}, what must be the mass ratio m_0/m for a final velocity v of 8 km s^{-1} (about equal to the orbital velocity of an earth satellite)?

8–46

a) Show that the acceleration of a rocket fired vertically upward is given by

$$a = -\frac{v_r}{m}\frac{dm}{dt} - g.$$

b) Suppose that the rate of ejection of mass by the rocket is constant, that is, the rate of decrease of mass is $dm/dt = -km_0$, where k is a positive constant and m_0 is the initial mass. What is the numerical value of k, and in what units is it expressed, for the rocket in Prob. 8–45?

c) Show that the mass at any time t is given by the equation

$$m = m_0(1 - kt).$$

d) Show that the acceleration at any time is equal to

$$\frac{v_r k}{1 - kt} - g.$$

e) Find the initial acceleration of the rocket in Prob. 8–45, in terms of the acceleration of gravity, g.

f) Find the acceleration 15 s after the motion starts.

g) Sketch the acceleration–time graph.

8–47 At what speed does the momentum of a particle differ by 1 percent from the value obtained using the nonrelativistic expression mv? Is the correct relativistic value greater or less than that obtained from the nonrelativistic expression?

8–48 By combining the relativistic energy and momentum expressions, Eqs. (7–22) and (8–16), show that these two quantities are related by

$$E^2 = (mc^2)^2 + (pc)^2,$$

where p is the relativistic momentum and E the *total* energy, including rest energy. Hence, show that when the total energy of a particle is much greater than its rest energy, the energy–momentum relation is approximately the same as for a massless particle. Note that for this to occur the speed of the particle must be very close to c.

8–49 Construct a right triangle in which one of the angles is α, where $\sin \alpha = v/c$. (v is the speed of a particle, c the speed of light.) If the base of the triangle (the side adjacent to α) is the rest energy mc^2, show that (a) the hypotenuse is the *total energy* and (b) the side opposite α is c times the relativistic momentum.

c) Describe a simple graphical procedure for finding the kinetic energy E_k.

8–50 The mass of an electron is 9.11×10^{-31} kg. Comparing the classical definition of momentum with its relativistic generalization, by how much is the classical expression in error if (a) $v = 0.01c$; (b) $v = 0.5c$; (c) $v = 0.9c$?

8–51 A radioactive isotope of cobalt, ^{60}Co, emits an electromagnetic photon (γ ray) of wavelength 0.932×10^{-12} m. The cobalt nucleus contains 27 protons and 33 neutrons, each with a mass of about 1.66×10^{-27} kg. If the nucleus is at rest before emission, what is the speed afterward? Is it necessary to use the relativistic generalization of momentum?

8–52 In Problem 8–51, suppose the cobalt atom is in a metallic crystal containing 0.01 moles of cobalt (about 6.02×10^{21} atoms) and that the entire crystal recoils as a unit, rather than just the single nucleus. Find the recoil velocity. (This recoil of the entire crystal rather than a single nucleus is called the *Mössbauer effect*, in honor of its discoverer, who first observed it in 1958.)

Chapter 9

Rotation

9-1 INTRODUCTION

Thus far our discussion of the principles of mechanics has been concerned primarily with particles. In problems involving a body whose size is not negligibly small, we nevertheless have assumed that its motion can be described in terms of the motion of a *point*. That is, the point mass serves as a *model* to represent a body whose size and shape are not relevant in the particular problem under discussion.

There are many problems in which this model is inadequate. One important class of problems involves *rigid bodies*, which can undergo both translational and rotational motion. The rigid body, a body with a perfectly definite and unchanging shape, is itself an idealized model, since real materials always deform somewhat when forces are exerted, but it is a useful model in cases where such deformation may be neglected.

The most general type of motion which a rigid body can undergo is a combination of *translation* and *rotation*. In preceding chapters we have discussed translational motion of a point, along a straight line or along a curve. We now discuss motion of rotation about a fixed axis, that is, motion of *rotation without translation*. We shall see that many of the equations describing rotation about a fixed axis are exactly analogous to those encountered in rectilinear motion.

If the axis is *not* fixed, the problem becomes more complicated, and we shall not attempt to give a complete discussion of the general case of translation plus rotation.

9–2 ANGULAR VELOCITY

Figure 9–1 represents a rigid body of arbitrary shape rotating about a fixed axis through point O and perpendicular to the plane of the diagram. Line OP is a line fixed with respect to the body and rotating with it. The position of the entire body is evidently completely specified by the angle θ which the line OP makes with some reference line fixed in space, such as Ox. The motion of the body is therefore analogous to the rectilinear motion of a particle whose position is completely specified by a single coordinate such as

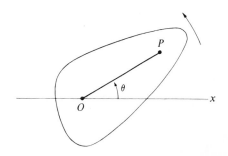

Fig. 9–1 *Body rotating about a fixed axis through point O.*

157

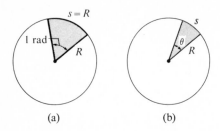

(a)   (b)

Fig. 9–2 *An angle* θ *in radians is defined as the ratio of the arc s to the radius R.*

Fig. 9–3 *Angular displacement* Δθ *of a rotating body.*

x or *y*. The equations of motion are greatly simplified if the angle θ is expressed in *radians*.

One radian (1 rad) is the angle subtended at the center of a circle by an arc of length equal to the radius of the circle (Fig. 9–2(a)). Since the radius is contained 2π times ($2\pi = 6.28\ldots$) in the circumference, there are 2π or $6.28\ldots$ rad in one complete revolution or 360°. Hence,

$$1 \text{ rad} \quad = \frac{360°}{2\pi} = 57.3 \text{ degrees},$$
$$360° \quad = 2\pi \text{ rad} = 6.28 \text{ rad},$$
$$180° \quad = \pi \text{ rad} = 3.14 \text{ rad},$$
$$90° \quad = \pi/2 \text{ rad} = 1.57 \text{ rad},$$
$$60° \quad = \pi/3 \text{ rad} = 1.05 \text{ rad},$$
$$45° \quad = \pi/4 \text{ rad} = 0.79 \text{ rad},$$

and so on.

In general (Fig. 9–2(b)), if θ represents any arbitrary angle subtended by an arc of length *s* on the circumference of a circle of radius *R*, then θ (in radians) is equal to the length of the arc *s* divided by the radius *R*:

$$\theta = \frac{s}{R}, \qquad s = R\theta. \qquad (9\text{–}1)$$

An angle in radians, being defined as the ratio of a length to a length, is a pure number. If $s = 1.5\ m$ and $R = 1$ m, the angle is usually described as $\theta = 1.5$ rad, but it would be equally correct to say simply $\theta = 1.5$.

In Fig. 9–3, a reference line *OP* in a rotating body makes an angle θ_1 with the reference line *Ox*, at a time t_1. At a later time t_2 the angle has increased to θ_2. The *average angular velocity* of the body, $\bar{\omega}$, in the time interval between t_1 and t_2, is defined as the ratio of the *angular displacement* $\theta_2 - \theta_1$, or $\Delta\theta$, to the elapsed time $t_2 - t_1$ or Δt:

$$\bar{\omega} = \frac{\Delta\theta}{\Delta t}.$$

The *instantaneous angular velocity* ω is defined as the limit approached by this ratio as Δt approaches zero:

$$\omega = \lim_{\Delta t \to 0} \frac{\Delta\theta}{\Delta t} = \frac{d\theta}{dt}. \qquad (9\text{–}2)$$

Since the body is rigid, *all* lines in it rotate through the same angle in the same time, and the angular velocity is characteristic of the body as a whole. If the angle θ is in radians, the unit of angular velocity is one radian per second (1 rad s^{-1}). Other units, such as the revolution per minute (rev min^{-1}), are in common use.

9–3 ANGULAR ACCELERATION

If the angular velocity of a body changes, it is said to have an angular acceleration. If ω_1 and ω_2 are the instantaneous angular velocities at times t_1 and t_2, the *average angular acceleration* $\bar{\alpha}$ is defined as

$$\bar{\alpha} = \frac{\omega_2 - \omega_1}{t_2 - t_1} = \frac{\Delta\omega}{\Delta t},$$

and the *instantaneous angular acceleration* α is defined as the limit of this ratio when Δt approaches zero:

$$\alpha = \lim_{\Delta t \to 0} \frac{\Delta\omega}{\Delta t} = \frac{d\omega}{dt}. \qquad (9\text{–}3)$$

The unit of angular acceleration is 1 rad s^{-2} or 1 s^{-2}. Angular velocity and angular acceleration are exactly analogous to linear velocity and acceleration.

Since $\omega = d\theta/dt$, the angular acceleration can be written

$$\alpha = \frac{d}{dt}\frac{d\theta}{dt} = \frac{d^2\theta}{dt^2}.$$

Also, by the chain rule,

$$\alpha = \frac{d\omega}{d\theta}\frac{d\theta}{dt} = \omega\frac{d\omega}{d\theta}. \qquad (9\text{–}4)$$

9–4 ROTATION WITH CONSTANT ANGULAR ACCELERATION

The simplest type of accelerated rotational motion is that in which the angular acceleration is constant. When this is the case, the expressions for the angular velocity and angular coordinate can readily be found by integration. We have

$$d\omega/dt = \alpha = \text{constant},$$
$$\int d\omega = \int \alpha\, dt,$$
$$\omega = \alpha t + C_1.$$

If ω_0 is the angular velocity when $t = 0$, the integration constant $C_1 = \omega_0$ and

$$\omega = \omega_0 + \alpha t. \qquad (9\text{–}5)$$

Then, since $\omega = d\theta/dt$,

$$\int d\theta = \int \omega_0\, dt + \int \alpha t\, dt, \qquad \theta = \omega_0 t + \tfrac{1}{2}\alpha t^2 + C_2.$$

The integration constant C_2 is the value of θ when $t = 0$, say θ_0.

$$\theta = \theta_0 + \omega_0 t + \tfrac{1}{2}\alpha t^2. \qquad (9\text{–}6)$$

If we write the angular acceleration as

$$\alpha = \omega\frac{d\omega}{d\theta},$$

then

$$\int \alpha\, d\theta = \int \omega\, d\omega + C_3, \qquad \alpha\theta = \tfrac{1}{2}\omega^2 + C_3.$$

If the angle θ again has the value θ_0 when $t = 0$, and if the initial angular velocity is ω_0, then $C_3 = \theta_0 - \tfrac{1}{2}\omega_0^2$ and

$$\omega^2 = \omega_0^2 + 2\alpha(\theta - \theta_0). \qquad (9\text{–}7)$$

Equations (9–5), (9–6), and (9–7) are exactly analogous to the corresponding equations for linear motion with constant acceleration:

$$v = v_0 + at,$$
$$x = x_0 + v_0 t + \tfrac{1}{2}at^2,$$
$$v^2 = v_0^2 + 2a(x - x_0)$$

Table 9–1 shows the similarity between the equations for motion with constant linear acceleration and those for motion with constant angular acceleration.

Example The angular velocity of a body is 4 rad s^{-1} at time $t = 0$, and its angular acceleration is constant and equal to 2 s^{-2}. A line OP in the body is horizontal at time $t = 0$. a) What angle does this line make with the horizontal at time $t = 3 \text{ s}$? b) What is the angular velocity at this time?

a) $\quad \theta = \theta_0 + \omega_0 t + \tfrac{1}{2}\alpha t^2$

$\quad = 0 + (4 \text{ rad s}^{-1})(3 \text{ s}) + \tfrac{1}{2}(2 \text{ rad s}^{-2})(3 \text{ s})^2$

$\quad = 21 \text{ rad} = \dfrac{21}{2\pi} \text{ rev} = 3.34 \text{ rev}.$

b) $\quad \omega = \omega_0 + \alpha t$

$\quad = 4 \text{ rad s}^{-1} + (2 \text{ rad s}^{-2})(3 \text{ s}) = 10 \text{ s}^{-1}.$

Table 9–1

Motion with constant linear acceleration	Motion with constant angular acceleration
$a = \text{constant}$	$\alpha = \text{constant}$
$v = v_0 + at$	$\omega = \omega_0 + \alpha t$
$x = x_0 + v_0 t + \frac{1}{2}at^2$	$\theta = \theta_0 + \omega_0 t + \frac{1}{2}\alpha t^2$
$x = x_0 + \dfrac{v_0 + v}{2} t$	$\theta = \theta_0 + \dfrac{\omega_0 + \omega}{2} t$
$v^2 = v_0^2 + 2a(x - x_0)$	$\omega^2 = \omega_0^2 + 2\alpha(\theta - \theta_0)$

Alternatively, from Eq. (9–7),

$$\omega^2 = \omega_0^2 + 2\alpha(\theta - \theta_0)$$

$$= (4 \text{ rad s}^{-1})^2 + 2(2 \text{ rad s}^{-2})(21 \text{ rad})$$

$$= 100 \text{ rad}^2 \text{ s}^{-2},$$

$$\omega = 10 \text{ rad s}^{-1}.$$

9–5 RELATION BETWEEN ANGULAR AND LINEAR VELOCITY AND ACCELERATION

In Section 6–6 we discussed the linear velocity and acceleration of a *particle* revolving in a circle. When a *rigid body* rotates about a fixed axis, every point in the body moves in a circle whose center is on the axis and which lies in a plane perpendicular to the axis. There are some useful and simple relations between the angular velocity and acceleration of the rotating body and the linear velocity and acceleration of points within it.

Let r be the distance from the axis to some point P in the body, so that the point moves in a circle of radius r, as in Fig. 9–4. When the radius makes an angle θ with the reference axis, the distance s to the point P, measured along the circular path, is

$$s = r\theta. \qquad (9\text{–}8)$$

Differentiating both sides of this equation with respect to t and noting that r is constant, we find

$$\frac{ds}{dt} = r\frac{d\theta}{dt}.$$

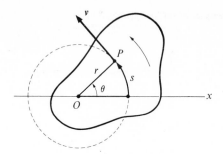

Fig. 9–4 *The distance s moved through by point P equals rθ.*

But ds/dt is the magnitude of the linear velocity v of point P, and $d\theta/dt$ is the angular velocity ω of the rotating body. Hence

$$v = r\omega \qquad (9\text{–}9)$$

and the magnitude v of the linear velocity equals the product of the angular velocity ω and the distance r of the point from the axis.

Differentiating Eq. (9–9) with respect to t gives

$$\frac{dv}{dt} = r\frac{d\omega}{dt}.$$

But dv/dt is the magnitude of the tangential component of acceleration $a_\parallel$ of point P, and $d\omega/dt$ is the angular acceleration α of the rotating body, so

$$a_\parallel = r\alpha \qquad (9\text{–}10)$$

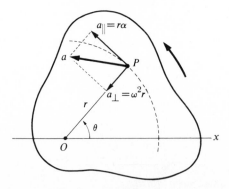

Fig. 9–5 *Nonuniform rotation about a fixed axis through point O. The tangential component of acceleration of point P equals rα; the radial component equals ω²r.*

9–6

Torque and angular acceleration. Moment of inertia   161

and the tangential component of acceleration equals the product of the angular acceleration and the distance from the axis.

The *radial* component of acceleration $a_\perp = v^2/r$ of the point P can also be expressed in terms of the angular velocity,

$$a_\perp = \frac{v^2}{r} = \omega^2 r. \qquad (9\text{–}11)$$

This is true at each instant *even when ω and v are not constant.*

The tangential and radial components of acceleration of any arbitrary point P in a rotating body are shown in Fig. 9–5.

9–6 TORQUE AND ANGULAR ACCELERATION. MOMENT OF INERTIA

We are now ready to consider the *dynamics* of rotation about a fixed axis, that is, the relation between the forces on a pivoted body and its angular acceleration.

Figure 9–6 represents a rigid body pivoted about a fixed axis through point O, perpendicular to the plane of the diagram.

The solid circle represents one of the particles of the body, of mass m_i. The particle is acted on by an external force $\mathbf{F}_i$, and by an internal force $\mathbf{f}_i$, the resultant of the forces exerted on it by all of the other particles of the body. It will suffice to consider only the case in which the forces $\mathbf{F}_i$ and $\mathbf{f}_i$ lie in a plane perpendicular to the axis. From Newton's second law,

$$\mathbf{F}_i + \mathbf{f}_i = m_i \mathbf{a}_i .$$

Let us resolve the forces and accelerations into radial and tangential components. Then

$$F_i \cos \theta_i + f_i \cos \phi_i = m_i a_{i\perp} = m_i r_i \omega^2,$$
$$F_i \sin \theta_i + f_i \sin \phi_i = m_i a_{i\parallel} = m_i r_i \alpha.$$

The first of these equations does not concern us further. When both sides of the second are multiplied by the distance r_i of the particle from the axis, we obtain

$$F_i r_i \sin \theta_i + f_i r_i \sin \phi_i = m_i r_i^2 \alpha. \qquad (9\text{–}12)$$

The first term on the left is the moment Γ_i of the external force about the axis, and the second is the moment of the internal force.

Equations corresponding to Eq. (9–12) can be written for all the particles of the body. When these equations are *added*, the moments of the *internal* forces cancel, since the resultant moment of every internal action-reaction force pair is zero. The sum of the left sides of the equations is then simply the resultant moment Γ of the *external* forces about the axis, or

$$\Gamma = \sum \Gamma_i = \sum F_i r_i \sin \theta_i .$$

Since the body is rigid, all particles have the same angular acceleration α and therefore

$$\Gamma = (m_1 r_1^2 + m_2 r_2^2 + \cdots)\alpha = (\sum m_i r_i^2)\alpha. \qquad (9\text{–}13)$$

The sum $\sum m_i r_i^2$ is called the *moment of inertia* of the body about the axis through point O, and is represented by I:

$$I = \sum m_i r_i^2. \qquad (9\text{–}14)$$

Equation (9–13) then becomes

$$\Gamma = I\alpha = I\frac{d\omega}{dt}. \qquad (9\text{–}15)$$

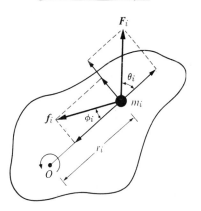

Fig. 9–6 *An external force $\mathbf{F}_i$ and an internal force $\mathbf{f}_i$ acting on a particle of mass m_i in a rigid body rotating about a fixed axis.*

That is, *when a rigid body is pivoted about a fixed axis, the resultant external torque about the axis equals the product of the moment of inertia of the body about the axis, and the angular acceleration.*

Thus the *angular* acceleration of a rigid body about a fixed axis is given by an equation having exactly the same form as that for the *linear* acceleration of a particle:

$$F = ma = m\frac{dv}{dt}.$$

The resultant torque Γ, about the axis corresponds to the resultant force F, the angular acceleration α corresponds to the linear acceleration a, and the moment of inertia I about the axis corresponds to the mass m.

The concept of moment of inertia can be thought of either as the sum of the product of the mass of each particle in a rigid body and the square of its distance from the axis, $I = \Sigma m_i r_i^2$, or as the ratio of the resultant torque to the angular acceleration, $I = \Gamma/\alpha$.

The analogy between translational and rotational quantities is displayed in Table 9-2.

Example A wheel of radius R, mass m_2, and moment of inertia I is mounted on an axle supported in fixed bearings, as in Fig. 9-7. A light flexible cord is wrapped around the rim of the wheel and carries a body of mass m_1. Friction in the bearings can be neglected. Discuss the motion of the system.

We must consider the resultant *force* on the suspended body and the resultant *torque* on the wheel. Let T represent the tension in the cord and P the upward force exerted on the shaft of the wheel by the bearings.

The resultant force on the suspended body is $w_1 - T$, and from Newton's second law for linear motion,

$$w_1 - T = m_1 a.$$

We have taken the downward direction as positive in order that a positive (counterclockwise) angular displacement of the wheel shall correspond to a positive linear displacement of the suspended body.

The forces P and w_2 have no moment about the axis of the wheel. The resultant torque on the wheel, about the axis, is TR, and from Newton's second law

TABLE 9-2 ANALOGY BETWEEN TRANSLATIONAL AND ROTATIONAL QUANTITIES

Concept	Translation	Rotation	Comments
Displacement	s	θ	$s = r\theta$
Velocity	$v = ds/dt$	$\omega = d\theta/dt$	$v = r\omega$
Acceleration	$a = dv/dt$	$\alpha = d\omega/dt$	$a_T = r\alpha$
Resultant force, moment	F	Γ	$\Gamma = Fr$
Equilibrium	$F = 0$	$\Gamma = 0$	
Acceleration constant	$v = v_0 + at$ $s = v_0 t + \frac{1}{2}at^2$ $v^2 = v_0^2 + 2as$	$\omega = \omega_0 + \alpha t$ $\theta = \omega_0 t + \frac{1}{2}\alpha t^2$ $\omega^2 = \omega_0^2 + 2\alpha\theta$	
Mass, moment of inertia	m	I	$I = \Sigma m_i r_i^2$
Newton's second law	$F = ma$	$\Gamma = I\alpha$	
Work	$W = \int F\, ds$	$W = \int \Gamma\, d\theta$	
Power	$P = Fv$	$P = \Gamma\omega$	
Potential energy	$E_p = mgy$		
Kinetic energy	$E_k = \frac{1}{2}mv^2$	$E_k = \frac{1}{2}I\omega^2$	
Impulse	$\int F\, dt$	$\int \Gamma\, dt$	
Momentum	mv	$L = I\omega$	

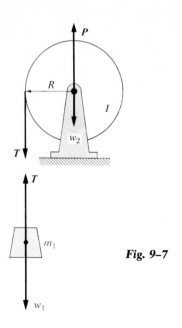

Fig. 9–7

for rotation,

$$TR = I\alpha$$

Since the linear acceleration of the suspended body equals the tangential acceleration of the rim of the wheel, we have

$$a = R\alpha.$$

Simultaneous solution of these equations gives

$$a = g\frac{1}{1 + (I/m_1R^2)}.$$

If the system starts from rest, the linear speed v of the suspended body, after descending a distance y (the acceleration is constant) is given by

$$v^2 = 2ay = 2\left[g\frac{1}{1 + (I/m_1R^2)}\right]y.$$

9–7 CALCULATION OF MOMENTS OF INERTIA

The moment of inertia of a body about an axis can be found *experimentally* by pivoting the body about the axis, applying a measured torque Γ to the body, and measuring the resulting angular acceleration α.

The moment of inertia is then given by

$$I = \frac{\Gamma}{\alpha}.$$

The moment of inertia can be *calculated* from the defining equation $I = \Sigma m_i r_i^2$, for any system consisting of discrete point masses.

Example 1   Three small bodies, which can be considered as particles, are connected by light rigid rods, as in Fig. 9–8. What is the moment of inertia of the system (a) about an axis through point A, perpendicular to the plane of the diagram, and (b) about an axis coinciding with the rod BC?

a)   The particle at point A lies on the axis. Its distance *from* the axis is zero and it contributes nothing to the moment of inertia. Therefore

$$I = \Sigma m_i r_i^2 = (10 \text{ g})(5 \text{ cm})^2 + (20 \text{ g})(4 \text{ cm})^2$$
$$= 570 \text{ g cm}^2.$$

b)   The particles at B and C both lie on the axis. The moment of inertia is

$$I = \Sigma m_i r_i^2 = (30 \text{ g})(4 \text{ cm})^2$$
$$= 480 \text{ g cm}^2.$$

This illustrates the important fact that the moment of inertia of a body, unlike its mass, is not a unique property of the body but depends on the axis about which it is computed.

For a body which is not composed of discrete point masses but is a continuous distribution of matter, the summation expressed in the definition of moment of inertia, $I = \Sigma m_i r_i^2$, must be evaluated by the methods of calculus. The body is imagined to be subdivided into volume elements, each of mass Δm.

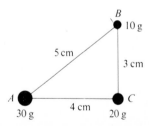

Fig. 9–8

Let r be the distance from any element to the axis of rotation. If each mass Δm is multiplied by the square of its distance from the axis and all the products $r^2\,\Delta m$ summed over the whole body, the moment of inertia is

$$I = \lim_{\Delta m \to 0} \Sigma r^2 \Delta m = \int r^2\,dm. \qquad (9\text{–}16)$$

If dV and dm represent the volume and mass, respectively, of an element, the density ρ is defined by the relation $dm = \rho\,dV$. Equation (9–16) may therefore be written

$$I = \int r^2\,dV.$$

If the density of a body is the same at all points, the body is said to be *uniform* or *homogeneous*, in which case

$$I = \rho \int r^2\,dV.$$

In using this equation, any convenient volume element may be chosen, *provided that all points within the element are the same distance r from the axis.*

The evaluation of integrals of this type may present considerable difficulty if the body is irregular, but for bodies of simple shape the integration can be carried out relatively easily. Three examples are given below.

Example 2 *Uniform slender rod, axis perpendicular to length.* Figure 9–9 shows a slender uniform rod of mass m and length ℓ. We wish to compute its moment of inertia about an axis through A, at an arbitrary distance h from one end. Select as an element of volume a short section of length dx and cross-sectional area S, at a distance x from point A. Then

$$dm = \rho\,dV = \rho S\,dx = \frac{\rho S \ell}{\ell}\,dx = \frac{m}{\ell}\,dx.$$

Fig. 9–9

Fig. 9–10

Now

$$\begin{aligned}
I_A &= \int x^2\,dm = \frac{m}{\ell}\int_{-h}^{\ell-h} x^2\,dx \\
&= \frac{m}{\ell}\frac{x^3}{3}\Big]_{-h}^{\ell-h} = \tfrac{1}{3}m(\ell^2 - 3\ell h + 3h^2).
\end{aligned}$$

From this general expression, we can find the moment of inertia about an axis through any point on the rod. For example, if the axis is at the left end, $h = 0$ and

$$I = \tfrac{1}{3}m\ell^2. \qquad (9\text{–}17)$$

If the axis is at the right end, $h = \ell$ and

$$I = \tfrac{1}{3}m\ell^2,$$

as would be expected. If the axis passes through the center,

$$h = \ell/2 \quad \text{and} \quad I = \tfrac{1}{12}m\ell^2. \qquad (9\text{–}18)$$

Example 3 *Hollow or solid cylinder, axis of symmetry* Figure 9–10 shows a hollow cylinder of length ℓ and inner and outer radii R_1 and R_2. We choose as the most convenient volume element the thin cylindrical shell of radius r, thickness dr, and length ℓ. If ρ is the density of the material, that is, the mass per unit volume, then

$$dm = \rho\,dV = \rho(2\pi r\,dr) \times \ell.$$

The moment of inertia is given by

$$I = \int r^2\,dm = 2\pi\ell \int_{R_1}^{R_2} \rho r^3\,dr.$$

If the body were of nonuniform density, one would have to know ρ as a function of r before the integration could be carried out. For a homogeneous solid, however, ρ is constant, and

$$I = 2\pi\ell\rho\int_{R_1}^{R_2} r^3\, dr$$

$$= \frac{\pi\ell\rho}{2}(R_2{}^4 - R_1{}^4).$$

The mass m of the entire cylinder is the product of its density and its volume. The volume is given by

$$\pi\ell(R_2{}^2 - R_1{}^2).$$

Hence

$$m = \pi\ell\rho(R_2{}^2 - R_1{}^2),$$

and the moment of inertia is therefore

$$I = \tfrac{1}{2}m(R_1{}^2 + R_2{}^2). \tag{9–19}$$

If the cylinder is solid, $R_1 = 0$, and if we let R represent the outer radius,

$$I = \tfrac{1}{2}mR^2. \tag{9–20}$$

If the cylinder is very thin-walled (like a stovepipe), $R_1 = R_2$ (very nearly) and if R represents this common radius,

$$I = mR^2.$$

Note that the moment of inertia of a cylinder about an axis coinciding with its axis of symmetry does not depend on the length ℓ. Two hollow cylinders of the same inner and outer radii, one of wood and one of brass, but having the same mass m, have equal moments of inertia even though the length of the former is much greater. Moment of inertia depends only on the *radial* distribution of mass, not on its distribution along the axis. Thus Eq. (9–19) holds for a very short cylinder like a washer, and Eq. (9–20) for a thin disk.

Example 4 *Uniform sphere, axis through center.* Divide the sphere into thin disks, as indicated in Fig. 9–11. The radius r of the disk shown is

$$r = \sqrt{R^2 - x^2}.$$

Its volume is

$$dV = \pi r^2\, dx = \pi(R^2 - x^2)\, dx$$

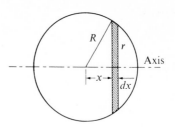

Fig. 9–11

and its mass is

$$dm = \rho\, dV.$$

Hence from Eq. (9–20) its moment of inertia is

$$dI = \frac{\pi\rho}{2}(R^2 - x^2)^2\, dx,$$

and for the whole sphere,

$$I = (2)\frac{\pi\rho}{2}\int_0^R (R^2 - x^2)^2\, dx,$$

since by symmetry the right hemisphere has the same moment of inertia as the left. Carrying out the integration, we obtain

$$I = \frac{8\pi\rho}{15}R^5.$$

The mass m of the sphere is

$$m = \rho V = \frac{4\pi\rho R^3}{3}.$$

Hence

$$I = \tfrac{2}{5}mR^2.$$

The moments of inertia of a few simple bodies are shown in Fig. 9–12.

Whatever the shape of a body, it is always possible to find a radial distance from any given axis at which the mass of the body could be concentrated without altering the moment of inertia of the body about that axis. This distance is called the *radius of gyration* of the body about the given axis, and is represented by k.

If the mass m of the body actually were concentrated at this distance, the moment of inertia would be that of a particle of mass m at a distance k from an axis, or mk^2. Since this equals the moment of inertia

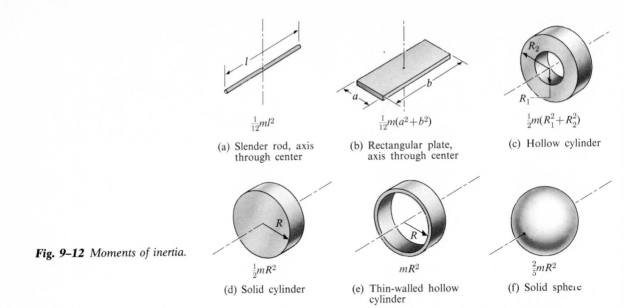

Fig. 9–12 Moments of inertia.

$\frac{1}{12}ml^2$

(a) Slender rod, axis through center

$\frac{1}{12}m(a^2+b^2)$

(b) Rectangular plate, axis through center

$\frac{1}{2}m(R_1^2+R_2^2)$

(c) Hollow cylinder

$\frac{1}{2}mR^2$

(d) Solid cylinder

mR^2

(e) Thin-walled hollow cylinder

$\frac{2}{5}mR^2$

(f) Solid sphere

I, then

$$mk^2 = I, \qquad k = \sqrt{I/m}. \qquad (9\text{--}21)$$

Example 5 What is the radius of gyration of a slender rod of mass m and length ℓ about an axis perpendicular to its length and passing through the center?

The moment of inertia about an axis through the center is

$$I_0 = \tfrac{1}{12}m\ell^2.$$

Hence

$$k_0 = \sqrt{\frac{\tfrac{1}{12}m\ell^2}{m}} = \frac{\ell}{2\sqrt{3}} = 0.289\ell.$$

The radius of gyration, like the moment of inertia, depends on the location of the axis.

Note carefully that, in general, the mass of a body cannot be considered as concentrated at its center of gravity for the purpose of computing its moment of inertia. For example, when a thin rod is pivoted about an axis through one end, perpendicular to one end, the moment of inertia is $I = mL^2/3$ while, if all the mass were concentrated at the center, the moment of inertia about this axis would be $m(L/2)^2$ or $mL^2/4$.

9–8 KINETIC ENERGY, WORK, AND POWER

When a rigid body rotates about a fixed axis, the velocity v_i of a particle at a perpendicular distance r_i from the axis equals $r_i\omega$, where ω is the angular velocity. The kinetic energy of the particle is then

$$\tfrac{1}{2}m_iv_i^2 = \tfrac{1}{2}m_ir_i^2\omega^2,$$

and the total kinetic energy of the body is

$$E_k = \sum \tfrac{1}{2}m_ir_i^2\omega^2 = \tfrac{1}{2}\left(\sum m_ir_i^2\right)\omega^2.$$

But $\sum m_ir_i^2$ equals the moment of inertia I about the axis, so

$$\boxed{E_k = \tfrac{1}{2}I\omega^2.} \qquad (9\text{--}22)$$

Thus the kinetic energy of a rigid body rotating about a fixed axis is given by an expression exactly analogous to that for the kinetic energy of a particle in linear motion, the moment of inertia I corresponding to the mass m and the angular velocity ω corresponding to the linear velocity v. (See Table 9–2.)

Example 1 Let us consider the motion of the system in Fig. 9–7 from the energy standpoint. Looking at the system as a whole, the external forces are the forces $\boldsymbol{P}$ and $\boldsymbol{w}_2$, which do no work, and the force $\boldsymbol{w}_1$, which is conservative. We can therefore apply the principle of conservation of energy, setting the decrease in potential energy of the body, as it descends a distance y, equal to the sum of the increase in *translational* kinetic energy of the body and the increase in *rotational* kinetic energy of the wheel:

$$m_1 gy = \tfrac{1}{2} m_1 v^2 + \tfrac{1}{2} I \omega^2 .$$

But

$$v = \omega R,$$

and again we find that

$$v^2 = 2 \left[g \frac{1}{1 + (I/m_1 R^2)} \right] y.$$

In Fig. 9-13, an external force F is applied at point P of a rigid body rotating about a fixed axis through O, perpendicular to the plane of the diagram. As the body rotates through a small angle $d\theta$, point P moves a distance $ds = r\, d\theta$ and the work done by the force F is

$$W = \int F_{\parallel}\, ds = \int F_{\parallel} r\, d\theta.$$

But $F_{\parallel} r$ is the moment Γ of the force about the axis, so

$$W = \int_{\theta_1}^{\theta_2} \Gamma\, d\theta. \qquad (9\text{--}23)$$

If more than one force acts on the body, the total work equals the work of the resultant moment.

From Eq. (9-15),

$$\Gamma = I\alpha = I\omega \frac{d\omega}{d\theta}.$$

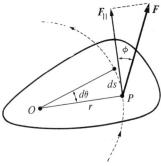

Fig. 9-13 *The work done by the force* F *in an angular displacement* $d\theta$.

Hence

$$\Gamma\, d\theta = I\omega\, d\omega,$$

and

$$W = \int_{\theta_1}^{\theta_2} \Gamma\, d\theta = \int_{\omega_1}^{\omega_2} I\omega\, d\omega$$

$$= \tfrac{1}{2} I\omega_2^2 - \tfrac{1}{2} I\omega_1^2. \qquad (9\text{--}24)$$

That is, *the work of the resultant moment equals the increase in kinetic energy*, in analogy with the work–energy equation for linear motion.

The power developed by the force Γ in Fig. 9-13, if v is the velocity of its point of application, is

$$P = F_{\parallel} v = F_{\parallel} r\omega,$$

and since $F_{\parallel} r = \Gamma$,

$$\boxed{P = \Gamma\omega,} \qquad (9\text{--}25)$$

the rotational analog of $P = F_{\parallel} v$. The power output of an engine therefore equals the product of torque and angular velocity. The transmission of an automobile engine is often called a "torque converter." An automobile engine can deliver full power only when running at its rated angular velocity and delivering the corresponding torque. When accelerating, or climbing a steep hill, the torque converter permits the engine to run at rated speed, torque, and power, while the same power is transmitted to the drive shaft at a larger torque and a smaller angular velocity.

Example 2 The manufacturers of an automobile state that its engine develops 345 hp and a torque of 475 lb ft. What is the corresponding angular velocity?

$$\omega = \frac{P}{\Gamma} = \frac{(345)(550 \text{ ft lb s}^{-1})}{475 \text{ lb ft}}$$

$$= 400 \text{ rad s}^{-1} \approx 3800 \text{ rev min}^{-1}.$$

9-9 ANGULAR MOMENTUM

The equation

$$\Gamma = I\alpha$$

can be written as

$$\Gamma = I\frac{d\omega}{dt} = \frac{d}{dt}(I\omega). \qquad (9\text{--}26)$$

The product of moment of inertia I and angular velocity ω is the rotational analog of the product of mass m and linear velocity v. The latter product is the linear momentum, and by analogy we call the product $I\omega$ the *angular momentum L:*

$$L = I\omega.$$ (9–27)

(Although in this special case of a rigid body rotating about a fixed axis the angular momentum is equal to $I\omega$, this is not the general definition of this quantity.)

Equation (9–26) can now be written

$$\Gamma = \frac{dL}{dt}.$$ (9–28)

The resultant external torque is equal to the rate of change of angular momentum, just as the resultant external force equals the rate of change of linear momentum.

Multiplying by dt and integrating, we get

$$J = \int_0^t \Gamma\, dt = L - L_0.$$ (9–29)

The integral $J = \int \Gamma\, dt$ is called the *angular impulse* of the torque and is analogous to the impulse of a force, $\int F\, dt$. The preceding equation is therefore the analog of the impulse–momentum principle in linear motion. That is, *the resultant angular impulse of the torque on a body is equal to the change in angular momentum of the body.*

9–10 CONSERVATION OF ANGULAR MOMENTUM

Here is an example of the usefulness of the concepts of angular impulse and angular momentum. Figure 9–14 shows two disks with moments of inertia I and I', rotating with initial constant angular velocities ω_0 and ω_0', respectively. At some time t_0, the disks are pushed together by a force directed along the axis, without any torque on either disk. After a short time interval, say Δt, the disks reach a common final angular velocity ω.

During the interval Δt, the larger disk exerts a torque Γ' on the smaller, and the smaller exerts a torque Γ on the larger. Both Γ and Γ' vary during the

contact, becoming zero after the common final angular velocity is reached. At each instant the two torques are equal in magnitude and opposite in direction, because of Newton's third law. That is, at each instant $\Gamma = -\Gamma'$; the total impulses on the two bodies are thus related by $J = -J'$.

Now according to Eq. (9–29), the impulse on each disk equals the change of angular momentum of that disk:

$$J = I\omega - I\omega_0,$$

$$J' = I'\omega - I'\omega_0'.$$

But because $J = -J'$, these are equal and opposite:

$$I\omega - I\omega_0 = -(I'\omega - I'\omega_0')$$

or

$$I\omega_0 + I'\omega_0' = (I + I')\omega.$$ (9–30)

(a)

(b)

Fig. 9–14 *An impulsive torque acts when two rotating disks engage.*

The left side of Eq. (9–30) is the total angular momentum of the system (consisting of both disks) before contact, and the right side is the total angular momentum after contact. We have therefore derived the important result that the total angular momentum of the whole system is unaltered. When both disks are regarded as one system, then the torques Γ and Γ' are *internal* torques; during the engaging process no *external* torque acts. *When the resultant external torque on a system is zero, the angular momentum of the system remains constant*; hence any internal interaction between the parts of a system cannot alter its total angular momentum. This is the *principle of conservation of angular momentum*, and it ranks with the principles of conservation of linear momentum and conservation of energy as one of the most fundamental of physical laws.

A circus acrobat, a diver, or a skater performing a pirouette on the toe of one skate, all take advantage of the principle. Suppose an acrobat has just left a swing as in Fig. 9–15, with arms and legs extended and with a counterclockwise angular momentum. When he pulls his arms and legs in, his moment of inertia I becomes much smaller. Since his angular momentum $I\omega$ remains constant and I decreases, his angular velocity ω *increases*.

Fig. 9–15 *Conservation of angular momentum.*

Example 1 In the situation of Fig. 9–14, suppose the large disk has mass 2 kg, radius 0.2 m, and initial angular velocity 50 rad s^{-1}, and the small disk mass 4 kg, radius 0.1 m, and initial angular velocity 200 rad s^{-1} (about 1900 rpm). Find the common final angular velocity after the disks are pushed into

contact. Is kinetic energy conserved during this process?

The moments of inertia of the two disks are

$$I = \tfrac{1}{2}(2 \text{ kg})(0.2 \text{ m})^2 = 0.04 \text{ kg m}^2,$$

and

$$I' = \tfrac{1}{2}(4 \text{ kg})(0.1 \text{ m})^2 = 0.02 \text{ kg m}^2.$$

From conservation of angular momentum, we have

$$(0.04 \text{ kg m}^2)(50 \text{ rad s}^{-1}) + (0.02 \text{ kg m}^2)(200 \text{ rad s}^{-1})$$

$$= (0.04 \text{ kg m}^2 + 0.02 \text{ kg m}^2)\omega,$$

$$\omega = 100 \text{ rad s}^{-1}.$$

The initial kinetic energy is

$$E_{k_0} = \tfrac{1}{2}(0.04 \text{ kg m}^2)(50 \text{ rad s}^{-1})^2$$

$$+ \tfrac{1}{2}(0.02 \text{ kg m}^2)(200 \text{ rad s}^{-1})^2 = 450 \text{ J}.$$

The final kinetic energy is

$$E_k = \tfrac{1}{2}(0.04 \text{ kg m}^2 + 0.02 \text{ kg m}^2)(100 \text{ rad s}^{-1})^2$$

$$= 300 \text{ J}.$$

One-third of the kinetic energy was lost during this "angular collision," which is the rotational analog of an inelastic collision. We should not expect kinetic energy to be conserved, even though the resultant external force and torque are zero, since nonconservative (frictional) internal forces act during the contact.

Example 2 A man stands at the center of a turntable, holding his arms extended horizontally, with a 5-kg mass in each hand, as in Fig. 9–16. He is set rotating about a vertical axis with an angular velocity of one revolution in 2 s. Find his new angular velocity if he drops his hands to his sides. The moment of inertia of the man may be assumed constant and equal to 6 kg m^2. The original distance of the weights from the axis is 1 m, and their final distance is 0.2 m.

If friction in the turntable is neglected, no external torques act about a vertical axis and the angular momentum about this axis is constant. That

Fig. 9–16 *Conservation of angular momentum about a fixed axis.*

is,

$$I\omega = (I\omega)_0 = I_0\omega_0,$$

where I and ω are the final moment of inertia and angular velocity, and I_0 and ω_0 are the initial values of these quantities:

$$I = I_{man} + I_{weights},$$

$$I_0 = 6 \text{ kg m}^2 + 2(5 \text{ kg})(1.0 \text{ m})^2 = 16 \text{ kg m}^2,$$

$$I = 6 \text{ kg m}^2 + 2(5 \text{ kg})(0.2 \text{ m})^2 = 6.4 \text{ kg m}^2,$$

$$\omega_0 = 2\pi(1/2) \text{ rad s}^{-1},$$

$$\omega = \omega_0\left(\frac{I_0}{I}\right) = \pi\frac{16 \text{ kg m}^2}{6.4 \text{ kg m}^2} = 2.5\pi \text{ rad s}^{-1}.$$

That is, the angular velocity is more than doubled.

The initial kinetic energy is

$$E_{k0} = \tfrac{1}{2}(16 \text{ kg m}^2)(\pi \text{ rad s}^{-1})^2 = 79 \text{ J}.$$

The final kinetic energy is

$$E_k = \tfrac{1}{2}(6.4 \text{ kg m}^2)(2.5\pi \text{ rad s}^{-1})^2 = 197 \text{ J}.$$

Where did the extra energy come from?

9–11 VECTOR REPRESENTATION OF ANGULAR QUANTITIES

A rotational quantity associated with an axis, such as angular velocity, angular momentum, torque, etc., can be represented by a *vector* along the axis. Thus angular velocity is a vector with magnitude equal to the number of radians through which the body turns in unit time (our original definition of angular velocity), with direction along the axis of rotation in the direction a righthand-thread screw would advance if turned with the body, as shown in Fig. 9–17. Thus far we have considered only problems of rotation about a fixed axis, for which the angular-velocity vector always lies along the axis of rotation. For motion where the direction of the axis changes, the vector nature of angular velocity is an essential feature of the description of motion. The concepts of moment and angular momentum must also be generalized to vector quantities.

We consider first the vector generalization of moment. In Fig. 9–18, a force $\boldsymbol{F}$ lying in the xy-plane is applied at point P of a rigid body. Point O is at a perpendicular distance r from point P, and $\boldsymbol{r}$ is the vector from O to P. There may or may not be an axis through point O. If there *were* an axis through O, perpendicular to the plane of $\boldsymbol{r}$ and $\boldsymbol{F}$, the moment Γ of the force $\boldsymbol{F}$ about this axis would be

$$\Gamma = rF.$$

We define the *vector moment* of the force $\boldsymbol{F}$, *about the* **point** O, as a vector Γ whose magnitude equals the moment rF and whose direction is perpendicular to the plane of $\boldsymbol{r}$ and $\boldsymbol{F}$, as shown. The sense of Γ along the axis is specified by the *right-hand screw rule:* Rotate the first vector ($\boldsymbol{r}$) through the smaller angle (θ) that will bring it into parallelism with the second vector ($\boldsymbol{F}$), as shown by the broken line in the

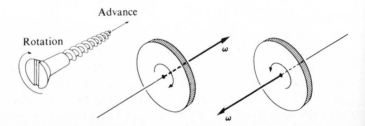

Fig. 9–17 *Vector angular velocity of a rotating body.*

Fig. 9–18 (a) Vector $\boldsymbol{\Gamma}$ is the vector moment of the force $\boldsymbol{F}$ about the point O: $\boldsymbol{\Gamma} = \boldsymbol{r} \times \boldsymbol{F}$.

diagram. The vector $\boldsymbol{\Gamma}$ then points in the direction of *advance* of a right-hand screw when it is *rotated* through the same angle θ. The vector moment of $\boldsymbol{F}$ about point O has the same magnitude and direction whatever the direction of an actual axis through O, *or even if there is no actual axis through this point.*

Since $\boldsymbol{r}$ and $\boldsymbol{F}$ are vectors, the definition above suggests that we define a new product of two vectors called the *vector product* or *cross product*. The vector product of any two vectors $\boldsymbol{A}$ and $\boldsymbol{B}$ is written $\boldsymbol{A} \times \boldsymbol{B}$ and is defined as a *vector* of magnitude $AB \sin \theta$, where θ is the angle between $\boldsymbol{A}$ and $\boldsymbol{B}$. Then if $\boldsymbol{C}$ represents this product

$$\boldsymbol{C} = \boldsymbol{A} \times \boldsymbol{B}, \qquad C = AB \sin \theta. \qquad (9\text{–}31)$$

The direction and sense of the vector product is specified by the right-hand screw rule given above. The vector movement of the force $\boldsymbol{F}$ about the point O can therefore be written in general as

$$\boldsymbol{\Gamma} = \boldsymbol{r} \times \boldsymbol{F}, \qquad (9\text{–}32)$$

where

$$\Gamma = rF \sin \theta$$

For the special case in Fig. 9–18, $\theta = 90°$ and $\sin \theta = 1$.

Angular momentum is generalized similarly. Figure 9–19 shows a horizontal flat plate rotating about a fixed vertical axis through point O with an angular velocity ω. A particle of the plate of mass m, at a distance r from the axis, has a linear velocity

of magnitude $v = r\omega$ and a linear momentum of magnitude mv. The angular momentum $\boldsymbol{L}$ of the particle is now defined as a vector quantity equal to the vector product of $\boldsymbol{r}$ and $m\boldsymbol{v}$:

$$\boldsymbol{L} = \boldsymbol{r} \times m\boldsymbol{v}. \qquad (9\text{–}33)$$

As an example, if a particle of mass m moves counterclockwise in a circle of radius r, in the plane of this page, with speed v, its angular momentum is a vector pointing out of the page, with magnitude mvr. This can also be expressed in terms of the angular velocity; since $v = r\omega$, we have

$$L = mvr = m(r\omega)r = mr^2\omega = I\omega. \qquad (9\text{–}34)$$

We have used the fact that the moment of inertia I of the particle about an axis through the center of the circle is mr^2.

The total angular momentum of a rigid body is the vector sum of the angular momenta of the particles composing it. If the body lies in a single plane, as in Fig. 9–19, and rotates about an axis perpendicular to this plane, then all these vectors lie along the axis. Then, the *vector* sum becomes an *arithmetic* sum and the magnitude of the angular momentum of the plate is

$$L = \sum rmv.$$

But $v = r\omega$, and ω has the same value for all particles. Hence

$$L = \left(\sum mr^2\right)\omega = I\omega,$$

and this definition of angular momentum leads to the same expression for L as that stated in Sec. 9–9.

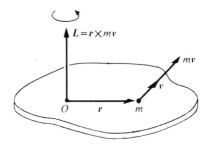

Fig. 9–19 Vector $\boldsymbol{L}$ represents the moment of momentum, or angular momentum, of a particle of mass m about point O: $\boldsymbol{L} = \boldsymbol{r} \times m\boldsymbol{v}.$

Now suppose that an external torque Γ is exerted on the plate. In a small time interval Δt the angular momentum of the plate changes by ΔL, where $\Delta L = \Gamma \Delta t$.

Since the *vectors* ΔL and Γ are both along the axis of rotation, this can be written as a *vector* equation:

$$\Delta L = \Gamma \Delta t. \qquad (9\text{--}35)$$

An analysis of the general case, where L and Γ may *not* be in the same direction, shows that the result above is always true. That is, *the **vector change** in angular momentum of a body, ΔL, is equal in magnitude and direction to the vector impulse $\Gamma \Delta t$ of the resultant external torque on the body.*

Let us now apply these vector concepts of torque and angular momentum to a specific example. In Fig. 9–20, a disk is mounted on a shaft through its center, the shaft being supported by fixed bearings. If the disk is rotating as indicated, its angular momentum vector L points toward the right, along the axis.

Suppose a cord is wrapped around the rim of the disk and a force F is exerted on the cord. The mag-

Fig. 9–20 Vector ΔL is the change in angular momentum produced in time Δt by the moment Γ of the force F. Vectors ΔL and Γ are in the same direction.

nitude of the resultant torque on the disk is $\Gamma = FR$, and the torque vector Γ also points along the axis. In a time Δt, the torque produces a vector change ΔL in the angular momentum, equal to $\Gamma \Delta t$ and having the same direction as Γ. When this change is added vectorially to the original angular momentum L, the resultant is a vector of length $L + \Delta L$, in the same direction as L. In other words, the *magnitude* of the angular momentum is increased, its *direction* remaining the same. An increase in the magnitude of the angular momentum simply means that the body rotates more rapidly.

The lengthy argument above appears at first to be nothing more than a difficult way of solving an easy problem in rotation about a fixed axis. However, the vector nature of torque and angular momentum are essential to an understanding of the gyroscope, to be discussed in the next section.

Example Two forces P and P', each of magnitude 20 N, are applied for 3 s to a disk of radius 0.5 m and moment of inertia 40 kg m², pivoted about an axis through its center, as in Fig. 9–21. The initial angular velocity of the disk is 5 rad s⁻¹. Show in a vector diagram the torque, the initial angular momentum, and the final angular momentum.

From the preceding discussion, the vectors representing the torque Γ due to the couple, and the initial angular momentum L_0, are directed as in Fig. 9–21. The magnitude of the initial angular momentum is

$$L_0 = I\omega_0 = (40 \text{ kg m}^2)(5 \text{ rad s}^{-1})$$
$$= 200 \text{ kg m}^2 \text{ s}^{-1}.$$

The angular impulse is the product of the constant torque Γ and the time interval Δt. Since this must equal the change in angular momentum, we have

$$\Gamma \Delta t = I\omega - I\omega_0 = \Delta(I\omega) = \Delta L.$$

But $\Gamma = (20 \text{ N})(0.5 \text{ m}) = 10 \text{ N m}$, and $\Delta t = 3 \text{ s}$. Hence

$$\Delta L = (10 \text{ N m})(3 \text{ s})$$
$$= 30 \text{ N m s} = 30 \text{ kg m}^2 \text{ s}^{-1}.$$

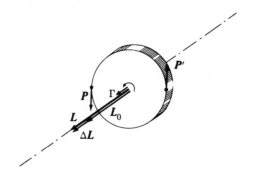

Fig. 9–21 *Vector* $\Delta\mathbf{L}$ *is the change in angular momentum produced by the couple* $\mathbf{P} - \mathbf{P}'$.

This increase has been represented by the vector $\Delta\mathbf{L}$ in Fig. 9–21.

The final angular momentum, $\mathbf{L}$, is the vector sum of $\mathbf{L}_0$ and $\Delta\mathbf{L}$. Since both are in the same direction, the vector sum is simply the arithmetic sum. That is

$$L = L_0 + \Delta L = 230 \text{ kg m}^2 \text{ s}^{-1}.$$

9–12 THE TOP AND THE GYROSCOPE

A symmetrical body rotating about an axis, one point of which is fixed, is called a *top*. If the fixed point is at the center of gravity, the body is called a *gyroscope*. The axis of rotation of a top or gyroscope can itself rotate about the fixed point, so the *direction* of the angular momentum vector can change.

Figure 9–22 illustrates the usual mounting of a toy gyroscope (more properly called a top, since the fixed point O is not at the center of mass). The top is spinning about its axis of symmetry, and if the axis is initially set in motion in the direction shown, with the proper angular velocity, the system continues to rotate uniformly about the pivot at O, the spin axis remaining horizontal.

If the axis of the top were fixed in space, its angular momentum would equal the product of its moment of inertia about the axis and its angular velocity about the axis, and would point along the axis. Because the axis itself is rotating, the angular momentum vector no longer lies on the axis. However, if the angular velocity of the axis is small com-

pared with the angular velocity about the axis, the component of angular momentum arising from the former effect is small, and we shall neglect it. The angular momentum vector $\mathbf{L}$, about the fixed point O, can then be drawn along the axis as shown and, as the top rotates about O, its angular momentum vector rotates with it.

The upward force $\mathbf{P}$ at the pivot has no moment about O. The resultant external moment is that due to the weight $\mathbf{w}$; its magnitude is

$$\Gamma = wR.$$

The direction of Γ is *perpendicular* to the axis, as shown. In a time Δt (compare with the analysis of Fig. 9–20), this torque produces a change $\Delta\mathbf{L}$ in the angular momentum, having the same direction as Γ and given by

$$\Delta\mathbf{L} = \Gamma\,\Delta t.$$

The angular momentum $\mathbf{L} + \Delta\mathbf{L}$, after a time Δt, is the vector sum of $\mathbf{L}$ and $\Delta\mathbf{L}$. Since $\Delta\mathbf{L}$ is perpendicular to $\mathbf{L}$, the new angular momentum vector has the same *magnitude* as the old but a different *direction*. The tip of the angular momentum

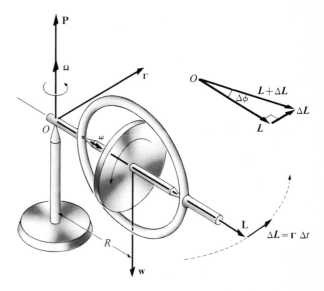

Fig. 9–22 *Vector* $\Delta\mathbf{L}$ *is the change in angular momentum produced in time* Δt *by the moment* Γ *of the force* $\mathbf{w}$. *Vectors* $\Delta\mathbf{L}$ *and* Γ *are in the same direction. (Compare with Fig. 9–20.)*

(a) Uniform circular motion of a particle

(b) Precessional motion of a top

$$v_1 = v_2 = v$$

$$\mathbf{v}_2 = \mathbf{v}_1 + \Delta\mathbf{v}$$

$$\Delta v \simeq v\,\Delta\theta$$

$$\lim_{\Delta t \to 0} \frac{\Delta v}{\Delta t} = v \lim_{\Delta t \to 0} \frac{\Delta\theta}{\Delta t}$$

$$a = v\omega \left(= \frac{v^2}{R} = R\omega^2 \right)$$

$$F = ma$$

$$F = mv\omega$$

$$\omega_1 = \omega_2 = \omega$$

$$\boldsymbol{\omega}_2 = \boldsymbol{\omega}_1 + \Delta\boldsymbol{\omega}$$

$$\Delta\omega \simeq \omega\,\Delta\phi$$

$$\lim_{\Delta t \to 0} \frac{\Delta\omega}{\Delta t} = \omega \lim_{\Delta t \to 0} \frac{\Delta\phi}{\Delta t}$$

$$\alpha = \omega\Omega$$

$$\Gamma = I\alpha$$

$$\Gamma = I\omega\Omega$$

Fig. 9–23 *Analogy between uniform circular motion of a particle and precessional motion of a top.*

vector moves as shown, and as time goes on it swings around a horizontal circle. Since the angular momentum vector lies along the gyroscope axis, the axis turns also, rotating in a horizontal plane about the point O. This motion of the axis of rotation is called *precession.*

The angle $\Delta\phi$ turned through by the vector $\mathbf{L}$ in time Δt (see the small inset diagram) is

$$\Delta\phi = \frac{\Delta L}{L} = \frac{\Gamma \Delta t}{L}.$$

The *angular velocity of precession* Ω is $\Delta\phi/\Delta t$, and therefore

$$\boxed{\Omega = \frac{\Gamma}{L}.} \qquad (9\text{–}36)$$

The angular velocity of precession is therefore *inversely* proportional to the angular momentum. If this is large, the precessional angular velocity will be small.

From a purely kinematic point of view, precessional motion of a top (see Fig. 9–22) is the rotational analog of uniform circular motion of a particle. The analogy is illustrated in Fig. 9–23, where the diagram and calculation for uniform circular motion are shown on the left and the corresponding diagram and calculation for precessional motion of a top are displayed on the right.

Why doesn't the top in Fig. 9–22 fall? The answer is that the upward force $\boldsymbol{P}$ exerted on it by the pivot is just equal in magnitude to its weight $\boldsymbol{w}$, so that the resultant vertical *force* is zero and the vertical acceleration of the center of gravity is zero. In other words, the vertical component of its linear momentum remains zero, since there is no resultant vertical force. However, the resultant *torque* of these forces is *not* zero and the angular momentum changes.

If the top were not rotating, it would have no angular momentum **L** initially. Its angular momentum Δ**L** after a time Δ*t* would be that acquired from the torque acting on it and would have the same direction as the torque. In other words, the top would rotate about an axis through *O* in the direction of the vector **Γ.** But if the top is originally rotating, the change in its angular momentum produced by the torque adds vectorially to the large angular momentum it already has, and since Δ**L** is horizontal and perpendicular to **L,** the result is a motion of precession, with both the angular momentum vector and the axis remaining horizontal.

To understand why the vertical force *P* should equal *w*, we must look further into the way in which the precessional motion in Fig. 9–22 originated. If the frame of the top is initially held at rest, say by supporting the projecting portion of the frame opposite *O* with one's finger, the upward forces exerted by the finger and by the pivot are equal to *w*/2. If the finger is suddenly removed, the upward force at *O*, at the first instant, is still *w*/2. The resultant vertical *force* is not zero, and the center of gravity has an initial downward acceleration. At the same time, precessional motion begins, although with a smaller angular velocity than that in the final steady state. The result of this motion is that the end of the frame at *O* presses down on the pivot with a greater force, so that the upward force *O* increases and eventually becomes *greater* than *w*. When this happens, the center of gravity starts to accelerate *upward*. The process repeats itself, and the motion consists of a precession together with an up-and-down oscillation of the axis, called *nutation*.

To start the top off with pure precession (without nutation), it is necessary to give the outer end of the axis a push in the direction in which it would normally precess. This causes the end of the frame at *O* to bear down on the pivot so that the upward force at *O* increases. When this force equals *w*, the vertical forces are in equilibrium, the outer end can be released, and precession goes on as in Fig. 9–22.

Fig. 9-24 *Photograph of a gimbal-mounted gyro.* [*Courtesy of Sperry Gyroscope Co.*]

Figure 9–24 is a photograph of a gyroscope mounted in gimbals. This mounting scheme permits the gyro axis *x* to take any possible orientation in space while the outer frame *C* remains stationary. If the center of mass of the gyro wheel is at the intersection of axes *x*, *y*, and *z*, no torque is exerted on it except for the small frictional torques at the pivot bearings.

PROBLEMS

9–1

a) What angle in radians is subtended by an arc 6 ft in length, on the circumference of a circle whose radius is 4 ft?

b) What angle in radians is subtended by arc of length 78.54 cm on the circumference of a circle of diameter 100 cm? What is this angle in degrees?

c) The angle between two radii of a circle is 0.60 radian. What length of arc is intercepted on the circumference of a circle of radius 200 cm? of radius 200 ft?

9–2 Compute the angular velocity, in rad s^{-1}, of the crankshaft of an automobile engine rotating at 4800 rev min^{-1}.

9–3 An electric motor running at 1800 rev min^{-1} has on its shaft three pulleys, of diameters 5, 10, and 15 cm, respectively. Find the linear velocity of the surface of each pulley in cm s^{-1}. The pulleys may be connected by a belt to a similar set on a countershaft, the 5-cm to the 15-cm, the 10-cm to the 10-cm, or the 15-cm to the 5-cm. Find the three possible angular velocities of the countershaft in rev min^{-1}.

9–4 A wheel 0.6 m in diameter starts from rest and accelerates uniformly to an angular velocity of 100 rad s^{-1} in 20 s. Find the angular acceleration and the angle turned through.

9–5 The angular velocity of a flywheel decreases uniformly from 1000 rev min^{-1} to 400 rev min^{-1} in 5 s. Find the angular acceleration and the number of revolutions made by the wheel in the 5-s interval. How many more seconds are required for the wheel to come to rest?

9–6 A flywheel requires 3 s to rotate through 234 radians. Its angular velocity at the end of this time is 108 rad s^{-1}. Find its constant angular acceleration.

9–7 A flywheel whose angular acceleration is constant and equal to 2 rad s^{-2}, rotates through an angle of 100 radians in 5 s. How long had it been in motion at the beginning of the 5-s interval if it started from rest?

9–8

a) Distinguish clearly between tangential and radial acceleration.

b) A flywheel rotates with constant angular velocity. Does a point on its rim have a tangential acceleration? a radial acceleration?

c) A flywheel is rotating with constant angular acceleration. Does a point on its rim have a tangential acceleration? a radial acceleration? Are these accelerations constant in magnitude?

9–9 A wheel 1.0 m in diameter is rotating about a fixed axis with an initial angular velocity of 2 rev s^{-1}. The acceleration is 3 rev s^{-2}.

a) Compute the angular velocity after 6 s.

b) Through what angle has the wheel turned in this time interval?

c) What is the tangential velocity of a point on the rim of the wheel at $t = 6$ s ?

d) What is the resultant acceleration of a point on the rim of the wheel at $t = 6$ s ?

9–10 A wheel having a diameter of 1 ft starts from rest and accelerates uniformly to an angular velocity of 900 rev min^{-1} in 5 s.

a) Find the position at the end of 1 s of a point originally at the top of the wheel.

b) Compute and show in diagram the magnitude and direction of the acceleration at the end of 1 s.

9–11 A flywheel of radius 30 cm starts from rest and accelerates with a constant angular acceleration of 0.50 rad s^{-2}. Compute the tangential acceleration, the radial acceleration, and the resultant acceleration, of a point on its rim (a) at the start, (b) after it has turned through 120°, (c) after it has turned through 240°.

9–12 A wheel starts from rest and accelerates uniformly to an angular velocity of 900 rev min^{-1} in 20 s. At the end of 1 s, (a) find the angle through which the wheel has rotated; and (b) compute and show in a diagram the magnitude and direction of the tangential and radial components of acceleration of a point 6 in from the axis.

9–13 Find the required angular velocity of an ultracentrifuge, in rpm, in order that the radial acceleration of a point 1 cm from the axis shall equal 300,000g (i.e., 300,000 times the acceleration of gravity).

9–14 An automobile engine is idling at 500 rev min^{-1}. When the accelerator is depressed, the angular velocity increases to 3000 rev min^{-1} in 5 s. Assume a constant angular acceleration.

a) What are the initial and final angular velocities, expressed in radians per second?

b) What was the angular acceleration, in radians per second squared?

c) How many revolutions did the engine make during the acceleration period?

d) The flywheel of the engine is 0.5 m in diameter. What is the linear speed of a point at its rim when the angular speed is 3000 rev min^{-1} ?

e) What was the tangential acceleration of the point during the acceleration period?

f) What is the radial acceleration of the point when the angular speed is 3000 rev min^{-1} ?

9–15 (a) Prove that when a body starts from rest and rotates about a fixed axis with constant angular acceleration, the radial acceleration of a point in the body is directly proportional to its angular displacement. (b) Through what angle will the body have turned when the resultant acceleration makes an angle of 60° with the radial acceleration?

9–16 A particle moves in the xy-plane according to the law

$$x = R \cos \omega t, \qquad y = R \sin \omega t,$$

where x and y are the coordinates of the body, t is the time, and R and ω are constants. (a) Eliminate t between these equations to find the equation of the curve in which the body moves. [*Hint:* Square each equation.] What is this curve? (b) Differentiate the original equations to find the x- and y-components of the velocity of the particle. Combine these expressions to obtain the magnitude and direction of the resultant velocity. (c) Differentiate again to obtain the magnitude and direction of the resultant acceleration. (This problem illustrates an alternative method of deriving the expressions for the speed and radial acceleration of a particle moving in a circle.)

9–17 Find the moment of inertia of a rod 4 cm in diameter and 2 m long, of mass 8 kg, (a) about an axis perpendicular to the rod and passing through its center, (b) about an axis perpendicular to the rod and passing through one end, (c) about a longitudinal axis through the center of the rod.

9–18 The four bodies shown in Fig. 9–25 have equal masses m. Body A is a solid cylinder of radius R. Body B is a hollow thin cylinder of radius R. Body C is a solid square with length of side $= 2R$. Body D is the same size as C, but hollow (i.e., made up of four thin sticks). The bodies haves axes of rotation perpendicular to the page and through the center of gravity of each body.

a) Which body has the smallest moment of inertia?

b) Which body has the largest moment of inertia?

Fig. 9-25

9–19 Small blocks, each of mass m, are clamped at the ends and at the center of a light rigid rod of length L. Compute the moment of inertia and the radius of gyration of the system about an axis perpendicular to the rod and passing through a point one-quarter of the length from one end. Neglect the moment of inertia of the rod.

9–20 The inner radius of a hollow cylinder is 10 cm, the outer radius is 20 cm, and the length is 30 cm. What is the radius of gyration of the cylinder about its axis?

9–21 The uniform thin rectangular plate in Fig. 9–26 has a length a, a width b, and a mass m. Find its moment of inertia (a) about the axis AA through its center O, (b) about the axis BB at one edge.

Fig. 9–26

9–22 (a) Prove that the moment of inertia of the thin flat plate in Fig. 9–27, about the z-axis, equals the sum of its moments of inertia about the x- and y-axes. (b) Given that the moment of inertia of a disk about an axis through its center and perpendicular to its plane is $mR^2/2$, use the relation above to find its moment of inertia about a diameter. (c) Derive the result of part (b) by direct integration of the defining equation, $I = \int r^2 \, dm$. (d) What is the moment of inertia of a disk about an axis tangent to its edge?

Fig. 9–27

9–23 A 60-kg grindstone is 1 m in diameter and has a radius of gyration of 0.25 m. A tool is pressed down on the rim with a normal force of 50 N. The coefficient of sliding friction between the tool and stone is 0.6, and there is a constant friction torque of 5 N m between the axle of the stone and its bearings.

a) How much force must be applied normally at the end of a crank handle 0.5 m long to bring the stone from rest to 120 rev min^{-1} in 9 s ?

b) After attaining a speed of 120 rev min^{-1}, what must the normal force at the end of the handle become to maintain a constant speed of 120 rev min^{-1} ?

c) How long will it take the grindstone to come from 120 rev min^{-1} to rest if it is acted on by the axle friction alone?

9–24 A flywheel consists of a solid disk 1 ft in diameter and 1 in. thick, and two projecting hubs 4 in. in diameter and 3 in. long. If the material of which it is constructed weighs 480 lb ft^{-3}, find (a) its moment of inertia, and (b) its radius of gyration about the axis of rotation.

9–25 A grindstone 1.0 m in diameter, of mass 50 kg, is rotating at 900 rev min^{-1}. A tool is pressed normally against the rim with a force of 200 N, and the grindstone comes to rest in 10 s. Find the coefficient of friction between the tool and the grindstone. Neglect friction in the bearings.

9–26 A constant torque of 20 N m is exerted on a pivoted wheel for 10 s, during which time the angular velocity of the wheel increases from zero to 100 rev min^{-1}. The external torque is then removed and the wheel is brought to rest by friction in its bearings in 100 s. Compute (a) the moment of inertia of the wheel, (b) the friction torque, (c) the total number of revolutions made by the wheel.

9–27 A cord is wrapped around the rim of a flywheel 0.5 m in radius, and a steady pull of 50 N is exerted on the cord, as in Fig. 9–28(a). The wheel is mounted in frictionless bearings on a horizontal shaft through its center. The moment of inertia of the wheel is 4 kg m^2.

a) Compute the angular acceleration of the wheel.

b) Show that the work done in unwinding 5 m of cord equals the gain in kinetic energy of the wheel.

c) If a mass having a weight of 50 N hangs from the cord, as in Fig. 9–28(b), compute the angular acceleration of the wheel. Why is this not the same as in part (a)?

50 N   50 N

(a)   (b)

Fig. 9–28

9–28 A solid cylinder of mass 15 kg, 0.3 m in diameter, is pivoted about a horizontal axis through its center, and a rope wrapped around the surface of the cylinder carries at its end a block of mass 8 kg.

a) How far does the block descend in 5 s, starting from rest?

b) What is the tension in the rope?

c) What is the force exerted on the cylinder by its bearings?

9–29 A bucket of water weighing 64 lb is suspended by a rope wrapped around a windlass in the form of a solid cylinder 1 ft in diameter, also weighing 64 lb. The bucket is released from rest at the top of a well and falls 64 ft to the water.

a) What is the tension in the rope while the bucket is falling?

b) With what velocity does the bucket strike the water?

c) What was the time of fall? Neglect the weight of the rope.

9–30 A 16-lb block rests on a horizontal frictionless surface. A cord attached to the block passes over a pulley, whose diameter is 6 in., to a hanging block also weighing 16 lb. The system is released from rest, and the blocks are observed to move 16 ft in 2 s.

a) What was the moment of inertia of the pulley?

b) What was the tension in each part of the cord?

9–31 Figure 9–29 represents an *Atwood's machine*. Find the linear accelerations of blocks *A* and *B*, the angular acceleration of the wheel *C*, and the tension in each side of the cord (a) if the surface of the wheel is frictionless; (b) if there is no slipping between the cord and the surface of the wheel. Let the masses of blocks *A* and *B* be 4 kg and 2 kg, respectively, the moment of inertia of the wheel about its axis be 0.2 kg m^2, and the radius of the wheel be 0.1 m.

C

A   *Fig. 9–29*

B

9–32 A flywheel 1.0 m in diameter is pivoted on a horizontal axis. A rope is wrapped around the outside of the flywheel, and a steady pull of 50 N is exerted on the rope. It is found that 10 m of rope are unwound in 4 s.

a) What was the angular acceleration of the flywheel?

b) What is its final angular velocity?

c) What is its final kinetic energy?

d) What is its moment of inertia?

9–33 A light rigid rod 1.0 m long has a small block of mass 0.05 kg attached at one end. The other end is pivoted, and the rod rotates in a vertical circle. At a certain instant, the rod makes an angle of 53° with the vertical, and the tangential speed of the block is 4 m s^{-1}.

a) What are the horizontal and vertical components of the velocity of the block?

b) What is the moment of inertia of the system?

c) What is the radial acceleration of the block?

d) What is the tangential acceleration of the block?

e) What is the tension or compression in the rod?

9–34

a) Compute the torque developed by an airplane engine whose output is 2000 hp at an angular velocity of 2400 rev min^{-1}.

b) If a drum 0.5 m in diameter were attached to the motor shaft, and the power output of the motor were used to raise a weight hanging from a rope wrapped around the shaft, how large a weight could be lifted?

c) With what velocity would it rise?

9–35 A block of mass $m = 5$ kg slides down a surface inclined 37° to the horizontal, as shown in Fig. 9–30. The coefficient of sliding friction is 0.25. A string attached to the block is wrapped around a flywheel on a fixed axis at O. The flywheel has a mass $M = 20$ kg, an outer radius $R = 0.2$ m, and a radius of gyration with respect to the axis of $k_0 = 0.1$ m.

a) What is the acceleration of the block down the plane?

b) What is the tension in the string?

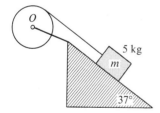

Fig. 9–30

9–36 A body of mass m is attached to a light cord wound around the shaft of a wheel, as in Fig. 9–31. The radius of the shaft is r, and the shaft is supported in fixed, frictionless bearings. When released from rest, the body descends a distance of 1.75 m in 5 s. Find the moment of inertia of the wheel and shaft, in terms of m and r.

9–37 A grinding wheel 0.2 m in diameter, of mass 3 kg, is rotating at 3600 rev min^{-1}.

a) What is its kinetic energy?

b) How far would it have to fall to acquire the same kinetic energy?

9–38 A disk of mass m and radius R is pivoted about a horizontal axis through its center, and a small body of

Fig. 9–31

mass m is attached to the rim of the disk. If the disk is released from rest with the small body at the end of a horizontal radius, find the angular velocity when the small body is at the bottom.

9–39 The flywheel of a gasoline engine is required to give up 380 ft lb of kinetic energy while its angular velocity decreases from 600 rev min^{-1} to 540 rev min^{-1}. What moment of inertia is required?

9–40 The flywheel of a punch press has a moment of inertia of 25 kg m^2 and it runs at 300 rev min^{-1}. The flywheel supplies all the energy needed in a quick punching operation.

a) Find the speed in rev min^{-1} to which the flywheel will be reduced by a sudden punching operation requiring 4000 J of work.

b) What must be the constant power supply to the flywheel in horsepower to bring it back to its initial speed in 5 s?

9–41 A magazine article described a passenger bus in Zurich, Switzerland, which derived its motive power from the energy stored in a large flywheel. The wheel was brought up to speed periodically, when the bus stopped at a station, by an electric motor which could then be attached to the electric power lines. The flywheel was a solid cylinder of mass 1000 kg, diameter 1.8 m, and its top speed was 3000 rev min^{-1}.

a) At this speed, what is the kinetic energy of the flywheel?

b) If the average power required to operate the bus is 25 hp, how long can it operate between stops?

9–42 A grindstone in the form of a solid cylinder has a radius of 0.5 m and a mass of 50 kg.

a) What torque will bring it from rest to an angular velocity of 300 rev min^{-1} in 10 s?

b) What is its kinetic energy when rotating at 300 rev min^{-1} ?

9–43 The flywheel of a motor has a mass of 300 kg and a radius of gyration of 1.5 m. The motor develops a constant torque of 2000 N m and the flywheel starts from rest.

a) What is the angular acceleration of the flywheel?

b) What will be its angular velocity after making 4 revolutions?

c) How much work is done by the motor during the first 4 revolutions?

9–44 The flywheel of a stationary engine has a moment of inertia of 20 slug ft^2.

a) What constant torque is required to bring it up to an angular velocity of 900 rev min^{-1} in 10 s, starting from rest?

b) What is its final kinetic energy?

9–45 In Fig. 9–32, the steel balls A and B have a mass of 500 g each, and are rotating about the vertical axis with an angular velocity of 4 rad s^{-1} at a distance of 15 cm from the axis. Collar C is now forced down until the balls are at a distance of 5 cm from the axis. How much work must be done to move the collar down?

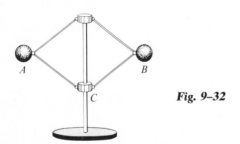

Fig. 9–32

9–46 A man sits on a piano stool holding a pair of dumbbells at a distance of 0.6 m from the axis of rotation of the stool. He is given an angular velocity of 5 rad s^{-1}, after which he pulls the dumbbells in until they are only 0.2 m distant from the axis. The moment of inertia of the man about the axis of rotation is 5 kg m^2 and may be considered constant. Each dumbbell has a mass of 5 kg and may be considered a point mass. Neglect friction.

a) What is the initial angular momentum of the system?

b) What is the angular velocity of the system after the dumbbells are pulled in toward the axis?

c) Compute the kinetic energy of the system before and after the dumbbells are pulled in. Account for the difference, if any.

9–47 A block of mass 0.05 kg is attached to a cord passing through a hole in a horizontal frictionless surface, as in Fig.

9–33. The block is originally revolving at a distance of 0.2 m from the hole, with an angular velocity of 3 rad s^{-1}. The cord is then pulled from below, shortening the radius of the circle in which the block revolves to 0.1 m. The block may be considered a point mass.

a) What is the angular velocity?

b) Find the change in kinetic energy of the block.

Fig. 9–33

9–48 A small block weighing 8 lb is attached to a cord passing through a hole in a horizontal frictionless surface. The block is originally revolving in a circle of radius 2 ft about the hole, with a tangential velocity of 12 ft s^{-1}. The cord is then pulled slowly from below, shortening the radius of the circle in which the block revolves. The breaking strength of the cord is 144 lb. What will be the radius of the circle when the cord breaks?

9–49 A block of mass M rests on a turntable which is rotating at constant angular velocity ω. A smooth cord runs from the block through a hole in the center of the table down to a hanging block of mass m. The coefficient of friction between the first block and the turntable is μ. (See Fig. 9–34.) Find the largest and smallest values of the radius r for which the first block will remain at rest relative to the turnable.

Fig. 9–34

9–50 A uniform rod of mass 0.03 kg and 0.2 m long rotates in a horizontal plane about a fixed vertical axis through its center. Two small bodies, each of mass 0.02 kg, are mounted so that they can slide along the rod. They are initially held by catches at positions 0.05 cm on each side of the center of the rod, and the system is rotating at 15 rev min^{-1}. Without otherwise changing the system, the catches are released and the masses slide outward along the rod and fly off at the ends.

a) What is the angular velocity of the system at the instant when the small masses reach the ends of the rod?

b) What is the angular velocity of the rod after the small masses leave it?

9–51 A turntable rotates about a fixed vertical axis, making one revolution in 10 s. The moment of inertia of the turntable about this axis is 1200 kg m². A man of mass 80 kg, initially standing at the center of the turntable, runs out along a radius. What is the angular velocity of the turntable when the man is 2 m from the center?

9–52 Disks A and B are mounted on a shaft SS and may be connected or disconnected by a clutch C, as in Fig. 9–35. The moment of inertia of disk A is one-half that of disk B. With the clutch disconnected, A is brought up to an angular velocity ω_0. The accelerating torque is then removed from A and it is coupled to disk B by the clutch. (Bearing friction may be neglected.) It is found that 2000 J of heat are developed in the clutch when the connection is made. What was the original kinetic energy of disk A?

Fig. 9–35

9–53 A man of mass 100 kg stands at the rim of a turntable of radius 2 m and moment of inertia 4000 kg m², mounted on a vertical frictionless shaft at its center. The whole system is initially at rest. The man now walks along the outer edge of the turntable with a velocity of 1 m s⁻¹, relative to the earth.

a) With what angular velocity and in what direction does the turntable rotate?

b) Through what angle will it have rotated when the man reaches his initial position on the turntable?

c) Through what angle will it have rotated when he reaches his initial position relative to the earth?

9–54 A man of mass 60 kg runs around the edge of a horizontal turntable mounted on a vertical frictionless axis through its center. The velocity of the man, relative to the earth, is 1 m s⁻¹. The turntable is rotating in the opposite direction with an angular velocity of 0.2 rad s⁻¹. The radius of the turntable is 2 m and its moment of inertia about the axis of rotation is 400 kg m². Find the final

angular velocity of the system if the man comes to rest, relative to the turntable.

9–55 Two flywheels, A and B, are mounted on shafts which can be connected or disengaged by a friction clutch C (see Fig. 9–35). The moment of inertia of wheel A is 4 slug ft². With the clutch disengaged, wheel A is brought up to an angular velocity of 600 rev min⁻¹. Wheel B is initially at rest. The clutch is now engaged, accelerating B and decelerating A, until both wheels have the same angular velocity. The final angular velocity of the system is 400 rev min⁻¹.

a) What was the moment of inertia of wheel B?

b) How much energy was lost in the process? (Neglect all bearing friction.)

9–56 The stabilizing gyroscope of a ship has a mass of 50,000 kg, its radius of gyration is 2 m, and it rotates about a vertical axis with an angular velocity of 900 rev min⁻¹.

a) How long a time is required to bring it up to speed, starting from rest, with a constant power input of 100 hp?

b) Find the torque needed to cause the axis to precess in a vertical fore-and-aft plane at the rate of 1 degree s⁻¹.

9–57 The mass of the rotor of a toy gyroscope is 150 g and its moment of inertia about its axis is 1500 g cm². The mass of the frame is 30 g. The gyroscope is supported on a single pivot, as in Fig. 9–36, with its center of gravity distant 4 cm horizontally from the pivot, and is precessing in a horizontal plane at the rate of one revolution in 6 s.

a) Find the upward force exerted by the pivot.

b) Find the angular velocity with which the rotor is spinning about its axis, expressed in rev min⁻¹.

c) Copy the diagram, and show by vectors the angular momentum of the rotor and the torque acting on it.

Fig. 9–36

9–58 The moment of inertia of the front wheel of a bicycle is 0.4 kg m², its radius is 0.5 m, and the forward speed of the bicycle is 5 m s⁻¹. With what angular velocity must the front wheel be turned about a vertical axis to counteract the capsizing torque due to a mass of 60 kg, 0.02 m horizontally to the right or left of the line of contact of wheels and ground? (Bicycle riders: Compare your own experience and see if your answer seems reasonable.)

9–59 The rotor of a small control gyro can be accelerated from rest to an angular velocity of 50,000 rev min^{-1} in 0.2 s. The moment of inertia of the rotor about its spin axis is 385 g cm^2.

a) What is the angular acceleration (assumed constant) as the rotor is brought up to speed?

b) What torque is needed to produce this angular acceleration?

c) Suppose the data above refer to the gyro in Fig. 9–24, and that the rotor is spinning clockwise when viewed from the left along the x-axis. Which way will the spin axis turn if ring B is forced to rotate clockwise, as seen from above, about the z-axis?

d) Suppose that rings A and B are locked in place. What torque must be exerted about the y-axis to rotate the outer frame about this axis at 1 deg s^{-1}?

9–60 A demonstration gyro wheel is constructed by removing the tire from a bicycle wheel 1.0 m in diameter, wrapping lead wire around the rim, and taping it in place. The shaft projects 0.2 m at each side of the wheel and a man holds the ends of the shaft in his hands. The mass of the system is 5 kg and its entire mass may be assumed to be located at its rim. The shaft is horizontal and the wheel is spinning about the shaft at 5 rev s^{-1}. Find the magnitude and direction of the force each hand exerts on the shaft under the following conditions:

a) The shaft is at rest.

b) The shaft is rotating in a horizontal plane about its center at 0.04 rev s^{-1}.

c) The shaft is rotating in a horizontal plane about its center at 0.20 rev s^{-1}.

d) At what rate must the shaft rotate in ordér that it may be supported at one end only?

Chapter 10

Elasticity

10–1 STRESS

The preceding chapter dealt with the motion of a rigid body, an idealized model used to represent the motion of a body when rotational motion may not be neglected but when the changes in the *shape* of the body are so small as to be negligible. Real materials always yield to some extent under the influence of applied forces; we shall now develop models to describe the relation between deformation of real materials and the forces responsible for these deformations.

Ultimately, the change in shape or volume of a body when outside forces act on it is determined by the forces between its molecules. Although molecular theory is at present not sufficiently advanced to enable us to calculate the elastic properties of, say, a

block of copper starting from the properties of a copper atom, the study of the solid state is an active subject in many research laboratories and our knowledge of it is steadily increasing. In this chapter we shall, however, confine ourselves to quantities that are directly measurable, and not attempt any detailed molecular explanation of the observed behavior.

Figure 10–1(a) shows a bar of uniform cross-sectional area A subjected to equal and opposite pulls F at its ends. The bar is said to be in *tension*. Consider a section through the bar at right angles to its length, as indicated by the dotted line. Since every portion of the bar is in equilibrium, that portion at the right of the section must be pulling on the portion at the left with a force F, and vice versa. If the section is not too near the ends of the bar, these pulls are

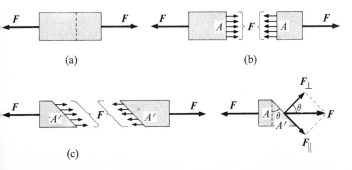

(a)

(b)

(c)

Fig. 10–1 (a) A bar in tension. (b) The stress at a perpendicular section equals F/A. (c) and (d) The stress at an inclined section can be resolved into a normal stress $F_\perp/A'$, and a tangential or shearing stress $F_\parallel/A'$.

183

uniformly distributed over the cross-sectional area A, as indicated by the short arrows in Fig. 10–1(b). We define the *stress S* at the section as the ratio of the force F to the area A:

$$\text{Stress} = \frac{F}{A}. \qquad (10\text{–}1)$$

The stress is called a *tensile* stress, meaning that each portion *pulls* on the other, and it is also a *normal* stress because the distributed force is perpendicular to the area. Units of stress are 1 newton per square meter (1 N m^{-2}), 1 dyne per square centimeter (1 dyn cm^{-2}), and 1 pound per square foot (1 lb ft^{-2}). The hybrid unit 1 lb in^{-2} is also commonly used.

Example　A human biceps (upper arm muscle) may exert a force of the order of 600 N on the bones to which it is attached. If the muscle has a cross-sectional area at its center of 50 cm^2 = 0.005 m^2, and the tendon attaching its lower end to the bones below the elbow joint has a cross section of 0.5 cm^2 = 5×10^{-5} m^2, find the tensile stress in each of these.

In each case the stress is the force per unit area. For the muscle,

$$\text{Tensile stress} = \frac{600 \text{ N}}{0.005 \text{ m}^2} = 120{,}000 \text{ N m}^{-2}$$
$$= 1.2 \times 10^5 \text{ N m}^{-2}.$$

For the tendon,

$$\text{Tensile stress} = \frac{600 \text{ N}}{5 \times 10^{-5} \text{ m}^2} = 1.2 \times 10^7 \text{ N m}^{-2}.$$

The maximum force a muscle can exert depends on its cross-sectional area, but the maximum *stress* is nearly the same for a wide variety of muscles.

Returning to Fig. 10–1, we consider next a section through the bar in some arbitrary direction, as in Fig. 10–1(c). The resultant force exerted on the portion at either side of this section, by the portion at the other, is equal and opposite to the force F at the end of the section. Now, however, the force is distributed over a larger area A' and is not at right

angles to the area. If we represent the resultant of the distributed forces by a single vector of magnitude F, as in Fig. 10–1(d), this vector can be resolved into a component $F_\perp$ normal to the area A', and a component $F_\parallel$ tangent to the area. The *normal* stress is defined, as before, as the ratio of the component $F_\perp$ to the area A'. The ratio of the component $F_\parallel$ to the area A' is called the *tangential* stress or, more commonly, the *shearing* stress at the section:

$$\text{Normal stress} = \frac{F_\perp}{A'}$$
$$\text{Tangential (shearing) stress} = \frac{F_\parallel}{A'}. \qquad (10\text{–}2)$$

Stress is not a vector quantity since, unlike a force, we cannot assign to it a specific direction. The *force* acting on the portion of the body on a specified side of a section has a definite direction. Stress is one of a class of physical quantities called *tensors*.

A bar subjected to *pushes* at its ends, as in Fig. 10–2, is said to be in *compression*. The stress on the dotted section, illustrated in part (b), is also a normal stress but is now a *compressive* stress, since each portion pushes on the other. It should be evident that if we take a section in some arbitrary direction, it will be subject to both a tangential (shearing) and a normal stress, the latter now being a compression.

As another example of a body under stress, consider the block of square cross section in Fig. 10–3(a), acted on by two equal and opposite couples produced by the pairs of forces F_x and F_y distributed over its surfaces. The block is in equilibrium, and any portion of it is in equilibrium also. Thus the distributed forces over the diagonal face in part (b) must have a resultant F whose components are equal to F_x and F_y. The stress at this section is therefore a pure compression, although the stresses at the right face and the bottom are both shearing stresses. Similarly, we see from Fig. 10–3(c) that the other diagonal face is in pure tension.

Fig. 10–2 A bar in compression.

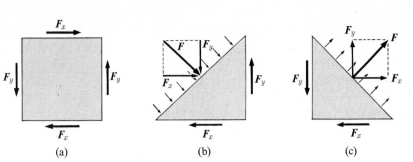

Fig. 10–3 (a) A body in shear. The stress on one diagonal, part (b), is a pure compression; that on the other, part (c), is a pure tension.

Consider next a fluid under pressure. The term "fluid" means a substance that can flow; hence the term applies to both liquids and gases. If there is a shearing stress at any point in a fluid, the fluid slips sidewise so long as the stress is maintained. Hence in a fluid at rest, the shearing stress is everywhere zero. Figure 10–4 represents a fluid in a cylinder provided with a piston, on which is exerted a downward force. The triangle is a side view of a wedge-shaped portion of the fluid. If for the moment we neglect the weight of the fluid, the only forces on this portion are those exerted by the rest of the fluid, and since these forces can have no shearing (or tangential) component, they must be normal to the surfaces of the wedge. Let F_x, F_y, and F represent the forces against the three faces. Since the fluid is in equilibrium, it follows that

$$F \sin \theta = F_x, \qquad F \cos \theta = F_y.$$

Also,

$$A \sin \theta = A_x, \qquad A \cos \theta = A_y.$$

Dividing the upper equations by the lower, we find

$$\frac{F}{A} = \frac{F_x}{A_x} = \frac{F_y}{A_y}.$$

Hence the force per unit area is the *same*, regardless of the direction of the section, and is always a compression. Any one of the preceding ratios defines the *hydrostatic pressure p* in the fluid,

$$\boxed{p = \frac{F}{A}, \qquad F = pA.} \qquad (10\text{–}3)$$

Units of pressure are 1 N m^{-2}, 1 dyn cm^{-2}, or 1 lb ft^{-2}. Like other types of stress, pressure is not a vector quantity and no direction can be assigned to it. *The force against any area within (or bounding) a fluid at rest and under pressure is normal to the area, regardless of the orientation of the area.* This is what is meant by the common statement that "the pressure in a fluid is the same in all directions."

The stress within a solid can also be a hydrostatic pressure, provided the stress at all points of the surface of the solid is of this nature. That is, the force per unit area must be the same at *all* points of the surface, and the force must be normal to the surface and directed inward. This is *not* the case in Fig. 10–2, where forces are applied at the ends of the bar only, but it is automatically the case if a solid is immersed in a fluid under pressure.

Fig. 10–4 A fluid under hydrostatic pressure. The force on a surface in any direction is normal to the surface.

10–2 STRAIN

The term *strain* refers to the relative change in dimensions or shape of a body which is subjected to stress. Associated with each type of stress described in the preceding section is a corresponding type of strain.

Figure 10–5 shows a bar whose natural length is ℓ_0 and which elongates to a length ℓ when equal and opposite pulls are exerted at its ends. The elongation, of course, does not occur at the ends only; every element of the bar stretches in the same proportion as does the bar as a whole. The *tensile strain* in the bar is defined as the ratio of the increase in length to the original length:

Fig. 10–5 *The longitudinal strain is defined as* $\Delta\ell/\ell_0$.

$$\text{Tensile strain} = \frac{\ell - \ell_0}{\ell_0} = \frac{\Delta\ell}{\ell_0}. \qquad (10\text{–}4)$$

The *compressive strain* of a bar in compression is defined in the same way, as the ratio of the decrease in length to the original length.

Figure 10–6(a) illustrates the nature of the deformation when shearing stresses act on the faces of a block, as in Fig. 10–3. The dotted outline *abcd* represents the unstressed block, and the full lines $a'b'c'd'$ represent the block under stress. The centers of the stressed and unstressed block coincide in part (a). In part (b), the edges *ad* and $a'd'$ coincide. The lengths of the faces under shear remain very nearly constant, while all dimensions parallel to the diagonal *ac* increase in length, and those parallel to the diagonal *bd* decrease in length. Note that this is to be expected in view of the nature of the corresponding internal stresses (see Fig. 10–3). This type of strain is called a *shear strain*, and is defined as the ratio of the displacement *x* of corner *b* to the transverse dimension *h*:

$$\text{Shearing strain} = \frac{x}{h}. \qquad (10\text{–}5)$$

Like other types of strain, shearing strain is a pure number.

The strain produced by a hydrostatic pressure, called a *volume strain*, is defined as the ratio of the change in volume, ΔV, to the original volume V. It also is a pure number:

$$\text{Volume strain} = \frac{\Delta V}{V}. \qquad (10\text{–}6)$$

(a)

(b)

Fig. 10–6 *Change in shape of a block in shear. The shearing strain is defined as* x/h.

10–3 ELASTICITY AND PLASTICITY

The relation between each of the three kinds of stress and its corresponding strain plays an important role in the branch of physics called the *theory of elasticity*,

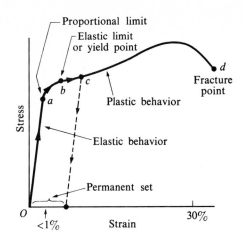

Fig. 10–7 *Typical stress–strain diagram for a ductile metal under tension.*

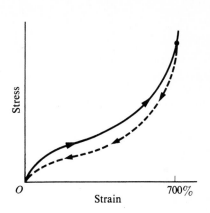

Fig. 10–8 *Typical stress–strain diagram for vulcanized rubber, showing elastic hysteresis.*

or its engineering counterpart, *strength of materials.* When any stress is plotted against the appropriate strain, the resulting stress–strain diagram is found to have several different shapes, depending on the kind of material. Two of the most important materials of present-day science and technology are metal and vulcanized rubber.

There are wide variations in the elastic behavior of metals. A typical stress–strain diagram for a ductile metal is shown in Fig. 10–7. The stress is a simple tensile stress and the strain is the percentage elongation. During the first portion of the curve (up to a strain of less than 1 percent), the stress and strain are proportional until the point *a*, the *proportional limit*, is reached. The fact that there is a region in which stress and strain are proportional is called *Hooke's Law.* (Robert Hooke (1635–1703) was a contemporary of Newton.)

From *a* to *b*, stress and strain are not proportional, but nevertheless, if the load is removed at any point between *O* and *b*, the curve will be retraced and the material will return to its original length. In the region *Ob*, the material is said to be *elastic* or to exhibit *elastic behavior*, and the point *b* is called the *elastic limit*, or the *yield point*. Up to this point, the forces exerted by the material are *conservative*; when the material returns to its original shape, work done in producing the deformation is recovered.

If the material is loaded further, the strain increases rapidly, but when the load is removed at some point beyond *b*, say *c*, the material does not come back to its original length but traverses the dashed line in Fig. 10–7. The length at zero stress is now greater than the original length and the material is said to have a *permanent set.* Further increase of load beyond *c* produces a large increase in strain until a point *d* is reached at which *fracture* takes place. From *b* to *d*, the metal is said to undergo *plastic flow* or *plastic deformation.* If large plastic deformation takes place between the elastic limit and the fracture point, the metal is said to be *ductile.* If, however, fracture occurs soon after the elastic limit is passed, the metal is said to be *brittle.*

Figure 10–8 shows a stress–strain curve for a typical sample of vulcanized rubber which has been stretched to over seven times its original length. During no portion of this curve is the stress proportional to the strain. The substance, however, is elastic, in the sense that when the load is removed, the rubber is restored to its original length. On decreasing the load, the stress–strain curve is *not* retraced but follows the dashed curve of Fig. 10–8.

The lack of coincidence of the curves for increasing and decreasing stress is known as *elastic hysteresis* (accent on the *third* syllable). (An analogous phenomenon observed with magnetic materials is called *magnetic hysteresis.*) When the stress–strain relation exhibits this behavior, the associated forces are *not* conservative, since the work done by the material in returning to its original shape is *less* than the work required to deform it. It can be shown that the area bounded by the two curves, that is, the area of the *hysteresis loop*, is equal to the energy dissipated within the elastic or magnetic material.

The large elastic hysteresis of some types of rubber makes these materials very valuable as vibration absorbers. If a block of such material is placed between a piece of vibrating machinery, and, let us say, the floor, elastic hysteresis takes place during every cycle of vibration. Mechanical energy is converted to a form known as internal energy which evidences itself by a rise in temperature. As a result, only a small amount of energy of vibration is transmitted to the floor.

The stress required to cause actual fracture of a material is called the *breaking stress* or the *ultimate strength.* Two materials, such as two steels, may have very similar elastic constants but vastly different breaking stresses.

10–4 ELASTIC MODULUS

The stress required to produce a given strain depends on the nature of the material under stress. The ratio of stress to strain, or the *stress per unit strain,* is called an *elastic modulus* of the material. The larger the elastic modulus, the greater the stress needed for a given strain.

Consider first longitudinal (tensile or compressive) stresses and strains. Experiment shows that up to the proportional limit, a given longitudinal stress produces a strain of the same magnitude whether the stress is a tension or a compression. Hence, the ratio of tensile stress to tensile strain, for a given material, equals the ratio of compressive stress to compressive strain. This ratio is called the *stretch modulus* or *Young's modulus* of the material, and will be denoted by Y:

$$Y = \frac{\text{tensile stress}}{\text{tensile strain}} = \frac{\text{compressive stress}}{\text{compressive strain}}$$
$$= \frac{F_\perp/A}{\Delta\ell/\ell_0} = \frac{\ell_0\,F_\perp}{A\,\Delta\ell}. \tag{10–7}$$

If the proportional limit is not exceeded, the ratio of stress to strain is constant, and Hooke's law is therefore equivalent to the statement that *within the proportional limit, the elastic modulus of a given material is constant,* depending only on the nature of the material.

Since a strain is a pure number, the units of Young's modulus are the same as those of stress, namely, force per unit area. Some typical values are listed in Table 10–1.

Table 10–1 APPROXIMATE ELASTIC CONSTANTS

Material	Young's modulus, Y		Shear modulus, S		Bulk modulus, B		Poisson's ratio, σ
	$10^{11}\,\text{N m}^{-2}$	$10^6\,\text{lb in}^{-2}$	$10^{11}\,\text{N m}^{-2}$	$10^6\,\text{lb in}^{-2}$	$10^{11}\,\text{N m}^{-2}$	$10^6\,\text{lb in}^{-2}$	
Aluminum	0.70	10	0.30	3.4	0.70	10	0.16
Brass	0.91	13	0.36	5.1	0.61	8.5	0.26
Copper	1.1	16	0.42	6.0	1.4	20	0.32
Glass	0.55	7.8	0.23	3.3	0.37	5.2	0.19
Iron	1.9	26	0.70	10	1.0	14	0.27
Lead	0.16	2.3	0.056	0.8	0.077	1.1	0.43
Nickel	2.1	30	0.77	11	2.6	34	0.36
Steel	2.0	29	0.84	12	1.6	23	0.19
Tungsten	3.6	51	1.5	21	2.0	29	0.20

When a material elongates under tensile stress, the dimensions *perpendicular* to the direction of stress become shorter by an amount proportional to the fractional change in length. If w is the original width and Δw the change in width, then it is found that

$$\frac{\Delta w}{w} = -\sigma \frac{\Delta \ell}{\ell}, \qquad (10\text{–}8)$$

where σ is a dimensionless constant characteristic of the material, called *Poisson's ratio*. For many common materials σ has a value between 0.1 and 0.3. Similarly, a material under compressive stress "bulges" at the sides, and again the fractional change in width is given by Eq. (10–8).

The *shear modulus S* of a material, within the Hooke's law region, is defined as the ratio of a shearing stress to the shearing strain it produces:

$$S = \frac{\text{shearing stress}}{\text{shearing strain}} \\ = \frac{F_\parallel/A}{x/h} = \frac{h}{A}\frac{F_\parallel}{x}. \qquad (10\text{–}9)$$

(Refer to Fig. 10–6 for the meaning of x and of h.)

The shear modulus of a material is also expressed as force per unit area. For most materials it is one-half to one-third as great as Young's modulus. The shear modulus is also called the *modulus of rigidity* or the *torsion modulus*.

The more general definition of shear modulus is

$$S = \frac{dF_\parallel/A}{dx/h} = \frac{h}{A}\frac{dF_\parallel}{dx}, \qquad (10\text{–}10)$$

where dx is the increase in x when the shearing force increases by $dF_\parallel$.

The shear modulus has a significance for *solid* materials only. A liquid or gas flows under the influence of a shearing stress, and cannot *permanently* support such a stress.

The modulus relating a hydrostatic pressure to the volume strain it produces is called the *bulk modulus*, and we shall represent it by B. The general definition of bulk modulus is the (negative) ratio of a change in pressure dp to the volume strain dV/V (fractional change in volume) produced by it:

$$B = -\frac{dp}{dV/V} = -V\frac{dp}{dV}. \qquad (10\text{–}11)$$

The minus sign is included in the definition of B because an *increase* of pressure always causes a *decrease* in volume. That is, if dp is positive, dV is negative. By including a minus sign in its definition, we make the bulk modulus itself a positive quantity.

The change in volume of a *solid* or *liquid* under pressure is so small that the volume V in Eq. (10–11) can be considered constant. Provided the pressure is not too great, the ratio dp/dV is constant also, the bulk modulus is constant, and we can replace dp and dV by finite changes in pressure and volume. The volume of a *gas*, however, changes markedly with pressure, and the general definition of B must be used for gases.

The reciprocal of the bulk modulus is called the *compressibility k*. From its definition,

$$k = \frac{1}{B} = -\frac{dV/V}{dp} = -\frac{1}{V}\frac{dV}{dp}. \qquad (10\text{–}12)$$

The compressibility of a material thus equals the *fractional decrease in volume*, $-dV/V$, *per unit increase* dp *in pressure*.

The units of a bulk modulus are the same as those of pressure, and the units of compressibility are those of *reciprocal pressure*. Thus the statement that the compressibility of water (see Table 10–2) is 50×10^{-6} atm^{-1} means that the volume decreases by

Table 10–2 COMPRESSIBILITIES OF LIQUIDS

Liquid	Compressibility, k		
	$(\text{N m}^{-2})^{-1}$	$(\text{lb in}^{-2})^{-1}$	atm^{-1}
Carbon disulfide	64×10^{-11}	45×10^{-7}	66×10^{-6}
Ethyl alcohol	110	78	115
Glycerine	21	15	22
Mercury	3.7	2.6	3.8
Water	49	34	50

Table 10-3 STRESSES AND STRAINS

Type of stress	Stress	Strain	Elastic modulus	Name of modulus
Tension or compression	$\dfrac{F_\perp}{A}$	$\dfrac{\Delta\ell}{\ell_0}$	$Y = \dfrac{F_\perp/A}{\Delta\ell/\ell_0}$	Young's modulus
Shear	$\dfrac{F_\parallel}{A}$	$\tan\phi \approx \phi$	$S = \dfrac{F_\parallel/A}{\phi}$	Shear modulus
Hydrostatic pressure	$p\left(=\dfrac{F_\perp}{A}\right)$	$\dfrac{\Delta V}{V_0}$	$B = -\dfrac{p}{\Delta V/V_0}$	Bulk modulus

50 millionths of the original volume for each atmosphere increase in pressure. ($1\,\text{atm} = 1.02 \times 10^5\,\text{N}\,\text{m}^{-2}$ $= 14.7\,\text{lb}\,\text{in}^{-2}$.)

The various types of stress, strain, and elastic modulus are summarized in Table 10-3.

Example 1 In an experiment to measure Young's modulus, a load of 500 kg, hanging from a steel wire of length 3 m and cross section $0.20\,\text{cm}^2$, was found to stretch the wire 0.4 cm above its no-load length. What were the stress, the strain, and the value of Young's modulus for the steel of which the wire was composed?

$$\text{Stress} = \frac{F_\perp}{A} = \frac{(500\,\text{kg})(10\,\text{m}\,\text{s}^{-2})}{0.00002\,\text{m}^2}$$
$$= 250 \times 10^6\,\text{N}\,\text{m}^{-2};$$

$$\text{Strain} = \frac{\Delta\ell}{\ell_0} = \frac{0.004\,\text{m}}{3\,\text{m}} = 0.00133;$$

$$Y = \frac{\text{stress}}{\text{strain}} = \frac{250 \times 10^6\,\text{N}\,\text{m}^{-2}}{0.00133}$$
$$= 1.9 \times 10^{11}\,\text{N}\,\text{m}^{-2}.$$

Example 2 Suppose the object in Fig. 10-6 is a brass plate 1.0 m square and 0.5 cm thick. How large a force F must be exerted on each of its edges if the displacement x in Fig. 10-6(b) is 0.02 cm? The shear modulus of brass is $0.36 \times 10^{11}\,\text{N}\,\text{m}^{-2}$.

The shearing stress on each edge is

$$\text{Shearing stress} = \frac{F_\parallel}{A}$$
$$= \frac{F}{(1.0\,\text{m})(0.005\,\text{m})} = 200\,F\,\text{N}\,\text{m}^{-2}.$$

The shearing strain is

$$\text{Shearing strain} = \frac{x}{h} = \frac{2 \times 10^{-4}\,\text{m}}{1.0\,\text{m}} = 2.0 \times 10^{-4}.$$

$$\text{Shear modulus } S = \frac{\text{stress}}{\text{strain}}$$
$$= 0.36 \times 10^{11}\,\text{N}\,\text{m}^{-2}$$
$$= \frac{200\,F\,\text{N}\,\text{m}^{-2}}{2.0 \times 10^{-4}},$$

$$F = 3.6 \times 10^4\,\text{N}.$$

Example 3 The volume of oil contained in a certain hydraulic press is $0.2\,\text{m}^3$. Find the decrease in volume of the oil when subjected to a pressure of $2.04 \times 10^7\,\text{N}\,\text{m}^{-2}$. The compressibility of the oil is $20 \times 10^{-6}\,\text{atm}^{-1}$.

The volume decreases by 20 parts per million for a pressure increase of 1 atm. Since $2.04 \times 10^7\,\text{N}\,\text{m}^{-2}$ $= 200\,\text{atm}$, the volume decrease is $200 \times 20 = 4000$ parts per million. Since the original volume is $0.2\,\text{m}^3$, the actual decrease is

$$\left(\frac{4000}{1,000,000}\right)(0.2\,\text{m}^3) = 8 \times 10^{-4}\,\text{m}^3 = 800\,\text{cm}^3;$$

or, from Eq. (10-12),

$$\Delta V = -kV\,\Delta p$$
$$= -(20 \times 10^{-6}\,\text{atm}^{-1})(0.2\,\text{m}^3)(200\,\text{atm})$$
$$= -800 \times 10^{-6}\,\text{m}^3.$$

Although the three elastic moduli and Poisson's ratio have been discussed separately, they are not completely independent. For materials having no distinction between various directions (i.e., *isotropic* materials), only two of these are really independent. For example, the bulk and shear moduli may be expressed in terms of Young's modulus and Poisson's ratio:

$$B = \frac{Y}{3(1 - 2\sigma)}, \qquad S = \frac{Y}{2(1 + \sigma)}.$$

For materials having directional properties, such as wood (which has a grain direction) and single crystals of materials, these relations do not hold, and the elastic behavior is more complex.

10–5 THE FORCE CONSTANT

The various elastic moduli are quantities which describe the elastic properties of a particular *material* and do not directly indicate how much a given rod, cable, or spring constructed of the material will distort under load. If Eq. (10–7) is solved for $F_\perp$, one obtains

$$F_\perp = \frac{YA}{\ell_0} \Delta \ell;$$

or, if YA/ℓ_0 is replaced by a single constant k, and the elongation $\Delta \ell$ is represented by x, then

$$F_\perp = kx. \qquad (10\text{–}13)$$

In other words, the elongation of a body in tension above its no-load length is directly proportional to the stretching force. Hooke's law was originally stated in this form, rather than in terms of stress and strain.

When a helical wire spring is stretched, the stress in the wire is practically pure shear. The elongation of the spring as a whole is directly proportional to the stretching force. That is, an equation of the form $F = kx$ still applies, the constant k depending on the shear modulus of the wire, its radius, the radius of the coils, and the number of coils.

The constant k, or the ratio of the force to the elongation, is called the *force constant* or the *stiffness* of the spring, and is expressed in newtons per meter or pounds per foot. It is equal numerically to the force required to produce unit elongation.

The ratio of elongation to the force, or the elongation per unit force, is called the *compliance* of the spring. The compliance equals the reciprocal of the force constant and is expressed in meters per newton or feet per pound. It is numerically equal to the elongation produced by unit force.

PROBLEMS

10–1 A steel rod 4 m long and 0.5 cm^2 in cross section is found to stretch 0.2 cm under a tension of 12,000 N. What is Young's modulus for this steel?

10–2 The elastic limit of a steel elevator cable is 40,000 lb in^{-2}. Find the maximum upward acceleration which can be given a 2-ton elevator when supported by a cable whose cross section is $\frac{1}{2}$ in^2 if the stress is not to exceed $\frac{1}{4}$ of the elastic limit.

10–3 A copper wire 12 ft long and 0.036 in. in diameter was given the test below. A load of 4.5 lb was originally hung from the wire to keep it taut. The position of the lower end of the wire was read on a scale.

Added load, lb	Scale reading, in.
0	3.02
2	3.04
4	3.06
6	3.08
8	3.10
10	3.12
12	3.14
14	3.65

a) Make a graph of these values, plotting the increase in length horizontally and the added load vertically.

b) Calculate the value of Young's modulus.

c) What was the stress at the proportional limit?

10–4 A steel wire has the following properties:

Length = 10 ft
Cross section = 0.01 in^2
Young's modulus = 30,000,000 lb in^{-2}
Shear modulus = 10,000,000 lb in^{-2}
Proportional limit = 60,000 lb in^{-2}
Breaking stress = 120,000 lb in^{-2}

The wire is fastened at its upper end and hangs vertically. (a) How great a load can be supported without exceeding the proportional limit? (b) How much will the wire stretch under this load? (c) What is the maximum load that can be supported?

10–5 A nylon rope used by mountaineers elongates 1.5 m under the weight of an 80-kg climber. If the rope is 50 m in length and 9 mm in diameter, what is Young's modulus for this material? If Poisson's ratio for nylon is 0.2, find the change in diameter under this stress.

10–6 A relaxed biceps muscle requires a force of 25 N for an elongation of 5 cm, and the same muscle under maximum tension requires a force of 500 N for the same elongation. Regarding the muscle as a uniform cylinder of length 0.2 m and cross-sectional area 50 cm^2, find Young's modulus for the muscle tissue under each of these conditions.

10–7

a) What is the maximum load that can be supported by an aluminum wire 0.1 cm in diameter without exceeding the proportional limit of $8 \times 10^7 \, \text{N m}^{-2}$?

b) If the wire was originally 5 m long, how much will it elongate under this load?

c) How much does the *diameter* change under this load?

10–8 A 5-kg mass hangs on a vertical steel wire 0.5 m long and 0.004 cm^2 in cross section. Hanging from the bottom of this weight is a similar steel wire which supports a 10-kg mass. Compute (a) the longitudinal strain, and (b) the elongation of each wire.

10–9 A 15-kg mass, fastened to the end of a steel wire of unstretched length 0.5 m, is whirled in a vertical circle with an angular velocity of 2 rev s^{-1} at the bottom of the circle. The cross section of the wire is 0.02 cm^2. Calculate the elongation of the wire when the weight is at the lowest point of its path.

10–10 A copper wire 8 m long and a steel wire 4 m long, each of cross section 0.5 cm^2, are fastened end-to-end and stretched with a tension of 500 N.

a) What is the change in length of each wire?

b) What is the elastic potential energy of the system?

10–11 A copper rod of length 2 m and cross-sectional area 2.0 cm^2 is fastened end-to-end to a steel rod of length L and cross-sectional area 1.0 cm^2. The compound rod is subjected to equal and opposite pulls of magnitude $3 \times 10^4 \, \text{N}$ at its ends.

a) Find the length L of the steel rod if the elongations of the two rods are equal.

b) What is the stress in each rod?

c) What is the strain in each rod?

10–12 A rod 1.05 m long, whose weight is negligible, is supported at its ends by wires A and B of equal length, as shown in Fig. 10–9. The cross section of A is 1 mm^2, that of B is 2 mm^2. Young's modulus for wire A is $2.4 \times 10^{11} \, \text{N m}^{-2}$, and for B it is $1.6 \times 10^{11} \, \text{N m}^{-2}$. At what point along the bar should a weight w be suspended in order to produce (a) equal stresses in A and B, (b) equal strains in A and B?

105 cm

Fig. 10–9

10–13 A bar of length L, cross-sectional area A, and Young's modulus Y, is subjected to a tension F. Represent the stress in the bar by Q and the strain by P. Derive the expression for the elastic potential energy, per unit volume, of the bar in terms of Q and P.

10–14 A certain elevator cable is to have a maximum stress of 10,000 lb in^{-2}, to allow for appropriate safety factors. If it is to support a loaded elevator with total weight of 4000 lb, which has a maximum upward acceleration of 5 ft s^{-2}, what should be the diameter of the cable?

10–15 A steel bar 0.2 cm square and 5 m long is stretched with a force of 400 N at each end. Find the stress, the strain, the total elongation, and the fractional change in thickness of the bar.

10–16 Two round rods, one of steel, the other of brass, are joined end to end. Each rod is 0.5 m long and 2 cm in diameter. The combination is subjected to tensile forces of 5000 N.

a) What is the strain in each rod?
b) What is the elongation of each rod?
c) What is the change in diameter of each rod?

10–17 A specimen of oil having an initial volume of 1000 cm^3 is subjected to a pressure of $12 \times 10^5 \, \text{N m}^{-2}$, and the volume is found to decrease by 0.3 cm^3. What is the bulk modulus for the material? The compressibility?

10–18 The compressibility of sodium is to be measured by observing the displacement of the piston in Fig. 10–4 when a force is applied. The sodium is immersed in an oil which fills the cylinder below the piston. Assume that the piston and walls of the cylinder are perfectly rigid, that there is no friction, and no oil leak. Compute the compressibility of the sodium in terms of the applied force F, the piston displacement x, the piston area A, the initial volume of the oil V_0, the initial volume of the sodium v_0, and the compressibility of the oil k_0.

10–19 Two strips of metal are riveted together at their ends by four rivets, each of diameter 0.5 cm. What is the maximum tension that can be exerted by the riveted strip if the shearing stress on the rivets is not to exceed 6×10^8 N m^{-2}? Assume each rivet to carry one-quarter of the load.

10–20 Find the density of ocean water at a depth of 500 m where the pressure is about 5.0×10^6 N m^{-2}. The density at the surface is 1.03×10^3 kg m^{-3}.

10–21 Compute the compressibility of steel, in reciprocal atmospheres, and compare with that of water. Which material is the more readily compressed?

10–22 A steel post 6 in. in diameter and 10 ft long is placed vertically and is required to support a load of 20,000 lb.

a) What is the stress in the post?

b) What is the strain in the post?

c) What is the change in length of the post?

10–23 A hollow cylindrical steel column 10 ft high shortens 0.01 in under a compression load of 72,000 lb. If the inner radius of the cylinder is 0.80 of the outer one, what is the outer radius?

10–24 A bar of cross section A is subjected to equal and opposite tensile forces F at its ends. Consider a plane through the bar making an angle θ with a plane at right angles to the bar (Fig. 10–10).

a) What is the tensile (normal) stress at this plane, in terms of F, A, θ?

b) What is the shearing (tangential) stress at the plane, in terms of F, A, and θ?

c) For what value of θ is the tensile stress a maximum?

d) For what value of θ is the shearing stress a maximum?

Fig. 10–10

10–25 Suppose the block in Fig. 10–3 is rectangular instead of square, but is in equilibrium under the action of shearing stresses on those outside faces perpendicular to the plane of the diagram. (Note that then $F_x \neq F_y$.)

a) Show that the shearing stress is the same on all outside faces perpendicular to the plane of the diagram.

b) Show that on all sections perpendicular to the plane of the diagram and making an angle of 45° with an end face, the stress is still a pure tension or compression.

Chapter 11

Harmonic Motion

11–1 INTRODUCTION

The motion of a body when acted upon by a constant force was considered in detail in Chapters 4 and 5. The motion is one of constant acceleration, and it was found useful to derive expressions for the position and velocity of the body at any time, and for its velocity at any position. In the present chapter, we are to study the motion of a body when the resultant force on it is *not* constant, but varies during the motion. Naturally, there is an infinite number of ways in which a force may vary; hence no general expressions can be given for the motion of a body when acted on by a variable force, except that the acceleration at each instant must equal the force at that instant divided by the mass of the body.

There is however a particular class of problems that occur so often in practical situations that it is useful to develop formulas for this special case. Such problems involve a situation where a body has an equilibrium position such that when displaced from equilibrium it experiences a restoring force, which causes it to undergo a back-and-forth motion past the equilibrium position. A familiar example is the pendulum, a mass suspended from a string. In the equilibrium position, the mass hangs straight down; when displaced from this position the mass does not simply return to the equilibrium position, but instead swings back and forth in a regular, repetitive manner.

Such a repetitive motion is said to be *oscillatory* or *periodic*. Many other examples might be cited; the balance wheel of a watch, the pistons in a gasoline engine, the strings in a musical instrument. The molecules of a solid body vibrate with oscillatory motion about their equilibrium positions in the crystal lattice, although of course this motion cannot be observed directly. The beating of the human heart is a periodic motion; Galileo is alleged to have used his own heartbeat for timing observations of motion.

In many kinds of *wave motion*, the particles of the medium in which the wave is traveling oscillate with periodic motion. This is true even for light waves and radio waves in empty space, except that instead of material particles the quantities that oscillate are the electric and magnetic fields associated with the wave. As a final example, an electrical circuit in which there is an alternating current is described in terms of voltages, currents, and electrical charges that oscillate with time. It can be seen that a study of periodic motion will lay the foundation for future work in many different fields of physics.

11–2 ELASTIC RESTORING FORCES

It has been shown in Chapter 10 that when a body is caused to change its shape, the distorting force is proportional to the amount of change, provided the proportional limit of elasticity is not exceeded. The change may be in the nature of an increase in length, as of a rubber band or a coil spring, or a decrease in length, or a bending as of a flat spring, or a twisting of a rod about its axis, or of many other forms. The term "force" is to be interpreted liberally as the force, or torque, or pressure, or whatever may be producing the distortion. If we restrict the discussion to the case of a push or a pull, where the distortion is simply the displacement of the point of application of the force, the force and displacement are related by Hooke's law,

$$F = kx,$$

where k is a proportionality constant called the *force constant* and x is the displacement from the equilibrium position.

In this equation, F stands for the force which must be exerted *on* an elastic body to produce the displacement x. The force with which the elastic body pulls back on an object to which it is attached is called the *restoring force* and is equal to $-kx$. The negative sign indicates that the force *on* the body is always opposite in sign to the displacement, so that no matter which way the body is displaced, the force is always toward the equilibrium position.

11-3 DEFINITIONS

To fix our ideas, suppose that a flat strip of steel such as a hacksaw blade is clamped vertically in a vise and a small body is attached to its upper end, as in Fig. 11-1. Assume that the strip is sufficiently long and the displacement sufficiently small so that the motion is essentially along a straight line. The mass of the strip itself is negligible.

Let the top of the strip be pulled to the right a distance A, as in Fig. 11-1, and released. The attached body is then acted on by a restoring force exerted by the steel strip and directed toward the equilibrium position O. It therefore accelerates in the direction of this force, and moves in toward the center with increasing speed. The *rate* of increase (i.e., the acceleration) is not constant, however, since the accelerating force becomes smaller as the body approaches the center.

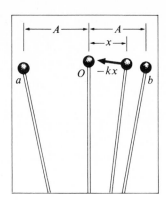

Fig. 11-1 *Motion under a restoring force.*

When the body reaches the center the restoring force has decreased to zero, but because of the velocity which has been acquired, the body "overshoots" the equilibrium position and continues to move toward the left. As soon as the equilibrium position is passed, the restoring force again comes into play, directed now toward the right. The body therefore decelerates, and at a rate which increases with increasing distance from O. It will therefore be brought to rest at some point to the left of O, and repeat its motion in the opposite direction.

Both experiment and theory show that the motion will be confined to a range $\pm A$ on either side of the equilibrium position, each to-and-fro movement taking place in the same interval of time. If there were no loss of energy by friction, the motion would continue indefinitely, once it had been started. This type of motion, under the influence of an elastic restoring force proportional to displacement and in the absence of all friction, is called *simple harmonic motion*, abbreviated SHM.

A *complete vibration* or *complete cycle* means one round trip, say from a to b and back to a, or from O to b to O to a and back to O.

The *periodic time*, or simply the *period* of the motion, represented by T, is the time required for one complete vibration.

The *frequency*, f, is the number of complete vibrations per unit time. Evidently the frequency is the reciprocal of the period, or $T = 1/f$. The mks unit, one cycle per second, is called one *hertz* (1 Hz = 1 s^{-1}).

The *coordinate*, x, at any instant, is the distance away from the equilibrium position or center of the path at that instant. Its sign indicates the direction of the displacement.

The *amplitude*, A, is the maximum value of $|x|$. The total range of the motion is therefore $2A$.

11–4 EQUATIONS OF SIMPLE HARMONIC MOTION

The motion of a point mass m under a restoring force given by $-kx$, where k is the force constant and x the displacement from equilibrium, is an idealized model constituting the simplest possible example of periodic motion. In more complex examples the force may depend on displacement in a more complicated way, but when the force is *directly proportional* to displacement, the resulting motion is called *simple harmonic motion*, abbreviated SHM. Since many more complex motions are approximately simple harmonic and may therefore be described approximately by this model, it is useful to analyze simple harmonic motion in some detail. We shall obtain expressions for the coordinate, velocity, and acceleration of a body moving with simple harmonic motion, just' as we found those for a body moving with constant acceleration. It must be emphasized that the equations of motion with *constant* acceleration cannot be applied, since the acceleration is continually changing.

Figure 11–2 represents the vibrating body of Fig. 11–1 at some instant when its displacement from the equilibrium position O is described by the coordinate x. The particle's mass is m, and the resultant force acting on it is simply the elastic restoring force $-kx$. From Newton's second law,

$$F = -kx = ma,$$

or

$$a = -\frac{k}{m}x. \qquad (11\text{--}1)$$

Thus, an alternate description of the essential feature of simple harmonic motion is that the acceleration at each instant is proportional to the negative of the displacement at that instant. When x has its maximum positive value A, the acceleration has its maximum negative value $-kA/m$; and at the instant the particle passes the equilibrium position ($x = 0$), the acceleration is zero. Its velocity is, of course, *not* zero at this point.

The energy principle provides a convenient basis for analysis of some aspects of simple harmonic motion. The elastic restoring force is a *conservative* force, and the work done by this force can be represented in terms of a potential energy $E_p = \frac{1}{2}kx^2$, just as in Section 7–5. The kinetic energy is $E_k = \frac{1}{2}mv^2$ and, according to the principle of conservation of energy, the total energy $E = E_k + E_p$ is constant. That is,

$$E = \tfrac{1}{2}mv^2 + \tfrac{1}{2}kx^2 = \text{constant}. \qquad (11\text{--}2)$$

The total energy E is also closely related to the amplitude A of the motion. When the particle reaches its maximum displacement $\pm A$, it stops and turns back toward equilibrium. At this instant, $v = 0$, so there is no kinetic energy; the (constant) total energy is thus equal to the potential energy at this point, which is $\frac{1}{2}kA^2 = E$. Thus,

$$\tfrac{1}{2}mv^2 + \tfrac{1}{2}kx^2 = \tfrac{1}{2}kA^2,$$

or

$$v = \pm\sqrt{\frac{k}{m}}\sqrt{A^2 - x^2}. \qquad (11\text{--}3)$$

This relation permits us to obtain the velocity (apart from a sign) for any given position, and Eq. (11–3) is analogous to the relation $v^2 - v_0^2 = 2a(x - x_0)$ for straight-line motion with constant acceleration.

The significance of Eq. (11–2) can be brought out by the graph shown in Fig. 11–3, in which energy

Equilibrium position

$O \quad F = -kx$

Fig. 11–2

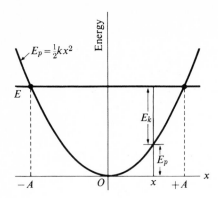

Fig. 11–3 *Relation between total energy E, potential energy E_p, and kinetic energy E_k, for a body oscillating with SHM.*

is plotted vertically and the coordinate x horizontally. First, the curve representing the potential energy, $E_p = \frac{1}{2}kx^2$, is constructed. (This curve is a parabola.) Next, a horizontal line is drawn at a height equal to the total energy E. We see at once that the motion is restricted to values of x lying between the points at which the horizontal line intersects the parabola since, if x were outside this range, the potential energy would exceed the total energy, and that is impossible. The motion of the vibrating body is analogous to that of a particle released at a height E on a frictionless track shaped like the potential energy curve, and the motion is said to take place in a "potential energy well."

If a vertical line is constructed at any value of x within the permitted range, the length of the segment between the x-axis and the parabola represents the potential energy E_p at that value of x, and the length of the segment between the parabola and the horizontal line at height E represents the corresponding kinetic energy E_k. At the endpoints, therefore, the energy is all potential and at the midpoint it is all kinetic. The velocity at the midpoint has its maximum (absolute) value v_{max}:

$$\tfrac{1}{2}mv_{max}^2 = E,$$

$$v_{max} = \pm\sqrt{2E/m}, \tag{11–4}$$

the sign of v being positive or negative depending on the direction of motion. The fact that both the maximum potential energy and maximum kinetic

energy are equal to the total energy (and thus to each other) may be used to relate the maximum velocity to the amplitude:

$$\tfrac{1}{2}kA^2 = \tfrac{1}{2}mv_{max}^2,$$

$$v_{max} = \sqrt{\frac{k}{m}}\,A. \tag{11–5}$$

The position–velocity relation given by Eq. (11–3) is very useful, but it does not tell us where the particle is at any given time. To have a complete description of the motion we need to know the position, velocity, and acceleration for any given time. The expression for the coordinate x as a function of time can be obtained by replacing v with dx/dt, in Eq. (11–3), and integrating. This gives

$$\int \frac{dx}{\sqrt{A^2 - x^2}} = \sqrt{\frac{k}{m}} \int dt,$$

$$\sin^{-1}\frac{x}{A} = \sqrt{\frac{k}{m}}\,t + C. \tag{11–6}$$

Let x_0 be the value of x when $t = 0$. The integration constant C is then

$$C = \sin^{-1}\frac{x_0}{A}.$$

That is, C is the *angle* (in radians) whose sine equals x_0/A. Let us represent this angle by θ_0:

$$\sin\theta_0 = \frac{x_0}{A}, \qquad \theta_0 = \sin^{-1}\frac{x_0}{A}. \tag{11–7}$$

Equation (11–6) can now be written

$$\sin^{-1}\frac{x}{A} = \sqrt{\frac{k}{m}}\,t + \theta_0,$$

$$x = A\sin\left(\sqrt{\frac{k}{m}}\,t + \theta_0\right). \tag{11–8}$$

The coordinate x is therefore a *sinusoidal* function of the time t. The term in parentheses is an *angle*, in radians. It is called the *phase angle*, or simply the *phase* of the motion. The angle θ_0 is the *initial phase angle*.

The period T is the time required for one complete oscillation. That is, the coordinate x has the same value at the times t and $t + T$. In other words, the phase angle $((\sqrt{k/m})t + \theta_0)$ increases by 2π radians in the time T:

$$\sqrt{k/m}\,(t + T) + \theta_0 = ((\sqrt{k/m})t + \theta_0) + 2\pi,$$

$$T = 2\pi\sqrt{\frac{m}{k}}. \qquad (11\text{-}9)$$

The period T therefore depends only on the mass m and the force constant k. It does not depend on the amplitude (or on the total energy). For given values of m and k, the time of one complete oscillation is the same whether the amplitude is large or small. A motion that has this property is said to be *isochronous* (equal time).

The frequency f, or the number of complete oscillations per unit time, is the reciprocal of the period T:

$$f = \frac{1}{T} = \frac{1}{2\pi}\sqrt{\frac{k}{m}}.$$

The *angular frequency* ω is defined as $\omega = 2\pi f$ and is expressed in radians per second. It follows from the two preceding equations that

$$\omega = \sqrt{k/m},$$

and Eqs. (11-8) and (11-3) can be written more compactly as

$$x = A\sin(\omega t + \theta_0), \qquad (11\text{-}10)$$

$$v = \omega\sqrt{A^2 - x^2}. \qquad (11\text{-}11)$$

Expressions for the velocity and acceleration as functions of time can now be obtained by differentiating Eq. (11-10):

$$v = \frac{dx}{dt} = \omega A\cos(\omega t + \theta_0), \qquad (11\text{-}12)$$

$$a = \frac{dv}{dt} = -\omega^2 A\sin(\omega t + \theta_0). \qquad (11\text{-}13)$$

Since $A\sin(\omega t + \theta_0) = x$, Eq. (11-13) may also be written

$$a = -\omega^2 x. \qquad (11\text{-}14)$$

We note that this is consistent with Eq. (11-1).

It follows from Eq. (11-12) that if v_0 is the velocity

when $t = 0$, then

$$\cos\theta_0 = \frac{v_0}{\omega A}. \qquad (11\text{-}15)$$

The equation, together with Eq. (11-7), $\sin\theta_0 = x_0/A$, completely determines the initial phase angle θ_0. That is, the angle θ_0 depends both on the initial position x_0 and the initial velocity v_0.

Figure 11-4 shows how the initial phase angle depends on the initial position and initial velocity. It follows from this triangle that

$$A = \sqrt{x_0^2 + (v_0/\omega)^2}, \qquad (11\text{-}16)$$

a relation that can also be obtained by setting the initial total energy, $\frac{1}{2}kx_0^2 + \frac{1}{2}mv_0^2$, equal to the potential energy at maximum displacement. The motion is therefore completely determined, for given values of m and k, when the initial position and initial velocity are known.

Fig. 11-4 *Relation between the initial phase angle θ_0, initial position x_0, and initial velocity v_0.*

The equations of simple harmonic motion may be summarized by comparing them with similar equations for rectilinear motion with constant acceleration (Table 11-1).

Figure 11-5 shows corresponding graphs of the coordinate x, velocity v, and acceleration a of a body oscillating with simple harmonic motion, plotted as functions of the time t (or of the angle ωt). The initial

TABLE 11-1

Rectilinear motion with constant acceleration	Simple harmonic motion (in terms of ω and θ_0)
$a = $ constant	$a = -\omega^2 x$
	$a = -\omega^2 A\sin(\omega t + \theta_0)$
$v^2 = v_0^2 + 2a(x - x_0)$	$v = \pm\omega\sqrt{A^2 - x^2}$
$v = v_0 + at$	$v = \omega A\cos(\omega t + \theta_0)$
$x = x_0 + v_0 t + \frac{1}{2}at^2$	$x = A\sin(\omega t + \theta_0)$
$\omega = 2\pi/T = 2\pi f = \sqrt{k/m},$	$A = \sqrt{x_0^2 + (v_0/\omega)^2},$
$\sin\theta_0 = x_0/A,$	$\cos\theta_0 = v_0/\omega A.$

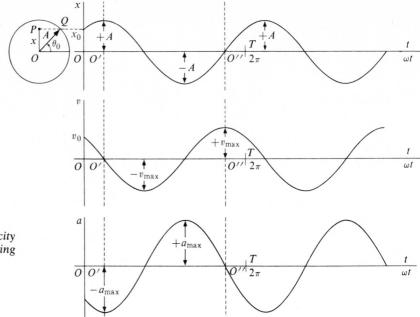

Fig. 11–5 *Graphs of coordinate x, velocity v, and acceleration a of a body oscillating with simple harmonic motion.*

coordinate is x_0 and the initial velocity is v_0. The angle θ_0 has been taken as $\pi/4$ rad. Each curve repeats itself in a time interval equal to the period T, during which the angle ωt increases by 2π rad. Note that when the body is at either end of its path, i.e., when the coordinate x has its maximum positive or negative value ($\pm A$), the velocity is zero and the acceleration has its maximum negative or positive value ($\mp a_{max}$). Note also that when the body passes through its equilibrium position ($x = 0$), the velocity has its maximum positive or negative value($\pm v_{max}$) and the acceleration is zero.

The equations of motion take a simpler form if we set $t = 0$ when the body is at the midpoint, or is at one end of its path. For example, suppose we let $t = 0$ when the body has its *maximum positive* displacement. Then $x_0 = +A$, $\sin \theta_0 = 1$, $\theta_0 = \pi/2$, and

$$x = A \sin (\omega t + \pi/2) = A \cos \omega t,$$
$$v = -\omega A \sin \omega t, \qquad (11\text{–}17)$$
$$a = -\omega^2 A \cos \omega t.$$

This corresponds to moving the origin from the point O, in Fig. 11–5, to the point O'. The graph of x versus t becomes a cosine curve, that of v versus t a

negative sine curve, and that of a a negative cosine curve.

If we set $t = 0$ when the body is at the midpoint and moving toward the right, then

$$x_0 = 0, \qquad \sin \theta_0 = 0, \qquad \theta_0 = 0$$

and

$$x = A \sin \omega t,$$
$$v = \omega A \cos \omega t, \qquad (11\text{–}18)$$
$$a = -\omega^2 A \sin \omega t.$$

This corresponds to displacing the origin to the point O'' in Fig. 11–5.

Example   Let the mass of the body in Fig. 11–2 be 0.025 kg, the force constant k be 0.4 N m^{-1}, and let the motion be started by displacing the body 0.1 m to the right of its equilibrium position and imparting to it a velocity toward the right of 0.4 m s^{-1}. Compute (a) the period T, (b) the frequency f, (c) the angular frequency ω, (d) the total energy E, (e) the amplitude A, (f) the angle θ_0, (g) the maximum velocity v_{max}, (h) the maximum acceleration a_{max}, (i) the coordinate, velocity, and acceleration at a time $\pi/8$ s after the start of the motion.

a) $\quad T = 2\pi\sqrt{\dfrac{m}{k}} = 2\pi\sqrt{\dfrac{0.025 \text{ kg}}{0.4 \text{ N m}^{-1}}} = \dfrac{\pi}{2}\,\text{s} = 1.57\text{ s}.$

b) $\quad f = \dfrac{1}{T} = \dfrac{2}{\pi}\,\text{Hz} = 0.638\text{ Hz}.$

c) $\quad \omega = 2\pi f = 4\text{ rad s}^{-1}.$

d) $\quad E = \tfrac{1}{2}mv_0^2 + \tfrac{1}{2}kx_0^2 = 0.004\text{ J}.$

e) $\quad A = \sqrt{2E/k} = 0.1\sqrt{2}\text{ m}.$

f) $\quad \sin\theta_0 = x_0/A = 1/\sqrt{2}, \qquad \theta_0 = \pi/4\text{ rad}.$

g) $\quad |v_{max}| = \sqrt{2E/m} = 0.4\sqrt{2}\text{ m s}^{-1} = 0.566\text{ m s}^{-1}.$

The maximum velocity occurs at the midpoint, where $x = 0$. Hence, from Eq. (11–12),

$$|v_{max}| = \omega A = 0.4\sqrt{2}\text{ m s}^{-1}.$$

h) The maximum acceleration occurs at the ends of the path where the force is a maximum. From Eq. (11–15),

$$|a_{max}| = \omega^2 x_{max} = 1.6\sqrt{2}\text{ m s}^{-2}.$$

i) The equations of motion are:

$$x = 0.1\sqrt{2}\sin\left(4t + \dfrac{\pi}{4}\right),$$

$$v = 0.4\sqrt{2}\cos\left(4t + \dfrac{\pi}{4}\right),$$

$$a = -1.6\sqrt{2}\sin\left(4t + \dfrac{\pi}{4}\right).$$

When $t = (\pi/8)$s, the phase angle is

$$\left(4t + \dfrac{\pi}{4}\right) = \dfrac{3\pi}{4}\text{ rad},$$

$$x = 0.1\sqrt{2}\sin(3\pi/4) = 0.1\text{ m},$$

$$v = 0.4\sqrt{2}\cos(3\pi/4) = -0.4\text{ m s}^{-1}.$$

$$a = -1.6\sqrt{2}\sin(3\pi/4) = -1.6\text{ m s}^{-2}.$$

The curves of Fig 11–5 represent the motion of the body in this example, if the scales of x, v, a, and t are chosen appropriately.

The equations of simple harmonic motion can be given a simple geometrical interpretation, as follows: In Fig 11–6(a), the point Q moves in a circle of radius equal to the amplitude A, with constant angular velocity ω. Then the position vector of this point, the line segment OQ, of length A, rotates with constant angular velocity ω about point O. Let the vector OQ make an angle with the horizontal axis, at time $t = 0$, equal to the initial phase angle θ_0. Point P is the *projection* of point Q onto the vertical axis, and as OQ rotates, point P oscillates along the axis.

We now show that the equations of motion of P are the same as those of a body oscillation with simple harmonic motion of amplitude A, angular frequency ω, and initial phase angle θ_0. Let x represent the length of OP. At any time t, the angle between the radius OQ and the horizontal axis is $\omega t + \theta_0$, and

$$x = A\sin(\omega t + \theta_0).$$

The *velocity* of point Q (Fig. 11–6(b)) is ωA, and its vertical component, equal to the velocity of P, is

$$v = \omega A\cos(\omega t + \theta_0).$$

(a)   (b)   (c)

Fig. 11–6 *Representation of simple harmonic motion by the projection onto the vertical axis of the tip of rotor OQ.*

The *acceleration* of Q is its radial acceleration $\omega^2 A$ (Fig. 11–6(c)) and its vertical component, equal to the acceleration of P, is

$$a = -\omega^2 A \sin(\omega t + \theta_0).$$

The negative sign must be included because the acceleration is negative whenever the sine of the phase angle is positive, and vice versa. These equations are, of course, just the general equations of harmonic motion. In the special case corresponding to Eqs. (11–17), the inital phase angle is 90° and the reference point Q is at the top of the circle when $t = 0$. If the reference point is at the right-hand end of the horizontal diameter when $t = 0$, then $\theta_0 = 0$ and the motion is described by Eqs. (11–18).

11–5 MOTION OF A BODY SUSPENDED FROM A COIL SPRING

Figure 11–7(a) shows a coil spring of force constant k and no-load length ℓ. When a body of mass m is attached to the spring as in part (b), it hangs in equilibrium with the spring extended by an amount $\Delta \ell$ such that the upward force P exerted by the spring is equal to the weight of the body, mg. But $P = k \Delta \ell$, so

$$k \Delta \ell = mg.$$

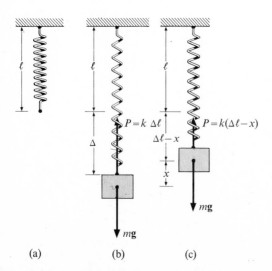

(a)   (b)   (c)

Fig. 11–7 *The restoring force on a body suspended by a spring is proportional to the coordinate measured from the equilibrium position.*

Now suppose the body is at a distance x *above* its equilibrium position, as in part (c). The extension of the spring is now $\Delta \ell - x$, the upward force it exerts on the body is $k(\Delta \ell - x)$, and the resultant force F on the body is

$$F = k(\Delta \ell - x) - mg = -kx.$$

The resultant force is therefore proportional to the displacement of the body *from its equilibrium position*, and if set in vertical motion the body oscillates with an angular frequency $\omega = \sqrt{k/m}$.

Except in the idealized case of a spring of zero mass, allowance must be made for the fact that the spring also oscillates. However, we cannot simply add the mass of the spring to that of the suspended body because not all portions of the spring oscillate with the same amplitude; the amplitude at the lower end equals that of the suspended body, while that at the upper end is zero. The correction term can be computed as follows.

Let L represent the length of the spring when the body is in its equilibrium position, and let m_s be the mass of the spring. Let us calculate the kinetic energy of the spring at an instant when the velocity of the lower end is v. Consider an element of the spring of length dy, at a distance y below the fixed upper end. The mass of the element, dm_s, is

$$dm_s = \frac{m_s}{L}\, dy.$$

All portions of the spring will be assumed to oscillate in phase and the velocity of the element, v_s, to be proportional to its distance from the fixed end: $v_s = (y/L)v$.

The kinetic energy of the element is

$$dE_k = \frac{1}{2} \cdot dm_s \cdot v_s^2 = \frac{1}{2} \cdot \frac{m_s}{L}dy \cdot \left(\frac{y}{L}v\right)^2,$$

and the total kinetic energy of the spring is

$$E_k = \frac{1}{2} \cdot \frac{m_s v^2}{L^3} \cdot \int_0^L y^2\, dy = \frac{1}{2}\left(\frac{1}{3}m_s\right)v^2.$$

This equals the kinetic energy of a body of mass one-third the mass of the spring, moving with the same velocity as that of the suspended body. In other words, the equivalent mass of the *vibrating system* equals that of the suspended body plus one-third the mass of the spring.

Example   A body of mass 1 kg is suspended from a coil spring whose mass is 0.09 kg and whose force constant is 66 N m^{-1}. Find the frequency and amplitude of the ensuing motion if the body is displaced 0.03 m below its equilibrium position and given a downward velocity of 0.4 m s^{-1}.

The angular frequency is

$$\omega = \sqrt{\frac{k}{m + m_s/3}} = \sqrt{\frac{66\,\text{N}\,\text{m}^{-1}}{1.03\,\text{kg}}} = 8.00\,\text{rad}\,\text{s}^{-1}.$$

The amplitude is expressed in terms of the initial coordinate and velocity by means of Eq. (11–16). Thus,

$$A = \sqrt{x_0^2 + (v_0/\omega)^2}$$
$$= \sqrt{(0.03\,\text{m})^2 + (0.4/8\,\text{m})^2}$$
$$= 0.0582.$$

11–6 THE SIMPLE PENDULUM

A simple pendulum, an idealized model for a more complex system, consists of a point mass suspended by an inextensible weightless string. When pulled to one side of its equilibrium position and released, the pendulum bob vibrates about this position. We wish to analyze this motion, asking in particular whether it is simple harmonic.

The necessary condition for simple harmonic motion is that the restoring force F shall be directly proportional to the coordinate x and oppositely directed. The path of the bob is not a straight line, but the arc of a circle of radius L, where L is the length of the supporting cord. The coordinate x refers to distances measured along this arc. (See Fig. 11–8.) Hence, if $F = -kx$, the motion will be simple harmonic; since $x = L\theta$, the requirement may also be written $F = -kL\theta$.

Figure 11–8 shows the forces on the bob at an instant when its coordinate is x. Choose axes tangent to the circle and along the radius, and resolve the weight into components. The restoring force F is

$$F = -mg \sin \theta. \qquad (11\text{–}19)$$

The restoring force is therefore proportional *not* to θ but to $\sin \theta$, so the motion is *not* simple harmonic.

However, *if the angle θ is small*, $\sin \theta$ is very nearly equal to θ. For example, when $\theta = 0.1$ rad, about 6°, $\sin \theta = 0.0998$, a difference of only 0.2 percent. With this approximation, Eq. (11–19) becomes

$$F = -mg\theta = -mg\frac{x}{L},$$

or

$$F = -\frac{mg}{L}x. \qquad (11\text{–}20)$$

The restoring force is then proportional to the coordinate *for small displacements*, and the constant mg/L represents the force constant k. The period of a simple pendulum when its amplitude is small is therefore

$$T = 2\pi\sqrt{m/k} = 2\pi\sqrt{m/(mg/L)},$$

or

$$T = 2\pi\sqrt{L/g}. \qquad (11\text{–}21)$$

The corresponding frequency relations are

$$f = \frac{1}{2\pi}\sqrt{\frac{g}{L}},$$
$$\omega = \sqrt{\frac{g}{L}}. \qquad (11\text{–}22)$$

We note that these expressions do not contain the *mass* of the particle; this is because the restoring force, a component of the particle's weight, is proportional to m. Thus the mass appears on both sides

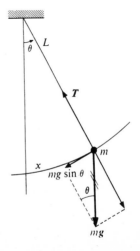

Fig. 11–8 *Forces on the bob of a simple pendulum.*

of $F = ma$ and may be cancelled out. For small oscillations the period of a pendulum for a given value of g is determined entirely by its length.

We emphasize again that the motion of a pendulum is only *approximately* simple harmonic, and when the amplitude is not small the departures from simple harmonic motion can be substantial. What constitutes a "small" amplitude? It can be shown that the general equation for the time of swing, when the maximum angular displacement is α, is

$$T = 2\pi\sqrt{L/g}\left(1 + \frac{1^2}{2^2}\sin^2\frac{\alpha}{2} + \frac{1^2 \cdot 3^2}{2^2 \cdot 4^2}\sin^4\frac{\alpha}{2} + \cdots\right).$$
$$(11\text{–}23)$$

The time may be computed to any desired degree of precision by taking enough terms in the infinite series. When $\alpha = 15°$ (on either side of the central position), the true period differs from that given by the approximate Eq. (11–21) by less than 0.5%.

The utility of the pendulum as a timekeeper is based on the fact that the period is practically independent of the amplitude. Thus, as a clock runs down and the amplitude of the swings becomes slightly smaller, the clock will still keep very nearly correct time.

The simple pendulum is also a precise and convenient method of measuring the acceleration of gravity g without actually resorting to free fall, since L and T may readily be measured. More complicated pendulums find considerable application in the field of geophysics. Local deposits of ore or oil, if their density differs from that of their surroundings, affect the local value of g, and precise measurements of this quantity over an area which is being prospected often furnish valuable information regarding the nature of underlying deposits.

11–7 ANGULAR HARMONIC MOTION

Angular harmonic motion occurs when a body which is pivoted about an axis experiences a restoring *torque* proportional to the angular displacement from its equilibrium position. This type of vibration is very similar to linear harmonic motion, and the corresponding equations may be written down immediately from the analogies between linear and angular quantities. The oscillatory motion of the balance

wheel of a watch is an example of angular harmonic motion.

A restoring torque proportional to angular displacement is expressed by

$$\Gamma = -k'\theta, \qquad (11\text{–}24)$$

where k' is a proportionality constant called the *torque constant*. The moment of inertia of the pivoted body corresponds to the mass of a body in linear motion. Hence, the formula for the period of angular harmonic motion is

$$T = 2\pi\sqrt{I/k'}, \qquad (11\text{–}25)$$

where k' is the constant in Eq. (11–24).

The equations for angular displacement, angular velocity, and angular acceleration can be obtained by comparison with the corresponding equations in Section 11–4, replacing x by θ, v by ω, and a by α.

The balance wheel of a watch is a familiar example of angular harmonic motion. If the hairspring behaves according to Eq. (11–24), the motion is *isochronous*; the period is constant even though the amplitude decreases somewhat as the mainspring unwinds.

11–8 THE PHYSICAL PENDULUM

A *physical* pendulum is any real pendulum, as contrasted with a simple pendulum in which all the mass is assumed to be concentrated at a point. In Fig. 11–9, a body of irregular shape is pivoted about a horizontal frictionless axis O and displaced from the vertical by an angle θ. The distance from the pivot to the center of gravity is h, the moment of inertia of the pendulum about an axis through the pivot is I, and the mass of the pendulum is m. The weight mg causes a restoring torque

$$\Gamma = -mgh \sin\theta. \qquad (11\text{–}26)$$

When released, the body will oscillate about its equilibrium position; as in the case of the simple pendulum, the motion is not simple harmonic, since the torque Γ is proportional not to θ but to $\sin\theta$. However, if θ is small, we can again approximate $\sin\theta$ by θ, and the motion is approximately harmon-

Fig. 11-9 A physical pendulum.

ic. With this approximation,

$$\Gamma \approx -(mgh)\theta,$$

and the effective torque constant is

$$k' = -\frac{\Gamma}{\theta} = mgh.$$

The period T is

$$T \equiv 2\pi\sqrt{\frac{I}{k'}} = 2\pi\sqrt{\frac{I}{mgh}}. \qquad (11\text{-}27)$$

Example 1 Let the body in Fig. 11-9 be a meterstick pivoted at one end. Then if L stands for the total length (1 m),

$$I = \tfrac{1}{3}mL^2, \qquad h = \frac{L}{2}, \qquad g = 9.8 \text{ m s}^{-2},$$

$$T = 2\pi\sqrt{\tfrac{1}{3}mL^2/(mgL/2)} = 2\pi\sqrt{(2/3)(L/g)}$$

$$= 2\pi\sqrt{(2/3)(1 \text{ m})/9.8 \text{ m s}^{-2}} = 1.65 \text{ s}.$$

Example 2 Equation (11-27) may be solved for the moment of inertia I, giving

$$I = \frac{T^2 mgh}{4\pi^2}.$$

The quantities on the right of the equation are all directly measurable. Hence the moment of inertia of a body of any complex shape may be found by suspending the body as a physical pendulum and measuring its period of vibration. The location of the

center of gravity can be found by balancing. Since T, m, g, and h are known, I can be computed. For example, Fig. 11-10 illustrates a connecting rod pivoted about a horizontal knife-edge. The connecting rod has a mass of 2 kg, and its center of gravity has been found by balancing to be 0.2 m below the knife-edge. When set into oscillation, it is found to make 100 complete vibrations in 120 s, so that $T = 120 \text{ s}/100 = 1.2 \text{ s}$. Therefore,

$$I = \frac{(1.2 \text{ s})^2 (2 \text{ kg})(9.8 \text{ m s}^{-2})(0.2 \text{ m})}{4\pi^2} = 0.143 \text{ kg m}^2.$$

Fig. 11-10

11-9 CENTER OF OSCILLATION

It is always possible to find an *equivalent* simple pendulum whose period is equal to that of a given physical pendulum. If L is the length of the equivalent simple pendulum,

$$T = 2\pi\sqrt{\frac{L}{g}} = 2\pi\sqrt{\frac{I}{mgh}}$$

or

$$L = \frac{I}{mh}. \qquad (11\text{-}28)$$

Thus, so far as its period of vibration is concerned, the mass of a physical pendulum may be considered to be concentrated at a point whose distance from the pivot is $L = I/mh$. This point is called the *center of oscillation* of the pendulum.

Example A slender uniform rod of length a is pivoted at one end and swings as a physical pendulum. Find the center of oscillation of the pendulum.

The moment of inertia of the rod about an axis through one end is

$$I = \tfrac{1}{3}ma^2.$$

The distance from the pivot to the center of gravity is $h = a/2$. The length of the equivalent simple pendulum is

$$L = \frac{I}{mh} = \frac{\tfrac{1}{3}ma^2}{m(a/2)}$$

$$= \tfrac{2}{3}a,$$

and the center of oscillation is at a distance $2a/3$ from the pivot.

Figure 11–11 shows a body, whose center of oscillation is at point C, pivoted about an axis through O. The center of oscillation and the point of support have the following interesting property: If the pendulum is pivoted about a new axis through point C, its period is unchanged and point O becomes the new center of oscillation. The point of support and the center of oscillation are said to be *conjugate* to each other.

The center of oscillation has another important property. Figure 11–12 shows a baseball bat pivoted at O. If a ball strikes the bat at its center of

Fig. 11–12 *The center of percussion coincides with the center of oscillation.*

Center of gravity

Center of percussion

oscillation, no impulsive force is exerted on the pivot and hence no "sting" is felt if the bat is held at that point. Because of this property, the center of oscillation is called the *center of percussion*.

In Fig. 11–13, a series of multiflash photographs, the *center of gravity* is marked by a black band. In (a), the body is struck at its center of percussion relative to a pivot at the upper end of the cord, and it starts to swing smoothly about this pivot. In (b), the body is struck at its center of gravity. Note that it does not start to rotate about the pivot, but that its initial motion is one of pure translation. That is, the center of percussion does not coincide with the center of gravity. In (c), the body is struck above, and in (d) below its center of percussion.

PROBLEMS

11–1 The general equation of simple harmonic motion,

$$x = A \sin(\omega t + \theta_0),$$

can be written in the equivalent form

$$x = B \sin \omega t + C \cos \omega t.$$

a) Find the expressions for the amplitudes B and C in terms of the amplitude A and the initial phase angle θ_0.

b) Interpret these expressions in terms of a rotating vector diagram.

11–2 A body of mass 0.25 kg is acted on by an elastic restoring force of force constant $k = 25 \text{ N m}^{-1}$.

a) Construct the graph of elastic potential energy E_p as a function of displacement x, over a range of x from −0.3 m to +0.3 m. Let 1 in. = 0.25 J vertically, and 1 in. = 0.1 m horizontally.

Fig. 11–11 *Center of oscillation. The length L equals that of the equivalent simple pendulum.*

Fig. 11–13 *The motion of a body suspended by a cord when the body is struck a horizontal blow.*

The body is set into oscillation with an initial potential energy of 0.6 J and an initial kinetic energy of 0.2 J. Answer the following questions by reference to the graph.

b) What is the amplitude of oscillation?

c) What is the potential energy when the displacement is one-half the amplitude?

d) At what displacement are the kinetic and potential energies equal?

e) What is the speed of the body at the midpoint of its path?

Find

f)   the period T,

g)   the frequency f, and

h)   the angular frequency ω.

i)   What is the initial phase angle θ_0 if the amplitude $A = 15$ cm, the initial displacement $x_0 = 7.5$ cm, and the initial velocity v_0 is negative?

11–3 A body is vibrating with simple harmonic motion of amplitude 15 cm and frequency 4 Hz. Compute (a) the maximum values of the acceleration and velocity, (b) the acceleration and velocity when the coordinate is 9 cm, and (c) the time required to move from the equilibrium position to a point 12 cm distant from it.

11–4 A body of mass 10 g moves with simple harmonic motion of amplitude 24 cm and period 4 s. The coordinate is $+24$ cm when $t = 0$. Compute (a) the position of the body when $t = 0.5$ s, (b) the magnitude and direction of the force acting on the body when $t = 0.5$ s, (c) the minimum time required for the body to move from its initial position to the point where $x = -12$ cm, (d) the velocity of the body when $x = -12$ cm.

11–5 The motion of the piston of an automobile engine is approximately simple harmonic.

a)   If the stroke of an engine (twice the amplitude) is 4 in., and the angular velocity is 3600 rev min^{-1}, compute the acceleration of the piston at the end of its stroke.

b)   If the piston weighs 1 lb, what resultant force must be exerted on it at this point?

c)   What is the velocity of the piston, in miles per hour, at the midpoint of its stroke?

11–6 A body of mass 2 kg is suspended from a spring of negligible mass, and is found to stretch the spring 20 cm.

a)   What is the force constant of the spring?

b)   What is the period of oscillation of the body, if pulled down and released?

c)   What would be the period of a body of mass 4 kg hanging from the same spring?

11–7 The scale of a spring balance reading from zero to 180 N is 9 cm long. A body suspended from the balance is observed to oscillate vertically at 1.5 Hz. What is the mass of the body? Neglect the mass of the spring.

11–8 A body weighing 8 lb is attached to a coil spring and oscillates vertically in simple harmonic motion. The amplitude is 2 ft, and at the highest point of the motion the spring has its natural unstretched length. Calculate the elastic potential energy of the spring, the kinetic energy of the body, its gravitational potential energy relative to the lowest point of the motion, and the sum of these three energies, when the body is (a) at its lowest point, (b) at its equilibrium position, and (c) at its highest point.

11–9 A mass of 100 kg suspended from a wire whose unstretched length ℓ_0 is 4 m is found to stretch the wire by 0.004 m. The cross-sectional area of the wire, which can be assumed constant, is 0.1 cm^2.

a)   If the load is pulled down a small additional distance and released, find the frequency at which it will vibrate.

b)   Compute Young's modulus for the wire.

11–10 A small block is executing simple harmonic motion in a horizontal plane with an amplitude of 10 cm. At a point 6 cm away from equilibrium, the velocity is 24 cm s^{-1}.

a)   What is the period?

b)   What is the displacement when the velocity is ± 12 cm s^{-1}?

c)   If a small object placed on the oscillating block is just on the verge of slipping at the endpoint of the path, what is the coefficient of friction?

11–11 A force of 30 N stretches a vertical spring 15 cm.

a)   What weight must be suspended from the spring so that the system will oscillate with a period of $(\pi/4)$ s?

b)   If the amplitude of the motion is 5 cm, where is the body and in what direction is it moving $(\pi/12)$ s after it has passed the equilibrium position, moving downward?

c)   What force does the spring exert on the body when it is 3 cm below the equilibrium position, moving upward?

11–12 A body of mass m is suspended from a coil spring and the time for 100 complete oscillations is measured for the following values of m:

m(g)	100	200	400	1000
Time of 100 oscillations (s)	23.4	30.6	41.8	64.7

Plot graphs of the measured values of (a) T vs. m, and (b) T^2 vs. m.

c)   Are the experimental results in agreement with theory?

d)   Is either graph a straight line?

e) Does the straight line pass through the origin?

f) What is the force constant of the spring?

g) What is the mass of the spring?

11–13 A body of mass 100 g hangs from a long spiral spring. When pulled down 10 cm below its equilibrium position and released, it vibrates with a period of 2 s.

a) What is its velocity as it passes through the equilibrium position?

b) What is its acceleration when it is 5 cm above the equilibrium position?

c) When it is moving upward, how long a time is required for it to move from a point 5 cm below its equilibrium position to a point 5 cm above it?

d) How much will the spring shorten if the body is removed?

11–14 A body of mass 5 kg hangs from a spring and oscillates with a period of 0.5 s. How much will the spring shorten when the body is removed?

11–15 Four passengers whose combined weight is 600 lb are observed to compress the springs of an automobile by 2 in. when they enter the automobile. If the total load supported by the springs is 1800 lb, find the period of vibration of the loaded automobile.

11–16

a) A block suspended from a spring vibrates with simple harmonic motion. At an instant when the displacement of the block is equal to one-half the amplitude, what fraction of the total energy of the system is kinetic and what fraction is potential?

b) When the block is in equilibrium, the length of the spring is an amount s greater than in the unstretched state. Prove that $T = 2\pi\sqrt{s/g}$.

11–17

a) With what additional force must a vertical spring carrying an 8-lb body in equilibrium be stretched so that, when released, it will perform 48 complete oscillations in 32 s with an amplitude of 3 in.?

b) What force is exerted by the spring on the body when it is at the lowest point, the middle, and the highest point of the path?

c) What is the kinetic energy of the system when the body is 1 in. below the middle of the path? its potential energy?

11–18 A force of 60 N stretches a certain spring 30 cm. A body of mass 4 kg is hung from this spring and allowed to come to rest. It is then pulled down 10 cm and released.

a) What is the period of the motion?

b) What are the magnitude and direction of the acceleration of the body when it is 5 cm above the equilibrium position, moving upward?

c) What is the tension in the spring when the body is 5 cm above the equilibrium position?

d) What is the shortest time required to go from the equilibrium position to the point 5 cm above?

e) If a small object were placed on the oscillating body, would it remain or leave?

f) If a small object were placed on the oscillating body and its amplitude doubled, where would the object and the oscillating body begin to separate?

11–19 Two springs with the same unstretched length but different force constants k_1 and k_2 are attached to a block of mass m on a level frictionless surface. Calculate the effective force constant in each of the three cases (a), (b), and (c) shown in Fig. 11–14.

d) A body of mass m, suspended from a spring with a force constant k, vibrates with a frequency f_1. When the spring is cut in half and the same body is suspended from one of the halves, the frequency is f_2. What is the ratio f_2/f_1?

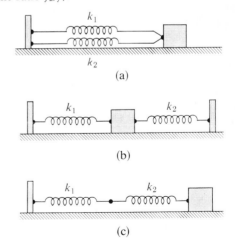

(a)

(b)

(c)

Fig. 11–14

11–20 Two springs, each of unstretched length 0.2 m but having different force constants k_1 and k_2, are attached to opposite ends of a block of mass m on a level frictionless surface. The outer ends of the springs are now attached to two pins P_1 and P_2, 10 cm from the original positions of the

springs. Let

$$k_1 = 1\,\mathrm{N\,m^{-1}}, \qquad k_2 = 3\,\mathrm{N\,m^{-1}}, \qquad m = 0.1\,\mathrm{kg}.$$

(See Fig. 11–15.)

a) Find the length of each spring when the block is in its new equilibrium position, after the springs have been attached to the pins.

b) Find the period of vibration of the block if it is slightly displaced from its new equilibrium position and released.

Fig. 11–15

11–21 The block in Problem 11–20 and Fig. 11–15 is oscillating with an amplitude of 0.05 m. At the instant it passes through its equilibrium position, a lump of putty of mass 0.1 kg is dropped vertically onto the block and sticks to it.

a) Find the new period and amplitude.

b) Was there a loss of energy and, if so, where did it go?

c) Would the answers be the same if the putty had been dropped on the block when it was at one end of its path?

11–22 A simple pendulum 8 ft long swings with an amplitude of 1 ft.

a) Compute the velocity of the pendulum at its lowest point.

b) Compute its acceleration at the ends of its path.

11–23 Find the length of a simple pendulum whose period is exactly 1 s at a point where $g = 9.80\ \mathrm{m\,s^{-2}}$.

11–24

a) What is the change ΔT in the period of a simple pendulum when the acceleration of gravity g changes by Δg? (*Hint*: The new period $T + \Delta T$ is obtained by substituting $g + \Delta g$ for g:

$$T + \Delta T = 2\pi\sqrt{\frac{L}{g + \Delta g}}.$$

To obtain an approximate expression, expand the $(g + \Delta g)^{1/2}$ using the binomial theorem and keeping only the first two terms:

$$(g + \Delta g)^{-1/2} = g^{-1/2} - \tfrac{1}{2}g^{-3/2}\Delta g + \cdots$$

The other terms contain higher powers of Δg and are very small if Δg is small.)

b) Find the *fractional* change in period $\Delta T/T$ in terms of the fractional change $\Delta g/g$.

c) The result of part (b) may also be obtained by taking differentials of both sides of Eq. (11–21). Use this procedure to derive the required relationship. The correct result is $dT/T = -\tfrac{1}{2}dg/g$.

d) A pendulum clock, which keeps correct time at a point where $g = 9.8000\ \mathrm{m\,s^{-1}}$, is found to lose 10 seconds each day at a higher elevation. Use the above results to find approximately the value of g at this new location.

11–25 A certain simple pendulum has a period on earth of 2.0 s. What is its period on the surface of the moon, where $g = 1.7\ \mathrm{m\,s^{-2}}$? (*Note*: The approximate method of Problem 11–24 is *not* adequate here because Δg is not small compared to g.)

11–26 The balance wheel of a watch vibrates with an angular amplitude of π radians and with a period of 0.5 s.

a) Find its maximum angular velocity.

b) Find its angular velocity when its displacement is one-half its amplitude.

c) Find its angular acceleration when its displacement is 45°.

11–27 A certain alarm clock ticks four times each second, each tick representing half a period. The balance wheel consists of a thin rim of radius 1.5 cm, connected to the balance staff by thin spokes of negligible mass. The total mass is 0.8 g.

a) What is the torque constant of the hairspring?

b) If the balance wheel is brass, what temperature increase is required for the clock to lose 10 s per day due to expansion of the balance wheel, assuming the torque constant does not change?

11–28 At one time it was proposed to define the second as the time required for a simple pendulum exactly one meter long to complete one side-to-side swing (one-half period).

a) By how much is this definition in error if $g = 9.8000\ \mathrm{m\,s^{-2}}$?

b) For what value of g would this definition be exactly correct?

11–29 Show that the moment of inertia of a thin uniform rod about a perpendicular axis through the rod, at a distance s from the center of mass, is equal to the moment of inertia about a parallel axis through the center of mass, plus ms^2, where m is the total mass. (This result is a special case of the theorem cited in Problem 11–32.)

11–30 A thin uniform rod is pivoted on a perpendicular axis through the rod, and its period as a physical pendulum is measured. By trial and error, a second pivot point is found so that the period is the same as for the first point. Show that, in this case, the period depends only on the distance L between the two points and is given by $T = 2\pi(L/g)^{1/2}$. (The result of Problem 11–29 will be useful.)

11–31 A monkey wrench is pivoted at one end and allowed to swing as a physical pendulum. The period is 0.9 s and the pivot is 20 cm from the center of gravity.

a) What is the radius of gyration of the wrench about an axis through the pivot?

b) If the wrench was initially displaced 0.1 rad from its equilibrium position, what is the angular velocity of the wrench as it passes through the equilibrium position?

11–32 It is shown in textbooks on mechanics that the moment of inertia I of a body about *any* axis through any point is given by

$$I = I_G + mh^2,$$

where I_G is the moment of inertia about a *parallel axis* through the center of gravity, m is the mass, and h is the perpendicular distance between the two parallel axes. A solid disk of radius $R = 12$ cm oscillates as a physical pendulum about an axis perpendicular to the plane of the disk at a distance r from its center. (See Fig. 11–16.)

a) Calculate the period of oscillation (for small amplitudes) for the following values of r: 0, $R/4$, $R/2$, $3R/4$, R.

b) Let T_0 represent the period when $r = R$, and T the period at any other value of r. Construct a graph of the dimensionless ratio T/T_0 as a function of the dimensionless ratio r/R. (Note that the graph then describes the behavior of *any* solid disk, whatever its radius.)

c) Prove by the methods of calculus that the period is a minimum when $r = R/\sqrt{2}$. Does this result agree with your graph?

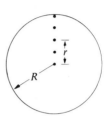

Fig. 11–16

11–33 It is desired to construct a pendulum of period 10 s.

a) What is the length of a *simple* pendulum having this period?

b) Suppose the pendulum must be mounted in a case not over 0.5 m high. Can you devise a pendulum, having a period of 10 s, that will satisfy this requirement?

11–34 A meter stick hangs from a horizontal axis at one end and oscillates as a physical pendulum. A body of small dimensions, and of mass equal to that of the meter stick, can be clamped to the stick at a distance h below the axis. Let T represent the period of the system with the body attached, and T_0 the period of the meter stick alone. Find the ratio T/T_0, (a) when $h = 0.5$ m, (b) when $h = 1.0$ m.

c) Is there any value of h for which $T = T_0$? If so, find it and explain why the period is unchanged when h has this value.

11–35 A meter stick is pivoted at one end. At what distance below the pivot should it be struck in order that it start swinging smoothly about the pivot?

Chapter 12

Hydrostatics

12–1 INTRODUCTION

The term "hydrostatics" is applied to the study of fluids at rest, and "hydrodynamics" to fluids in motion. The special branch of hydrodynamics relating to the flow of gases and of air in particular is called "aerodynamics."

A fluid is a substance which can flow. Hence the term includes both liquids and gases. Liquids and gases differ markedly in their compressibilities; a gas is easily compressed, while a liquid is practically incompressible. The small volume changes of a liquid under pressure can usually be neglected in this part of the subject.

The *density* of a homogeneous material is defined as its mass per unit volume. Units of density are one kilogram per cubic meter (1 kg m^{-3}) in the mks system, and one gram per cubic centimeter (1 g cm^{-3}) in the cgs system. We shall represent density by the greek letter ρ (rho):

$$\rho = \frac{m}{V}, \qquad m = \rho V. \qquad (12\text{–}1)$$

Typical values of density at room temperature are given in Table 12–1. The conversion factor

$$1 \text{ g cm}^{-3} = 1000 \text{ kg m}^{-3}$$

is useful in connection with this table.

The *specific gravity* of a material is the ratio of its density to that of water and is therefore a pure number. "Specific gravity" is an exceedingly poor term, since it has nothing to do with gravity. "Relative density" would describe the concept more precisely.

Table 12–1 DENSITIES   *CGS*

Material	Density, g cm^{-3}	Material	Density, g cm^{-3}
Aluminum	2.7	Silver	10.5
Brass	8.6	Steel	7.8
Copper	8.9	Mercury	13.6
Gold	19.3	Ethyl alcohol	0.81
Ice	0.92	Benzene	0.90
Iron	7.8	Glycerin	1.26
Lead	11.3	Water	1.00
Platinum	21.4	Sea water	1.03

Density measurements are an important analytical technique in a wide variety of circumstances. The condition of an automobile storage battery can be judged by measuring the density of the electrolyte, a sulfuric acid solution. As the battery discharges, the sulfuric acid (H_2SO_4) combines with lead in the battery plates to form insoluble lead sulfate ($PbSO_4$), decreasing the concentration of the solution. The density varies from about 1.30 g cm^{-3} for a fully charged battery to 1.15 g cm^{-3} for a discharged battery. Similarly, permanent-type antifreeze is usually a solution of ethylene glycol (density 1.12 g cm^{-3}) in water, with small quantities of additives to retard corrosion. The glycol concentration, which determines the freezing point of the solution, can be determined from a simple density measurement. Both these measurements are performed routinely in service stations with the aid of a simple hydrometer, which measures density by observation of the level at which a calibrated body floats in a sample of the solution. The hydrometer is discussed in Section 12–6.

Medical science finds many uses for density measurements; examples include tests on body fluids such as blood and urine. The density of normal human blood is about 1.04 to 1.06 g cm^{-3}. Since density increases with the concentration of red cells, an unusually low density may indicate anemia. Similarly, the usual density of urine is about 1.02 g cm^{-3}. Certain diseases lead to increased excretion of salts and a corresponding increase in urine density.

12–2 PRESSURE IN A FLUID

When the concept of hydrostatic pressure was introduced in Section 10–1, the weight of the fluid was neglected and the pressure was assumed the same at all points. It is a familiar fact, however, that atmospheric pressure decreases with increasing altitude and that the pressure in a lake or in the ocean decreases with increasing distance from the bottom. We therefore generalize the definition of pressure and define the pressure *at any point* as the ratio of the normal force dF exerted on a small area dA including the point, to the area dA:

$$ p = \frac{dF}{dA}, \qquad dF = p\, dA. \qquad (12\text{–}2) $$

If the pressure is the same at all points of a finite plane surface of area A, these equations reduce to Eq. (10–3):

$$ p = \frac{F}{A}, \qquad F = pA. $$

Let us find the general relation between the pressure p at any point in a fluid and the elevation of the

point y. If the fluid is in equilibrium, every volume element is in equilibrium. Consider an element in the form of a thin slab, shown in Fig. 12–1, whose thickness is dy and whose faces have an area A. If ρ is the density of the fluid, the mass of the element is $\rho A\, dy$ and its weight dw is $\rho g A\, dy$. The force exerted on the element by the surrounding fluid is everywhere normal to its surface. By symmetry, the resultant horizontal force on its rim is zero. The upward force on its lower face is pA, and the downward force on its upper face is $(p + dp)A$. Since it is in equilibrium,

$$ \sum F_y = 0, $$
$$ pA - (p + dp)A - \rho g A\, dy = 0, $$

and therefore

$$ \frac{dp}{dy} = -\rho g. \qquad (12\text{–}3) $$

Since ρ and g are both positive quantities, it follows that a positive dy (an increase of elevation) is

(a)

(b)

Fig. 12–1 *Forces on an element of fluid in equilibrium.*

Fig. 12–2

Fig. 12–3

accompanied by a negative dp (decrease of pressure). If p_1 and p_2 are the pressures at elevations y_1 and y_2 above some reference level, then integration of Eq. (12–3), when ρ and g are constant, gives

$$p_2 - p_1 = -\rho g(y_2 - y_1).$$

Let us apply this equation to a liquid in an open vessel, such as that shown in Fig. 12–2. Take point 1 at any level and let p represent the pressure at this point. Take point 2 at the top where the pressure is atmospheric pressure p_a. Then

$$p_a - p = -\rho g(y_2 - y_1),$$
$$p = p_a + \rho g h. \qquad (12\text{–}4)$$

Note that the shape of the containing vessel does not affect the pressure, and that the pressure is the same at all points at the same depth. It also follows from Eq. (12–4) that if the pressure p_a is increased in any way, say by inserting a piston on the top surface and pressing down on it, the pressure p at any depth must increase by exactly the same amount. This fact was recognized in 1653 by the French scientist Blaise Pascal (1623–1662) and is called "Pascal's law." It is often stated: "*Pressure applied to an enclosed fluid is transmitted undiminished to every portion of the fluid and the walls of the containing vessel.*" We can see now that it is not an independent principle but a necessary consequence of the laws of mechanics.

Pascal's law is illustrated by the operation of a hydraulic press, shown in Fig. 12–3. A piston of small cross-sectional area a is used to exert a small

force f directly on a liquid such as oil. The pressure $p = f/a$ is transmitted through the connecting pipe to a larger cylinder equipped with a larger piston of area A. Since the pressure is the same in both cylinders,

$$p = \frac{f}{a} = \frac{F}{A} \qquad \text{and} \qquad F = \frac{A}{a} f.$$

It follows that the hydraulic press is a force-multiplying device with a multiplication factor equal to the ratio of the areas of the two pistons. Barber chairs, dentist chairs, car lifts and jacks, and hydraulic brakes are all devices that make use of the principle of the hydraulic press.

12–3 THE HYDROSTATIC "PARADOX"

If a number of vessels of different shapes are interconnected, as in Fig. 12–4(a), it is found that a liquid poured into them stands at the same level in each. Before the principles of hydrostatics were completely understood, this seemed a very puzzling phenomenon and was called the "hydrostatic paradox." It would appear at first sight, for example, that vessel C should develop a greater pressure at its base than should B, and hence that liquid would be forced from C into B.

However, Eq. (12–4) states that the pressure depends only on the depth below the liquid surface and not at all on the shape of the containing vessel. Since the depth of the liquid is the same in each vessel, the pressure at the base of each is the same

(a)   (b)

Fig. 12–4 (a) The hydrostatic paradox. The top of the liquid stands at the same level in each vessel. (b) Forces on the liquid in vessel C.

and hence the system is in equilibrium. Thus there is no real paradox.

A more detailed explanation may be helpful in understanding the situation. Consider vessel C in Fig. 12–4(b). The forces exerted against the liquid by the walls are shown by arrows, the force being everywhere perpendicular to the walls of the vessel. The inclined forces at the sloping walls may be resolved into horizontal and vertical components. The weight of the liquid in the sections lettered A is supported by the vertical components of these forces. Hence the pressure at the base of the vessel is due only to the weight of the liquid in the cylindrical column B. Any vessel, regardless of its shape, may be treated in the same way.

12–4 PRESSURE GAUGES

The simplest type of pressure gauge is the open-tube manometer, illustrated in Fig. 12–5(a). It consists of a U-shaped tube containing a liquid, one end of the tube being at the pressure p which it is desired to measure, while the other end is open to the atmosphere.

The pressure at the bottom of the left column is $p + \rho g y_1$, while that at the bottom of the right column (the same point) is $p_a + \rho g y_2$, where ρ is the density of the manometric liquid. Since these pressures must be equal,

$$p + \rho g y_1 = p_a + \rho g y_2,$$

and

$$p - p_a = \rho g(y_2 - y_1) = \rho g h. \qquad (12\text{–}5)$$

The pressure p is called the *absolute pressure*, whereas the difference $p - p_a$ between this and the

atmospheric pressure is called the *gauge pressure*. It is seen that the gauge pressure is proportional to the difference in height of the liquid columns.

The mercury barometer is a long glass tube that has been filled with mercury and then inverted in a dish of mercury, as shown in Fig. 12–5(b). The space above the mercury column contains only mercury vapor, whose pressure, at room temperature, is so small that it may be neglected. It is easily seen that

$$p_a = \rho g(y_2 - y_1) = \rho g h. \qquad (12\text{–}6)$$

Because mercury manometers and barometers are used so frequently in laboratories, it is customary to express atmospheric pressure and other pressures as so many "inches of mercury," "centimeters of mercury," or "millimeters of mercury." Although these are not real units of pressure, they are so descriptive that they are widely used. The pressure exerted by a column of mercury one millimeter high is commonly called *one Torr*, after the Italian physi-

(a)   (b)

Fig. 12–5 (a) The open-tube manometer. (b) The barometer.

cist Torricelli (1608–1647), who first investigated the mercury barometric column.

One type of blood pressure gauge commonly used by physicians, called a *sphygmomanometer*, includes a form of manometer similar to Fig. 12–5(a). Blood-pressure readings, such as 130/80, refer to the maximum and minimum pressures, measured in millimeters of mercury or *Torrs*. Because of height differences, the hydrostatic pressure varies at different points in the body; the standard reference point is the upper arm, level with the heart. Pressure is also affected by the viscous nature of blood flow and by valves throughout the vascular system which act as pressure regulators.

Example Compute the atmospheric pressure on a day when the height of the barometer is 76.0 cm.

The height of the mercury column depends on ρ and g as well as on the atmospheric pressure. Hence both the density of mercury and the local acceleration of gravity must be known. The density varies with the temperature, and g with the latitude and elevation above sea level. All accurate barometers are provided with a thermometer and with a table or chart from which corrections for temperature and elevation can be found. If we assume $g = 9.8 \text{ m s}^{-2}$ and $\rho = 13.6 \times 10^3 \text{ kg m}^{-3}$

$$p_a = \rho g h = (13.6 \times 10^3 \text{ kg m}^{-3})(9.8 \text{ m s}^{-2})(0.76 \text{ m})$$

$$= 101{,}300 \text{ N m}^{-2} = 1.013 \times 10^5 \text{ Pa}.$$

(One pascal, abbreviated Pa, is defined as one newton per square meter; $1 \text{ Pa} = 1 \text{ N m}^{-2}$.)

In English units,

$$0.76 \text{ m} = 30 \text{ in.} = 2.5 \text{ ft},$$

$$\rho g = 850 \text{ lb ft}^{-3},$$

$$p_a = 2120 \text{ lb ft}^{-2} = 14.7 \text{ lb in}^{-2}.$$

A pressure of $1.013 \times 10^5 \text{ Pa} = 14.7 \text{ lb in}^{-2}$, is called *one atmosphere* (1 atm). A pressure of exactly 10^5 Pa is called one *bar*, and a pressure one one-thousandth as great is one *millibar*. Atmospheric pressures are of the order of 1000 millibars, and are now stated in terms of this unit by the National Weather Service.

The Bourdon-type pressure gauge is more convenient for most purposes than a liquid manometer. It consists of a flattened brass tube closed at one end and bent into a circular form. The closed end of the tube is connected by a gear and pinion to a pointer which moves over a scale. The open end of the tube is connected to the apparatus, the pressure within which is to be measured. When pressure is exerted within the flattened tube, it straightens slightly, just as a bent rubber hose straightens when water is admitted. The resulting motion of the closed end of the tube is transmitted to the pointer.

12–5 PUMPS

Many devices of modern life include glass or metal containers from which nearly all the air has been exhausted; examples include electric light bulbs, TV picture tubes, thermos bottles, and many others. Successful operation of these devices requires air pressures as low as 10^{-8} mm of mercury, and special pumps are used to produce these "high" vacuums. Two commonly used types are the *rotary oil pump* for pressures as low as 10^{-4} mm of mercury, and the *mercury or oil diffusion pump* for pressures as low as 10^{-8} mm Hg.

Figure 12–6 depicts schematically a common type of rotary oil pump. The vessel to be exhausted

Fig. 12–6 *The rotary oil pump.*

Fig. 12–7 *The diffusion pump.*

is connected to the tube *A*, which communicates directly with the space marked *B*. As an eccentric cylinder *C* rotates in the direction shown, the point of contact between it and the inner walls of the stationary cylinder moves around in a clockwise direction, thereby trapping some air in the space marked *E*. The sliding vane *D* is kept in contact with the rotating cylinder by the pressure of the rod *F*. When the air in *E* is compressed enough to increase the pressure slightly above atmospheric, the valve *G* opens and the air bubbles through the oil and leaves through an opening *H* in the upper plate. The cylinder is caused to rotate by means of a small electric motor.

For pressures below about 10^{-3} or 10^{-4} mm of mercury, a *diffusion pump* is usually employed. In this type of pump, oil or mercury is heated to produce vapor, which is formed by the shape of the pump into a jet of rapidly moving vapor. As air molecules from the vessel to be exhausted diffuse into the jet, they are swept away from this vessel. A common type of diffusion pump is shown in Fig. 12–7. A rotary oil pump is used to reduce the pressure within the dif-

fusion pump to the low value necessary to ensure a well-defined jet, to prevent air molecules from diffusing from the jet back to the vessel. The air molecules swept away by the jet are removed by the rotary oil pump, which is called the *forepump*, and the mercury or oil condenses on the cool walls of the pump and returns to the well at the bottom.

Pumps are also used to compress gases and circulate liquids. Compressors for home refrigerators and air conditioners frequently use rotary pumps similar in principle to that shown in Fig. 12–6; reciprocating-piston pumps, equipped with pistons similar to those in gasoline engines, are also used for refrigerator compressors, and as air-compressors. In these pumps, the essential principle is a varying volume and a valve arrangement to ensure one-way flow of the fluid being pumped.

In situations where it is important to prevent contamination of the fluid, the varying volume is often provided by a flexible diaphragm, as shown in Fig. 12–8. The center portion of the diaphragm moves up and down, driven by a motor. When it moves up, fluid is pushed into the chamber from the left by gravity or external air pressure; when it moves down, the fluid in the chamber is forced out the right side. Backflow is prevented by the two valves. Many automobile-engine fuel pumps operate on this principle, and the human heart comprises four such pumps, with the expansion and contraction of the four chambers by the heart muscles playing the role of the diaphragm in Fig. 12–8. Some artificial heart pumps use diaphragm pumps.

Fig. 12–8 *A diaphragm pump.*

12–6 ARCHIMEDES' PRINCIPLE

Buoyancy is a familiar phenomenon; a body immersed in water seems to have less weight than when immersed in air, and a body whose average density is less than that of the fluid in which it is immersed can float in that fluid. Examples are the human body in water, or a helium-filled balloon in air.

Archimedes' principle states that *when a body is immersed in a fluid, the fluid exerts an upward force on the body equal to the weight of the fluid which is displaced by the body.* To prove this principle, we consider an arbitrary portion of fluid at rest. In Fig. 12–9, the irregular outline is the surface bounding this portion of fluid. The short arrows represent the forces exerted by the surrounding fluid against small elements of the boundary surface.

Since the entire fluid is at rest, the x-component of the resultant of these surface forces is zero. The y-component of the resultant, F_y, must equal the weight mg of the fluid inside the arbitrary surface, and its line of action must pass through the center of gravity of this fluid.

Now suppose that the fluid inside the surface is removed and replaced by a solid body having exactly the same shape. The pressure at every point is exactly the same as before; the force exerted on the body by the surrounding fluid is unaltered, and is therefore equal to the weight mg of fluid displaced. The line of action of this force again passes through the center of gravity of the displaced fluid.

The submerged body need not be in equilibrium. Its weight may be greater or less than F_y, and if it is not homogeneous, its center of gravity may not lie on the line of F_y. Therefore, in general, it will be acted on by a resultant force through its own center of gravity and by a couple, and will rise or fall and also rotate.

The weight of a balloon floating in air, or of a submarine floating at some depth below the surface of the water, is just equal to the weight of a volume of air, or water, that is equal to the volume of the balloon or submarine. That is, the average density of the balloon equals that of air, and the average density of the submarine equals the density of water.

A body whose average density is less than that of a liquid can float partially submerged at the free

Fig. 12–9 *Archimedes' principle. The buoyant force F_y equals the weight of the displaced fluid.*

upper surface of the liquid. A familiar example is the hydrometer, shown in Fig. 12–10. The instrument sinks in the fluid until the weight of fluid it displaces is exactly equal to its own weight. In a fluid of greater density, less fluid is displaced for the same weight, so less of the instrument is immersed; i.e., in denser fluids it floats *higher*. This corresponds to the familiar fact that when swimming in sea water (density 1.03 g cm^{-3}), one's body floats higher than in fresh water.

The hydrometer is weighted at its bottom end so the upright position is stable, and a scale in the stem at the top permits direct density readings. Fig. 12–10(b) shows a hydrometer commonly used to measure density of battery acid or anti-freeze, as discussed in Section 12–1. When the tube at the bottom is immersed in the fluid and the bulb squeezed to expel air and then released (like a giant medicine dropper) the resulting pressure difference causes the fluid to rise into the outer tube, and the hydrometer floats in this sample.

When making precise "weighings" with a sensitive analytical balance, correction must be made for the buoyant force of the air if the density of the body being "weighed" is very different from that of the standard "weights," which are usually of brass. For example, suppose a block of wood of density 0.4 g cm^{-3} is balanced on an equal-arm balance by brass "weights" of mass 20 g, density 8.0 g cm^{-3}. The

(a)   (b)

Fig. 12–10 *(a) A simple hydrometer. (b) Hydrometer used as a tester for battery acid or antifreeze.*

apparent weight of each body is the difference between its true weight and the buoyant force of the air. If ρ_w, ρ_b, and ρ_a are the densities of the wood, brass, and air, and V_w and V_b are the volumes of the wood and brass, the apparent weights, which are equal, are

$$\rho_w V_w g - \rho_a V_w g = \rho_b V_b g - \rho_a V_b g.$$

The true mass of the wood is $\rho_w V_w$, and the true mass of the standard is $\rho_b V_b$. Hence,

$$\text{True mass} = \rho_w V_w = \rho_b V_b + \rho_a(V_w - V_b)$$

$$= \text{mass of standard} + \rho_a(V_w - V_b).$$

In the specific example cited,

$$V_w = \frac{20}{0.4} = 50 \text{ cm}^3 \quad \text{(very nearly)},$$

$$V_b = \frac{20}{8} = 2.5 \text{ cm}^3,$$

$$\rho_a = 0.0013 \text{ g cm}^{-3}.$$

Hence,

$$\rho_a(V_w - V_b) = 0.0013 \times 47.5$$

$$= 0.062 \text{ g}.$$

MASS OF V_a DIFFERENCE

Therefore,

$$\text{True mass} = 20.062 \text{ g}.$$

If measurements are being made to one one-thousandth of a gram (0.001 g), the correction of 0.062 g is essential.

Example   A tank containing water is placed on a spring scale, which registers a total weight W. A stone of weight w is hung from a string and lowered into the water without touching the sides or bottom of the tank (Fig. 12–11(a)). What will be the reading on the spring scale?

First, for the stone alone, the forces are as shown in Fig. 12–11(b), where B is the buoyant force and T is the tension in the string. Since $\sum F_y = 0$,

$$T + B = w.$$

Next, for the tank with the water and stone in it, the forces are as shown in Fig. 12–11(c), where S is the force exerted by the spring scale on the isolated system and, by Newton's third law, is equal in

(a)   (b)   (c)

Fig. 12–11

magnitude and opposite in direction to the force exerted on the scale. The condition for equilibrium yields the equation

$$T + S = w + W.$$

Subtracting the first equation from the second, we get

$$S = W + B.$$

That is, the reading of the spring scale has been *increased* by an amount equal to the buoyant force.

An interesting medical application of buoyancy is found in the technique of *hydrotherapy*. A patient may be unable to lift a limb because of disease or injury in the associated muscles or joints. When the body is immersed in water it becomes, in effect, nearly weightless because its average density is only a little greater than that of water. As a result, the forces required to move the limb are greatly reduced, and therapeutic exercise becomes possible.

12–7 FORCES AGAINST A DAM

Water stands at a depth H behind the vertical upstream face of a dam (Fig. 12–12). It exerts a certain resultant horizontal force on the dam, tending to slide it along its foundation, and a certain moment tending to overturn the dam about the point O. We wish to find the horizontal force and its moment.

Figure 12–12(b) is a view of the upstream face of the dam. The pressure at an elevation y is

$$p = \rho g(H - y).$$

(Atmospheric pressure can be omitted, since it also acts upstream against the other face of the dam.) The

force against the shaded strip is

$$dF = p \, dA$$
$$= \rho g(H - y)L \, dy.$$

The total force is

$$F = \int dF = \int_0^H \rho g L(H - y) dy$$
$$= \tfrac{1}{2}\rho g L H^2.$$

The moment of the force dF about an axis through O is

$$d\Gamma = y \, dF = \rho g L y(H - y) \, dy.$$

The total torque about O is

$$\Gamma = \int d\Gamma = \int_0^H \rho g L y(H - y) \, dy$$
$$= \tfrac{1}{6}\rho g L H^3.$$

If $\bar{H}$ is the height above O at which the total force F would have to act to produce this torque,

$$F\bar{H} = \tfrac{1}{2}\rho g L H^2 \times \bar{H} = \tfrac{1}{6}\rho g L H^3,$$
$$\bar{H} = \tfrac{1}{3}H.$$

Hence the line of action of the resultant force is at 1/3 of the depth above O, or at 2/3 of the depth below the surface.

12–8 SURFACE TENSION

A liquid flowing slowly from the tip of a medicine dropper emerges not as a continuous stream but as a succession of drops. A sewing needle, if placed carefully on a water surface, makes a small depression in the surface and rests there without sinking,

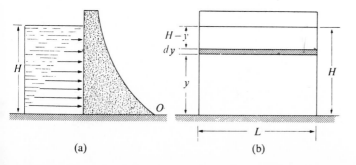

Fig. 12–12 Forces on a dam.

(a)   (b)

even though its density may be as much as 10 times that of water. When a clean glass tube of small bore is dipped into water, the water rises in the tube, but if the tube is dipped in mercury, the mercury is depressed. All these phenomena, and many others of a similar nature, are associated with the existence of a *boundary surface* between a liquid and some other substance.

All surface phenomena indicate that the surface of a liquid can be considered to be in a state of stress such that, for any line lying in or bounding the surface, the material on either side of the line exerts a *pull* on the material on the other side. This pull lies in the plane of the surface and is perpendicular to the line. In Fig. 12–13, a wire ring has a loop of thread attached to it as shown. When the ring and thread are dipped in a soap solution and removed, a thin film of liquid is formed in which the thread "floats" freely, as shown in part (a). If the film inside the loop of thread is punctured, the thread springs out into a circular shape as in part (b), as if the surfaces of the liquid were pulling radially outward on it, as shown by the arrows. Presumably, the same forces were acting before the film was punctured, but since there was film on *both* sides of the thread the net force exerted by the film on every portion of the thread was zero.

Fig. 12–14 *The horizontal slide wire is in equilibrium under the action of the upward surface force $2\gamma\ell$ and the downward pull $w_1 + w_2$.*

Another simple apparatus for demonstrating surface effects is shown in Fig. 12–14. A piece of wire is bent into the shape of a U and a second piece of wire is used as a slider. When the apparatus is dipped in a soap solution and removed, the slider (if its weight w_1 is not too great) is quickly pulled up to the top of the U. It may be held in equilibrium by adding a second weight w_2. Surprisingly, the same total force $F = w_1 + w_2$ will hold the slider at rest in *any* position, regardless of the area of the liquid film, provided the film remains at constant temperature. That is, the force does not increase as the surface is stretched farther. This is very different from the elastic behavior of a sheet of rubber, for which the force would be greater as the sheet was stretched.

Although a soap film like that in Fig. 12–14 is very thin, its thickness is still enormous compared with the size of a molecule. Hence it can be considered as made up chiefly of bulk liquid, bounded by two surface layers a few molecules thick. When the crossbar in Fig. 12–14 is pulled down and the area of the film is increased, molecules formerly in the main body of the liquid move into the surface layers. That is, these layers are not "stretched" as a rubber sheet would be; more surface is created by molecules moving from the bulk liquid.

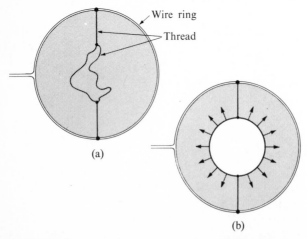

Fig. 12–13 *A wire ring with a flexible loop of thread, dipped in a soap solution, (a) before and (b) after puncturing the surface films inside the loop.*

Let ℓ be the length of the wire slider. Since the film has two surfaces, the total length along which the surface force acts is 2ℓ. The *surface tension* in the film, γ, is defined as *the ratio of the surface force to the length* (perpendicular to the force) *along which the force acts.* Hence, in this case,

$$\gamma = \frac{F}{2\ell}. \qquad (12\text{–}8)$$

The mks unit of surface tension is the newton per meter (1 N m^{-1}). This unit is not in common use; the usual unit is the cgs unit, the dyne per centimeter (1 dyn cm^{-1}). The conversion factor is

$$1 \text{ N m}^{-1} = 1000 \text{ dyn cm}^{-1}.$$

Another example is shown in Fig. 12–15. A circular wire of circumference ℓ is lifted out from the body of a liquid. The additional force F needed to balance the surface forces $2\gamma\ell$ due to the two surface films on each side is measured either by the stretch of a delicate spring or by the twist of a torsion wire. The surface tension is then given by

$$\gamma = \frac{F}{2\ell}.$$

Other methods of measuring surface tension will be apparent in what is to follow. Some typical values are shown in Table 12–2.

Fig. 12–15 *Lifting a circular wire of length ℓ out of a liquid requires an additional force **F** to balance the surface forces $2\gamma\ell$. This method is commonly used to measure surface tension.*

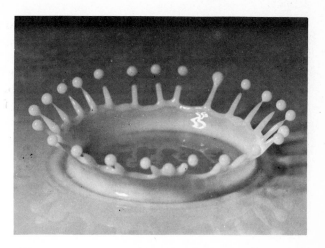

Fig. 12–16 *A drop of milk splashes on a hard surface.* (*Reproduced from* Flash, *courtesy of Ralph S. Hale & Co.*)

The surface tension of a liquid surface in contact with its own vapor or with air is found to depend *only* on the nature of the liquid and on the temperature. Surface tension usually decreases as temperature increases; Table 12–2 illustrates this behavior for water.

A surface under tension tends to contract until it has the minimum area consistent with any fixed boundaries that may be present and with pressure

Table 12–2 EXPERIMENTAL VALUES OF SURFACE TENSION

Liquid in contact with air	T, °C	Surface tension, dyn cm^{-1}
Benzene	20	28.9
Carbon tetrachloride	20	26.8
Ethyl alcohol	20	22.3
Glycerin	20	63.1
Mercury	20	465
Olive oil	20	32.0
Soap solution	20	25.0
Water	0	75.6
Water	20	72.8
Water	60	66.2
Water	100	58.9
Oxygen	−193	15.7
Neon	−247	5.15
Helium	−269	0.12

the surface. For example, a drop of liquid under no external forces, or in free fall in vacuum, is always spherical in shape because the sphere has smaller surface area for a given volume than has any other geometric shape. Figure 12–16 is a beautiful example of formation of spherical droplets in a very complex phenomenon, the impact of a drop on a rigid surface. This was taken by Dr. Edgerton of M.I.T., one of the pioneers in the development of high-speed photographic technique.

12–9 PRESSURE DIFFERENCE ACROSS A SURFACE FILM

A soap bubble consists of two spherical surface films very close together, with liquid between. Surface tension makes the films tend to contract, but as the bubble contracts it compresses the inside air, increasing the interior pressure to a point which prevents further contraction. To obtain a relation between surface tension and this excess pressure, we apply the principles of statics to one-half of the bubble.

As a preliminary, consider first a small element ΔA of the surface, shown in Fig. 12–17. Suppose the air pressure on the left of this element is p and that on the right is p_a. The force normal to the element is therefore $(p - p_a)\Delta A$. The component of this force in the x-direction is

$$(p - p_a)\Delta A \cos \theta.$$

But $\Delta A \cos \theta$ is the area projected on a plane perpendicular to the x-axis. The force in the x-direction is therefore the difference of pressure multiplied by the *projected area* in the x-direction.

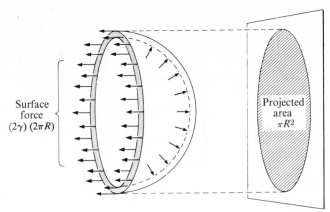

Fig. 12–18 *Equilibrium of half a soap bubble. The force exerted by the outer half is $(2\gamma)(2\pi R)$, and the net force exerted by the air inside and outside the bubble is the pressure difference times projected area, or $(p - p_a)\pi R^2$.*

Now consider the half-bubble shown in Fig. 12–18. The other half exerts a force to the left equal to twice the surface tension times the perimeter or

$$F(\text{to the left}) = (2\gamma)(2\pi R).$$

The force to the right on this half-bubble is equal to the pressure difference $p - p_a$ multiplied by the area obtained by projecting the half-bubble on a plane perpendicular to the direction in question. Since this projected area is πR^2,

$$F(\text{to the right}) = (p - p_a)\pi R^2.$$

Since the half-bubble is in equilibrium,

$$(p - p_a)\pi R^2 = 4\pi R\gamma,$$

or

$$\boxed{p - p_a = \frac{4\gamma}{R} \quad \text{(soap bubble).}} \quad (12\text{–}9)$$

If the surface tension remains constant (this means constant temperature), the pressure difference is greater for small bubbles than for large ones. If two bubbles are blown at opposite ends of a pipe, the smaller of the two will force air into the larger; the smaller one will get still smaller, and the larger will increase in size.

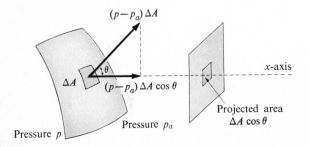

Fig. 12–17 *The force in the x-direction is the difference of pressure multiplied by the projected area in the x-direction.*

For a liquid drop, which has only *one* surface film, the difference between pressure of the liquid and that of the outside air is half that for a soap bubble:

$$p - p_a = \frac{2\gamma}{R} \quad \text{(liquid drop).} \quad (12\text{–}10)$$

Example Calculate the excess pressure inside a drop of mercury of diameter 4 mm at 20°C.

$$p - p_a = \frac{2\gamma}{R}$$
$$= \frac{(2)(465 \times 10^{-3}\,\mathrm{N\,m^{-1}})}{0.002\,\mathrm{m}}$$
$$= 465\,\mathrm{N\,m^{-2}} \cong 0.005\,\mathrm{atm.}$$

12–10 CONTACT ANGLE AND CAPILLARITY

The preceding sections have been concerned with surface films lying in the boundary between a liquid and a gas. Surface films also exist between a solid wall and a liquid, and between a solid and a vapor. The three boundaries and their accompanying films are shown schematically in Fig. 12–19. The films are only a few molecules thick. Associated with each film is an appropriate surface tension. Thus,

γ_{SL} = surface tension of the solid-liquid film,

γ_{SV} = surface tension of the solid-vapor film,

γ_{LV} = surface tension of the liquid-vapor film.

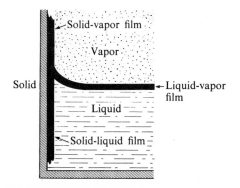

Fig. 12–19 *Surface films exist at the solid–vapor boundary as well as the liquid–vapor boundary.*

Table 12–3 CONTACT ANGLES

Liquid	Wall	Contact angle
α-Bromonaphthalene ($C_{10}H_7Br$)	Soda-lime glass	5°
	Lead glass	6°45′
	Pyrex	20°30′
	Fused quartz	21°
Methylene iodide (CH_2I_2)	Soda-lime glass	29°
	Lead glass	30°
	Pyrex	29°
	Fused quartz	33°
Water	Paraffin	107°
Mercury	Soda-lime glass	140°

The symbol γ without subscripts, defined and used in the preceding sections, now appears as γ_{LV}.

The curvature of the surface of a liquid near a solid wall depends upon the difference between γ_{SV} and γ_{SL}. Consider a portion of a glass wall in contact with methylene iodide, as shown in Fig. 12–20(a). When γ_{SV} is greater than γ_{SL}, the line along which the three films meet is pulled upward, as in Figs. 12–19 and 12–20(a). Angle θ in Fig. 12–20 is called the *contact angle*, and we see that when γ_{SV} is greater than γ_{SL}, the angle of contact is between zero and 90°. Figure 12–20(b) shows a case where γ_{SV} is *less* than γ_{SL}, the line of intersection of the films is pulled *downward*, and the contact angle is between 90° and 180°. In Fig. 12–20(c), γ_{SV} and γ_{SL} are very nearly equal, and the contact angle is 90°. See Table 12–3 for contact angles of representative liquids in containers of several materials.

Impurities and adulterants present in or added to a liquid may alter the contact angle considerably. Wetting agents or detergents change the contact angle from a large value, greater than 90°, to a value much *smaller* than 90°. Conversely, waterproofing agents applied to cloth cause the contact angle of water in contact with the cloth to be larger than 90°. The effect of a detergent on a drop of water resting on a block of paraffin is shown in Fig. 12–21.

An important phenomenon resulting from surface tension is the elevation of a liquid in an open tube of small cross section. The term *capillarity*, used to describe effects of this sort, originates from the

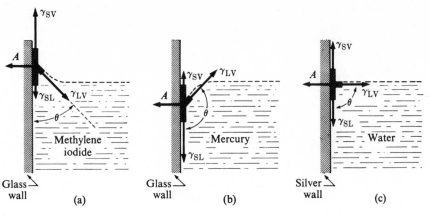

Fig. 12–20 *The surface of a liquid near a solid wall is curved if the solid–vapor surface tension γ_{SV} differs from the solid–liquid surface tension γ_{SL}.*

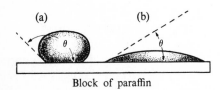

Fig. 12–21 *Effect of decreasing contact angle by a wetting agent.*

description of such tubes as "capillary" or "hairlike." In the case of a liquid that wets the tube, the contact angle is less than 90° and the liquid rises until an equilibrium height y is reached, as shown in Fig. 12–22(a). (The curved liquid surface in the tube is called a *meniscus*.)

If the tube is a cylinder of radius r, the liquid makes contact with the tube along a line of length $2\pi r$. When we isolate the cylinder of liquid of height y and radius r, along with its liquid-vapor film, the total upward force is

$$F = 2\pi r \gamma_{LV} \cos \theta.$$

The downward force is the weight w of the cylinder; this is equal to the weight-density ρg times the volume, which is approximately $\pi r^2 y$, neglecting the small volume of the meniscus. Thus

$$w = \rho g \pi r^2 y.$$

Since the cylinder is in equilibrium,

$$\rho g \pi r^2 y = 2\pi r \gamma_{LV} \cos \theta,$$

or

$$\boxed{y = \frac{2\gamma_{LV} \cos \theta}{\rho g r}.} \qquad (12\text{–}11)$$

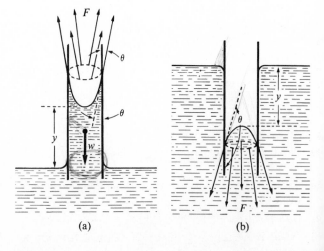

Fig. 12–22 *Surface tension forces on a liquid in a capillary tube. The liquid rises if $\theta < 90°$ and is depressed if $\theta > 90°$.*

The same equation holds for capillary depression, shown in Fig. 12–22(b). Capillarity accounts for the rise of ink in blotting paper, the rise of lighter fluid in the wick of a cigarette lighter, and many other common phenomena.

Capillarity is very important in a variety of life processes. A familiar example is the rising of water (actually a dilute aqueous solution) from the roots of a plant to its foliage, due partly to capillarity and partly to osmotic pressure developed in the roots. In the higher animals, including man, blood is pumped through the arteries and veins, but capillarity is still important in the smallest blood vessels, which indeed are called capillaries.

PROBLEMS

12–1 The piston of a hydraulic automobile lift is 30 cm in diameter. What pressure in pascals (newtons per square meter) is required to lift a car of mass 150 kg? Also express this pressure in atmospheres.

12–2 The expansion tank of a household hot-water heating system is open to the atmosphere and is 10 m above a pressure gauge attached to the furnace. What is the gauge pressure at the furnace, in newtons per square meter? in atmospheres?

12–3 Why can't a skin diver obtain an air supply at any desired depth by breathing through a "snorkel," a tube connected to his face mask and having its upper end above the water surface?

12–4 Suppose the door of a room makes an airtight but frictionless fit in its frame. Do you think you could open the door if the air pressure on one side were standard atmospheric pressure and that on the other side differed from standard by 1%?

12–5 The liquid in the open-tube manometer in Fig. 12–5(a) is mercury, and $y_1 = 3$ cm, $y_2 = 8$ cm. Atmospheric pressure is 970 millibars.

a) What is the absolute pressure at the bottom of the U-tube?

b) What is the absolute pressure in the open tube, at a depth of 5 cm below the free surface?

c) What is the absolute pressure of the gas in the tank?

d) What is the gauge pressure of the gas, in "cm of mercury"?

e) What is the gauge pressure in "cm of water"?

12–6 What gauge pressure must a pump produce in order to pump water from the bottom of Grand Canyon (2400-ft. elevation) to Indian Gardens (4500-ft.)? Express your results in pounds per square inch and in atmospheres.

12–7 A diving bell is to be designed to withstand the pressure of water at a depth of 2000 ft.

a) If sea water weighs 64 lb per cubic foot, what is the pressure at this depth?

b) What is the total force on a glass window 6 inches in diameter?

12–8 A common method for testing gold for purity is to measure its density by weighing it in air and then in water, as discussed in Section 12–6. In an era of rising gold prices, a swindler proposed to make a fake gold ingot by using a hollow slab of iridium (density 22.5 g cm^{-3}) and plating it with a thin layer of gold (density 19.3 g cm^{-3}).

a) To make a fake ingot of total mass 0.5 kg, what should be the total volume?

b) What is the volume of the interior air space?

c) Is iridium really any cheaper than gold?

12–9 According to advertising claims, a certain small car will float in water.

a) If the mass of the car is 1000 kg and its interior volume 4.0 m^3, what fraction of the car is immersed when it floats? The buoyancy of steel and other materials may be neglected.

b) As water gradually leaks in and displaces the air in the car, what fraction of the interior volume is water when the car sinks?

12–10

a) A small test tube, partially filled with water, is inverted in a large jar of water and floats, as shown in Fig. 12–23. The lower end of the test tube is open and the top of the large jar is covered by a tightly fitting rubber membrane. When the membrane is pressed down the test tube sinks, and when the membrane is released it rises again. Explain. (A hollow glass figure in human form is often used instead of the test tube, and is called a "Cartesian diver.")

Fig. 12–23

b) A torpedoed ship sinks below the surface of the ocean. If the depth is sufficiently great, is it possible for the ship to remain suspended in equilibrium at some point above the ocean bottom?

12–11 A tube 1 cm^2 in cross section is attached to the top of a vessel 1 cm high and of cross section 100 cm^2. Water is poured into the system, filling it to a depth of 100 cm above the bottom of the vessel, as in Fig. 12–24.

a) What is the force exerted by the water against the bottom of the vessel?

b) What is the weight of the water in the system?

c) Explain why (a) and (b) are not equal.

Fig. 12–24

1 cm^2

99 cm

1 cm

100 cm^2

12–12 A piece of gold-aluminum alloy weighs 10 lb. When suspended from a spring balance and submerged in water, the balance reads 8 lb. What is the weight of gold in the alloy if the specific gravity of gold is 19.3 and the specific gravity of aluminum is 2.5?

12–13 What is the area of the smallest block of ice 1 ft thick that will just support a man weighing 180 lb? The specific gravity of the ice is 0.917, and it is floating in fresh water.

12–14 A cubical block of wood 10 cm on a side floats at the interface between oil and water as in Fig. 12–25, with its lower surface 2 cm below the interface. The density of the oil is 0.6 g cm^{-3}.

a) What is the mass of the block?

b) What is the gauge pressure at the lower face of the block?

Oil

Wood

10 cm

Water

10 cm

Fig. 12–25

12–15 The densities of air, helium, and hydrogen (at standard conditions) are, respectively, 1.29 kg m^{-3}, 0.178 kg m^{-3}, and 0.0899 kg m^{-3}. What is the volume in cubic meters displaced by a hydrogen-filled dirigible which has a total "lift" of 10,000 kg? What would be the "lift" if helium were used instead of hydrogen?

12–16 A piece of wood is 2 ft long, 1 ft wide, and 2 in. thick. Its specific gravity is 0.6. What volume of lead must be fastened underneath to sink the wood in calm water so that its top is just even with the water level?

12–17 A cubical block of wood 10 cm on a side and of density 0.5 g cm^{-3} floats in a jar of water. Oil of density 0.8 g cm^{-3} is poured on the water until the top of the oil layer is 4 cm below the top of the block.

a) How deep is the oil layer?

b) What is the gauge pressure at the lower face of the block?

12–18 A cubical block of steel (density = 7.8 g cm^{-3}) floats on mercury (density = 13.6 g cm^{-3}).

a) What fraction of the block is above the mercury surface?

b) If water is poured on the mercury surface, how deep must the water layer be so that the water surface just rises to the top of the steel block?

12–19 Block A in Fig. 12–26 hangs by a cord from spring balance D and is submerged in a liquid C contained in beaker B. The weight of the beaker is 2 lb, the weight of the liquid is 3 lb. Balance D reads 5 lb and balance E reads 15 lb. The volume of block A is 0.1 ft^3.

D

A   B
     C

E

Fig. 12–26

a) What is the mass per unit volume of the liquid?

b) What will each balance read if block A is pulled up out of the liquid?

12-20 A hollow sphere of inner radius 9 cm and outer radius 10 cm floats half submerged in a liquid of specific gravity 0.8.

a) Calculate the density of the material of which the sphere is made.

b) What would be the density of a liquid in which the hollow sphere would just float completely submerged?

12-21 Two spherical bodies having the same diameter are released simultaneously from the same height. If the mass of one is 10 times that of the other and if the air resistance on each is the same, show that the heavier body will arrive at the ground first.

12-22 When a life preserver having a volume of 0.03 m³ is immersed in sea water (specific gravity 1.03), it will just support a clothed 80-kg man (specific gravity 1.2) with 2/10 of his volume above water. What is the mass per unit volume of the material composing the life preserver?

12-23 A block of balsa wood placed in one scale pan of an equal-arm balance is found to be exactly balanced by a 100-g brass "weight" in the other scale pan. Find the true mass of the balsa wood, if its specific gravity is 0.15.

12-24 A 1500-kg cylindrical can buoy floats vertically in salt water (specific gravity = 1.03). The diameter of the buoy is 1.0 m. Calculate (a) the additional distance the buoy will sink when a 100-kg man stands on top, (b) the period of the resulting vertical simple harmonic motion when the man dives off.

12-25 A cubical block of wood 1 ft on a side is weighted so that its center of gravity is at the point shown in Fig. 12-27(a), and it floats in water with one-half its volume submerged. Compute the restoring torque when the block is "heeled" at an angle of 45° as in Fig. 12-27(b).

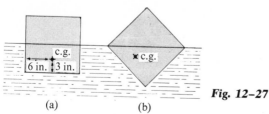

(a)   (b)   *Fig. 12-27*

12-26 A hydrometer consists of a spherical bulb and a cylindrical stem of cross section 0.4 cm². The total volume of bulb and stem is 13.2 cm³. When immersed in water the hydrometer floats with 8 cm of the stem above the water surface. In alcohol, 1 cm of the stem is above the surface. Find the density of the alcohol.

12-27 The following is quoted from a letter. How would you reply?

It is the practice of carpenters hereabouts, when laying out and leveling up the foundations of relatively long buildings, to use a garden hose filled with water, into the ends of the hose being thrust glass tubes 10 to 12 inches long.

The theory is that the water, seeking a common level, will be of the same height in both the tubes and thus effect a level. Now the question rises as to what happens if a bubble of air is left in the hose. Our greybeards contend the air will not affect the reading from one end to the other. Others say that it will cause important inaccuracies.

Can you give a relatively simple answer to this question, together with an explanation? I include a rough sketch (Fig. 12-28) of the situation that caused the dispute.

Fig. 12-28

12-28 A swimming pool measures 25 × 8 × 3 m deep. Compute the force exerted by the water against either end and against the bottom.

12-29 The upper edge of a vertical gate in a dam lies along the water surface. The gate is 6 ft wide and is hinged along the bottom edge, which is 10 ft below the water surface. What is the torque about the hinge?

12-30 The upper edge of a gate in a dam runs along the water surface. The gate is 6 ft high and 10 ft wide and is hinged along a horizontal line through its center. Calculate the torque about the hinge.

12-31 Figure 12-29 is a cross-sectional view of a masonry dam whose length perpendicular to the diagram is 100 ft. The depth of water behind the dam is 30 ft. The masonry of which the dam is constructed weighs 150 lb ft⁻³.

a) Find the dimensions x and 2x, if the weight of the dam is to be 10 times as great as the horizontal force exerted on it by the water.

b) Is the dam then stable with respect to its overturning about the edge through point O?

c) How does the size of the reservoir behind the dam affect the answers above?

Fig. 12-29

12-32 A U-tube of length ℓ (see Fig. 12-30) contains a liquid. What is the difference in height between the liquid columns in the vertical arms (a) if the tube has an acceleration a toward the right, and (b) if the tube is mounted on a horizontal turntable rotating with an angular velocity ω, with one of the vertical arms on the axis of rotation?

c) Explain why the difference in height does not depend on the density of the liquid, or on the cross-sectional area of the tube. Would it be the same if the vertical tubes did not have equal cross-sections? Would it be the same if the horizontal portion were tapered from one end to the other?

Fig. 12-30

12-33 Compare the tension of a soap bubble with that of a rubber balloon in the following respects:

a) Has each a surface tension?

b) Does the surface tension depend on area?

c) Is Hooke's law applicable?

12-34 Water can rise to a height y in a certain capillary. Suppose that this tube is immersed in water so that only a length $y/2$ is above the surface. Will you have a fountain or not? Explain your reasoning.

12-35 A capillary tube is dipped in water with its lower end 10 cm below the water surface. Water rises in the tube

to a height of 4 cm above that of the surrounding liquid, and the angle of contact is zero. What gauge pressure is required to blow a hemispherical bubble at the lower end of the tube?

12-36 A glass tube of inside diameter 1 mm is dipped vertically into a container of mercury, with its lower end 1 cm below the mercury surface.

a) What must be the gauge pressure of air in the tube to blow a hemispherical bubble at its lower end?

b) To what height will mercury rise in the tube if the air pressure in the tube is 3000 N m^{-2} below atmospheric? The angle of contact between mercury and glass is 140°.

12-37 On a day when the atmospheric pressure is 950 millibars, (a) what would be the height of the mercury column in a barometric tube of inside diameter 2 mm?

b) What would be the height in the absence of any surface tension effects?

c) What is the minimum diameter a barometric tube may have in order that the correction for capillary depression shall be less than 0.01 cm of mercury?

12-38

a) Derive the expression for the height of capillary rise in the space between two parallel plates dipping in a liquid.

b) Two glass plates, parallel to each other and separated by 0.5 mm, are dipped in water. To what height will the water rise between them? Assume zero angle of contact.

12-39 A tube of circular cross section and outer radius 0.14 cm is closed at one end. This end is weighted and the tube floats vertically in water, heavy end down. The total mass of the tube and weights is 0.20 g. If the angle of contact is zero, how far below the water surface is the bottom of the tube?

12-40 Find the gauge pressure, in dynes per square centimeter, in a soap bubble 5 cm in diameter. The surface tension is 25 dyn cm^{-1} = 25 × 10^{-3} N m^{-1}.

12-41 Two large glass plates are clamped together along one edge and separated by spacers a few millimeters thick along the opposite edge to form a wedge-shaped air film. These plates are then placed vertically in a dish of colored liquid. Show that the edge of the liquid forms an equilateral hyperbola.

12–42 When mercury is poured onto a flat glass surface which it does not wet, it spreads out into a pool of uniform thickness regardless of the size of the pool. Find the thickness of the pool.

12–43 Suppose the xylem tubes in the actively growing outer layer of a tree are uniform cylinders, and that the rising of sap is due entirely to capillarity, with a contact angle of 45° and surface tension of 0.05 N m^{-1}. What is the maximum radius of the tubes for a tree 20 m tall?

12–44 A tree 50 m tall has xylem tubes (sap-carrying pipes) in the form of uniform cylinders of radius 2 × 10^{-4} mm. If the surface tension is 0.05 N m^{-1} and the contact angle 45°, what minimum pressure must be developed in the roots in order for the sap to reach the top of the tree?

Chapter 13

Hydrodynamics and Viscosity

13–1 INTRODUCTION

Hydrodynamics is the study of fluids in motion. It is one of the most complex branches of mechanics, as illustrated by such familiar examples of fluid flow as a river in flood or a swirling cloud of cigarette smoke. While each drop of water or each smoke particle is governed by Newton's laws of motion, the resulting equations can be exceedingly complex. Fortunately, many situations of practical importance can be represented by idealized models which are simple enough to permit detailed analysis.

To begin, we shall consider only a so-called *ideal fluid*, that is, one which is incompressible and which has no internal friction or viscosity. The assumption of incompressibility is usually a good approximation for liquids. A gas can also be treated as incompressible provided the flow is such that pressure differences are not too great. Internal friction in a fluid gives rise to shearing stresses when two adjacent layers of fluid move relative to each other, or when the fluid flows inside a tube or around an obstacle. In some cases these shearing forces can be neglected in comparison with gravitational forces and forces arising from pressure differences.

The path followed by an element of a moving fluid is called a *line of flow*. In general, the velocity of

the element changes in both magnitude and direction along its line of flow. If every element passing through a given point follows the same line of flow as that of preceding elements, the flow is said to be *steady* or *stationary*. When any given flow is first started, it passes through a nonsteady state, but in many instances the flow becomes steady after a certain period of time has elapsed. In steady flow, the velocity at each point of space remains constant in time, although the velocity of a particular particle of the fluid may change as it moves from one point to another.

A *streamline* is defined as a curve whose tangent, at any point, is in the direction of the fluid velocity at that point. In steady flow, the streamlines coincide with the lines of flow.

If we construct all of the streamlines passing through the periphery of an element of area, such as the area A in Fig.13–1, these lines enclose a tube called a *flow tube* or *tube of flow*. From the definition of a streamline, no fluid can cross the side walls of a tube of flow; in steady flow there can be no mixing of the fluids in different flow tubes.

Figure 13–2 illustrates the nature of the flow around a number of obstacles, and in a channel of varying cross section. The photographs were made using an apparatus designed by Pohl, in which

Fig. 13–1 *A flow tube bounded by streamlines.*

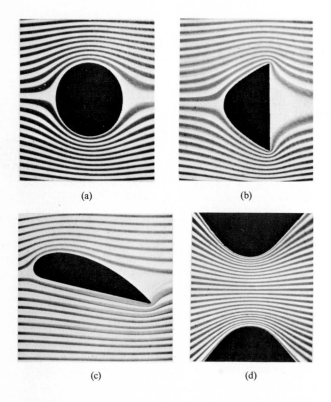

(a) (b)

(c) (d)

Fig. 13–2 *(a), (b), (c): Streamline flow around obstacles of various shapes. (d) Flow in a channel of varying cross-sectional area.*

alternate streams of clear and colored water flow between two closely spaced glass plates. The obstacles and the channel walls are opaque flat plates which fit between the glass plates. It will be noted that each obstacle is completely surrounded by a tube of flow. The tube splits into two portions at a so-called *stagnation point* on the upstream side of the obstacle. These portions rejoin at a second stagna-

tion point on the downstream side. The velocity at the stagnation points is zero. It will also be noted that the cross sections of all flow tubes decrease at a constriction and increase again when the channel widens.

The patterns of Fig. 13–2 are typical of *streamline* or *laminar* flow, in which adjacent layers of fluid slide smoothly past each other. At sufficiently high flow rates, or when boundary surfaces cause abrupt changes in velocity, the flow becomes irregular and much more complex and is called *turbulent* flow. In turbulent flow there is, strictly speaking, no steady-state pattern, since the flow pattern continuously changes.

13–2 THE EQUATION OF CONTINUITY

Let us consider any stationary, closed surface in a moving fluid; in general, fluid flows into the volume enclosed by the surface at some points and flows out at other points. The *equation of continuity* is a mathematical statement of the fact that the *net* rate of flow of mass *inward* across any closed surface is equal to the rate of increase of the mass within the surface.

For an incompressible fluid in steady flow, the equation takes the following form. Figure 13–3 represents a portion of a tube of flow, between two fixed cross sections of areas A_1 and A_2. Let v_1 and v_2 be the speeds at these sections. There is no flow across the side walls of the tube. The volume of fluid that flows into the tube across A_1 in a time interval dt is that contained in the short cylindrical element of base A_1 and height $v_1 \, dt$, or is $A_1 v_1 \, dt$. If the density of the fluid is ρ, the *mass* flowing in is $\rho A_1 v_1 \, dt$. Similarly, the mass

Fig. 13–3 *Flow into and out of a portion of a tube of flow.*

that flows out across A_2 in the same time is $\rho A_2 v_2\, dt$. The volume between A_1 and A_2 is constant, and since the flow is steady, the mass flowing out equals that flowing in. Hence

$$\rho A_1 v_1\, dt = \rho A_2 v_2\, dt,$$

or

$$A_1 v_1 = A_2 v_2, \qquad (13\text{–}1)$$

and the product Av is constant along any given tube of flow. It follows that when the cross section of a flow tube decreases, as in the constriction in Fig. 13–2(d), the velocity increases. This can readily be shown by introducing small particles into the fluid and observing their motion.

13–3 BERNOULLI'S EQUATION

When an incompressible fluid flows along a horizontal flow tube of varying cross section, its velocity changes; that is, it accelerates or decelerates. It must therefore be acted on by a resultant force, and this means that the pressure must vary *along* the flow tube even though the elevation does not change. For two points at *different* elevations, the pressure difference depends not only on the difference in level but also on the difference between the velocities at the points. The general expression for the pressure difference can be obtained directly from Newton's second law, but it is simpler to make use of the work-energy theorem. The problem was first solved by Daniel Bernoulli in 1738.

Figure 13–4 represents a portion of a tube of flow. We are to follow a small element of the fluid, indicated by shading, as it moves from one point to another along the tube. Let y_1 be the elevation of the first point above some reference level, v_1 the speed at that point, A_1 the cross-sectional area of the tube, and p_1 the pressure. All these quantities may vary from point to point: y_2, v_2, A_2, and p_2 are their values at the second point.

To derive the Bernoulli equation, we apply the work-energy theory to the fluid in a section of a flow tube. At some initial time, this element of fluid lies between two cross-sections a and c of the flow tube, in Fig. 13–4. The width of the tube is exaggerated for

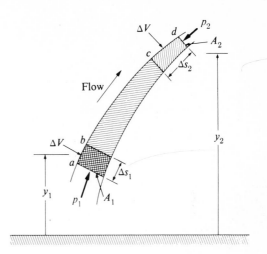

Fig. 13–4 *The net work done on the shaded element equals the increase in its kinetic and potential energy.*

clarity. In a time interval Δt, the element moves, the ends undergoing displacements Δs_1 and Δs_2 as shown. Because of the continuity relation, $\Delta V = A_1 \Delta s_1 = A_2 \Delta s_2$. The force on the cross-section at a is $p_1 A_1$, and that at b is $p_2 A_2$. The net *work* done on the element during this displacement is therefore

$$W = p_1 A_1 \Delta s_1 - p_2 A_2 \Delta s_2 = (p_1 - p_2)\, \Delta V. \quad (13\text{–}2)$$

The negative sign in the second term arises because the force at c is opposite in direction to the displacement.

We now equate this work to the total change in energy, kinetic and potential, of the element. During Δt, a volume of fluid $\Delta V = A_1 \Delta s_1$ with a mass $\Delta m = \rho \Delta V$ enters the tube past section a, bringing with it kinetic energy $\frac{1}{2}\Delta mv^2 = \frac{1}{2}\rho\, \Delta V v_1{}^2$. Similarly, during this interval, an equal mass leaves the tube past section c, taking with it kinetic energy $\frac{1}{2}\rho\, \Delta V v_2{}^2$. Thus the net change in kinetic energy is

$$\Delta E_k = \tfrac{1}{2}\rho\, \Delta V(v_2{}^2 - v_1{}^2). \qquad (13\text{–}3)$$

The change in potential energy is obtained similarly. The potential energy of the mass entering at a in time Δt is $\Delta mgy_1 = \rho\, \Delta Vgy_1$, and that of the mass leaving at c is $\Delta mgy_2 = \rho\, \Delta Vgy_2$.

The net change in potential energy is

$$\Delta E_p = \rho \, \Delta V g(y_2 - y_1). \qquad (13\text{–}4)$$

Combining Eqs. (13–2), (13–3), and (13–4) in the work-energy theorem $W = \Delta E_k + \Delta E_p$, we obtain

$$(p_1 - p_2) \Delta V = \tfrac{1}{2}\rho \, \Delta V(v_2{}^2 - v_1{}^2) + \rho \, \Delta V g(y_2 - y_1)$$

or

$$\boxed{p_1 - p_2 = \tfrac{1}{2}\rho(v_2{}^2 - v_1{}^2) + \rho g(y_2 - y_1).} \qquad (13\text{–}5)$$

In this form, Bernoulli's equation represents the equality of the work per unit volume of fluid $(p_1 - p_2)$ to the sum of the changes in kinetic and potential energies per unit volume that occur during the flow. Or we may interpret Eq. (13–5) in terms of pressures. The second term on the right is the pressure difference arising from the weight of the fluid and the difference in elevation of the two ends of the fluid element. The first term is the additional pressure difference associated with the change of velocity of the fluid.

Equation (13–5) can also be written

$$p_1 + \rho g y_1 + \tfrac{1}{2}\rho v_1{}^2 = p_2 + \rho g y_2 + \tfrac{1}{2}\rho v_2{}^2, \qquad (13\text{–}6)$$

and since the subscripts 1 and 2 refer to *any* two points along the tube of flow, Bernoulli's equation may also be written

$$z + h + \frac{v^2}{2g} = z_2 + h_2 + \frac{v^2}{2g}$$

$$\boxed{p + \rho g y + \tfrac{1}{2}\rho v^2 = \text{constant.}} \qquad (13\text{–}7)$$

$$h = \frac{p}{d}$$

Note carefully that p is the *absolute* (not gauge) pressure, and that a consistent set of units must be used. In the mks system, pressure is expressed in N m^{-2}, density in kg m^{-3}, and velocity in m s^{-1}.

Example Water enters a house through a pipe 2.0 cm in inside diameter, at an absolute pressure of 4×10^5 Pa (about 4 atm). The pipe leading to the second floor bathroom 5 m above is 1.0 cm in diameter. When the flow velocity at the inlet pipe is 4 m s^{-1}, find the flow velocity and pressure in the bathroom.

Solution. The flow velocity is obtained from the continuity equation:

$$v_2 = \frac{A_1}{A_2} v_1 = \frac{(2.0 \text{ cm})^2}{(1.0 \text{ cm})^2} (4 \text{ m s}^{-1})$$

$$= 16 \text{ m s}^{-1}.$$

The pressure is now obtained from Bernoulli's equation:

$$p_2 = p_1 - \tfrac{1}{2}\rho(v_2{}^2 - v_1{}^2) - \rho g(y_2 - y_1)$$

$$= 4 \times 10^5 \text{ Pa} - \tfrac{1}{2}(1.0 \times 10^3 \text{ kg m}^{-3})$$

$$\times (256 \text{ m}^2 \text{ s}^{-2} - 16 \text{ m}^2 \text{ s}^{-2})$$

$$- (1.0 \times 10^3 \text{ kg m}^{-3})(9.8 \text{ m s}^{-2})(5 \text{ m})$$

$$= 2.3 \times 10^5 \text{ Pa} \simeq 2.3 \text{ atm}.$$

We note that when the water is turned off, the second term on the right vanishes and the pressure rises to 3.5×10^5 Pa.

13–4 APPLICATIONS OF BERNOULLI'S EQUATION

1. The equations of hydrostatics are special cases of Bernoulli's equation, when the velocity is everywhere zero. Thus, when v_1 and v_2 are zero, Eq. (13–5) reduces to

$$p_1 - p_2 = \rho g(y_2 - y_1),$$

which is the same as Eq. (12–3).

2. *Speed of efflux. Torricelli's theorem.* Figure 13–5 represents a tank of cross-sectional area A_1, filled to a depth h with a liquid of density ρ. The space above the top of the liquid contains air at pressure p, and the liquid flows out of an orifice of area A_2. Let us consider the entire volume of moving fluid as a single tube of flow, and let v_1 and v_2 be the speeds at points 1 and 2. The quantity v_2 is called the *speed of efflux*. The pressure at point 2 is atmospheric, p_a. Applying Bernoulli's equation to points 1 and 2, and taking the bottom of the tank as our reference level, we get

$$p + \tfrac{1}{2}\rho v_1{}^2 + \rho g h = p_a + \tfrac{1}{2}\rho v_2{}^2,$$

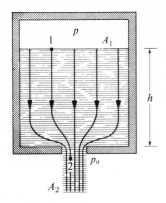

Fig. 13–5 *Flow of a liquid out of an orifice.*

or

$$v_2^2 = v_1^2 + 2\frac{p - p_a}{\rho} + 2gh. \qquad (13\text{–}8)$$

From the equation of continuity,

$$v_2 = \frac{A_1}{A_2} v_1. \qquad (13\text{–}9)$$

Because of the converging of the streamlines as they approach the orifice, the cross section of the stream continues to diminish for a short distance outside the tank. It is the area of smallest cross section, known as the *vena contracta*, which should be used in Eq. (13–9). For a sharp-edged circular opening, the area of the *vena contracta* is about 65 percent as great as the area of the orifice.

Now let us consider some special cases. Suppose the tank is open to the atmosphere, so that

$$p = p_a \qquad \text{and} \qquad p - p_a = 0.$$

Suppose also that $A_1 \gg A_2$. Then v_1^2 is very much less than v_2^2 and can be neglected, and from Eq. (13–8),

$$v_2 = \sqrt{2gh}. \qquad (13\text{–}10)$$

That is, *the speed of efflux is the same as that acquired by any body in falling freely through a height h.* This is *Torricelli's theorem.* It is not restricted to an opening in the bottom of a vessel, but applies also to a hole in the side walls at a depth h below the surface.

Now suppose again that the ratio of areas is such that v_1^2 is negligible and that the pressure p (in a closed vessel) is so large that the term $2gh$ in Eq. (13–8) can be neglected, compared with $2(p - p_a)/\rho$. The speed of efflux is then

$$v_2 = \sqrt{2(p - p_a)/\rho}. \qquad (13\text{–}11)$$

The density ρ is that of the fluid escaping from the orifice. If the vessel is partly filled with a liquid, as in Fig. 13–5, ρ is the density of the liquid. On the other hand, if the vessel contains only a gas, ρ is the density of the gas. The efflux speed of a gas may be very great, even for small pressures, since its density is small. However, if the pressure is too great, our idealized model is no longer adequate. Compressibility of the gas must be considered, and if the speed is too great, the motion may become turbulent. Bernoulli's equation can no longer be applied to the motion under these conditions.

A flow of fluid out of an orifice in a vessel gives rise to a *thrust* or *reaction force* on the remainder of the system. The mechanics of the problem are the same as those involved in rocket propulsion. The thrust can be computed as follows, provided conditions are such that Bernoulli's equation is applicable. If A is the area of the orifice, ρ the density of the escaping fluid, and v the speed of efflux, the mass of fluid flowing out in time Δt is $\rho A v \, \Delta t$, and its momentum (mass × velocity) is $\rho A v^2 \, \Delta t$. Since we are neglecting the relatively small velocity of the fluid in the container, we can say that the escaping fluid started from rest and acquired the momentum above in time Δt. Its *rate of change* of momentum was therefore $\rho A v^2$, and from Newton's second law this equals the force acting on it. By Newton's third law, an equal and opposite reaction force acts on the remainder of the system. Taking the expression for v^2 from Eq. (13–11), the reaction force can be written

$$F = \rho A v^2 = \rho A \frac{2(p - p_a)}{\rho},$$

or

$$F = 2A(p - p_a). \qquad (13\text{–}12)$$

Thus, while the *speed* of efflux is inversely proportional to the density, the *thrust* is independent of

Fig. 13–6 *A toy water rocket.*

the density and depends only on the area of the orifice and the gauge pressure $p - p_a$.

Example   Figure 13–6 shows a toy "water rocket." Above the water is air at a pressure $p = 2$ atm. (a) If the rocket is held at rest, what is the speed of efflux out of an opening in the base of the rocket? (b) What is the upward thrust if the area of the opening is 0.5 cm²?

a) The gauge pressure $p - p_a = 1$ atm $\approx 10^5$ Pa. The efflux speed is therefore

$$v \approx \sqrt{\frac{2 \times 10^5 \text{ Pa}}{10^3 \text{ kg m}^{-3}}}$$

$$\approx 14 \text{ m s}^{-1}.$$

The upward thrust, or reaction force, is

$$F \approx (2)(0.5 \times 10^{-4} \text{ m}^2)(10^5 \text{ Pa})$$

$$\approx 10 \text{ N}.$$

This force is much larger than the weight of the rocket and contents. Note that the reaction force, for the same gauge pressure, would be the same if the rocket initially contained air only. What is the reason for partially filling it with water?

It is interesting to treat this problem in the same way that the rocket was treated earlier. We have the rocket equation, Eq. (8–13)

$$m \frac{dv}{dt} = v_r \frac{dm}{dt} - mg.$$

Each term has a simple interpretation. The left-hand term is the resultant force which is seen to be the difference between the upward reaction force $v_r \, dm/dt$ and the weight mg downward. For the water rocket, $v_r \approx 14 \text{ m s}^{-1}$; $dm/dt = \rho v_r A \approx 0.7 \text{ kg s}^{-1}$. Hence the reaction force is

$$F = v_r \, dm/dt \approx (0.7 \text{ kg s}^{-1})(14 \text{ m s}^{-1}) \approx 10 \text{ N}.$$

3.   The *Venturi tube*, illustrated in Fig. 13–7, consists of a constriction or throat inserted in a pipeline and having properly designed tapers at inlet and outlet to avoid turbulence. Bernoulli's equation, applied to the wide and to the constricted portions of the pipe, becomes

$$p_1 + \tfrac{1}{2}\rho v_1{}^2 = p_2 + \tfrac{1}{2}\rho v_2{}^2.$$

From the equation of continuity, the speed v_2 is greater than the speed v_1, and hence the pressure p_2 in the throat is *less* than the pressure p_1. Thus a net force to the right acts to accelerate the fluid as it enters the throat, and a net force to the left decelerates it as it leaves. The pressures p_1 and p_2 can be measured by attaching vertical side tubes as shown in the diagram. From a knowledge of these pressures and of the cross-sectional areas A_1 and A_2, the velocities and the mass rate of flow can be computed. When used for this purpose, the device is called a *Venturi meter*.

The reduced pressure at a constriction finds a number of technical applications. Gasoline vapor is drawn into the manifold of an internal combustion engine by the low pressure produced in a Venturi throat to which the carburetor is connected. The *aspirator pump* is a Venturi throat through which

Fig. 13–7 *The Venturi tube.*

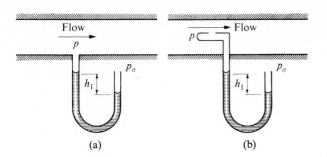

Fig. 13–8 *Pressure gauges for measuring the static pressure p in a fluid flowing in an enclosed channel.*

water is forced. Air is drawn into the low-pressure water rushing through the constricted portion.

4.   *Measurement of pressure in a moving fluid.* The pressure p in a fluid flowing in an enclosed channel can be measured with an open-tube manometer, as shown in Fig. 13–8. In (a), one arm of the manometer is connected to an opening in the channel wall. In (b), a *probe* is inserted in the stream. The probe should be small enough so that the flow is not appreciably disturbed and should be shaped so as to avoid turbulence. The difference h_1 in height of the liquid in the arms of the manometer is proportional to the difference between atmospheric pressure p_a and the fluid pressure p. That is,

$$p_a - p = \rho_m g h_1,$$
$$p = p_a - \rho_m g h_1, \qquad (13\text{–}13)$$

where ρ_m is the density of the manometer liquid.

The *Pitot tube,* shown in Fig. 13–9, is a probe with an opening at its upstream end. A stagnation point forms at the opening, where the pressure is p_2

Fig. 13–9 *The Pitot tube.*

Fig. 13–10 *The Prandtl tube.*

and the speed is zero. Applying Bernoulli's equation to the stagnation point, and to a point at a large distance from the probe where the pressure is p and the speed is v, we get

$$p_2 = p + \tfrac{1}{2}\rho v^2 \qquad (13\text{–}14)$$

(ρ is the density of the flowing fluid). The pressure p_2 at the stagnation point is therefore the sum of the pressure p and the quantity $\tfrac{1}{2}\rho v^2$.

The quantity $\tfrac{1}{2}\rho v^2$ is sometimes called the *dynamic pressure* and p the *static pressure*. These terms are somewhat misleading, inasmuch as p is always the true pressure (force per unit area) in both static and dynamic situations.

The instrument shown in Fig. 13–10 combines in a single device the functions of the instruments in Figs. 13–8 and 13–9. This device is called a *Prandtl tube*; the term Pitot tube is also used. The pressure at opening 1 is the "static pressure," corresponding to the pressure p measured in Fig. 13–8, and that at opening 2 is the quantity $p_2 = p + \tfrac{1}{2}\rho v^2$ measured in Fig. 13–9. The manometric height h_3 is proportional to the difference of these, or to $\tfrac{1}{2}\rho v^2$. Hence

$$\tfrac{1}{2}\rho v^2 = \rho_m g h_3. \qquad (13\text{–}15)$$

This instrument is self-contained, and its reading does not depend on atmospheric pressure. If held at rest, it can be used to measure the velocity of a stream of fluid flowing past it. If mounted on an aircraft, it indicates the velocity of the aircraft relative to the surrounding air and is known as an *airspeed indicator.*

Fig. 13–11 *Flow lines around an airfoil.*

5. *Lift on an aircraft wing.* Figure 13–11 shows flow lines around a section of an aircraft wing or an airfoil. The orientation of the wing relative to the flow direction causes the flow lines to crowd together above the wing, corresponding to increased flow velocity in this region, much as in the throat of a Venturi. Hence the region above the wing is one of increased velocity and reduced pressure, while below the wing the pressure remains nearly atmospheric. Because the upward force on the under side of the wing is greater than the downward force on the top side, there is a net upward force or *lift.*

This phenomenon can also be understood qualitatively on the basis of Newton's laws. In order for the fluid to exert a net upward force on the wing, the wing must exert a *downward* reaction force on the fluid, deflecting the stream downward. This effect is shown in Fig. 13–11; the fluid suffers a net change in the vertical component of momentum as it passes the wing, corresponding to the downward force the wing exerts on it. The force *on* the wing is thus *upward*, in agreement with the above discussion.

As the angle of the wing relative to the flow increases, turbulent flow occurs in a larger and larger region above the wing, and the pressure drop is no longer as great as that predicted by Bernoulli's principle. The lift on the wing decreases, and in extreme cases the airplane stalls.

6. *The curved flight of a spinning ball.* This effect is somewhat more subtle than the lift on an airplane wing, but the same basic principles are involved. Figure 13–12(a) represents a stationary ball in a blast of air moving from right to left. The motion of the air stream around and past the ball is the same as though the ball were moving through still air from left to right. Because of the large velocities ordinarily involved, there is a region of turbulent flow behind the ball, as shown.

When the ball is spinning, as in Fig. 13–12(b), the viscosity of air causes layers of air near the ball's surface to be pulled around in the direction of spin. The velocity of air relative to the ball surface is greater at the top than at the bottom. The region of turbulence becomes asymmetric, with turbulence occurring farther forward on the top side than on the bottom. This asymmetry gives rise to a pressure difference, the average pressure at the top becoming greater than that at the bottom. The corresponding net downward force deflects the ball as shown. In a baseball curve pitch, the actual deflection is sideways, and in that case Fig. 13–12 shows a *top* view of the situation.

Corresponding to the downward force on the ball in Fig. 13–12(b), the ball must exert an equal and opposite reaction force *upward* on the air. This deflects the air stream upward near the ball, as the figure shows, just as the lift force on an airplane wing corresponds to a downward deflection of the air stream.

Fig. 13–12 *(a) Air flow past a stationary ball, showing symmetric region of turbulence behind ball. (b) Air flow past a spinning ball, showing asymmetric region of turbulence and deflection of air stream. The net force on the ball is in the direction shown.*

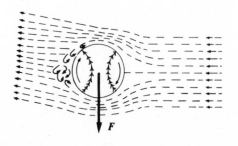

(a)   (b)

A similar effect occurs with golf balls, which always have "backspin" from impact with the slanted club face. The resulting pressure difference between top and bottom of the ball causes a "lift" force which keeps the ball in the air considerably longer than would be possible without spin. A well-hit drive appears from the tee to "float" or even curve *upward* during the initial portion of its flight, and this is a real effect, not an illusion. The dimples on the ball play an essential role; an undimpled ball has a much shorter trajectory than a dimpled one given the same initial velocity and spin. Some manufacturers even claim that the *shape* of the dimples is significant; one feels that polygonal dimples are superior to round ones!

13–5 VISCOSITY

Viscosity may be thought of as the internal friction of a fluid. Because of viscosity, a force must be exerted to cause one layer of a fluid to slide past another, or to cause one surface to slide past another if there is a layer of fluid between the surfaces. Both liquids and gases exhibit viscosity, although liquids are much more viscous than gases. In developing the fundamental equations of viscous flow, it will be seen that the problem is very similar to that of the shearing stress and strain in a solid.

Figure 13–13 illustrates one type of apparatus for measuring the viscosity of a liquid. A cylinder is pivoted on nearly frictionless bearings so as to rotate coaxially within a cylindrical vessel. The liquid whose viscosity is to be measured is poured into the annular space between the cylinders. A torque can be applied to the inner cylinder by the weight-pulley system. When the weight is released, the inner cylinder accelerates momentarily but very quickly reaches a constant angular velocity and continues to rotate at that velocity so long as the torque acts. It is obvious that this velocity will be smaller with a liquid such as glycerine in the annular space than it will be if the liquid is water or kerosene. From a knowledge of the torque, the dimensions of the apparatus, and the angular velocity, the viscosity of the liquid may be computed.

To reduce the problem to its essential terms, we replace the concentric cylinders by two parallel

Fig. 13–13 *Schematic diagram of one type of viscosimeter.*

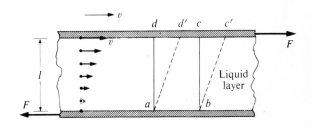

Fig. 13–14 *Laminar flow of a viscous fluid.*

plates, as in Fig. 13–14. The bottom plate is stationary, while the top plate moves with constant velocity v. It is found that the fluid in contact with each surface has the same velocity as that surface; thus at the top surface the fluid has velocity v, while the fluid adjacent to the bottom surface is at rest. The velocities of intermediate layers of fluid increase uniformly from one surface to the other, as shown by the arrows.

Flow of this type is called *laminar*. (A lamina is a thin sheet.) The layers of liquid slide over one another much as do the leaves of a book when it is placed flat on a table and a horizontal force applied to the top cover. As a consequence of this motion, a portion of the liquid which at some instant has the shape $abcd$, will a moment later take the shape $abc'd'$, and will become more and more distorted as the motion continues. That is, the liquid is in a state of continually increasing shearing strain.

In order to maintain the motion, a constant force must be exerted to the right on the upper, moving plate, and hence indirectly on the upper liquid surface. This force tends to drag the fluid and the lower plate as well to the right. Therefore an equal force must be exerted toward the left on the lower plate to hold it stationary. These forces are lettered F in Fig. 13–14. If A is the area of the fluid over which these forces are applied, the ratio F/A is the shearing stress exerted on the fluid.

When a shearing stress is applied to a *solid*, the effect of the stress is to produce a certain displacement of the solid, such as dd'. The shearing strain is defined as the ratio of this displacement to the transverse dimension ℓ, and within the elastic limit the shearing stress is proportional to the shearing strain. With a fluid, on the other hand, the shearing strain increases without limit so long as the stress is applied, and the stress is found by experiment to depend not on the shearing strain, but on its *rate of change*. The strain in Fig. 13–14, at the instant when the volume of fluid has the shape $abc'd'$, is dd'/ad, or dd'/ℓ. Since ℓ is constant, the rate of change of strain equals $1/\ell$ times the rate of change of dd'. But the rate of change of dd' is simply the velocity of point d', or the velocity v of the moving wall. Hence

$$\text{Rate of change of shearing strain} = \frac{v}{\ell}.$$

The *coefficient of viscosity* of the fluid, or simply its viscosity η, is defined as the ratio of the shearing stress, F/A, to the rate of change of shearing strain:

$$\eta = \frac{\text{shearing stress}}{\text{rate of change of shearing strain}} = \frac{F/A}{v/\ell},$$

or

$$F = \eta A \frac{v}{\ell} \qquad (13\text{–}16)$$

For a liquid which flows readily, such as water or kerosene, the shearing stress is relatively small for a given rate of change of shearing strain, and the viscosity also is relatively small. For a liquid such as molasses or glycerine, a greater shearing stress is

necessary for the same rate of change of shearing strain, and the viscosity is correspondingly greater. Viscosities of gases at ordinary pressures and temperatures are very much smaller than those of common liquids. Viscosities of all fluids are markedly dependent on temperature, increasing for gases and decreasing for liquids as the temperature is increased, hence the expression "as slow as molasses in January." An important consideration in the design of oils for engine lubrication is to reduce the temperature variation of viscosity as much as possible.

Equation (13–16) was derived for the special case in which the velocity increased at a uniform rate with increasing distance from the lower plate. The general term for the *space* rate of change of velocity, in a direction at right angles to the flow, is the *velocity gradient* in this direction. In this special case it is equal to v/ℓ. In the general case, the velocity gradient is not uniform and its value at any point can be written as dv/dy, where dv is the small difference in velocity between two points separated by a distance dy measured at right angles to the direction of flow. Hence the general form of Eq. (13–16) is

$$F = \eta A \frac{dv}{dy}. \qquad (13\text{–}17)$$

From Eq. (13–16), the unit of viscosity is that of force times distance, divided by area times velocity. In the mks system the unit is

$$1\,\text{N m m}^{-2}\,(\text{m s}^{-1})^{-1} = 1\,\text{N s m}^{-2}.$$

The corresponding cgs unit, $1\,\text{dyn s cm}^{-2}$, is the only viscosity unit in common use, and is called 1 *poise* in honor of the French scientist Poiseuille. Thus,

$$1\,\text{poise} = 1\,\text{dyn s cm}^{-2} = 10^{-1}\,\text{N s m}^{-2}.$$

Small viscosities are expressed in *centipoises* ($1\,\text{cp} = 10^{-2}$ poise) or *micropoises* ($1\,\mu\text{p} = 10^{-6}$ poise). A few typical values are given in Table 13–1.

Not all fluids behave according to the direct proportionality of force and velocity predicted by Eq. (13–16). An interesting exception is blood, for which velocity increases more rapidly than force. Thus doubling the force in Fig. 13–14 produces *more*

Table 13–1 TYPICAL VALUES OF VISCOSITY

Temperature, °C	Viscosity of castor oil, poise	Viscosity of water, centipoise	Viscosity of air, micropoise
0	53	1.792	171
20	9.86	1.005	181
40	2.31	0.656	190
60	0.80	0.469	200
80	0.30	0.357	209
100	0.17	0.284	218

than a twofold increase in velocity. This behavior may be understood on the basis that, on a microscopic scale, blood is not a homogeneous fluid but rather a suspension of solid particles in a liquid. The suspended particles have characteristic shapes; for example, red cells are roughly disk-shaped. At small velocities their orientations are random, but as velocity increases they tend to become oriented so as to facilitate flow. The fluids which provide lubrication in human joints exhibit similar behavior.

Fluids for which Eq. (13–16) holds are called *newtonian fluids*; we see that this description is an idealized model which not all fluids obey. In general, fluids which are suspensions or dispersions are often nonnewtonian in their viscous behavior. Nevertheless, Eq. (13–16) provides a useful model to describe approximately the properties of many pure substances.

13–6 POISEUILLE'S LAW

It is evident from the general nature of viscous effects that the velocity of a viscous fluid flowing through a tube is not the same at all points of a cross section. The outermost layer of fluid clings to the walls of the tube, and its velocity is zero. The tube walls exert a backward drag on this layer, which in turn drags backward on the next layer beyond it, and so on. Provided the velocity is not too great, the flow is laminar, with a velocity which is a maximum at the center of the tube and which decreases to zero at the walls. The flow is like that of a number of telescoping tubes sliding relative to one another, the central tube advancing most rapidly and the outer tube remaining at rest.

Let us consider the variation of velocity with radius for a cylindrical pipe of inner radius R. We consider the flow of a cylindrical element of fluid coaxial with the pipe, of radius r and length L, as shown in Fig. 13–15(a). The force on the left end is $p_1 \pi r^2$, and that on the right end $p_2 \pi r^2$, as shown. The net force is thus

$$F = (p_1 - p_2)\pi r^2.$$

This force must just balance the viscous retarding force at the surface of this element. This force is given by Eq. (13–17). The area over which the viscous force acts is $A = 2\pi rL$. Thus the viscous force is

$$F = \eta 2\pi rL \frac{dv}{dr}.$$

Equating this to the net force due to pressure on the ends and rearranging,

$$-\frac{dv}{dr} = \frac{(p_1 - p_2)r}{2\eta L}.$$

This shows that the velocity changes more and more rapidly as we go from the center ($r = 0$) to the pipe wall ($r = R$). The negative sign must be introduced because v decreases as r increases. Integrating, we get

$$-\int_v^0 dv = \frac{p_1 - p_2}{2\eta L}\int_r^R r\,dr,$$

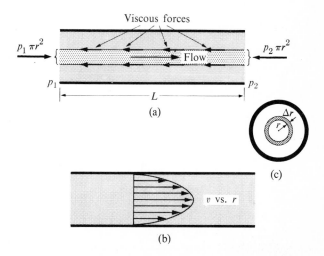

Fig. 13–15 (a) Forces on a cylindrical element of a viscous fluid. (b) Velocity distribution for viscous flow.

and therefore

$$v = \frac{p_1 - p_2}{4\eta L}(R^2 - r^2). \qquad (13\text{–}18)$$

We see that the velocity decreases from a maximum value $(p_1 - p_2)R^2/4\eta L$ at the center to zero at the wall. Thus the maximum velocity is proportional to the *square* of the pipe radius and is also proportional to the pressure change per unit length $(p_1 - p_2)/L$, called the *pressure gradient*. The curve in Fig. 13–15(b) is a graph of Eq. (13–18) with the v-axis horizontal and the r-axis vertical.

Equation (13–18) may be used to find the total rate of flow of fluid through the pipe. The velocity at each point is proportional to the pressure gradient $(p_1 - p_2)/L$, so the total flow rate must also be proportional to this quantity. Let us consider the thin-walled element in Fig. 13–15(c). The volume of fluid dV crossing the ends of this element in a time dt is $v\,dA\,dt$, where v is the velocity at the radius r and dA is the shaded area, equal to $2\pi r\,dr$. Taking the expression for v from Eq. (13–18), we get

$$dV = \frac{p_1 - p_2}{4\eta L}(R^2 - r^2)\,2\pi r\,dr\,dt.$$

The volume flowing across the entire cross section is obtained by integrating over all elements between $r = 0$ and $r = R$:

$$dV = \frac{\pi(p_1 - p_2)}{2\eta L}\int_0^R (R^2 - r^2)r\,dr\,dt$$

$$= \frac{\pi R^4}{8}\frac{p_1 - p_2}{\eta}\,dt.$$

The total volume of flow per unit time, denoted by Q, is given by

$$Q = \frac{dV}{dt} = \frac{\pi R^4}{8}\frac{p_1 - p_2}{\eta L}. \qquad (13\text{–}19)$$

This relation was first derived by Poiseuille* and is called *Poiseuille's law*. The volume rate of flow is

---

* Approximate pronounciation: Pwah-zoy'.

inversely proportional to viscosity, as might be expected. It is proportional to the pressure gradient along the pipe, and it varies as the fourth power of the radius. For example, if the radius is halved, the flow rate is reduced by a factor of 16. This relation is familiar to physicians in connection with the selection of needles for hypodermic syringes. Needle size is much more important than thumb pressure in determining the flow rate from the needle; doubling the needle diameter has the same effect as increasing the thumb force sixteenfold.

The difference between the flow of an ideal non-viscous fluid and one having viscosity is illustrated in Fig. 13–16, where fluid is flowing along a horizontal tube of varying cross section. The height of the fluid in the small vertical tubes is proportional to the gauge pressure.

In part (a), the fluid is assumed to have no viscosity. The pressure at b is very nearly the static pressure $\rho g y$, since the velocity is small in the large tank. The pressure at c is less than at b because the fluid must accelerate between these points. The pressures at c and d are equal, however, since the velocity and elevation at these points are the same. There is a further pressure drop between d and e, and between

(a)

(b)

Fig. 13–16 *Pressures along a horizontal tube in which is flowing (a) an ideal fluid, (b) a viscous fluid.*

f and *g*. The pressure at *g* is atmospheric and the gauge pressure at this point is zero.

Part (b) of the diagram illustrates the effect of viscosity. Again, the pressure at *b* is nearly the static pressure $\rho g y$. There is a pressure drop from *b* to *c*, due now in part to viscous effects, and also a further drop from *c* to *d*. The pressure gradient in this part of the tube is represented by the slope of the dotted line. The drop from *d* to *e* results in part from acceleration and in part from viscosity. The pressure gradient between *e* and *f* is greater than between *c* and *d* because of the smaller radius in this portion. Finally, the pressure at *g* is somewhat above atmospheric, since there is now a pressure gradient between this point and the end of the tube.

13–7 STOKES' LAW

When an ideal fluid of zero viscosity flows past a sphere, or when a sphere moves through a stationary fluid, the streamlines form a perfectly symmetrical pattern around the sphere, as shown in Fig. 13–2(a). The pressure at any point on the upstream hemispherical surface is exactly the same as that at the corresponding point on the downstream face, and the resultant force on the sphere is zero. If the fluid has viscosity, however, there will be a viscous drag on the sphere. (A viscous drag will of course be experienced by a body of any shape, but only for a sphere is the drag readily calculable.)

We shall not attempt to derive the expression for the viscous force directly from the laws of flow of a viscous fluid. The only quantities on which the force can depend are the viscosity η of the fluid, the radius *r* of the sphere, and its velocity *v* relative to the fluid. A complete analysis shows that the force *F* is given by

$$F = 6\pi\eta rv. \qquad (13\text{–}20)$$

This equation was first deduced by Sir George Stokes in 1845 and is called *Stokes' law*. We have already used it in Section 5–6 (Example 10) to study the motion of a sphere falling in a viscous fluid,

although at that point it was necessary to know only that the viscous force on a given sphere in a given fluid is proportional to the relative velocity.

It will be recalled that a sphere falling in a viscous fluid reaches a *terminal velocity* v_T at which the viscous retarding force plus the buoyant force equals the weight of the sphere. Let ρ be the density of the sphere and ρ' the density of the fluid. The weight of the sphere is then $(4/3)\pi r^3 \rho g$, and the buoyant force is $(4/3)\pi r^3 \rho' g$; when the terminal velocity is reached, the total force is zero, and

$$\tfrac{4}{3}\pi r^3 \rho' g + 6\pi\eta r v_T = \tfrac{4}{3}\pi r^3 \rho g,$$

or

$$v_T = \frac{2}{9}\frac{r^2 g}{\eta}(\rho - \rho'). \qquad (13\text{–}21)$$

By measuring the terminal velocity of a sphere of known radius and density, the viscosity of the fluid in which it is falling can be found from the equation above. Conversely, if the viscosity is known, the radius of the sphere can be determined by measuring the terminal velocity. This method was used by Millikan to determine the radius of very small electrically charged oil drops (used to measure the electrical charge of an individual electron) by observing free fall in air.

Even for nonspherical bodies, a relation of the form of Eq. (13–19) holds, with a different numerical coefficient. Biologists call the terminal velocity the *sedimentation velocity*, and experiments with sedimentation can give useful information concerning very small particles. It is often useful to increase the terminal velocity by spinning the sample in a centrifuge, which greatly increases the effective acceleration of gravity.

13–8 REYNOLDS NUMBER

When the velocity of a fluid flowing in a tube exceeds a certain critical value (which depends on the properties of the fluid and the diameter of the tube) the nature of the flow becomes extremely complicated. Within a very thin layer adjacent to the tube walls, called the *boundary layer*, the flow is still laminar. The flow velocity in the boundary layer is zero at the tube walls and increases uniformly throughout the layer. The properties of the boundary layer are of the greatest importance in determining the resistance to flow, and the transfer of heat to or from the moving fluid.

Beyond the boundary layer, the motion is highly irregular. Random local circular currents called *vortices* develop within the fluid, with a large increase in the resistance to flow. Flow of this sort is called *turbulent.*

Experiment indicates that a combination of four factors determines whether the flow of a fluid through a tube or pipe is laminar or turbulent. This combination is known as the *Reynolds number*, N_R, and is defined as

$$N_R = \frac{\rho v D}{\eta}, \qquad (13\text{–}22)$$

where ρ is the density of the fluid, v the average forward velocity, η the viscosity, and D the diameter of the tube. (The average velocity is defined as the uniform velocity over the entire cross section of the tube, which would result in the same volume rate of flow.) The Reynolds number, $\rho v D/\eta$, is a *dimensionless* quantity and has the same numerical value in any consistent system of units. For example, for water at 20°C flowing in a tube of diameter 1 cm with an average velocity of 10 cm s^{-1}, the Reynolds number is

$$N_R = \frac{\rho v D}{\eta} = \frac{(1\,\mathrm{g\,cm^{-3}})(10\,\mathrm{cm\,s^{-1}})(1\,\mathrm{cm})}{0.01\,\mathrm{dyn\,s\,cm^{-2}}} = 1000.$$

Had the four quantities been expressed originally in mks units, the same value of 1000 would have been obtained.

All the experiments show that when the Reynolds number is less than about 2000 the flow is laminar, whereas above about 3000 the flow is turbulent. In the transition region between 2000 and 3000

Fig. 13–7 *(a) Laminar flow. (b) Turbulent flow. (c) First laminar, then turbulent.*

(a)   (b)   (c)

the flow is unstable and may change from one type to the other. Thus for water at 20°C flowing in a tube 1 cm in diameter, the flow is laminar when

$$\frac{\rho v D}{\eta} \leq 2000,$$

or when

$$v \leq \frac{(2000)(0.01 \text{ dyn s cm}^{-2})}{(1 \text{ g cm}^{-3})(1 \text{ cm})} = 20 \text{ cm s}^{-1}.$$

Above about 30 cm s^{-1} the flow is turbulent. If air at the same temperature were flowing at 30 cm s^{-1} in the same tube, the Reynolds number would be

$$N_R = \frac{(0.0013 \text{ g cm}^{-3})(30 \text{ cm s}^{-1})(1 \text{ cm})}{181 \times 10^{-6} \text{ dyn s cm}^{-2}} = 216.$$

Since this is much less than 3000, the flow would be laminar and would not become turbulent unless the velocity were as great as 420 cm s^{-1}.

The distinction between laminar and turbulent flow is shown in the photographs of Fig. 13–17. In (a) and (b) the fluid is water and in (c) air and smoke particles.

The Reynolds number of a system forms the basis for the study of the behavior of real systems through the use of small scale models. A common example is the wind tunnel, in which one measures the aerodynamic forces on a scale model of an aircraft wing. The forces on a full-size wing are then deduced from these measurements.

Two systems are said to be *dynamically similar* if the Reynolds number, $\rho v D/\eta$, is the same for both. The letter D may refer, in general, to any dimension of a system, such as the span or chord of an aircraft wing. Thus the flow of a fluid of given density ρ and viscosity η, about a half-scale model, is dynamically similar to that around the full-size object if the velocity v is twice as great.

PROBLEMS

13–1 A circular hole 2 cm in diameter is cut in the side of a large standpipe, 10 m below the water level in the standpipe. Find (a) the velocity of efflux, and (b) the

volume discharged per unit time. Neglect the contraction of the streamlines after emerging from the hole.

13–2 Water stands at a depth H in a large open tank whose side walls are vertical (Fig. 13–18). A hole is made in one of the walls at a depth h below the water surface.

a) At what distance R from the foot of the wall does the emerging stream of water strike the floor?

b) At what height above the bottom of the tank could a second hole be cut so that the stream emerging from it would have the same range?

Fig. 13–18

13–3 A cylindrical vessel, open at the top, is 20 cm high and 10 cm in diameter. A circular hole whose cross-sectional area is 1 cm^2 is cut in the center of the bottom of the vessel. Water flows into the vessel from a tube above it at the rate of 140 cm^3 s^{-1}. How high will the water in the vessel rise?

13–4 At a certain point in a pipeline the velocity is 2 m s^{-1} and the gauge pressure is 10^4 Pa above atmospheric. Find the gauge pressure at a second point in the line 1 m lower than the first, if the cross section at the second point is one-half that at the first. The liquid in the pipe is water.

13–5 Water in an enclosed tank is subjected to a gauge pressure of 4 lb in^{-2}, applied by compressed air introduced into the top of the tank. There is a small hole in the side of the tank 16 ft below the level of the water. Calculate the speed with which water escapes from this hole.

13–6 What gauge pressure is required in the city mains in order that a stream from a fire hose connected to the mains may reach a vertical height of 60 ft?

13–7 A tank of large area is filled with water to a depth of 0.3 m. A hole of 5 cm^2 cross section in the bottom allows water to drain out in a continuous stream.

a) What is the rate at which water flows out of the tank, in m^3 s^{-1}?

b) At what distance below the bottom of the tank is the cross-sectional area of the stream equal to one-half the area of the hole?

13–8 A sealed tank containing sea water to a height of 2 m also contains air above the water at a gauge pressure of 40 atm. Water flows out from a hole at the bottom. The cross-sectional area of the hole is 10 cm².

a) Calculate the efflux velocity of the water.

b) Calculate the reaction force on the tank exerted by the water in the emergent stream.

13–9 A pipeline 6 in. in diameter, flowing full of water, has a constriction of diameter 3 in. If the velocity in the 6-in. portion is 4 ft s⁻¹, find (a) the velocity in the constriction, and (b) the discharge rate in cubic feet per second.

13–10 A horizontal pipe of 6 in² cross section tapers to a cross section of 2 in². If water is flowing with a velocity of 180 ft min⁻¹ in the large pipe where a pressure gauge reads 10.5 lb in⁻², what is the gauge pressure in the adjoining part of the small pipe? The barometer reads 30 in. of mercury.

13–11 At a certain point in a pipeline the velocity is 1 m s⁻¹ and the gauge pressure is 3×10^5 Pa. Find the gauge pressure at a second point in the line 20 m lower than the first, if the cross section at the second point is one-half that at the first. The liquid in the pipe is water.

13–12 Water stands at a depth of 1 m in an enclosed tank whose side walls are vertical. The space above the water surface contains air at a gauge pressure of 8×10^5 Pa. The tank rests on a platform 2 m above the floor. A hole of cross-sectional area 1 cm² is made in one of the side walls just above the bottom of the tank.

a) Where does the stream of water from the hole strike the floor?

b) What is the vertical force exerted on the floor by the stream?

c) What is the horizontal force exerted on the tank? Assume the water level and the pressure in the tank to remain constant, and neglect any effect of viscosity.

13–13 Water flows steadily from a reservoir, as in Fig. 13–19. The elevation of point 1 is 10 m; of points 2 and 3 it is 1 m. The cross section at point 2 is 0.04 m² and at point 3 it is 0.02 m². The area of the reservoir is very large compared with the cross sections of the pipe.

a) Compute the gauge pressure at point 2.

b) Compute the discharge rate in cubic feet per second.

13–14 Water flows steadily in a pipeline of constant cross section leading out of an elevated tank. At a point 2 m below the water level in the tank the gauge pressure in the flowing stream is 10^4 Pa.

Fig. 13–19

a) What is the velocity of the water at this point?

b) If the pipe rises to a point 3 m above the level of the water in the tank, what are the velocity and the pressure at the latter point?

13–15 Water flows through a horizontal pipe of cross-sectional area 10 cm². At one section the cross-sectional area is 5 cm². The pressure difference between the two sections is 300 Pa. How many cubic meters of water will flow out of the pipe in 1 min?

13–16 Two very large open tanks, A and F (Fig. 13–20), both contain the same liquid. A horizontal pipe BCD, having a constriction at C, leads out of the bottom of tank A, and a vertical pipe E opens into the constriction at C and dips into the liquid in tank F. Assume streamline flow and no viscosity. If the cross section at C is one-half that at D, and if D is at distance h_1 below the level of the liquid in A, to what height h_2 will liquid rise in pipe E? Express your answer in terms of h_1. Neglect changes in atmospheric pressure with elevation.

Fig. 13–20

13–17 At a certain point in a horizontal pipeline the gauge pressure is 6.24 lb in⁻². At another point the gauge pressure is 4.37 lb in⁻². If the areas of the pipe at these two points are 3 in² and 1.5 in², respectively, compute the number of cubic feet of water which flow across any cross section of the pipe per minute.

13–18 Water flowing in a horizontal pipe discharges at the rate of 0.004 m³ s⁻¹. At a point in the pipe where the cross section is 0.001 m², the absolute pressure is 1.2 $\times 10^5$ Pa. What must be the cross section of a constriction in the pipe such that the pressure there is reduced to 1.0 $\times 10^5$ Pa?

13–19 The pressure difference between the main pipeline and the throat of a Venturi meter is 10^5 Pa. The areas of the pipe and the constriction are 0.1 m² and 0.05 m². How many cubic meters per second are flowing through the pipe? The liquid in the pipe is water.

13–20 The section of pipe shown in Fig. 13–21 has a cross section of 40 cm² at the wider portions and 10 cm² at the constriction. The discharge of water from the pipe is 3000 cm³ s⁻¹.

a) Find the velocities at the wide and the narrow portions.

b) Find the pressure difference between these portions.

c) Find the difference in height between the mercury columns in the U-tube.

Fig. 13–21

13–21 Water is used as the manometric liquid in a Prandtl tube mounted in an aircraft to measure airspeed. If the maximum difference in height between the liquid columns is 0.1 m, what is the maximum airspeed that can be measured? The density of air is 1.3 kg m⁻³.

13–22 In Problem 13–21, suppose the manometric liquid is mercury. What is the maximum airspeed that can be measured?

13–23 In a wind-tunnel experiment, the pressure on the top surface of an airplane wing was 0.90×10^5 Pa, and the pressure on the bottom surface 0.91×10^5 Pa. If the area of each surface is 40 m², what is the net lift force on the wing?

13–24 An airplane of mass 6000 kg has a wing area of 60 m². If the pressure on the lower wing surface is 0.60×10^5 Pa during level flight at an elevation of 4000 m, what is the pressure on the upper wing surface?

13–25 Water at 20° C flows through a pipe of radius 1.0 cm. If the flow velocity at the center is 10 cm s⁻¹, find the pressure drop along a 2-m section of pipe due to viscosity.

13–26 The inner, rotating cylinder of the viscosimeter in Fig. 13–12 is 5 cm in diameter. The inner diameter of the outer, fixed cylinder is 5.4 cm, and the diameter of the pulley attached to the inner cylinder is 4 cm. A liquid whose viscosity is 6 poise fills the space between inner and outer cylinders to a depth of 8 cm. A body of mass 30 g is

supported by a thread wrapped around the pulley attached to the inner cylinder, and hangs vertically, as shown in Fig. 13–12. Find the speed of descent of the body after it has reached its terminal velocity.

13–27 A viscous liquid flows through a tube with laminar flow, as in Fig. 13–14(b). Prove that the volume rate of flow is the same as if the velocity were uniform at all points of a cross section and equal to half the velocity at the axis.

13–28

a) With what terminal velocity will an air bubble 1 mm in diameter rise in a liquid of viscosity 150 cp and density 0.90 g cm⁻³?

b) What is the terminal velocity of the same bubble in water?

13–29

a) With what velocity is a steel ball 1 mm in radius falling in a tank of glycerine at an instant when its acceleration is one-half that of a freely falling body?

b) What is the terminal velocity of the ball? The densities of steel and of glycerine are 8.5 g cm⁻³ and 1.32 g cm⁻³, respectively. Assume $\eta = 8.3$ poise.

13–30 Assume that air is streaming horizontally past an aircraft wing such that the velocity is 40 m s⁻¹ over the top surface and 30 m s⁻¹ past the bottom surface. If the wing has a mass of 300 kg and an area of 5 m², what is the net force on the wing? The density of air is 1.3 kg m⁻³.

13–31 Modern airplane design calls for a "lift" of about 20 lb per square foot of wing area. Assume that air flows past the wing of an aircraft with streamline flow. If the velocity of flow past the lower wing surface is 300 ft s⁻¹, what is the required velocity over the upper surface to give a "lift" of 20 lb ft⁻²? The density of air is 0.0025 slug ft⁻³.

13–32 Water at 20°C flows with a speed of 50 cm s⁻¹ through a pipe of diameter 3 mm.

a) What is the Reynolds number?

b) What is the nature of the flow?

13–33 Water at 20°C is pumped through a horizontal smooth pipe 15 cm in diameter and discharges into the air. If the pump maintains a flow velocity of 30 cm s⁻¹,

a) What is the nature of the flow?

b) What is the discharge rate in liters per second?

13–34 The tank at the left of Fig. 13–15(a) has a very large cross section and is open to the atmosphere. The depth y = 40 cm. The cross sections of the horizontal tubes leading out of the tank are respectively 1 cm^2, 0.5 cm^2, and 0.2 cm^2. The liquid is ideal, having zero viscosity.

a)   What is the volume rate of flow out of the tank?

b)   What is the velocity in each portion of the horizontal tube?

c)   What are the heights of the liquid in the vertical side tubes?

Suppose that the liquid in Fig. 13–15(b) has a viscosity of 0.5 poise, a density of 0.8 g cm^{-3}, and that the depth of liquid in the large tank is such that the volume rate of flow is the same as in part (a) above. The distance between the side tubes at c and d, and between those at e and f, is 20 cm. The cross sections of the horizontal tubes are the same in both diagrams.

d)   What is the difference in level between the tops of the liquid columns in tubes c and d?

e)   In tubes e and f?

f)   What is the flow velocity on the axis of each part of the horizontal tube?

13–35

a)   Is it reasonable to assume that the flow in the second part of Problem 13–34 is laminar?

b)   Would the flow be laminar if the liquid were water?

13–36 Oil having a viscosity of 300 centipoise and a density of 0.90 g cm^{-3} is to be pumped from one large open tank to another through 1 km of smooth steel pipe 15 cm in diameter. The line discharges into the air at a point 30 m above the level of the oil in the supply tank.

a)   What gauge pressure, in atmospheres, must the pump exert in order to maintain a flow of 0.05 m^3 s^{-1}?

b)   What is the power consumed by the pump?

Chapter 14

Relativistic Mechanics

14–1 INVARIANCE OF PHYSICAL LAWS

In previous chapters we have stressed the importance of inertial frames of reference. Newton's laws of motion are valid only in inertial frames, but they are valid in *all* inertial frames. Any frame moving with constant velocity with respect to an inertial frame is itself an inertial frame, and all such frames are equivalent with respect to expressing the basic principles of mechanics. The *laws of mechanics are the same in every inertial frame of reference.*

Einstein proposed in 1905 that this principle should be extended to include *all* the basic laws of physics. This innocent-sounding proposition has far-reaching and startling consequences, a few of which have already been mentioned. If the principle of conservation of momentum is to be valid in all inertial systems, for example, the definition of momentum, for particles moving at speeds comparable to the speed of light, must be changed from $m\boldsymbol{v}$ to $m\boldsymbol{v}/(1 - v^2/c^2)^{1/2}$, as discussed in Section 8–8. Even more fundamental are the modifications needed in the *kinematic* aspects of motion, as we shall see. Nevertheless, Einstein's *principle of relativity*, as it has come to be called, is now accepted as an essential requirement for a physical theory. The principle of relativity states that *the laws of physics are the same in every inertial frame of reference.*

The speed of light plays a special role in relativity theory. Light is an electromagnetic wave; it travels in vacuum with a speed c which is independent of the motion of the source. Its numerical value is known very precisely; to six significant figures it is

$$c = 2.99793 \times 10^8 \text{ m s}^{-1}.$$

The approximate value $c = 3.00 \times 10^8$ m s^{-1} is in error by less than one part in 1000 and is often used when greater precision is not required.

At one time it was thought that light traveled through a hypothetical medium called *the ether*, just as sound waves travel through air. Intensive experimental efforts to find direct evidence for its existence yielded consistently negative results, and it is now known that there is no ether. With this result in mind, let us suppose the speed of light is measured by two observers, one at rest with respect to the source, the other moving away from it. Both are in inertial frames of reference, and according to Einstein's principle of relativity, the laws of physics, and in particular the speed of light, must be the same in both frames.

For example, suppose a light source is located in a spaceship moving with respect to earth. An observer moving with the spaceship measures the speed of the light and obtains the value c. But after the light has left the source its motion cannot be influenced by the motion of the source, so an observer on earth measuring the speed of this same light must also obtain the value c, despite the fact that there is relative motion between the two observers. This conclusion may appear not to be consistent with common sense; but it is important to recognize that "common sense" is intuition based on everyday experience, and this does not usually include measurements of the speed of light. We must be prepared to accept results which seem to be in conflict with common sense when they involve realms far removed from everyday observation.

Thus the speed of light (in vacuum) is independent of the motion of the source and is the same in all frames of reference. To explore the consequences of this statement, we consider the newtonian relationship between two inertial frames, labeled S and S' in Fig. 14–1. Let the x axes of the two systems lie along the same line, but let the origin O' of S' move relative to the origin O of S with constant velocity u along the common x-axis. If the two origins coincide at time $t = 0$, then their separation at a later time t is ut.

A point P may be described by the coordinates (x, y, z) in S or by coordinates (x', y', z') in S'. Refer-

ence to the figure shows that these are related by

$$x = x' + ut, \qquad y = y', \qquad z = z'. \quad (14\text{–}1)$$

These equations are called the *galilean coordinate transformation*.

If the point P moves in the x-direction, its velocity v relative to S is given by $v = dx/dt$, and its velocity v' relative to S' is $v' = dx'/dt$. Intuitively it is clear that these are related by

$$v = v' + u. \qquad (14\text{–}2)$$

This relation may also be obtained formally from Eqs. (14–1). Suppose the particle is at a point described by coordinate x_1 or x_1' at time t_1, and at x_2 or x_2' at time t_2. Then $\Delta t = t_2 - t_1$, and, from Eq. (14–1),

$$\Delta x = x_2 - x_1 = (x_2' - x_1') + u(t_2 - t_1)$$

$$= \Delta x' + u \, \Delta t,$$

$$\frac{\Delta x}{\Delta t} = \frac{\Delta x'}{\Delta t} + u,$$

and in the limit as $\Delta t \to 0$,

$$v = v' + u,$$

in agreement with Eq. (14–2).

A fundamental problem now appears. Applied to the speed of light, Eq. (14–2) says $c = c' + u$. Einstein's principle of relativity, supported by experimental observations, says $c = c'$. This is a genuine inconsistency, not an illusion, and it demands resolution. If we accept the principle of relativity, we are forced to conclude that Eqs. (14–1) and (14–2), intuitively appealing as they are, cannot be correct, but need to be modified to bring them into harmony with this principle.

The resolution involves modifications of our basic kinematic concepts. The first of these involves an assumption so fundamental that it might seem unnecessary, namely the assumption that the same *time scale* is used in frames S and S'. This may be stated formally by adding to Eqs. (14–1) a fourth equation $t = t'$. Obvious though this assumption may seem, it is not strictly correct when the relative speed u of the two frames of reference is comparable

Fig. 14–1 *The position of point P can be described by the coordinates x and y in frame of reference S, or by x' and y' in S'. S' moves relative to S with constant velocity u along the common x-x' axis. The two origins O and O' coincide at time $t = t' = 0$.*

to the speed of light. The difficulty lies in the concept of *simultaneity*, which we examine next.

14–2 RELATIVE NATURE OF SIMULTANEITY

Measuring times and time intervals involves the concept of *simultaneity*. When a person says he awoke at seven o'clock he means that two *events*, his awakening and the arrival of the hour hand of his clock at the number seven, occurred *simultaneously*. The fundamental problem in measuring time intervals is that in general two events that appear simultaneous in one frame of reference *do not* appear simultaneous in a second frame which is moving relative to the first, even if both are inertial frames.

The following thought experiment, devised by Einstein, illustrates this point. Consider a long train moving with uniform velocity, as shown in Fig. 14–2(a). Two lightning bolts strike the train, one at each end. Each bolt leaves a mark on the train and one on the ground at the same instant. The points on the ground are labeled A and B in the figure, and the corresponding points on the train A' and B'. An observer on the ground is located at O, midway between A and B; another observer is at O', moving

with the train and midway between A' and B'. Both these observers use the light signals from the lightning to observe the events.

Suppose the two light signals reach the observer at O simultaneously; he concludes that the two events took place at A and B simultaneously. But the observer at O' is moving with the train, and the light pulse from B' reaches him before the light pulse from A' does; he concludes that the event at the front of the train happened *earlier* than that at the rear. Thus the two events appear simultaneous to one observer, not to the other. *Whether or not two events at different space points are simultaneous depends on the state of motion of the observer.* It follows that *the time interval between two events at different space points is in general different for two observers in relative motion.*

It might be argued that, in this example, the lightning bolts really *are* simultaneous, and that if the observer at O' could communicate with the distant points without time delay, he would realize this. But the finite speed of information transmission is not the problem. If O' is really midway between A' and B', then, in his frame of reference, the time for a signal to travel from A' to O' is the same as from B' to O'. Two signals arrive simultaneously at O' only if they were emitted simultaneously at A' and B'; in this example they *do not* arrive simultaneously at O', and so O' must conclude that the events at A' and B' were *not* simultaneous.

Furthermore, there is no basis for saying either that O is right and O' is wrong, or the reverse, since according to the principle of relativity, no inertial frame of reference is preferred over any other in the formulation of physical laws. Each observer is correct *in his own frame of reference*, but simultaneity is not an absolute concept. Whether or not two events are simultaneous depends on the frame of reference, and the time interval between two events also depends on the frame of reference.

14–3 RELATIVITY OF TIME

Fig. 14–2 (a) *To the stationary observer at point O, two lightning bolts appear to strike simultaneously.* (b) *The moving observer at point O' sees the light from the front of the train first and thinks that the bolt at the front struck first.* (c) *The two light pulses arrive at O simultaneously.*

To derive a quantitative relation between time intervals in different coordinate systems, we consider another thought experiment. As before, a frame of reference S' moves with velocity u relative to a frame

S. An observer in S' directs a source of light at a mirror a distance d away, as shown in Fig. 14–3, and measures the time interval $\Delta t'$ for light to make the "round trip" to the mirror and back. The total distance is $2d$, so the time interval is

$$\Delta t' = \frac{2d}{c}. \qquad (14\text{–}3)$$

As measured in frame S, the time for the round trip is a different interval Δt. During this time, the source moves relative to S a distance ut, and the total round-trip distance is not just $2d$ but is 2ℓ, where

$$\ell = \sqrt{d^2 + (u\,\Delta t/2)^2}\,.$$

The speed of light is the same for both observers, so the relation in S analogous to Eq. (14–3) is ·

$$\Delta t = \frac{2\ell}{c} = \frac{2}{c}\sqrt{d^2 + (u\,\Delta t/2)^2}\,. \qquad (14\text{–}4)$$

To obtain a relation between Δt and $\Delta t'$ which does not contain d, we solve Eq. (14–3) for d and substitute the result into Eq. (14–4), obtaining

$$\Delta t = \frac{2}{c}\sqrt{(c\,\Delta t'/2)^2 + (u\,\Delta t/2)^2}\,.$$

This may now be squared and solved for Δt; the result is

$$\boxed{\Delta t = \frac{\Delta t'}{\sqrt{1 - u^2/c^2}}\,.} \qquad (14\text{–}5)$$

We may generalize this important result: If a time interval $\Delta t'$ separates two events occurring at the same space point in a frame of reference S' (in this case, the departure and arrival of the light signal at O'), then the time interval Δt between these two events as observed in S is *larger* than $\Delta t'$, and the two are related by Eq. (14–5). Thus when the rate of a clock at rest in S' is measured by an observer in S, the rate measured in S is *slower* than the rate observed in S'. This effect is called *time dilation*.

It is important to note that the observer in S measuring the time interval Δt cannot do so with a

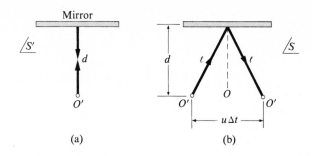

Fig. 14–3 (a) *Light pulse emitted from source at O' and reflected back along the same line, as observed in S'.* (b) *Path of the same light pulse, as observed in S. The positions of O' at the times of departure and return of the pulse are shown. The speed of the pulse is the same in S as in S', but the path is longer in S.*

single clock. In Fig. 14–3 the points of departure and return of the light pulse are different space points in S, although they are the same point in S'. If S tries to use a single clock, the finite time of communication between two points will becloud the issue. To avoid this, S may use two clocks at the two relevant points. There is no difficulty in synchronizing two clocks in the same frame of reference; one procedure is to send a light pulse simultaneously to two clocks from a point midway between them, with the two operators setting their clocks to a predetermined time when the pulses arrive. In thought experiments it is often helpful to imagine a large number of synchronized clocks distributed conveniently in a single frame of reference. Only when a clock is moving relative to a given frame of reference do ambiguities of synchronization or simultaneity arise.

Example   A spaceship flies past earth with a speed of $0.99c$ (about 2.97×10^8 m s^{-1}). A high-intensity signal light (perhaps a pulsed laser) blinks on and off, each pulse lasting 2×10^{-6} s. At a certain instant the ship appears to an earthling observer to be directly overhead at an altitude of 1000 km, and to be traveling perpendicular to the line of sight. What is the duration of each light pulse, as measured by this observer, and how far does the ship travel relative to earth during each pulse?

The observer does not see the pulse at the instant it is emitted, because the light signal requires a time equal to $(1000 \times 10^3 \text{ m})/(3 \times 10^8 \text{ m s}^{-1})$, or $(1/300)$ s, to travel from the ship to earth. But if the distance from the spaceship to observer is essentially constant during the emission of a pulse, the time delays at beginning and end of the pulse are equal and the time *interval* is not affected.

Let S be the earth's frame of reference, S' that of the spaceship. Then, in the notation of Eq. (14–5), $\Delta t' = 2 \times 10^{-6}$s. This interval refers to two events occurring at the same point relative to S', namely, the starting and stopping of the pulse. The corresponding interval in S is given by Eq. (14–5):

$$\Delta t = \frac{\Delta t'}{\sqrt{1 - u^2/c^2}} = \frac{2 \times 10^{-6}\,\text{s}}{\sqrt{1 - (0.99)^2}}$$

$$= 14.1 \times 10^{-6}\text{s}.$$

Thus the time dilation in S is about a factor of seven. The distance D traveled in S during this interval is

$$D = u\,\Delta t = (0.99)(3 \times 10^8\,\text{m s}^{-1})(14.1 \times 10^{-6}\,\text{s})$$

$$= 4190\,\text{m} = 4.19\,\text{km}.$$

If the spaceship is traveling directly *toward* the observer, the time interval cannot be measured directly by a single observer because the time delay is not the same at the beginning and end of the pulse. One possible scheme, at least in principle, is to use *two* observers at rest in S, with synchronized clocks, one at the position of the ship when the pulse starts, the other at its position at the end of the pulse. These observers will again measure a time interval in S of 14.1×10^{-6} s.

From the derivation of Eq. (14–5) and the spaceship example, it can be seen that a time interval between two events occurring *at the same point* in a given frame of reference is a more fundamental quantity than an interval between events at different points. The term *proper time* is used to denote an interval between two events occurring at the same space point. Thus Eq. (14–5) may be used *only* when $\Delta t'$ is a proper time interval in S', in which case Δt is *not* a proper time interval in S. If, instead, Δt is

proper in S, then Δt and $\Delta t'$ must be interchanged in Eq. (14–5).

When the relative velocity u of S and S' is very small, the factor $(1 - u^2/c^2)$ is very nearly equal to unity, and Eq. (14–5) approaches the newtonian relation $\Delta t = \Delta t'$ (i.e., the same time scale for all frames of reference). This assumption, therefore, retains its validity in the limit of small relative velocities.

14–4 RELATIVITY OF LENGTH

Just as the time interval between two events depends on the frame of reference, the *distance* between two points also depends on the frame of reference. To measure a distance one must, in principle, observe the positions of two points, such as the two ends of a ruler, simultaneously; but what is simultaneous in one frame is not in another.

To develop a relation between lengths in various coordinate systems we consider another thought experiment. We attach a source of light pulses to one end of a ruler and a mirror to the other, as shown in Fig. 14–4. Let the ruler be at rest in S' and the length in this frame be ℓ'. Then the time $\Delta t'$ required for a light pulse to make the round trip from source to

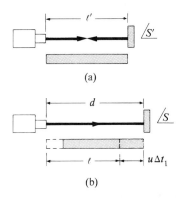

(a)

(b)

Fig. 14–4 (a) A light pulse is emitted from a source at one end of a ruler, reflected from a mirror at the opposite end, and returned to the source position. (b) Motion of the light pulse as seen by an observer in S. The distance traveled from source to mirror is greater than the length ℓ measured in S, by the amount $u\,\Delta t_1$, as shown.

mirror and back is given by

$$\Delta t' = \frac{2\ell'}{c}. \qquad (14\text{--}6)$$

This is a proper time interval, since departure and return occur at the same point in S'.

In S the ruler is displaced during this travel of the light pulse. Let the length of the ruler in S be ℓ, and let the time of travel from source to mirror, as measured in S, be Δt_1. During this interval the mirror moves a distance $u\,\Delta t_1$, and the total length of path d from source to mirror is not ℓ but

$$d = \ell + u\,\Delta t_1. \qquad (14\text{--}7)$$

But since the pulse travels with speed c, it is also true that

$$d = c\,\Delta t_1. \qquad (14\text{--}8)$$

Combining Eqs. (14–7) and (14–8) to eliminate d,

$$c\,\Delta t_1 = \ell + u\,\Delta t_1,$$

or

$$\Delta t_1 = \frac{\ell}{c - u}. \qquad (14\text{--}9)$$

In the same way it can be shown that the time Δt_2 for the return trip from mirror to source is

$$\Delta t_2 = \frac{\ell}{c + u}. \qquad (14\text{--}10)$$

The *total* time $\Delta t = \Delta t_1 + \Delta t_2$ for the round trip, as measured in S, is

$$\Delta t = \frac{\ell}{c - u} + \frac{\ell}{c + u} = \frac{2\ell}{c(1 - u^2/c^2)}. \qquad (14\text{--}11)$$

It is also known that Δt are $\Delta t'$ are related by Eq. (14–5), since $\Delta t'$ is proper in S'. Thus Eq. (14–6) becomes

$$\Delta t\sqrt{1 - u^2/c^2} = \frac{2\ell'}{c}. \qquad (14\text{--}12)$$

Finally, combining this with Eq. (14–11) to eliminate

Δt, and simplifying, we obtain

$$\boxed{\ell = \ell'\sqrt{1 - u^2/c^2}.} \qquad (14\text{--}13)$$

Thus the length measured in S, in which the ruler is moving, is *shorter* than in S', where it is at rest. A length measured in the rest frame of the body is called a *proper length*; thus, ℓ' above is a proper length in S', and the length measured in any other frame is less than ℓ'. This effect is called *contraction of length*.

Example   In the spaceship example of Section 14–3, what distance does the spaceship travel during emission of a pulse, as measured in its rest frame?

The question is somewhat ambiguous since, of course, in its own frame of reference the ship is at rest. But suppose it leaves markers in space, such as small smoke bombs, at the instants when the pulse starts and stops, and measures the distance between these markers, with the aid of observers behind the ship but moving with it, each with a clock synchronized with that of the ship. The distance d between the markers is a proper length in the earth's frame S. In the spaceship's frame S', the distance d' is contracted by the factor given in Eq. (14–13):

$$d' = d\sqrt{1 - u^2/c^2} = (4190\ \text{m})\sqrt{1 - (0.99)^2}$$
$$= 594\ \text{m}.$$

(Note that because d, not d', is a proper length, we must reverse the roles of ℓ and ℓ'.) An observer in the spaceship can calculate his speed relative to earth from this data:

$$u = \frac{d'}{\Delta t'} = \frac{594\ \text{m}}{2 \times 16^{-6}\ \text{s}} = 2.97 \times 10^8\ \text{m s}^{-1},$$

which agrees with the initial data.

When u is very small compared to c, the contraction factor in Eq. (14–13) approaches unity, and in the limit of small speeds we recover the newtonian relation $\ell = \ell'$. This and the corresponding result for time dilation show that Eqs. (14–1) retain their validity in the limit of small speeds; only at speeds comparable to c are modifications needed.

Lengths measured perpendicular to the direction of motion are *not* contracted; this may be verified by constructing a thought experiment for measuring in S and S' the length of a ruler oriented perpendicular to the direction of relative motion. The details of this discussion are not essential for our further work and will not be given here.

14–5 THE LORENTZ TRANSFORMATION

The galilean coordinate transformation given by Eqs. (14–1) is valid only in the limit when u is much smaller than c, but we are now in position to derive a more general transformation not subject to this limitation. The more general relations are called the *Lorentz transformation*. When u is small they reduce to the galilean transformation, but they may also be used when u is comparable to c.

The basic problem is this: When an event occurs at point (x, y, z) at time t, as observed in a frame of reference S, what are the coordinates (x', y', z') and time t' of the event as observed in a second frame S' moving relative to S with constant velocity u along the x-direction?

To derive the transformation equations, we return to Fig. 14–1. As before, we assume that the origins coincide at the initial time $t = t' = 0$. Then in S the distance from O to O' at time t is still ut. The coordinate x' is a proper length in S', so in S it appears contracted by the factor given in Eq. (14–13). Thus the distance x from O to P in S is given not simply by $x = ut + x'$ as in the galilean transformation, but by

$$x = ut + x'\sqrt{1 - u^2/c^2}. \qquad (14\text{–}14)$$

Solving this equation for x', we obtain

$$x' = \frac{x - ut}{\sqrt{1 - u^2/c^2}}. \qquad (14\text{–}15)$$

This is half of the Lorentz transformations; the other half is the equation giving t' in terms of x and t. To obtain this we note that the principle of relativity requires that the *form* of the transformation from S to S' must be identical to that from S' to S, the only difference being a change in the sign of the relative

velocity u. Thus, from Eq. (14–14), it must be true that

$$x' = -ut' + x\sqrt{1 - u^2/c^2}. \qquad (14\text{–}16)$$

We may now equate Eqs. (14–15) and (14–16) to eliminate x' from this expression, obtaining the desired relation for t' in terms of x and t. The algebraic details will be omitted; the result is

$$t' = \frac{t - ux/c^2}{\sqrt{1 - u^2/c^2}}. \qquad (14\text{–}17)$$

As remarked previously, lengths perpendicular to the direction of relative motion are not affected by the motion, so $y' = y$ and $z' = z$. Collecting all the transformation equations, we have

$$\boxed{\begin{aligned} x' &= \frac{x - ut}{\sqrt{1 - u^2/c^2}}, \\ y' &= y, \\ z' &= z, \\ t' &= \frac{t - ux/c^2}{\sqrt{1 - u^2/c^2}}. \end{aligned}} \qquad (14\text{–}18)$$

These are the *Lorentz transformation equations*, the relativistic generalization of the galilean transformation, Eqs. (14–1). When u is much smaller than c, the two transformations become identical.

Next we consider the relativistic generalization of the velocity transformation relation, Eq. (14–2), which, as previously noted, is valid only when u is very small. The relativistic expression can easily be obtained from the Lorentz transformation. Suppose that a body observed in S' is at point x'_1 at time t'_1 and point x'_2 at time t'_2. Then its speed v' in S' is given by

$$v' = \frac{x'_2 - x'_1}{t'_2 - t'_1} = \frac{\Delta x'}{\Delta t'}. \qquad (14\text{–}19)$$

To obtain the speed in S we use Eqs. (14–18) to translate this expression into terms of the corresponding positions x_1 and x_2 and times t_1 and t_2

observed in S. We find

$$x_2' - x_1' = \frac{x_2 - x_1 - u(t_2 - t_1)}{\sqrt{1 - u^2/c^2}} = \frac{\Delta x - u\,\Delta t}{\sqrt{1 - u^2/c^2}},$$

$$t_2' - t_1' = \frac{t_2 - t_1 - u(x_2 - x_1)/c^2}{\sqrt{1 - u^2/c^2}} = \frac{\Delta t - u\,\Delta x/c^2}{\sqrt{1 - u^2/c^2}}.$$

Using these results in Eq. (14–19), we find

$$v' = \frac{\Delta x - u\,\Delta t}{\Delta t - u\,\Delta x/c^2} = \frac{\dfrac{\Delta x}{\Delta t} - u}{1 - \dfrac{u}{c^2}\dfrac{\Delta x}{\Delta t}}.$$

Now $\Delta x/\Delta t$ is just the velocity v measured in S, so we finally obtain

$$\boxed{v' = \frac{v - u}{1 - uv/c^2}.} \qquad (14\text{–}20)$$

We note that when u and v are much smaller than c, the denominator becomes equal to unity, and we obtain the nonrelativistic result $v' = v - u$. The opposite extreme is the case $v = c$; then we find

$$v' = \frac{c - u}{1 - uc/c^2} = c.$$

That is, anything moving with speed c relative to S also has speed c relative to S', despite the relative motion of the two frames. This result demonstrates the consistency of Eq. (14–20) with our initial assumption that the speed of light is the same in all frames of reference.

Equation (14–20) may also be rearranged to give v in terms of v'. The algebraic details are left as a problem; the result is

$$\boxed{v = \frac{v' + u}{1 + uv'/c^2}.} \qquad (14\text{–}21)$$

Example A spaceship moving away from earth with a speed $0.9c$ fires a missile in the same direction

as its motion, with a speed $0.9c$ relative to the spaceship. What is the missile's speed relative to earth?

Let the earth's frame of reference be S, the spaceship's S'. Then $v' = 0.9c$ and $u = 0.9c$. The nonrelativistic velocity addition formula would give a velocity relative to earth of $1.8c$. The correct relativistic result, obtained from Eq. (14–21), is

$$v = \frac{0.9c + 0.9c}{1 + (0.9c)(0.9c)/c^2} = 0.994c.$$

When u is less than c, a body moving with a speed less than c in one frame of reference also has a speed less than c in *every other* frame of reference. This is one reason for thinking that no material body may travel with a speed greater than that of light, relative to any frame of reference. The relativistic generalizations of energy and momentum, to be considered next, give further support to this hypothesis.

14–6 MOMENTUM

We have discussed the fact that Newton's laws of motion are *invariant* under the galilean coordinate transformation, but that to satisfy the principle of relativity this transformation must be replaced by the more general Lorentz transformation. This requires corresponding generalizations in the laws of motion and the definitions of momentum and energy.

The principle of conservation of momentum states that *when two bodies collide, the total momentum is constant,* provided there is no interaction except that of the two bodies with each other. However, when one considers a collision in one coordinate system S, in which momentum is conserved, and then uses the Lorentz transformation to obtain the velocities in a second system S', it is found that if the newtonian definition of momentum ($p = mv$) is used, momentum is not conserved in the second system. Thus if the Lorentz transformation is correct, and if we believe in the principle of relativity (i.e., momentum conservation must hold in *all* systems), the only choice remaining is to modify the *definition* of momentum.

Deriving the correct relativistic generalization is beyond our scope, and we simply quote the result, which has already been mentioned in Section 8–8:

$$p = \frac{mv}{\sqrt{1 - v^2/c^2}}. \qquad (14\text{-}22)$$

We note that, as usual when the particle's speed v is much less than c, this reduces to the Newtonian expression $p = mv$, but that in general the momentum is greater in magnitude than mv.

In newtonian mechanics the second law of motion can be stated in the form

$$F = \frac{dp}{dt}. \qquad (14\text{-}23)$$

That is, force equals time rate of change of momentum. Experiment shows that this result is still valid in relativistic mechanics, provided we use the relativistic momentum given by Eq. (14–22). This has the effect that a body under the action of a constant force *does not* experience a constant acceleration; as the particle's speed increases the acceleration for a given force continuously *decreases*. As the speed approaches c, the acceleration approaches zero, no matter how great the force. Thus it is impossible to accelerate a particle from a state of rest to a speed equal to or greater than c, and the speed of light is sometimes referred to as "the ultimate speed."

14–7 WORK AND ENERGY

The work–energy relation developed in Chapter 7 made use of Newton's laws of motion. Since these must be generalized to bring them into accord with the principle of relativity, it is not surprising that the work–energy relation also requires generalization.

Because a constant force on a body no longer causes a constant acceleration (except at very small velocities), even the simplest dynamics problems require the use of the calculus. We may, however, follow the same pattern as that of Section 7–3 in deriving the relativistic generalization of the work–energy principle. We begin with the definition of work: $W = \int F\, dx$. According to Eq. (14–23), $F = dp/dt$. By repeated application of the chain rule for derivatives,

we obtain

$$F = \frac{dp}{dt} = \frac{dp}{dv}\frac{dv}{dx}\frac{dx}{dt} = \frac{dp}{dv}\frac{dv}{dx}v. \qquad (14\text{-}24)$$

Thus if the particle has speed v_1 at point x_1 and v_2 at x_2, the work done by F during the motion from x_1 to x_2 may be expressed as

$$W = \int_{x_2}^{x_1} F\, dx = \int_{v_1}^{v_2} \frac{dp}{dv} v\, dv. \qquad (14\text{-}25)$$

By differentiating Eq. (14–22), we obtain

$$\frac{dp}{dv} = \frac{m}{(1 - v^2/c^2)^{3/2}}. \qquad (14\text{-}26)$$

The reader should verify this calculation. Substituting this result in Eq. (14–25) and integrating, we find

$$W = \int_{v_1}^{v_2} \frac{mv\, dv}{(1 - v^2/c^2)^{3/2}}$$

$$= \frac{mc^2}{\sqrt{1 - v_2^2/c^2}} - \frac{mc^2}{\sqrt{1 - v_1^2/c^2}}, \qquad (14\text{-}27)$$

where v_1 and v_2 are the initial and final velocities of the particle.

This result suggests defining kinetic energy as

$$E = \frac{mc^2}{\sqrt{1 - v^2/c^2}}. \qquad (14\text{-}28)$$

But this expression is not zero when $v = 0$; instead it becomes equal to mc^2. Thus the correct relativistic generalization of kinetic energy E_k is:

$$E_k = \frac{mc^2}{\sqrt{1 - v^2/c^2}} - mc^2. \qquad (14\text{-}29)$$

This expression, if correct, must reduce to the newtonian expression $E_k = \frac{1}{2}mv^2$ when v is much smaller than c. It is not obvious that this is the case; to demonstrate that it is so, we can expand the radical using the binomial theorem:

$$\left(1 - \frac{v^2}{c^2}\right)^{-1/2} = 1 + \frac{1}{2}\frac{v^2}{c^2} + \frac{3}{8}\frac{v^4}{c^4} + \frac{5}{16}\frac{v^6}{c^6} + \cdots$$

Combining this with Eq. (14–25),

$$E_k = mc^2\left(1 + \frac{1}{2}\frac{v^2}{c^2} + \frac{3}{8}\frac{v^4}{c^4} + \cdots\right) - mc^2$$

$$= \frac{1}{2}mv^2 + \frac{3}{8}m\frac{v^4}{c^2} + \cdots \tag{14–30}$$

In each expression the dots stand for omitted terms. When v is much smaller than c, all terms in the series except the first are negligibly small, and we obtain the classical $\frac{1}{2}mv^2$.

But what is the significance of the term mc^2 that had to be subtracted in Eq. (14–29)? Although Eq. (14–28) does not give the *kinetic energy* of the particle, perhaps it represents some kind of *total* energy, including both the kinetic energy and an additional energy mc^2 which the particle possesses even when it is not moving. This hypothetical energy associated with mass rather than motion may be called the *rest energy* of the particle. This speculation does not prove that the concept of rest energy is meaningful, but it points the way toward further investigation.

There is in fact direct experimental evidence of the existence of rest energy. The simplest example is the decay of the π° meson, an unstable particle that "decays"; in the decay process, the particle disappears and electromagnetic radiation appears. When the particle is at rest (and therefore with no kinetic energy) before its decay, the total energy of the radiation produced is found to be exactly equal to mc^2. There are many other examples of fundamental particle transformations in which the total mass of the system changes, and in every case there is a corresponding energy change consistent with the assumption of a rest energy mc^2 associated with a mass m.

Although the principles of conservation of mass and of energy originally developed quite independently, the theory of relativity shows that they are but two special cases of a single broader conservation principle, the *principle of conservation of mass and energy*. There are physical phenomena where neither mass nor energy is separately conserved, but where the changes in these quantities are governed by the more general relation that a change m in the mass of the system must be accompanied by an opposite change mc^2 in its energy.

The term *mass* as used here always means the rest mass of a particle, the inertial mass m measured through Eq. (14–22). For a given particle, m is a constant, independent of the state of motion of the particle. The concept of a variable, velocity-dependent relativistic mass is unnecessary and is not used in this discussion.

The possibility of conversion of mass into energy is the fundamental principle involved in the generation of power through nuclear reactions, a subject to be discussed in later chapters. When a uranium nucleus undergoes fission in a nuclear reactor, the total mass of the resulting fragments is *less* than that of the parent nucleus, and the total kinetic energy of the fragments is equal to this mass deficit times c^2. This kinetic energy can be used to produce steam to operate turbines for electric power generators or in a variety of other ways.

The total energy (kinetic plus rest) of a particle is related simply to its momentum, as shown by combining Eqs. (14–22) and (14–28) to eliminate the particle's velocity. This is most easily accomplished by rewriting these equations in the following forms:

$$\left(\frac{E}{mc^2}\right)^2 = \frac{1}{1 - v^2/c^2}; \qquad \left(\frac{p}{mc}\right)^2 = \frac{v^2/c^2}{1 - v^2/c^2}.$$

Subtracting the second of these from the first and rearranging, we find

$$\boxed{E^2 = (mc^2)^2 + (pc)^2.} \tag{14–31}$$

Again we see that for a particle at rest ($p = 0$), $E = mc^2$. Equation (14–31) also suggests that a particle may have energy and momentum even when it has no rest mass. In such a case, $m = 0$ and

$$E = pc. \tag{14–32}$$

Massless particles, including photons, the quanta of electromagnetic radiation, and others, were mentioned in Section 8–8. The existence of such particles is well established, and they will be discussed in greater detail in later chapters. These particles always travel with the speed of light; they are emitted and absorbed during changes of state of atomic systems, accompanied by corresponding changes in the energy and momentum of these systems.

14–8 RELATIVITY AND NEWTONIAN MECHANICS

The sweeping changes required by the principle of relativity go to the very roots of newtonian mechanics, including the concepts of length and time, the equations of motion, and the conservation principles. Thus it may appear that foundations on which Newton's mechanics are built have been destroyed. While this is true in one sense, it is essential to keep in mind that the Newtonian formulation still retains its validity whenever speeds are small compared with the speed of light. In such cases time dilation, length contraction, and the modifications of the laws of motion do not appear. In fact, every one of the principles of newtonian mechanics survives as a special case of the more general relativistic formulation.

Relativity does not *contradict* the older mechanics but *generalizes* it. After all, Newton's laws rest on a very solid base of experimental evidence, and it would be very strange indeed to advance a new theory inconsistent with this evidence. So it always is with the development of physical theory. Whenever a new theory is in partial conflict with an older, established theory, it nevertheless must yield the same predictions as the old in areas where the old theory is supported by experimental evidence. Every new physical theory must pass this test, called the *correspondence principle*, which has come to be regarded as a fundamental procedural rule in all physical theory. There are many problems for which newtonian mechanics is clearly inadequate, including all situations where particle speeds approach that of light or there is direct conversion of mass to energy. But there is still a large area, including nearly all the behavior of macroscopic bodies in mechanical sys-

tems, in which Newtonian mechanics is still perfectly adequate.

At this point it is legitimate to ask whether the relativistic mechanics just discussed is the final word on this subject or whether *further* generalizations are possible or necessary. For example, inertial frames of reference have occupied a privileged position in all our discussion thus far. Should the principle of relativity be extended to noninertial frames as well?

Here is an example to illustrate some implications of this question. A man decides to go over Niagara Falls while enclosed in a large wooden box. During his free fall over the falls he can in principle perform experiments inside the box. An object released inside the box does not fall to the floor because both the box and the object are in free fall with a downward acceleration of 9.8 m s^{-2}. But an alternative interpretation, from this man's point of view, is that the force of gravity has suddenly been turned off. Provided he remains in the box and it remains in free fall, he cannot tell whether he is indeed in free fall or whether the force of gravity has vanished. A similar problem appears in a space station in orbit around the earth. Objects in the spaceship appear weightless, but without going outside the ship there is no way to determine whether gravity has disappeared or the spaceship is in an accelerated (i.e., noninertial) frame of reference.

These considerations form the basis of Einstein's *general theory of relativity.* If one cannot distinguish experimentally between a gravitational field and an accelerated reference system, then there can be no real distinction between the two. Pursuing this concept, we may try to represent *any* gravitational field in terms of special characteristics of the coordinate system. This turns out to require even more sweeping revisions of our space-time concepts than the special theory of relativity did, and we find that, in general, the geometric properties of the space are noneuclidean.

The basic ideas of the general theory of relativity are now well established, but some of the details remain speculative in nature. Its chief application is in cosmological investigations of the structure of the universe, the formation and evolution of stars, and related matters. It is not believed to have any relevance for atomic or nuclear phenomena or macroscopic mechanical problems of less than astronomical dimensions.

PROBLEMS

14–1 The π^+ meson, an unstable particle, lives, on the average, about 2.6×10^{-8}s (measured in its own frame of reference) before decaying.

a) If such a particle is moving with respect to the laboratory with a speed of $0.8c$, what lifetime is measured in the laboratory?

b) What distance, measured in the laboratory, does the particle move before decaying?

14–2 The μ^+ meson (or positive muon) is an unstable particle with a lifetime of about 2.3×10^{-6}s (measured in the rest frame of the muon).

a) If a muon travels with a speed $0.99c$ relative to a laboratory, what is the lifetime as measured in the laboratory?

b) What distance, measured in the laboratory, does the particle travel during its lifetime?

14–3 For the two trains discussed in Section 14–2, suppose the two lightning bolts appear simultaneous to an observer on the train. Show that they *do not* appear simultaneous to an observer on the ground. Which appears to come first?

14–4 Solve Eqs. (14–18) to obtain x and t in terms of x' and t', and show that the resulting transformation has the same form as the original one except for a change of sign for u.

14–5 A light pulse is emitted at the origin of a frame of reference S' at time $t' = 0$. Its distance x' from the origin after a time t' is given by $x'^2 = c^2 t'^2$. Use the Lorentz transformation to transform this equation to an equation in x and t, and show that the result is $x^2 = c^2 t^2$; that is, the motion appears exactly the same in the frame of reference S of x and t, as it does in S'.

14–6 Two events observed in a frame of reference S have positions and times given by (x_1, t_1) and (x_2, t_2), respectively. Show that in a frame S' moving just fast enough so the two events occur at the same point in S', the time interval $\Delta t'$ between the two events is given by

$$\Delta t' = \sqrt{(\Delta t)^2 - (\Delta x/c)^2} \,,$$

where $\Delta x = x_2 - x_1$, and $\Delta t = t_2 - t_1$. Hence show that, if

$\Delta x \geq c \, \Delta t$, there is *no* frame S' in which the two events occur at the same point. The interval $\Delta t'$ is sometimes called the *proper time interval* for the events; is this term appropriate?

14–7 For the two events in Problem 14–6, show that if $\Delta x > c \, \Delta t$ there is a frame of reference S' in which the two events occur *simultaneously*. Find the distance between the two events in S'. This distance is sometimes called a *proper length*; is this term appropriate?

14–8 Two events are observed in a frame of reference S to occur at the same space point, the second occurring 2 s after the first. In a second frame S' moving relative to S, the second event is observed to occur 3 s after the first. What is the distance between the positions of the two events as measured in S'?

14–9 Two events are observed in a frame of reference S to occur simultaneously, at points separated by a distance of 1 m. In a second frame S' moving relative to S along the line joining the two points in S, the two events appear to be separated by 2 m. What is the time interval between the events, as measured in S'?

14–10 A particle is said to be in the *extreme relativistic range* when its kinetic energy is much larger than its rest energy.

a) What is the speed of a particle (expressed as a fraction of c) such that the total energy is ten times the rest energy?

b) For such a particle, what percent error in the energy–momentum relation of Eq. (14–31) results if the term $(mc^2)^2$ is neglected?

14–11 A photon of energy E is emitted by an atom of mass m, which recoils in the opposite direction. Assuming the atom can be treated nonrelativistically, compute the recoil velocity of the atom. Hence, show that the recoil velocity is much smaller than c whenever E is much smaller than the rest energy mc^2 of the atom.

14–12 Two particles emerge from a high-energy accelerator in opposite directions, each with a speed $0.6c$. What is the relative velocity of the particles?

14–13 At what speed is the momentum of a particle twice as great as the result obtained from the nonrelativistic expression mv?

14–14 At what speed does the momentum of a particle differ from the value obtained using the nonrelativistic expression mv by 1 percent? Is the correct relativistic value greater or less than that obtained from the nonrelativistic expression?

14–15 The mass of an electron is 9.11×10^{-31}kg. Comparing the classical definition of momentum with its relativistic generalization, by how much is the classical expression in error if (a) $v = 0.01c$; (b) $v = 0.5c$; (c) $v = 0.9c$?

14–16 A radioactive isotope of cobalt, ^{60}Co, emits an electromagnetic photon (γ ray) of wavelength 0.932×10^{-12} m. The cobalt nucleus contains 27 protons and 33 neutrons, each with a mass of about 1.66×10^{-27}kg. If the nucleus is at rest before emission, what is the speed afterward? Is it necessary to use the relativistic generalization of momentum?

14–17 In Problem 14–16, suppose the cobalt atom is in a metallic crystal containing 0.01 moles of cobalt (about 6.02×10^{21} atoms) and that the entire crystal recoils as a unit, rather than just the single nucleus. Find the recoil velocity, (This recoil of the entire crystal rather than a single nucleus is called the *Mössbauer effect*, in honor of its discoverer, who first observed it in 1958.)

14–18 What is the speed of a particle whose kinetic energy is equal to its rest energy?

14–19 At what speed is the kinetic energy of a particle equal to $10\ mc^2$?

14–20 How much work must be done to accelerate a particle from rest to a speed $0.1c$? From a speed $0.9c$ to a speed $0.99c$?

14–21 In *positron annihilation*, an electron and a positron (a positively-charged electron) collide and disappear, producing electromagnetic radiation. If each particle has a mass of 9.1×10^{-31}kg and they are at rest just before the annihilation, find the total energy of the radiation.

14–22 The total consumption of electrical energy per year in the United States is of the order 10^{19} joules. If matter could be converted completely into energy, how many kilograms of matter would have to be converted to produce this much energy?

14–23 Compute the kinetic energy of an electron (mass 9.11×10^{-31} kg) using both the nonrelativistic and relativistic expressions, and compare the two results, for speeds of

a) 1.0×10^8 m s^{-1};

b) 2.0×10^8 m s^{-1}.

14–24 In a hypothetical nuclear-fusion reactor, two deuterium nuclei combine or "fuse" to form one helium nucleus. The mass of a deuterium nucleus, expressed in atomic mass units (u), is 2.0147 u; that of a helium nucleus is 4.0039 u. (1 u = 1.66×10^{-27} kg.)

a) How much energy is released when 1 kg of deuterium undergoes fusion?

b) The annual consumption of electrical energy in the United States is of the order of 10^{19} J. How much deuterium must react to produce this much energy?

14–25 A nuclear bomb containing 20 kg of plutonium explodes. The rest mass of the products of the explosion is less than the original rest mass by one part in 10^4.

a) How much energy is released in the explosion?

b) If the explosion takes place in 1 μs, what is the average power developed by the bomb?

c) How much water could the released energy lift to a height of 1 km?

14–26 Construct a right triangle in which one of the angles is α, where $\sin \alpha = v/c$. (v is the speed of a particle, c the speed of light.) If the base of the triangle (the side adjacent to α) is the rest energy mc^2, show that (a) the hypotenuse is the total energy and (b) the side opposite α is c times the relativistic momentum.

c) Describe a simple graphical procedure for finding the kinetic energy E_k.

Chapter 15

Temperature and Expansion

15–1 CONCEPT OF TEMPERATURE

To describe the equilibrium states of mechanical systems, as well as to study and predict the motions of rigid bodies and fluids, only three fundamental indefinables were needed: length, mass, and time. All other physical quantities of importance in mechanics could be expressed in terms of these three indefinables. We come, now, however, to a series of phenomena, called *thermal effects* or *heat phenomena*, which involve aspects that are essentially nonmechanical and which require for their description a fourth fundamental indefinable, the *temperature*.

The familiar sensations of hotness and coldness are described with adjectives such as cold, cool, tepid, warm, hot, etc. When we touch an object, we use our *temperature sense* to ascribe *to the object* a property called *temperature*, which determines whether it will feel hot or cold to the touch. The hotter it feels, the higher the temperature. This procedure plays the same role in "qualitative science" that hefting a body does in determining its weight or that kicking an object does in estimating its mass. To determine the mass of an object *quantitatively*, we must first arrive at the concept of mass by means of *quantitative* operations such as measuring the acceleration imparted to the object by meas-

ured force, and then taking the ratio of F to a. Similarly, the quantitative determination of temperature requires a set of operations that are independent of our sense perceptions of hotness or coldness, and which involve quantities that can be measured objectively. How this is done will be explained in the following paragraphs.

Even before treating the concept of *temperature* in a precise, quantitative manner, we can note that there are numerous simple systems in which a quantity characterizing the state of the system varies with the hotness or coldness of the system. A simple example is a liquid such as mercury or alcohol in a bulb attached to a very thin tube, as in Fig. 15–1(a). The significant quantity characterizing the state of this system is the length L of the liquid column, measured from some arbitrary fixed point. Another simple system is a quantity of gas in a constant-volume container, shown in Fig. 15–1(b). Here the varying quantity, which we may refer to in these examples as a *state coordinate*, is the pressure, which varies as the gas becomes hotter or colder. A third example is the electrical resistance of a wire, which also varies with hotness and coldness.

Let A stand for the liquid-in-capillary system, with state coordinate L, and let B stand for the gas at constant volume, with state coordinate p. If A and

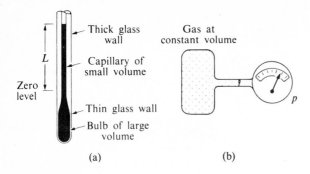

(a)   (b)

Fig. 15–1 (a) A system whose state is specified by the value of L. (b) A system whose state is given by the value of p.

B are brought into contact, their state coordinates, in general, are found to change. When A and B are separated, however, the change is slower, and when thick walls of various materials such as wood, plaster, felt, asbestos, etc., are used to separate A and B, the values of the respective state coordinates L and p are almost independent of each other. Generalizing from these observations, we postulate the existence of an ideal partition, called an **adiabatic wall**, which, when used to separate two systems, allows their state coordinates to vary over a large range of values **independently.** An adiabatic wall is an idealization that cannot be realized perfectly but may be approximated closely. In Fig. 15–2(a), such a wall is represented as a thick cross-shaded region.

When systems A and B are first put into actual contact or are separated by a thin metallic partition, their state coordinates may or may not change. A wall which enables a state coordinate of one system to influence that of another is called a **diathermic wall.** A thin sheet of copper is an example of a diathermic wall. In Fig. 15–2(b), a diathermic wall is depicted as a thin, darkly shaded region. Eventually, a time will be reached when no further change in the coordinates of A and B takes place. The joint state of both systems that exists when all changes in the coordinates have ceased is called **thermal equilibrium.**

Imagine two systems A and B separated from each other by an adiabatic wall but each in contact with a third system C through diathermic walls, the whole assembly being surrounded by an adiabatic wall as shown in Fig. 15–3(a). Experiment shows that the two systems will come to thermal equilibrium with the third and that no further change will occur if the adiabatic wall separating them is then replaced by a diathermic wall (Fig. 15–3(b)). If, instead of allowing both systems A and B to come to equilibrium with C at the same time, we first have equilibrium between A and C and then equilibrium between B and C (the state of system C being the same in both cases), then, when A and B are brought into communication through a diathermic wall, they will be found to be in thermal equilibrium. We shall use the expression "two systems are in thermal equilibrium"

Adiabatic
wall

(a)

Diathermic
wall

(b)

Fig. 15–2 System A, a liquid column, and system B, a gas at constant volume, separated by (a) an adiabatic wall, p and L independent, and (b) a diathermic wall, p and L dependent.

Fig. 15–3 *The zeroth law of thermodynamics. (a) If A and B are each in thermal equilibrium with C, then (b) A and B are in thermal equilibrium with each other.*

to mean that the two systems are in states such that if the two *were* to be connected through a diathermic wall, the combined system *would be* in thermal equilibrium.

These experimental facts may then be stated concisely in the following form: *Two systems in thermal equilibrium with a third are in thermal equilibrium with each other.* Following R. H. Fowler, we shall call this postulate the *zeroth law of thermodynamics.* At first thought it might seem that the zeroth law is obvious, but its truth must be verified by experiment.

When two systems *A* and *B* are first put in contact through a diathermic wall, they may or may not be in thermal equilibrium. One is entitled to ask, "What is there about *A* and *B* that determines whether or not they are in thermal equilibrium?" *We are led to infer the existence of a new property called the* **temperature.** *The temperature of a system is that property which determines whether or not it will be in thermal equilibrium with other systems.* When two or more systems are in thermal equilibrium, they are said to have the same temperature.

The temperature of all systems in thermal equilibrium may be represented by a number. The establishment of a temperature scale is merely the adoption of a set of rules for assigning numbers to temperatures. Once this is done, the condition for thermal equilibrium between two systems is that they have the same temperature. When the temperatures of two systems are different, we may be sure that they are *not* in thermal equilibrium.

The temperature of a material is directly related to the energies of its molecules; as temperature increases, molecular motion increases. The relation of temperature to microscopic mechanical energy will be explored in detail in Chapter 20. It is important to understand, however, that temperature can be defined *without* reference to molecular considerations. Indeed, temperature is inherently a macroscopic concept that has no meaning for an individual molecule. Temperature can be related to molecular motion only by considering the *average* energy of a large number of molecules.

15–2 THERMOMETERS

In defining a temperature scale, the simplest procedure is to choose a system such as one of those described above, arbitrarily assigning a numerical value of temperature to each value of the state coordinate of the system. This then defines quantitatively the temperature of this system, and of all systems in thermal equilibrium with it.

Although the system of Fig. 15–1(a) was one of the earliest thermometers, various other systems are now used. Important characteristics of a thermometer include *sensitivity* (an appreciable change in the state coordinate produced by a small change in temperature), *accuracy* in the measurement of the state coordinate, and *reproducibility*. Another often desirable property is *speed* in coming to thermal equilibrium with other systems. The thermometers which satisfy these requirements best will be described in the following paragraphs.

A thermometer widely used in research and engineering laboratories is the *thermocouple*, which consists of a junction of two different metals or alloys, such as *A* and *B*, labeled "test junction" in Fig. 15–4. The test junction is usually embedded in the material whose temperature is to be measured. Since the test junction is small and has a small mass, it can follow temperature changes rapidly and come to equilibrium quickly. The reference junction consists of two junctions: one of *A* and copper and the other of *B* and copper. These two junctions are maintained at any desired constant temperature, called the *reference temperature*. The state coordinate of this thermometer is an electrical quantity called

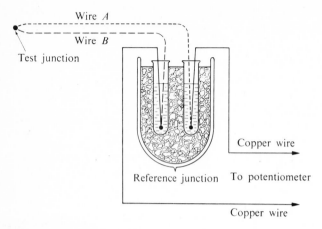

Fig. 15–4 *Thermocouple, showing the test junction and the reference junction.*

the emf (electromotive force) which is measured with an instrument known as a potentiometer. A thermocouple with one junction of pure platinum and the other of 90% platinum and 10% rhodium is often used. Copper and an alloy called constantan are also frequently used.

A different electrical thermometer, the *resistance thermometer*, consists of a fine wire, often enclosed in a thin-walled silver tube for protection. Copper wires lead from the thermometer unit to a resistance-measuring device such as a Wheatstone bridge. Since resistance may be measured with great precision, the resistance thermometer is one of the most precise instruments for the measurement of temperature. In the region of extremely low temperatures, a small carbon cylinder or a small piece of germanium crystal is used instead of a coil of platinum wire.

To measure temperatures above the range of thermocouples and resistance thermometers, an *optical pyrometer* is used. As shown in Fig. 15–5, it consists essentially of a telescope *T*, in the tube of which is mounted a filter *F* of red glass and a small electric lamp bulb *L*. When the pyrometer is directed toward a furnace, an observer looking through the telescope sees the dark lamp filament against the bright background of the furnace. The lamp filament is connected to a battery *B* and a rheostat *R*. By turning the rheostat knob, we may gradually increase the current in the filament, and hence the brightness, until the brightness of the filament just matches the

brightness of the background. From previous calibration of the instrument at known temperatures, the scale of the ammeter *A* in the circuit may be marked to read the unknown temperature directly. Since no part of the instrument needs to come into contact with the hot body, the optical pyrometer may be used at very high temperatures, above the melting points of metals.

Among all the state coordinates or, as they are often called, *thermometric properties*, the pressure of a gas whose volume is maintained constant is outstanding in its sensitivity, accuracy of measurement, and reproducibility. The constant-volume gas thermometer is illustrated schematically in Fig. 15–6. The

Fig. 15–5 *Principle of the optical pyrometer.*

Fig. 15–6 *The constant-volume gas thermometer.*

gas, usually helium, is contained in bulb C and the pressure exerted by it can be measured by the open-tube mercury manometer. As the temperature of the gas increases, the gas expands, forcing the mercury down in tube B and up in tube A. Tubes A and B communicate through a rubber tube D with a mercury reservoir R. By raising R, the mercury level in B may be brought back to reference mark E. The gas is thus kept at constant volume.

Gas thermometers are used mainly in bureaus of standards and in some university research laboratories. They are usually large, bulky, and slow in coming to thermal equilibrium.

15–3 THE ESTABLISHMENT OF A TEMPERATURE SCALE

Any one of the thermometers described in the preceding section may be used to indicate the constancy of a temperature if its state coordinate or thermometric property remains constant. By this means, it has been found that a system composed of a solid and a liquid of the same material maintained at constant pressure will remain in *phase equilibrium* (that is, the liquid and solid exist together without the liquid changing into solid or the solid changing into liquid) only at one definite temperature. Similarly, a liquid will remain in phase equilibrium with its vapor at only one definite temperature when the pressure is maintained constant.

The temperature at which a solid and liquid of the same material coexist in phase equilibrium *at atmospheric pressure* is called the *normal melting point,* abbreviated NMP. The temperature at which a liquid and its vapor exist in phase equilibrium at atmospheric pressure is called the *normal boiling point,* abbreviated NBP.

Phase equilibrium between a solid and its vapor is sometimes possible at atmospheric pressure. The temperature at which this takes place is the *normal sublimation point,* NSP.

It is possible for all three phases—solid, liquid, and vapor—to coexist in equilibrium, but only at one definite pressure and temperature known as the *triple point,* abbreviated TP. The triple-point of water occurs at 4.58 mm of mercury and 0.01°C.

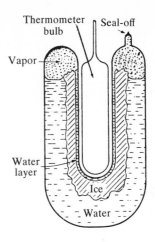

Fig. 15–7 *Triple-point cell with a thermometer in the well, which melts a thin layer of ice nearby.*

The NMP, NBP, NSP, or TP of any material can be chosen as a standard reference point for the purpose of setting up a temperature scale. Any temperature so chosen is called a *fixed point. The standard fixed point used in modern thermometry is the triple point of water,* to which is given the arbitrary* number

$$273.16 \text{ K},$$

read 273.16 kelvins.†

To achieve the triple point, water of the highest purity is distilled into a vessel like that shown schematically in Fig. 15–7. When all air has been removed, the vessel is sealed off. With the aid of a freezing mixture in the inner well, a layer of ice is formed around the well. When the freezing mixture

---

* By extrapolating backwards with the aid of the gas law formulas, the "absolute zero" point was "determined" to be −273.15°C. Since the kelvin has the same magnitude as one degree on the Celsius scale, the triple point on the Kelvin scale becomes 273.16 K.

† At a meeting of the Thirteenth General Conference on Weights and Measures on Oct. 13, 1967, the name of the unit of temperature was changed from *degree Kelvin* (symbol °K) to *kelvin* (symbol K). The kelvin, now the unit of temperature, is the fraction 1/273.16 of the thermodynamic temperature of the triple point of water.

is replaced by a thermometer bulb, a thin layer of ice is melted nearby. So long as the solid, liquid, and vapor phases coexist in equilibrium, the system is at the triple point.

We start our program of setting up a temperature scale by denoting with the letter X *any one* of the thermometric properties mentioned previously: the emf of a thermocouple, $\mathscr{E}$, the resistance of a wire, R, the pressure of a gas at constant volume, p, etc. *We define the ratio of two temperatures to be the same as the ratio of the two corresponding values of X.* Thus, if a thermometer with thermometric property X is put in thermal equilibrium with a system and registers a value X, and is then put in thermal equilibrium with another system and registers a value X_3, the ratio of the temperatures of those two systems is given by

$$\boxed{\frac{T(X)}{T(X_3)} = \frac{X}{X_3}.} \qquad (15\text{–}1)$$

If now, we let the subscript 3 stand for the standard fixed point, the triple point of water, then $T(X_3)$ = 273.16 K. Hence

$$T(X) = 273.16 \text{ K}\left(\frac{X}{X_3}\right). \qquad (15\text{–}2)$$

Example Suppose a gas thermometer registers a pressure of 15.0×10^4 Pa at the triple point of water and a pressure of 20.5×10^4 Pa at the normal boiling point; what is the temperature at the normal boiling point?

From Eq. (15–2),

$$T_B = 273.16 \text{ K} \frac{20.5 \times 10^4 \text{ Pa}}{15.0 \times 10^4 \text{ Pa}} = 373 \text{ K}.$$

The remaining problem in temperature measurement is that when different thermometers are used to measure the same temperature, the results do not always agree. Best agreement is found with gas thermometers; there is some variation with pressure for any gas, but it is found that, in the limit as pressure becomes smaller and smaller, all gas ther-

mometers at the same temperature approach the same reading, independent of the nature of the gas. Thus the usual procedure is to use the constant-volume thermometer (in the limit of very low pressure) to *define* a temperature scale, and then use this device to calibrate the other thermometers in terms of this scale.

To measure a low temperature, a gas that does not liquefy at the low temperature must be used. The lowest temperature that can be measured with a gas thermometer is about 1 K, provided low-pressure helium is used as the gas. *The temperature $T = 0$ remains as yet undefined.* In Chapter 19 we shall discuss the Kelvin temperature scale, which is independent of the properties of any particular substance. It can be shown that in the temperature region in which a gas thermometer can be used, the gas scale and the Kelvin scale are identical. In anticipation of this result, we write K after a gas temperature.

It will also be shown in Chapter 19 how the absolute zero of temperature is defined on the Kelvin scale. Until then, the term "absolute zero" will have no meaning. The statement made so often that all molecular activity ceases at the temperature $T = 0$ is entirely erroneous. When it is necessary in statistical mechanics to correlate temperature with molecular activity, it is found that classical statistical mechanics must be modified with the aid of quantum mechanics. When this modification is carried out, the molecules of a substance at absolute zero have a *finite* amount of kinetic energy known as the *zero-point energy*.

The body temperatures of warm-blooded animals are held constant to within a few tenths of one celsius degree by an elaborate temperature-control system. The thermometric property which senses the temperature of the blood is provided by a chemical equilibrium condition in a part of the brain called the hypothalamus, which then activates appropriate temperature-controlling mechanisms. The most important mechanisms are dilation or contraction of blood vessels near the surface, increasing or decreasing loss of body heat by conduction, and activation or deactivation of sweat glands to increase or decrease evaporative cooling.

Table 15–1 TEMPERATURES OF FIXED POINTS

Basic fixed points	T, K	t, °C	T_R, °R	t_F, °F
Standard: Triple point of water	**273.16**	**0.01**	**491.688**	**32.018**
NBP of oxygen	90.18	−182.97	162.32	−297.35
Equilibrium of ice and air-saturated water (ice point)	273.15	0.00	491.67	32.00
NBP of water (steam point)	373.15	100.00	671.67	212.00
NMP of zinc	692.66	419.51	1246.78	787.11
NMP of antimony	903.65	630.50	1626.57	1166.90
NMP of silver	1233.95	960.80	2221.11	1761.44
NMP of gold	1336.15	1063.00	2405.07	1945.40

15–4   THE CELSIUS, RANKINE, AND FAHRENHEIT SCALES[*]

The Celsius temperature scale (formerly called the *centigrade* scale in the United States and Great Britain) employs a degree of the same magnitude as that of the Kelvin scale, but its zero point is shifted so that *the Celsius temperature of the triple point of water is 0.01 degree Celsius*, abbreviated 0.01°C. Thus, if t denotes the Celsius temperature

$$t = T - 273.15 \text{ K}. \qquad (15\text{–}3)$$

The Celsius temperature t_S at which steam condenses at 1 atm pressure is

$$t_S = T_S - 273.15 \text{ K};$$

T_S is found to be 373.15 K, so

$$t_S = 373.15° - 273.15° \quad \text{or } t_S = 100.00°C.$$

There are two other scales in common use in engineering and in everyday life in the United States and in Great Britain. The Rankine temperature T_R (written °R) is proportional to the Kelvin temperature according to the relation

$$T_R = \frac{9}{5} T. \qquad (15\text{–}4)$$

A degree of the same size is used in the Fahrenheit

scale t_F (written °F), but with the zero point shifted according to the relation

$$t_F = T_R - 459.67°R. \qquad (15\text{–}5)$$

Substituting Eqs. (15–3) and (15–4) into Eq. (15–5), we get

$$t_F = \frac{9}{5}t + 32°F, \qquad (15\text{–}6)$$

from which it follows that the Fahrenheit temperature of the ice point ($t = 0°C$) is 32°F and of the steam point ($t = 100°C$) is 212°F. The 100 Celsius or Kelvin degrees between the ice point and the steam point correspond to 180 Fahrenheit or Rankine degrees, as shown in Fig. 15–8, where the four scales are compared.

Fig. 15–8 *Relations among Kelvin, Celsius, Rankine, and Fahrenheit temperature scales. Temperatures have been rounded off to the nearest degree.*

---

[*] Named after Anders Celsius (1701–1744). William John MacQuorn Rankine (1820–1872). Gabriel Fahrenheit (1686–1736).

The accurate measurement of a boiling point or melting point with the aid of a gas thermometer requires very careful laboratory work. Fortunately, this has been done for a large number of substances which are obtainable with high purity. Some of these results are shown in Table 15–1. With the aid of these basic fixed points other thermometers may be calibrated.

Suppose the temperature of a beaker of water is raised from 20°C to 30°C, through a temperature interval of 10 Celsius degrees. It is desirable to distinguish between such a temperature *interval* and the actual temperature of 10 degrees above the Celsius zero. Hence we shall use the phrase "10 degrees Celsius," or "10°C," when referring to an *actual temperature*, and "10 Celsius degrees," or "10C°" to mean a temperature *interval*. Thus there is an interval of 10 Celsius degrees between 20 degrees Celsius and 30 degrees Celsius.

15–5 EXPANSION OF SOLIDS AND LIQUIDS

With a few exceptions, the volumes of all bodies increase with increasing temperature if the external pressure on the body remains constant. Suppose that a solid or a liquid undergoes a small change of volume dV when the temperature is changed a small amount dT (or dt, since the kelvin and the Celsius degree are temperature intervals of equal magnitude). The *coefficient of volume expansion* β is defined as *the fractional change in volume dV/V divided by the change of temperature dT*, or

$$\beta = \frac{1}{V}\frac{dV}{dT} \text{(at constant pressure).} \quad (15\text{–}7)$$

The unit β is 1 *reciprocal degree*, or 1 $(C°)^{-1}$. The numerical value depends, of course, on the size of the degree. Since the kelvin and the Celsius degree are $\frac{9}{5}$ as large as the Rankine and Fahrenheit degrees, the fractional volume change per kelvin or Celsius degree is $\frac{9}{5}$ as great as that per Rankine or Fahrenheit degree.

The coefficient of volume expansion is insensitive to a change of pressure, but varies markedly with

Table 15–2 COEFFICIENT OF VOLUME EXPANSION (APPROXIMATE)

Solids	$\bar{\beta}$, $(C°)^{-1}$	Liquids	$\bar{\beta}$, $(C°)^{-1}$
Aluminum	7.2×10^{-5}	Alcohol, ethyl	75×10^{-5}
Brass	6.0	Carbon disulfide	115
Copper	4.2	Glycerin	49
Glass	1.2–2.7	Mercury	18
Steel	3.6		
Invar	0.27		
Quartz (fused)	0.12		

temperature. For many substances β decreases as the temperature is lowered, approaching zero as the Kelvin temperature approaches zero. It is also a peculiar circumstance that the higher the melting point of a metal, the lower the coefficient of volume expansion.

Calling $\bar{\beta}$ the average value of β within a moderate temperature interval ΔT, we may rewrite Eq. (15–7):

$$\Delta V = \bar{\beta} V_0 \Delta T \quad (15\text{–}8)$$

where V_0 is the original volume.

Some values of $\bar{\beta}$ in the neighborhood of room temperature are listed in Table 15–2. Note that the values for liquids are much larger than those for solids.

If there is a hole in a solid body, the volume of the hole increases when the body expands, just as if the hole were a solid of the same material as the body. Thus the volume enclosed by a thin-walled glass flask or thermometer bulb increases just as would a solid body of glass of the same size.

Example   A glass flask of volume 200 cm³ is just filled with mercury at 20°C. How much mercury will overflow when the temperature of the system is raised to 100°C? The coefficient of volume expansion of the glass is $1.2 \times 10^{-5}(C°)^{-1}$.

The increase in the volume of the flask is

$$\Delta V = (1.2 \times 10^5 (\text{C}°)^{-1})(200 \text{ cm}^3)(100° - 20°)$$
$$= 0.192 \text{ cm}^3.$$

The increase in the volume of the mercury is

$$\Delta V = (18 \times 10^{-5}(\text{C}°)^{-1})(200 \text{ cm}^3)(100° - 20°)$$
$$= 2.88 \text{ cm}^3.$$

The volume of mercury overflowing is therefore

$$2.88 - 0.19 = 2.69 \text{ cm}^3.$$

Water, in the temperature range from 0°C to 4°C, *decreases* in volume with increasing temperature, contrary to the behavior of most substances. That is, between 0°C and 4°C the coefficient of expansion of water is *negative*. Above 4°C, water expands when heated. Since the volume of a given mass of water is smaller at 4°C than at any other temperature, the density of water is a maximum at 4°C. Water also expands upon freezing, unlike most materials.

This anomalous behavior has an important effect on plant and animal life in lakes. When a lake cools, the cooled water at the surface flows to the bottom because of its greater density. But when the temperature reaches 4°C this flow ceases and the water near the surface remains colder (and less dense) than that at the bottom. As the surface freezes, the ice floats because it is less dense than water. The water at the bottom remains at 4°C until nearly the entire body is frozen. If water behaved like most substances, contracting continuously on cooling and freezing, lakes would freeze from the bottom up, circulation due to density differences would continuously carry warmer water to the surface for efficient cooling, and lakes would freeze solid much more easily, thus destroying all plant and animal life that can withstand cold water but not freezing.

The anomalous expansion of water in the temperature range 0°C to 10°C is shown in Fig. 15–9. Table 15–3 covers a wider range of temperatures.

For a body in the form of a rod or cable, we are often interested only in the change of *length* with temperature, and we define a *coefficient of linear*

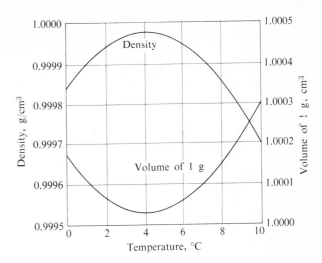

Fig. 15–9 *Density of water, and volume of 1 gram, in the temperature range from 0°C to 10°C.*

expansion α. If L is the length,

$$\alpha = \frac{1}{L}\frac{dL}{dT}, \qquad (15\text{–}9)$$

and, over a moderate temperature change ΔT,

$$\Delta L = L_0 \bar{\alpha} \Delta T, \qquad (15\text{–}10)$$

and

$$L = L_0 + \Delta L = L_0(1 + \bar{\alpha} \Delta T), \qquad (15\text{–}11)$$

where $\bar{\alpha}$ is the average coefficient within the temperature interval, and L_0 is the original length.

Table 15–3 DENSITY AND VOLUME OF WATER

t, °C	density, g cm^{-3}	Volume of 1 g, cm^3
0	0.9998	1.0002
4	1.0000	1.0000
10	0.9997	1.0003
20	0.9982	1.0018
50	0.9881	1.0121
75	0.9749	1.0258
100	0.9584	1.0434

The volume coefficient is related to the linear coefficient. To obtain the relation, we consider a solid body in the form of a rectangular parallelepiped with dimensions L_1, L_2, and L_3. Then the volume is

$$V = L_1 L_2 L_3,$$

and

$$\frac{dV}{dT} = L_2 L_3 \frac{dL_1}{dT} + L_1 L_3 \frac{dL_2}{dT} + L_1 L_2 \frac{dL_3}{dT}.$$

Dividing by $L_1 L_2 L_3$, we obtain

$$\frac{1}{V}\frac{dV}{dT} = \frac{1}{L_1}\frac{dL_1}{dT} + \frac{1}{L_2}\frac{dL_2}{dT} + \frac{1}{L_3}\frac{dL_3}{dT}.$$

If the solid has the same properties in each of the three directions, each of the three expressions on the right is the coefficient of linear expansion α, and hence

$$\boxed{\beta = 3\alpha.} \qquad (15\text{–}12)$$

15–6 THERMAL STRESSES

If the ends of a rod are rigidly fixed so as to prevent expansion or contraction and the temperature of the rod is changed, tensile or compressive stresses, called *thermal stresses*, will be set up in the rod. These stresses may become very large, sufficiently so to stress the rod beyond its elastic limit or even beyond its breaking strength. Hence in the design of any structure which is subject to changes in temperature, some provision must, in general, be made for expansion. In a long steam pipe this is accomplished by the insertion of expansion joints or a section of pipe in the form of a U. In bridges, one end may be rigidly fastened to its abutment while the other rests on rollers.

It is a simple matter to compute the thermal stress set up in a rod which is not free to expand or contract. Suppose that a rod at a temperature T has its ends rigidly fastened, and that while they are thus held the temperature is reduced to a lower value, T_0.

The fractional change in length if the rod were free to contract would be

$$\frac{\Delta L}{L_0} = \bar{\alpha}(T - T_0) = \bar{\alpha}\,\Delta T. \qquad (15\text{–}13)$$

Since the rod is *not* free to contract, the tension must increase by a sufficient amount to produce the same fractional change in length. But from the definition of Young's modulus,

$$Y = \frac{F/A}{\Delta L/L_0}, \qquad F = AY\frac{\Delta L}{L_0}.$$

Introducing the expression for $\Delta L/L_0$ from Eq. (15–13), we have

$$F = AY\bar{\alpha}\,\Delta T, \qquad (15\text{–}14)$$

which gives the tension F in the rod. The *stress* in the rod is

$$\frac{F}{A} = Y\bar{\alpha}\,\Delta T. \qquad (15\text{–}15)$$

If, instead, Δt represents an *increase* in temperature, F/A is a *compressive* stress of the same magnitude.

Thermal stresses can also be induced by nonuniform heating. Even if a solid body at uniform temperature has no internal stresses, stress may be induced by nonuniform expansion if it is heated unevenly. The breaking of a thick glass container when hot water is poured into it is a familiar phenomenon. Heat-resistant glasses such as Pyrex have exceptionally low expansion coefficients, and usually also have high strength to permit thin-wall construction to minimize temperature differences.

PROBLEMS

15–1 The limiting value of the ratio of the pressures of a gas at the melting point of lead and at the triple point of water, when the gas is kept at constant volume, is found to be 2.19816. What is the Kelvin temperature of the melting point of lead?

15–2

a) If you feel sick in France and are told you have a fever of 40°C, should you be concerned?

b)   What is normal body temperature on the Celsius scale?

c)   The normal boiling point of liquid oxygen is −182.97°C. What is this temperature on the Kelvin and Rankine scales?

d)   At what temperature do the Fahrenheit and Celsius scales coincide?

15–3 When the United States finally converts officially to metric units, the Celsius temperature scale will very likely replace the Fahrenheit scale in everyday use. As a familiarization exercise, find the Celsius temperatures corresponding to (a) a cool room (68°F); (b) a hot summer day (95°F); (c) a cold winter day (5°F).

15–4 In a rather primitive experiment with a constant-volume gas thermometer, the pressure at the triple point of water was found to be 4.0×10^4 Pa. and the pressure at the normal boiling point 5.4×10^4 Pa. According to these data, what is the temperature of absolute zero on the Celsius scale?

15–5 A gas thermometer of the type shown in Fig. 15–6 registered a pressure corresponding to 5.0 cm of mercury when in contact with water at the triple point. What pressure will it read when in contact with water at the normal boiling point?

15–6 The electrical resistance of some metals varies with temperature (as measured by a gas thermometer) approximately according to $R = R_0[1 + \beta(T - T_0)]$, where R_0 is the resistance at temperature T_0. For a certain metal, β is found to be 0.004 K^{-1}.

a)   If the resistance at 0°C is 100 ohms, what is the resistance at 20°C?

b)   At what temperature is the resistance 200 ohms?

15–7 The pressure p, volume V, number of moles n, and Kelvin temperature T of an ideal gas are related by the equation $pV = nRT$. Prove that the coefficient of volume expansion is equal to the reciprocal of the Kelvin temperature.

15–8

a)   For any material, density ρ, mass m, and volume V are related by $\rho = m/V$. Prove that

$$\beta = -\frac{1}{\rho}\frac{\partial \rho}{\partial T}.$$

b)   The density of rock salt between −193°C and −13°C is given by the empirical formula

$$\rho = 2.1680(1 - 11.2 \times 10^{-5}T - 0.5 \times 10^{-7}T^2),$$

with T measured on the Celsius scale. Calculate β at −100°C.

15–9 The total change of volume ΔV of a metal, when the temperature changes from room temperature ($T_r = 300$ K) to the melting point T_m, is still only a small fraction of the original volume, so that

$$\frac{\Delta V}{V} = \int_{T_r}^{T_m} \beta \, dT.$$

Given the data in Table 15–4, calculate the average value of $\Delta V/V$ for these three metals.

Table 15–4

	T_m, K	β, 10^{-6} K^{-1}
Copper	1360	$43 + 0.022T$
Palladium	1830	$30 + 0.015T$
Platinum	2050	$24 + 0.0086T$

15–10 A glass flask whose volume is exactly 1000 cm^3 at 0°C is filled level full of mercury at this temperature. When flask and mercury are heated to 100°C, 15.2 cm^3 of mercury overflow. If the coefficient of volume expansion of mercury is 0.000182 per Celsius degree, compute the coefficient of linear expansion of the glass.

15–11 At a temperature of 20°C, the volume of a certain glass flask, up to a reference mark on the stem of the flask, is exactly 100 cm^3. The flask is filled to this point with a liquid whose cubical coefficient of expansion is 120×10^{-5} $(C°)^{-1}$, with both flask and liquid at 20°C. The linear coefficient of expansion of the glass is 8×10^{-6} $(C°)^{-1}$. The cross section of the stem is 1 mm^2 and can be considered constant. How far will the liquid rise or fall in the stem when the temperature is raised to 40°C?

15–12 To ensure a tight fit, the aluminum rivets used in airplane construction are made slightly larger than the rivet holes and cooled by "dry ice" (solid CO_2) before being driven. If the diameter of a hole is 0.2500 in., what should be the diameter of a rivet at 20°C if its diameter is to equal that of the hole when the rivet is cooled to −78°C, the temperature of dry ice? Assume the expansion coefficient to remain constant at the value given in Table 15–2.

15–13 A metal rod 30.0 cm long expands by 0.075 cm when its temperature is raised from 0°C to 100°C. A rod of a different metal of the same length expands by 0.045 cm for the same rise in temperature. A third rod, also 30.0 cm long, is made up of pieces of each of the above metals placed end-to-end and expands 0.065 cm between 0°C and 100°C. Find the length of each portion of the composite bar.

15–14 A hole 1.000 in. in diameter is bored in a brass plate at a temperature of 20°C. What is the diameter of the hole when the temperature of the plate is increased to 200°C? Assume the expansion coefficient to remain constant.

15–15 Suppose that a steel hoop could be constructed around the earth's equator. just fitting it at a temperature of 20°C. What would be the space between the hoop and the earth if the temperature of the hoop were increased by 1 C°?

15–16 A clock whose pendulum makes one vibration in 2 s is correct at 25°C. The pendulum shaft is of steel and its mass may be neglected compared with that of the bob. (a) What is the fractional change in length of the shaft when it is cooled to 15°C? (b) How many seconds per day will the clock gain or lose at 15°C? [*Hint:* Use differentials.]

15–17 A clock with a brass pendulum shaft keeps correct time at a certain temperature.

a)   How closely must the temperature be controlled if the clock is not to gain or lose more than 1 second a day? Does the answer depend on the period of the pendulum?

b)   Will an increase of temperature cause the clock to gain or lose?

15–18 The length of a bridge is 2000 ft. (a) If it were a continuous span, fixed at one end and free to move at the other, about what would be the range of motion of the free end between a cold winter day (−20°F) and a hot summer day (100°F)? (b) If both ends were rigidly fixed on the summer day, what would be the stress on the winter day?

15–19 The cross section of a steel rod is 1.5 in². What is the least force that will prevent it from contracting while cooling from 520°C to 20°C?

15–20 A steel wire which is 10 ft long at 20°C is found to increase in length by $\frac{3}{4}$ in. when heated to 520°C. Compute its average coefficient of linear expansion. (b) Find the stress in the wire if it is stretched taut at 520°C and cooled to 20°C without being allowed to contract.

15–21 A steel rod of length 40 cm and a copper rod of length 36 cm, both of the same diameter, are placed end-to-

end between two rigid supports, with no initial stress in the rods. The temperature of the rods is now raised by 50 C°. What is the stress in each rod?

15–22 A heavy brass bar has projections at its ends, as in Fig. 15–10. Two fine steel wires fastened between the projections are just taut (zero tension) when the whole system is at 0°C. What is the tensile stress in the steel wires when the temperature of the system is raised to 300°C? Make any simplifying assumptions you think are justified, but state what they are.

Brass

Fig. 15–10

Steel wire

15–23 The steel rails of the missile testing track at Tularosa Basin, New Mexico, are welded into 10,000-ft lengths and each length is prestressed by stretching it 3 ft before it is fastened to the concrete base slab.

a)   What is the strain in the rails?

b)   If Young's modulus is 30×10^6 lb in^{-2}, what is the stress?

c)   What stretching force is required if the cross-sectional area of a rail is 50 in²?

d)   If the stretched rails are fastened to the base slab at a temperature of 60°F, what is the stress when the temperature decreases to 0°F?

e)   At what temperature will the stress change from tension to compression?

15–24 Prove that if a body under hydrostatic pressure is raised in temperature but not allowed to expand, the increase in pressure is

$$\Delta p = B\bar{\beta} \, \Delta t.$$

where the bulk modulus B and the average coefficient of volume expansion $\bar{\beta}$ are both assumed positive and constant.

15–25

a)   A block of metal at a pressure of 1 atm and a temperature of 20°C is kept at constant volume. If the temperature is raised to 32°C , what will be the final pressure?

b)   If the block is maintained at constant volume by rigid walls that can withstand a maximum pressure of 1200 atm, what is the highest temperature to which the system may be raised? Assume B and $\bar{\beta}$ to remain practically constant at the values 1.5×10^{11} Pa and $5.0 \times 10^{-5}(C°)^{-1}$, respectively.

15–26 What hydrostatic pressure is necessary to prevent a copper block from expanding when its temperature is increased from 20°C to 30°C?

15–27 Table 15–3 lists the density of water, and the volume of 1 g. at atmospheric pressure. A steel bomb is filled with water at 10°C and atmospheric pressure, and the system is heated to 75°C. What is then the pressure in the bomb? Assume the bomb to be sufficiently rigid so that its volume is not affected by the increased pressure.

15–28 A liquid is enclosed in a metal cylinder provided with a piston of the same metal. The system is originally at atmospheric pressure and at a temperature of 80°C. The piston is forced down until the pressure on the liquid is increased by 100 atm, and it is then clamped in this position. Find the new temperature at which the pressure of the liquid is again 1 atm. Assume that the cylinder is sufficiently strong so that its volume is not altered by changes in pressure, but only by changes in temperature.

Compressibility of liquid	$k = 50 \times 10^{-6}$ atm^{-1}.
Cubical coefficient of expansion of liquid	$\beta = 5.3 \times 10^{-4}$ $(C°)^{-1}$.
Linear coefficient of expansion of metal	$\alpha = 10 \times 10^{-6}$ $(C°)^{-1}$.

Chapter 16

Heat and Heat Measurements

16–1 HEAT TRANSFER

Suppose that system *A*, at a higher temperature than system *B*, is put in contact with *B*. When thermal equilibrium has been reached, *A* will be found to have undergone a temperature decrease and *B* a temperature increase. It was therefore quite natural for the early investigators in this field to assume that *A* lost something and that this "something" flowed into *B*. While the temperature changes are taking place, it is customary to refer to a *heat flow* or a *heat transfer* from *A* to *B*. The process of heat transfer was formerly thought to be a flow of an invisible weightless fluid called *caloric*, which was produced when a substance burned and which could flow from a region rich in caloric (where the temperature was high) to a region where there was less caloric (and a lower temperature). The abandonment of the caloric theory was a part of the general development of physics during the eighteenth and nineteenth centuries. Due to the experimental skill and physical insight of Count Rumford (1753–1814) and Sir James Prescott Joule (1818–1889), the idea slowly emerged that heat flow is an *energy transfer*.

When an energy transfer takes place by virtue of a temperature difference exclusively, it is called a *heat flow*. Thus, in Fig. 16–1(a), the hot flame of a Bunsen burner is in contact with a system consisting of water and water vapor at a lower temperature. Proceeding from left to right, water is converted to steam at a higher temperature and pressure. Under these conditions the steam is capable of doing more work (by pushing against a turbine blade, for example) than before, and therefore has received energy in a process involving a heat flow.

An energy transfer to a system may take place, however, without a flow of heat. Consider the three diagrams (from left to right) of Fig. 16–1(b). As an object hanging from a string attached to a cam descends, a piston is caused to descend and a gas is compressed. According to the definition given in Chapter 7, this compression is the result of the performance of *work*, since the point of application of a force has moved in the direction of the force. In its compressed state the gas is capable of doing more work than before and hence the gas has received energy in a process that involves the performance of work.

(a)

(b)

Fig. 16–1 *Distinction between flow of heat and performance of work.*

(c)

The three diagrams of Fig. 16–1(c) show how a process may take place involving the simultaneous flow of heat and performance of work.

16–2 QUANTITY OF HEAT

A system consisting of a quantity of water and a small piece of resistance wire is depicted in Fig. 16–2. In part (a), the system is caused to undergo a temperature rise ΔT in a process involving a heat flow through a diathermic wall (Section 15–1) at the bottom of the enclosure. In (b) the system is surrounded by adiabatic walls and there is no heat flow. The same temperature change ΔT is produced by permitting a falling body to turn an electric generator, causing a current in the resistance wire. Thus in (b) the method of energy transfer to the system is the performance of mechanical work.

Suppose, however, that we regard the water alone as our system in part (b). Then the flow of energy *into the water* is due to the presence of a resistance wire whose temperature is maintained a trifle higher than that of the water by means of the electric current in the wire. This process involves a heat flow from the wire to the water.

Two conclusions may be drawn from these considerations:

1. Whether a process is regarded as involving a flow of heat or a performance of work depends on the *choice of the system.*

2. Whether the energy transfer depicted in Fig. 16–2(b) is regarded as a flow of heat or as a performance of work has no bearing on the quantity of energy delivered to the water.

We may describe the results of the two processes depicted in Fig. 16–2 by saying that a heat flow and a performance of work are *equivalent.* Once either process is completed, the energy of the system is greater than before, and no experiment can tell whether this energy increase has been caused by a flow of heat or a performance of work. They are both methods of energy transfer; in each case the *time rate* of energy transfer is expressed in power units, such as watts (joules per second) or foot-pounds per second.

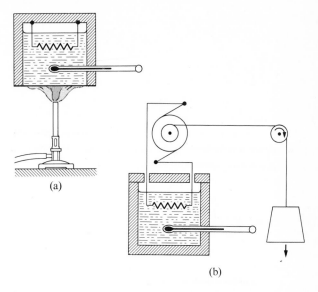

Fig. 16–2 *The same temperature change of the same system may be accomplished by either (a) a heat flow or (b) the performance of work.*

Up to this point, the word *heat* has been used only in conjunction with *flow* or *transfer.* Thus, a heat flow is an energy transfer brought about by a temperature difference only. Ideally, the word *heat* should be used *only* when referring to a *method* of energy *transfer* and, when the transfer is completed, to refer to the total amount of energy so transferred. The expression "quantity of heat," however, has played such a huge role in the development of the subject and is to be found in so many books and tables, that it is almost impossible to avoid.

It is essential to bear in mind that the concept "quantity of heat" has meaning *only* in the context of an interaction in which energy is *transferred* from one system to another as a result of a temperature difference. As we shall see later, to say that a given system *contains* a certain quantity of heat is meaningless. A careful discussion of this point leads to the concept of *internal energy,* to be discussed in detail in Chapter 19.

In the eighteenth century, a unit of quantity of heat, the *calorie,* was defined as that amount of heat required to raise the temperature of one gram of water one kelvin or one Celsius degree. It was found later that more heat was required to raise the temper-

c of steam = .48 cal/gr °C

ature of 1 g of water from, say, 90°C to 91°C than from 30°C to 31°C. The definition was then refined, and the chosen calorie became known as the "15° calorie," that is, the quantity of heat required to change the temperature of 1 g of water from 14.5°C to 15.5°C. A corresponding unit defined in terms of Fahrenheit degrees and British units is the *British thermal unit*, or Btu. By definition, 1 Btu is the quantity of heat required to raise the temperature of one pound of water from 63°F to 64°F. A third unit in common use, especially in measuring food energy, is the kilocalorie (kcal). The relations among these three units are   .2 cal = .0002

$$1 \text{ Btu} = 252 \text{ cal} = 0.252 \text{ kcal}.$$

It should also be noted that the unit of quantity of heat is fundamentally a unit of energy; thus there must be a relation between the above units and the familiar mechanical energy units such as the joule. It has been found experimentally that

$$1 \text{ cal} = 4.186 \text{ joules} = 4.186 \text{ J}.$$

The International Committee on Weights and Measures no longer recognizes the calorie as a fundamental unit, but recommends instead that the joule be used for quantity of heat as well as all other forms of energy. Although the calorie is convenient in problems involving water, it is awkward when both heat and other forms of energy are involved. There seems little doubt that ultimately the joule will become the universal unit of energy; many of the examples and problems of this and the following chapters will anticipate this usage. The relation between heat and mechanical energy will be explored in detail in Chapters 19 and 20.

JOULES

16-3 HEAT CAPACITY

Suppose that a small quantity of heat dQ is transferred between a system and its surroundings. If the system undergoes a small temperature change dT, the *specific heat capacity c of the system is defined as the ratio of the heat dQ to the product of the mass m and temperature change dT*; thus

$$c = \frac{dQ}{m \, dT}. \qquad (16\text{–}1)$$

= To MOLE

The specific heat capacity of water can be taken to be 4.19 J g^{-1} (C°)$^{-1}$, 1 cal g^{-1} (C°)$^{-1}$, or 1 Btu lb^{-1} (F°)$^{-1}$ for most practical purposes.

It is often convenient to use the *mole* to describe the amount of substance. By definition, one mole (1 mol) of any substance is a quantity of matter such that its mass in grams is numerically equal to the *molecular mass M* (often called *molecular weight*).* To calculate the number of moles n, we divide the mass in grams by the molecular mass; thus $n = m/M$. Replacing the mass m in Eq. (16–1) by the product nM, we get

$$Mc = \frac{dQ}{n \, dT}.$$

The product Mc is called the *molar heat capacity* and is represented by the symbol C. Hence, by definition

$$C = Mc = \frac{dQ}{n \, dT}. \qquad (16\text{–}2)$$

The molar heat capacity of water is approximately 75.3 J mol^{-1} (C°)$^{-1}$ or 18 cal mol^{-1} (C°)$^{-1}$.

The quantity defined by Eq. (16–1) is sometimes called simply *specific heat*, and the molar heat capacity defined by Eq. (16–2) is often called the *molar specific heat*. However, the terms *specific heat capacity* and *molar heat capacity* will be used throughout this book. Representative values of specific and molar heat capacity are given in Table 16–1.

From Eq. (16–1) the *total* quantity of heat Q that must be supplied to a body of mass m to change its temperature from T_1 to T_2 is

$$Q = m \int_{T_1}^{T_2} c \, dT. \qquad (16\text{–}3)$$

The specific heat capacities of all materials vary somewhat with temperature, and of course c must be expressed as a function of T in order to carry out the

---

* Although the term molecular *weight* is in common use, the term measures the *mass*, not the *weight* of a molecule. Thus the term *molecular mass* is preferable. This term is coming into more common use and will be used throughout this book.

cal

$$\frac{Joules}{gr\ °C} \times \frac{1\ cal}{4.186\ J.} = \frac{cal}{gr\ °C}$$

Table 16–1 MEAN SPECIFIC AND MOLAR HEAT CAPACITIES OF METALS

Metal	(Specific) $\bar{c}_p$, J g^{-1}(C°)$^{-1}$	Temperature range °C	M, g mol^{-1}	(Molar) $\bar{C}_p = M\bar{c}_p$, J mol^{-1}(C°)$^{-1}$
Beryllium	1.97 ,476	20–100	9.01	17.7
Aluminum	0.91 .22	17–100	27.0	24.5
Iron	0.47 .11	18–100	55.9	26.4
Copper	0.39 .094	15–100	63.5	24.7
Silver	0.234 ,056	15–100	108	25.3
Mercury	0.138 ,033	0–100	201	27.8
Lead	0.130 .031	20–100	207	26.9

integration. However, when the temperature interval is not too great, this variation may often be ignored, and c may be treated as a constant. In that case, the integral is trivial and Eq. (16–3) becomes

$$Q = mc(T_2 - T_1). \qquad (16\text{–}4)$$

Similarly, when the molar heat capacity C may be treated as constant, Eq. (16–2) may be integrated to give

$$Q = nC(T_2 - T_1). \qquad (16\text{–}5)$$

At very low temperatures, specific heat capacities always *decrease*, and the variation of heat capacity with temperature provides useful information concerning the microscopic structure of materials.

The specific or molar heat capacity of a substance is not the only physical property whose experimental determination requires the measurement of a quantity of heat. Heat conductivity, heat of fusion, heat of vaporization, heat of combustion, heat of solution, and heat of reaction are examples of other such properties which are called *thermal properties* of matter. The field of physics and physical chemistry concerned with the measurement of thermal properties is called *calorimetry*.

16–4 THE MEASUREMENT OF HEAT CAPACITY

To measure a heat capacity we need to add a measured quantity of heat to a measured quantity of a material and observe the resulting temperature change. For maximum precision the thermal measurements are usually made electrically. The heat input is provided by passing a current through a resistance wire in contact with (often wound around) the material and measuring the input of electrical energy. The thermometer is usually a small resistance thermometer or thermocouple embedded in the sample and is chosen for its rapidity of response and its sensitivity.

Readings are taken before the switch in the heater circuit is closed; these are shown on the left of Fig. 16–3. The fact that the temperature rises before energy is supplied to the heater indicates that the surroundings are at a temperature higher than that of the sample. The current is maintained in the heater for a time interval Δt, during which we do not measure the temperature. After the switch in the heater circuit is opened, temperature measurements are resumed, giving rise to the righthand part of Fig. 16–3, where the positive slope indicates that the surroundings are at still higher temperature than the sample.

If the potential difference across the heater is V and an electric current I flows, the power delivered to the heater is VI, and if the current is maintained for a time Δt, the total energy (heat) transferred to the sample is $VI\Delta t$. The molar heat capacity is therefore

$$C = \frac{VI\Delta t}{n\Delta T}, \qquad (16\text{–}6)$$

Fig. 16–3 *Temperature–time graph of the data taken during a heat-capacity measurement.*

where C is expressed in joules per mole-degree when I is expressed in amperes, V in volts, t in seconds, and ΔT is read from the graph, as in Fig. 16–3. The resulting molar heat capacity C is the value at the temperature T which is in the middle of the interval ΔT. In careful experiments, ΔT may be as small as 0.01 K.

Precise measurements of specific heats require great experimental skill, partly because of the difficulty of avoiding (and compensating for) unwanted heat transfer between the sample and its surroundings. Still, such measurements are very important. The temperature variation of specific heats provides the most direct approach to the understanding of molecular energies in matter. Low-temperature thermal properties in particular are of great interest in contemporary physics.

Figure 16–4 shows the variation of the specific heat capacity of water with temperature. It may be seen that the quantity of heat necessary to raise the temperature of 1 g of water from 14.5°C to 15.5°C is

$$1 \ 15° \ \text{cal} = 4.186 \ \text{J}.$$

Two other calories are frequently used. The *international table calorie* (IT cal) is *defined* to be

$$1 \ \text{IT cal} = \frac{1 \ \text{W hr}}{860} = \frac{3600 \ \text{J}}{860} = 4.186 \ \text{J},$$

and is almost identical with the 15° cal. The *thermochemical calorie*, however, is equal to 4.1840 J and may be seen from Fig. 16–4 to correspond to about a 17° calorie. In the absence of any additional information, the word *calorie* should be taken to mean 4.186 J. This variety of definitions of the calorie is an additional argument in favor of *eliminating* the calorie completely and using the joule as the fundamental unit of quantity of heat.

Mechanical engineers frequently use the British thermal unit (Btu), defined as the quantity of heat required to raise the temperature of 1 lb (mass) of water from 63°F to 64°F. The following relations hold:

$$1 \ \text{Btu} = 778.3 \ \text{ft lb} = 252.0 \ \text{cal} = 1055 \ \text{J}.$$

16–5 EXPERIMENTAL VALUES OF HEAT CAPACITIES

The amount of heat transferred to or from a system depends on the manner in which the system is controlled or constrained during the transfer, that is, whether the system is kept at *constant pressure* or at *constant volume*. The two corresponding specific (or molar) heat capacities are denoted by the respective symbols c_P (or C_P) and c_V (or C_V). In the electrical method of measuring heat capacities, the sample under investigation is maintained at constant pressure. As a matter of fact, it would be almost impossible to make a precise determination of c_V because there is no effective way of keeping the volume of a system constant and of correcting for the heat transferred to the containing walls.

A few average values of heat capacities of metals are listed in Table 16–1. The specific heat capacities are seen to be less than that of water and to decrease with increasing molecular mass. The last column reveals an interesting regularity, first noted in 1819

Fig. 16–4 *Specific heat capacity of water as a function of temperature.*

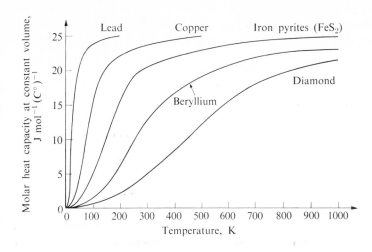

Fig. 16–5 *Temperature variation of C_V of solids.*

by two French physicists, Dulong and Petit. The average molar heat capacities (at constant pressure) for all metals except the very lightest are approximately the same and equal to about 25 J mol^{-1} (C°)$^{-1}$. This is known as the *Dulong and Petit law*; although only an approximate rule, it contains the germ of a very important idea. We know from chemistry that the number of molecules in one mole is the same for all substances. It follows that (nearly) the same amount of heat is required *per molecule* to raise the temperature of each of these metals by a given amount, even though the *mass* of a molecule of lead (for example) is nearly 10 times as great as that of a molecule of aluminum. To put it in a different way, the heat required to raise the temperature of a sample of metal depends only on *how many* molecules the sample contains, and not on the *mass* of an individual molecule. This is the first time in our study of physics that we have met a property of matter so directly related to its molecular structure.

The really important theoretical interpretation of heat-capacity measurements can be made only when the theoretician has at his disposal the complete temperature dependence, from the lowest to the highest possible temperatures, of the *molar heat capacity at constant volume*, C_V, because it is this quantity that is directly connected with the molecular energy which, in turn, may be calculated by statistical methods. Fortunately, there are relations which permit obtaining values of C_V from experimental

values of C_P. Figure 16–5 shows the resulting values of C_V and their temperature variation from about 4 K to 1000 K. These curves represent the extremes of behavior. All other metals and nonmetals lie within the boundaries of lead and diamond. Except for certain anomalies that are too complicated to dwell upon, all such curves approach about 6 cal mole^{-1} K^{-1} as the temperature approaches infinity. The Dulong and Petit value, therefore, is approached by all solids, lead arriving at this value at about 200 K (below room temperature) and diamond requiring well over 2000 K.

Each curve in Fig. 16–5 has the same shape; it starts at zero, rises rapidly at first, then bends over and approaches the value 25 J mol^{-1} K^{-1}. It was shown by Debye that the behavior of the molar heat capacity of *nonmetals* over the entire temperature range can be accounted for by the *vibrational motion of the molecules occupying regular positions in the crystal lattice*. There are, however, many substances whose temperature variation of heat capacity cannot be explained entirely on the basis of lattice vibrations. The low-temperature behavior of four "abnormal" types of substance is depicted in Fig. 16–6. There are many more different types of heat capacity curves; each of them indicates some special molecular or atomic or ionic property of the particles that occupy the regular positions within the crystal lattice. The theoretical understanding of heat capacity curves is among the most interesting areas of modern solid-state physics.

Fig. 16-6 Different ways in which the molar heat capacity varies with the temperature at low temperatures, indicating widely different atomic processes.

(a)   (b)   (c)   (d)

16-6 CHANGE OF PHASE

The term *phase* as used here relates to the fact that matter exists either as a solid, liquid, or gas. Thus the chemical substance H_2O exists in the *solid phase* as ice, in the *liquid phase* as water, and in the *gaseous phase* as steam. Provided they do not decompose at high temperatures, all substances can exist in any of the three phases under the proper conditions of temperature and pressure. Transitions from one phase to another are accompanied by the absorption or liberation of heat and usually by a change in volume, even when the transition occurs at constant temperature.

As an illustration, suppose that ice is taken from a freezer where its temperature was, say, $-25°C$. Let the ice be crushed quickly, placed in a container, and a thermometer inserted in the mass. Imagine the container to be surrounded by a heating coil which supplies heat to the ice at a uniform rate, and suppose that no other heat reaches the ice. The temperature of the ice would be observed to increase steadily, as shown by the portion of the graph (Fig. 16-7) from a to b, or until the temperature has risen to 0°C. In this temperature range the specific heat capacity of ice is approximately $2.3 \ J \ g^{-1} \ (C°)^{-1}$ or 0.55 $cal \ g^{-1} \ (C°)^{-1}$. As soon as this temperature is reached, some liquid water will be observed in the container. In other words, the ice begins to *melt*. The melting process is a *change of phase*, from the solid phase to the liquid phase. The thermometer, however, will show no *increase in temperature*, and even though heat is being supplied at the same rate as before, the temperature remains at 0°C until all the ice is melted (point c, Fig. 16-7), if the pressure is maintained constant at one atmosphere.

As soon as the last of the ice has melted, the temperature begins to rise again at a uniform rate (from c to d, Fig. 16-7) although this rate will be slower than that from a to b because the specific heat of water is greater than that of ice. When a temperature of 100°C is reached (point d), bubbles of steam (gaseous water or water vapor) start to escape from the liquid surface; the water begins to *boil*. The temperature remains constant at 100°C (at constant atmospheric pressure) until all the water has boiled away. Another change of phase has therefore taken place, from the liquid phase to the gaseous phase.

If all the water vapor had been trapped and not allowed to diffuse away (a very large container would be needed), the heating process could be continued as from e to f. The gas would now be called "superheated steam."

Although water was chosen as an example in the process just described, the same type of curve as in Fig. 16-7 is obtained for many other substances. Some, of course, decompose before reaching a melting or boiling point, and others, such as glass or

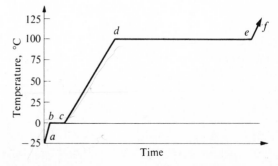

Fig. 16-7 The temperature remains constant during each change of phase, provided the pressure remains constant.

pitch, do not change phase at a definite temperature but become gradually softer as their temperature is raised. Crystalline substances, such as ice, or a metal, melt at a definite temperature. Glass and pitch behave like supercooled liquids of very high viscosity.

The quantity of heat per unit mass that must be supplied to a material at its melting point to convert it completely to a liquid *at the same temperature* is called the *heat of fusion* of the material. The quantity of heat per unit mass that must be supplied to a material at its boiling point to convert it completely to a gas at the same temperature is called the *heat of vaporization* of the material. Heats of fusion and vaporization are expressed in units of energy per unit mass, such as joules per kilogram, calories per gram, or Btu per pound-mass. Thus, the heat of fusion of ice is about 335,000 J kg^{-1}, 80 cal g^{-1}, or 144 Btu lb^{-1}. The heat of vaporization of water at 100°C is 2.26 × 10^6 J kg^{-1}, 539 cal g^{-1}, or 970 Btu lb^{-1}.

Some heats of fusion and vaporization are listed in Table 16–2. These quantities of heat are sometimes called *latent* heats because they change the phase of the material but not its temperature. This term is somewhat redundant and will not be used here.

When heat is removed from a gas, its temperature falls; at the same temperature at which it boiled,

it returns to the liquid phase, or *condenses*. In so doing it gives up to its surroundings the same quantity of heat which was required to vaporize it. The heat so given up, per unit mass, is called the *heat of condensation* and is equal to the heat of vaporization. Similarly, a liquid returns to the solid phase, or freezes, when cooled to the temperature at which it melted, and gives up heat called *heat of solidification* exactly equal to the heat of fusion. Thus the melting point and the freezing point are at the same temperature, and the boiling point and condensation point are at the same temperature.

Under some conditions a material can be cooled below the normal phase-change temperature without a phase change occurring. The resulting state is unstable and is described as *supercooled*. Very pure water can be cooled several degrees below the normal freezing point under ideal conditions: when a small ice crystal is dropped in or the water is agitated, it crystallizes very quickly. Similarly, supercooled water vapor condenses quickly into fog droplets when a disturbance, such as dust particles or ionizing radiation, is introduced. This phenomenon is used in the *cloud chamber*, where charged particles such as protons and electrons induce condensation of supercooled vapor along their path, thus making the path visible.

$$\frac{J}{g} \times \frac{1\,\text{cal}}{4.186\,J} = \frac{\text{cal}}{g}$$

Table 16–2 HEATS OF FUSION AND VAPORIZATION

Substance	Normal melting point		Heat of fusion, J g^{-1}	Normal boiling point		Heat of vaporization, J g^{-1}
	°K	°C		°K	°C	
Helium	3.5	−269.65	5.23	4.216	−268.93	20.9
Hydrogen	13.84	−259.31	58.6	20.26	−252.89	452
Nitrogen	63.18	−209.97	25.5	77.34	−195.81	201
Oxygen	54.36	−218.79	13.8	90.18	−182.97	213
Ethyl alcohol	159	−114	104.2	351	78	854
Mercury	234	−39	11.8	630	357	272
Water	273.15	0.00	333	373.15	100.00	2256
Sulfur	392	119	38.1	717.75	444.60	326
Lead	600.5	327.3	24.5	2023	1750	871
Antimony	903.65	630.50	165	1713	1440	561
Silver	1233.95	960.80	88.3	2466	2193	2336
Gold	1336.15	1063.00	64.5	2933	2660	1578
Copper	1356	1083	134	1460	1187	5069

LATENT HEAT OF FUSION
"   "   " VAPORIZATION

No SUBLIMATION

Whether a substance, at its melting point, is freezing or melting depends on whether heat is being supplied or removed. That is, if heat is supplied to a beaker containing both ice and water at 0°C, some of the ice will melt; if heat is removed, some of the water will freeze; the temperature in either case remains at 0°C so long as both ice and water are present. If heat is *neither supplied nor removed*, no change at all takes place and the relative amounts of ice and water, and the temperature, all remain constant. Such a system is said to be in *phase equilibrium*. This furnishes another point of view regarding the melting point. That is, the melting (or freezing) point of a substance is *that temperature at which both the liquid and solid phases can exist together*. At any higher temperature, the substance can be only a liquid; at any lower temperature, it can be only a solid.

The general term *heat of transformation* is applied to both heats of fusion and heats of vaporization, and both are designated by the letter L. Since L represents the heat absorbed or liberated in the change of phase of unit mass, the heat Q absorbed or liberated in the change of phase of a mass m is

$$Q = mL. \qquad (16\text{–}5)$$

The household steam-heating system makes use of a boiling–condensing process to transfer heat from the furnace to the radiators. Each kilogram of water which is turned to steam in the furnace absorbs 2.26×10^6 J (the heat of vaporization of water) from the furnace, and gives up this amount when it condenses in the radiators. (This figure is correct if the steam pressure is 1 atm. It is slightly smaller at higher pressures.) Thus the steam-heating system does not need to circulate as much water as a hot-water heating system. If water leaves a hot-water furnace at 60°C and returns at 40°C, dropping 20°C, about 27 kg of water must circulate to carry the same heat as is carried in the form of heat of vaporization by 1 kg of steam.

The temperature-control mechanisms of many warm-blooded animals operate on a similar princi-

ple. When the hypothalamus detects a slight rise in blood temperature, the sweat glands are activated. As sweat (chiefly water) evaporates from the surface of the body, it removes heat from the body as heat of vaporization. The heat is brought to the surface of the skin by the blood in the vascular system. This plays the same role as the pipes connecting the furnace to the radiators, with the heart playing the role of the circulating pump in a forced-circulation hot-water or steam system.

Under proper conditions of temperature and pressure, a substance can change directly from the solid to the gaseous phase without passing through the liquid phase. The transfer from solid to vapor is called *sublimation*, and the solid is said to *sublime*. "Dry ice" (solid carbon dioxide) sublimes at atmospheric pressure. Liquid carbon dioxide cannot exist at a pressure lower than about 5×10^5 Pa (about 5 atm). Heat is absorbed in the process of sublimation, and liberated in the reverse process. The quantity of heat per unit mass is called the *heat of sublimation*.

Definite quantities of heat are involved in chemical reactions. Perhaps the most familiar are those associated with combustion; complete combustion of one gram of gasoline produces about 46,000 J or about 11,000 cal, and the *heat of combustion* of gasoline is said to be 46,000 J g^{-1} or 46×10^6 J kg^{-1}. Energy values of foods are defined similarly; the unit of food energy, although called a calorie, is really a *kilocalorie*, equal to 1000 cal, or 4186 J. When we say that a gram of peanut butter "contains" 12 calories, we mean that, when it reacts with oxygen, with the help of enzymes, to convert the carbon and hydrogen completely to CO_2 and H_2O, the total energy liberated as heat is 12,000 cal, or 50,200 J. Not all of this energy is directly useful for mechanical work; the matter of efficiency of utilization of energy will be discussed in detail in Chapter 19.

16–7 EXAMPLES

As indicated above, the basic principle in calculations involving quantity of heat is that, when heat flow occurs between two bodies in thermal contact, the amount of heat lost by one body must equal that gained by the other. The following examples illustrate this principle in the context of phenomena discussed in this chapter.

Example 1 A copper cup of mass 0.1 kg, initially at 20°C, is filled with 0.2 kg of coffee initially at 70°C. What is the final temperature after the coffee and cup attain thermal equilibrium?

Let the final temperature be T. The heat lost by the coffee (which is assumed to have a specific heat capacity equal to that of water) is

$$Q = (0.2 \text{ kg})(4186 \text{ J kg}^{-1}(\text{C}°)^{-1})(70°C - T).$$

Similarly, the heat gained by the copper cup is

$$Q = (0.1 \text{ kg})(390 \text{ J kg}^{-1}(\text{C}°)^{-1})(T - 20°).$$

Equating these two quantities of heat yields an algebraic equation for T; solution of this equation gives $T = 67.8°C$. The final temperature is much closer to the initial temperature of the coffee than that of the cup because of the much larger specific heat capacity of water.

Example 2 How much ice at $-20°C$ must be dropped into 0.25 kg of water, initially at 20°C, in order for the final temperature to be 0°C with the ice all melted? The heat capacity of the container may be neglected.

The heat lost by the water is

$$Q = (0.25 \text{ kg})(4186 \text{ J kg}^{-1}(\text{C}°)^{-1})(20°C - 0°C)$$
$$= 20,930 \text{ J}.$$

The specific heat of ice is 2302 J kg^{-1} (C°)$^{-1}$. Let the mass of ice be m; then the heat needed to heat it from $-20°C$ to 0°C is

$$Q = m(2302 \text{ J kg}^{-1}(\text{C}°)^{-1})(0°C - (-20°C))$$
$$= m(46,046 \text{ J kg}^{-1}).$$

The additional heat needed to melt the ice is the latent heat of fusion times the mass:

$$Q = m(335,000 \text{ J kg}^{-1}).$$

The sum of these two quantities must equal the heat lost by the water:

$$m(381,000 \text{ J kg}^{-1}) = 20,930 \text{ J},$$

from which $m = 0.055 \text{ kg} = 55 \text{ g}$. This is roughly equivalent to two medium-size ice cubes.

Example 3 A certain gasoline camping lantern emits as much light as a 25-watt electric light bulb. Assuming the efficiency of conversion of heat into light is the same for the lantern and the bulb (which is not actually correct), how much gasoline does the lantern burn in 10 hours?

The rate of energy conversion in the lantern is 25 W, so in 10 hours (36,000 s), the total energy needed is

$$(25 \text{ J s}^{-1})(36,000 \text{ s}) = 0.9 \times 10^6 \text{ J}.$$

As mentioned in Section 16–6, combustion of one gram of gasoline produces 46,000 J, so the mass of gasoline required is

$$\frac{0.9 \times 10^6 \text{ J}}{46,000 \text{ J g}^{-1}} = 19.5 \text{ g}.$$

Actual lanterns are much less efficient than this, and typically require of the order of 300 to 400 g of gasoline (roughly one pint) to operate for ten hours.

PROBLEMS

[*Note.* When no information is given about the temperature variation of specific or molar heat capacity, assume it to be constant and use the average value given in Table 16–1.]

16–1 A combustion experiment is performed by burning a mixture of fuel and oxygen in a constant-volume "bomb" surrounded by a water bath. During the experiment the temperature of the water is observed to rise. Regarding the mixture of fuel and oxygen as the system, (a) has heat been transferred? (b) has work been done?

16–2 A liquid is irregularly stirred in a well-insulated container and thereby undergoes a rise in temperature. Regarding the liquid as the system, (a) has heat been transferred? (b) has work been done?

16–3 An automobile of mass 1500 kg is travelling at 5 m s^{-1}. How many calories are transferred in the brake mechanism when it is brought to rest?

16–4 A copper vessel of mass 200 g contains 400 g of water. The water is heated by a friction device which dissipates mechanical energy, and it is observed that the temperature of the system rises at the rate of 3 C° min^{-1}.

elec energy $I^2R = P$
$I^2RT = W$

Neglect heat losses to surroundings. What power in watts is dissipated in the water?

16–5 How long could a 2000-hp motor be operated on the heat energy liberated by 1 mi^3 of ocean water when the temperature of the water is lowered by 1 C° if all this heat were converted to mechanical energy? Why do we not utilize this tremendous reservoir of energy?

16–6

a) A certain house burns 10 tons of coal in a heating season. The heat of combustion of the coal is 11,000 Btu lb^{-1}. If stack losses are 15%, how many Btu were actually used to heat the house?

b) In some localities large tanks of water are heated by solar radiation during the summer and the stored energy is used for heating during the winter. Find the required dimensions of the storage tank, assuming it to be a cube, to store a quantity of energy equal to that computed in part (a). Assume that the water is raised to 120°F in summer and cooled to 80°F in the winter.

16–7 An artificial satellite, constructed of aluminum, encircles the earth at a speed of 9000 m s^{-1}.

a) Find the ratio of its kinetic energy to the energy required to raise its temperature by 600 C°. (The melting point of aluminum is 660°C.) Assume a constant specific heat capacity of 0.20 cal g^{-1} $(C°)^{-1}$.

b) Discuss the bearing of your answer on the problem of the re-entry of a satellite into the earth's atmosphere.

16–8 A calorimeter contains 100 g of water at 0°C. A 1000-g copper cylinder and a 1000-g lead cylinder, both at 100°C, are placed in the calorimeter. Find the final temperature if there is no loss of heat to the surroundings.

16–9

a) Compare the heat capacities (heat capacity is the heat per unit temperature change) of equal *masses* of water, copper, and lead.

b) Compare the heat capacities of equal *volumes* of water, copper, and lead.

16–10 An aluminum can of mass 500 g contains 117.5 g of water at a temperature of 20°C. A 200-g block of iron at 75°C is dropped into the can. Find the final temperature, assuming no heat loss to the surroundings.

16–11 A casting weighing 100 lb is taken from an annealing furnace where its temperature was 900°F and plunged

into a tank containing 800 lb of oil at a temperature of 80°F. The final temperature is 100°F, and the specific heat capacity of the oil is 0.5 Btu lb^{-1} $(F°)^{-1}$. What was the specific heat capacity of the casting? Neglect the heat capacity of the tank itself and any heat losses.

16–12 A lead bullet, traveling at 350 m s^{-1}, strikes a target and is brought to rest. What would be the rise in temperature of the bullet if there were no heat loss to the surroundings?

16–13 A copper calorimeter (mass 300 g) contains 500 g of water at a temperature of 15°C. A 560-g block of copper, at a temperature of 100°C, is dropped into the calorimeter and the temperature is observed to increase to 22.5°C. Neglect heat losses to the surroundings. Find the specific heat capacity of copper.

16–14 A 50-g sample of a material, at a temperature of 100°C, is dropped into a calorimeter containing 200 g of water initially at 20°C. The calorimeter is of copper and its mass is 100 g. The final temperature of the calorimeter is 22°C. Compute the specific heat capacity of the sample.

16–15 In Fig. 16–8 an electric heater is shown whose purpose is to provide a continuous supply of hot water. Water is flowing at the rate of 300 g min^{-1}, the inlet thermometer registers 15°C, the voltmeter reads 120 V and the ammeter 10 amp.

a) When a steady state is finally reached, what is the reading of the outlet thermometer?

b) Why is it unnecessary to take into account the heat capacity (mc or nC) of the apparatus itself?

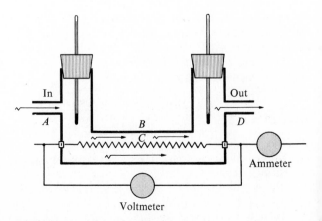

Fig. 16–8

w/ = 40 hp

RATE 4.5 gal/HR

E = gal

16–16 An automobile engine whose output is 40 hp uses 4.5 gallons of gasoline per hour. The heat of combustion is 12×10^7 joules per gallon. What is the efficiency of the engine? That is, what fraction of the heat of combustion is converted to mechanical work?

16–17 The electric power input to a certain electric motor is 0.5 kW, and the mechanical power output is 0.54 hp.

a) What is the efficiency of the motor? (That is, what fraction of the electrical power is converted to mechanical power?)

b) How many calories are developed in the motor in one hour of operation, assuming all electrical energy not converted to mechanical energy is converted to heat?

16–18 The molar heat capacity at constant pressure of a certain substance varies with temperature according to the empirical equation

$$C_p = 27.2 \text{ J mol}^{-1}\text{K}^{-1} + (4 \times 10^{-3}\text{ J mol}^{-1}\text{K}^{-2})T.$$

How much heat is necessary to change the temperature of 10 moles from 27°C to 527°C?

16–19 At very low temperatures, the molar heat capacity of rock salt varies with the temperature according to "Debye's T^3 law"; thus

$$C = k\frac{T^3}{\Theta^3},$$

where $k = 1940 \text{ J mol}^{-1}\text{K}^{-1}$ and $\Theta = 281$ K.

a) How much heat is required to raise the temperature of 2 moles of rock salt from 10 K to 50 K?

b) What is the mean molar heat capacity in this range?

c) What is the true molar heat capacity at 50 K?

16–20 Figure 16–9 shows a sketch of a continuous-flow calorimeter used to measure the heat of combustion of a gaseous fuel. Water is supplied at the rate of 12.5 lb min^{-1} and natural gas at 0.020 ft^3 min^{-1}. In the steady state, the inlet and outlet thermometers register 60°F and 76°F, respectively. What is the heat of combustion of natural gas in Btu ft^{-3}? Why should the gas flow be made as small as possible?

10060 BTU/ft³

16–21

a) Make a rough sketch of the type of temperature–time graph (such as that in Fig. 16–3) that would result if the surroundings were much *cooler* than the sample.

b) An electrical resistor is immersed in a liquid and electrical energy is dissipated for 100 s at a constant rate of 50 W. The mass of the liquid is 530 g, and its

Fig. 16–9

temperature increases from 17.64°C to 20.77°C. Find the mean specific heat capacity of the liquid in this temperature range.

16–22 How much heat is required to convert 1 g of ice at -10°C to steam at 100°C?

16–23 A beaker of very small mass contains 500 g of water at a temperature of 80°C. How many grams of ice at a temperature of -20°C must be dropped in the water so that the final temperature of the system will be 50°C?

16–24 An open vessel contains 500 g of ice at -20°C. The mass of the container can be neglected. Heat is supplied to the vessel at the constant rate of 1000 cal min^{-1} for 100 min. Plot a curve showing the elapsed time as abscissa and the temperature as ordinate.

16–25 A copper calorimeter of mass 100 g contains 150 g of water and 8 g of ice in thermal equilibrium at atmospheric pressure. 100 g of lead at a temperature of 200°C are dropped into the calorimeter. Find the final temperature if no heat is lost to the surroundings.

16–26 500 g of ice at -16°C are dropped into a calorimeter containing 1000 g of water at 20°C. The calorimeter can is of copper and has a mass of 278 g. Compute the final temperature of the system, assuming no heat losses.

16–27 A tube leads from a flask in which water is boiling under atmospheric pressure to a calorimeter. The mass of the calorimeter is 150 g, its heat capacity is 15 cal (C°)$^{-1}$ and it contains originally 340 g of water at 15°C. Steam is allowed to condense in the calorimeter until its temperature increases to 71°C, after which the total mass of calorimeter and contents is found to be 525 g. Compute the heat of condensation of steam from these data.

16–28 An aluminum canteen whose mass is 500 g contains 750 g of water and 100 g of ice. The canteen is dropped from an aircraft to the ground. After landing, the temperature of the canteen is found to be 25°C. Assuming that no energy is given to the ground in the impact, what was the velocity of the canteen just before it landed? State any additional assumptions needed.

16–29 A calorimeter contains 500 g of water and 300 g of ice, all at a temperature of 0°C. A block of metal of mass 1000 g is taken from a furnace where its temperature was 240°C and is dropped quickly into the calorimeter. As a result, all the ice is just melted. What would the final temperature of the system have been if the mass of the block had been twice as great? Neglect heat loss from and the heat capacity of the calorimeter.

16–30 An ice cube whose mass is 50 g is taken from a refrigerator where its temperature was −10°C and is dropped into a glass of water at 0°C. If no heat is gained or lost from outside, how much water will freeze onto the cube?

16–31 A copper calorimeter can ($mc = 30$ cal deg^{-1}) contains 50 g of ice. The system is initially at 0°C. 12 g of steam at 100°C and 1 atm pressure are run into the calorimeter. What is the final temperature of the calorimeter and its contents?

16–32 A vessel whose walls are thermally insulated contains 2100 g of water and 200 g of ice, all at a temperature of 0°C. The outlet of a tube leading from a boiler, in which water is boiling at atmospheric pressure, is inserted in the water. How many grams of steam must condense to raise the temperature of the system to 20°C? Neglect the heat capacity of the container.

16–33 A 2-kg iron block is taken from a furnace where its temperature was 650°C and placed on a large block of ice at 0°C. Assuming that all the heat given up by the iron is used to melt the ice, how much ice is melted?

16–34 In a household hot-water heating system, water is delivered to the radiators at 140°F and leaves at 100°F. The system is to be replaced by a steam system in which steam at atmospheric pressure condenses in the radiators, the condensed steam leaving the radiators at 180°F. How

many pounds of steam will supply the same heat as was supplied by 1 lb of hot water in the first installation?

16–35 A "solar house" has storage facilities for 4 million Btu. Compare the space requirements for this storage on the assumption (a) that the heat is stored in water heated from a minimum temperature of 80°F to a maximum at 120°F, and (b) that the heat is stored in Glauber salt ($Na_2SO_4 \cdot 10H_2O$) heated in the same temperature range.

PROPERTIES OF GLAUBER SALT

Specific heat capacity	
Solid	0.46 Btu lb^{-1} °)$^{-1}$
Liquid	0.68 Btu lb^{-1} °)$^{-1}$
Specific gravity	1.6
Melting point	90°F
Heat of fusion	104 Btu lb^{-1}

16–36 The nominal food-energy value of butter is about 6 kcal g^{-1}, where 1 kcal = 1000 cal = 41 6 J. If all this energy could be converted completely to mechanical energy, how much butter would be required to power an 80-kg mountaineer on his journey from Lupine Meadows (elevation 2070 m) to the summit of Grand Teton (4196 m)?

16–37 The capacity of air conditioners is typically expressed in Btu hr^{-1} or "tons," the latter being the number of tons of ice that can be frozen from water at 0° C in 24 hr by the unit.

a) What is the rating in Btu/hr of a one-ton air conditioner?

b) Express the capacity of a one-ton air conditioner in watts.

16–38 A 6000-Btu-hr^{-1} air conditioner typically consumes about 800 W of electrical power. What is the ratio of the rate of removal of heat to the rate of consumption of electrical energy? Express the two rates in the same units to obtain a dimensionless ratio.

16–39 Suppose 1 liter of gasoline will propel a car a distance of 10 km. The density of gasoline is about 0.7 g cm^{-3}, and its heat of combustion is about 4.6×10^4 J g^{-1}.

a) If the engine is 25% efficient, that is, if 1/4 of the heat of combustion is converted into useful mechanical work, what total work does the engine do during the 10-km displacement?

b) If this work is assumed to be done in opposing a constant resisting force F, find the magnitude of F.

Chapter 17

Transfer of Heat

17–1 CONDUCTION

If one end of a metal rod is placed in a flame while the other is held in the hand, that part of the rod one is holding will be felt to become hotter and hotter, although it was not itself in direct contact with the flame. Heat is said to reach the cooler end of the rod by *conduction* along or through the material of the rod. The molecules at the hot end of the rod increase the violence of their vibration as the temperature of the hot end increases. Then, as they collide with their more slowly moving neighbors farther out on the rod, some of their energy of motion is shared with these neighbors and they in turn pass it along to those still farther out from the flame. Hence energy of thermal motion is passed along from one molecule to the next, while each individual molecule remains at its original position.

It is well known that metals are good conductors of electricity and also good conductors of heat. The ability of a metal to conduct an electric current is due to the fact that there are within it so-called "free" electrons, that is, electrons that have become detached from their parent molecules. The free electrons also play a part in the conduction of heat, and the reason metals are such good heat conductors is that the free electrons provide an effective mecha-

nism for carrying thermal energy from the hotter to the cooler portions of the metal.

Conduction of heat can take place in a body only when different parts of the body are at different temperatures, and the direction of heat flow is always from points of higher to points of lower temperature.

Figure 17–1 represents a rod of material of cross section A and thickness L. Let the whole of the left face of the rod be kept at a temperature T_2, and the whole of the right face at a lower temperature T_1. The direction of the heat current is then from left to right through the rod. The sides of the rod are covered by an insulating material, so no heat transfer occurs at the sides. This is of course an idealized model; all real materials, even the best thermal insulators, conduct heat to some extent.

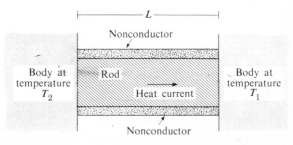

Fig. 17–1 *Steady-state heat flow in a uniform rod.*

293

After the ends of the rod have been kept at the temperatures T_1 and T_2 for a sufficiently long time, the temperature at points within the rod is found to decrease uniformly with distance from the hot to the cold face. At each point, however, the temperature remains constant with time. This condition is called "steady-state" heat flow.

Experiment shows that the rate of flow of heat through the rod in the steady state is proportional to the area A, proportional to the temperature difference $(T_2 - T_1)$, and inversely proportional to the length L. These proportions may be converted to an equation by introducing a constant k whose numerical value depends on the material of the rod. The quantity k is called the *thermal conductivity* of the material,

$$H = \frac{kA(T_2 - T_1)}{L}, \qquad (17\text{–}1)$$

where H is the quantity of heat flowing through the rod per unit time, also called the *heat current.*

Equation (17–1) may also be used to compute the rate of heat flow through a slab, or *any* homogeneous body having a uniform cross-section perpendicular to the direction of flow, provided the flow has attained steady-state conditions and the ends are kept at constant temperatures.

When the cross-section is not uniform, or when steady state conditions are not present, the temperature does not necessarily change uniformly along the direction of heat flow. Equation (17–1) may still be applied to a thin layer of material perpendicular to the direction of flow. If x is the coordinate measured along the flow path, dx the thickness of the layer, and A the cross section area perpendicular to that path, then Eq. (17–1) may be rewritten

$$H = -kA\frac{dT}{dx}, \qquad (17\text{–}2)$$

where dT is the temperature change across the distance dx. The negative sign is included because if the temperature *increases* in the direction of increas-

ing x (dx and dT both positive) the direction of heat flow is the direction of *decreasing x*, and conversely. The temperature change per unit length dT/dx is called the *temperature gradient;* Eq. (17–1) refers to the special case where the temperature gradient is constant, and equal to $(T_2 - T_1)/L$.

The mks unit of rate of heat flow is the joule per second, although other energy units such as the calorie are sometimes used. The units of k are revealed by solving Eq. (17–2) for k:

$$k = -\frac{H}{A}\frac{dx}{dT}.$$

This shows that in mks units, the unit of k is

$$\frac{(1\ \mathrm{J\ s^{-1}})(1\ \mathrm{m})}{(1\ \mathrm{m})^2(1\ \mathrm{C^\circ})} = 1\ \mathrm{J\ (s\,m\,C^\circ)^{-1}}.$$

Values of thermal conductivity are often tabulated using cgs units, with the calorie as the energy unit. The unit of k is then $1\ \mathrm{cal\ (s\,cm\,C^\circ)^{-1}}$. The conversion is

$$1\ \mathrm{cal\ (s\,cm\,C^\circ)^{-1}} = 419\ \mathrm{J\ (s\,m\,C^\circ)^{-1}}.$$

The thermal conductivity of most materials is a function of temperature, increasing slightly with increasing temperature, but the variation is small and can often be neglected. Some numerical values of k, at temperatures near room temperature, are given in Table 17–1. The properties of materials used commercially as heat insulators are expressed in a mixed system in which the unit of heat current is $1\ \mathrm{Btu\ hr^{-1}}$, the unit of area is $1\ \mathrm{ft^2}$, and the unit of temperature gradient is one fahrenheit degree per *inch* ($1\ \mathrm{F^\circ\ in^{-1}}$).

It is evident from Eq. (17–1) that the larger the thermal conductivity k, the larger the heat current, other factors being equal. A material for which k is large is therefore a good heat conductor, while if k is small, the material is a poor conductor or a good insulator. A "perfect heat conductor" ($k = \infty$) or a "perfect heat insulator" ($k = 0$) does not exist. However, it will be seen from Table 17–1 that the metals as a group have much greater thermal conductivities than the nonmetals, and that those of gases are extremely small.

Before steady-state conditions are reached, the temperature at each point may change with time. For

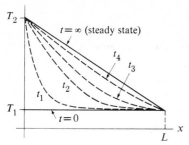

Fig. 17–2 *Transient and steady-state temperature distribution along a rod initially at temperature T_1. The transient distributions were computed for $t_2 = 5t_1$, $t_3 = 10t_1$, $t_4 = 20t_1$.*

example, suppose the rod in Fig. 17–1 is initially at the temperature T_1 throughout, and then at time $t = 0$ the ends are placed in contact with the bodies of temperature T_1 and T_2. The resulting temperature distributions at various times are shown in Fig. 17–2.

Table 17–1 THERMAL CONDUCTIVITIES

	k, J s^{-1} m^{-1} (C°)$^{-1}$	k, cal s^{-1} cm^{-1} (C°)$^{-1}$
Metals		
Aluminum	205	0.49
Brass	109	0.26
Copper	385	0.92
Lead	34.7	0.083
Mercury	8.3	0.020
Silver	406	0.97
Steel	50.2	0.12
Various solids		
(Representative values)		
Insulating brick	0.15	0.00035
Red brick	0.6	0.0015
Concrete	0.8	0.002
Cork	0.04	0.0001
Felt	0.04	0.0001
Glass	0.8	0.002
Ice	1.6	0.004
Rock wool	0.04	0.0001
Wood	0.12–0.04	0.0003–0.0001
Gases		
Air	0.024	0.000057
Argon	0.016	0.000039
Helium	0.14	0.00034
Hydrogen	0.14	0.00033
Oxygen	0.023	0.000056

At time t_1 the temperature changes much more rapidly with x near the left end than near the right, corresponding to the fact that the heat flow is much greater near that end to provide a net flow of heat into the middle region and raise its temperature. With increasing time the differences become smaller and smaller, until after a very long time ($t = \infty$) the steady-state uniform temperature distribution is attained. It is a remarkable fact that the final steady state is completely independent of the initial conditions at $t = 0$, and depends only on the temperatures at the ends.

In the steady state the heat current in the rod is the same in all cross sections; otherwise there would be a net flow of heat into certain regions and out of others and their temperatures would change, contradicting the assumption of a steady state. Thus steady-state heat flow is comparable to the flow of an incompressible fluid.

Since there are no restrictions on the relative dimensions of the rod in Fig. 17–1, Eq. (17–1) applies also when the length L becomes relatively small and the area A relatively large, as in the case of a sheet or slab of material such as a portion of the wall of a house, a refrigerator, or a furnace. (The dimension L would then ordinarily be described as the "thickness" rather than the "length.")

17–2 RADIAL HEAT FLOW IN A SPHERE OR CYLINDER

We next consider two examples of heat flow in which the temperature gradient is not uniform along the direction of flow, even in the steady state. Figure 17–3 represents a steam pipe surrounded by a layer

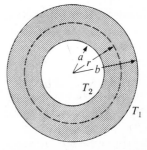

Fig. 17–3 *Radial heat flow in a cylinder or sphere.*

Table 17–2

Cylinder (length L)	Sphere
$A = 2\pi rL$	$A = 4\pi r^2$
$H = -k(2\pi rL)\dfrac{dT}{dr}$	$H = -k(4\pi r^2)\dfrac{dT}{dr}$
$H\dfrac{dr}{r} = -2\pi kL\,dT$	$H\dfrac{dr}{r^2} = -4\pi k\,dT$
$H\ln r = -2\pi kLT + C$	$-\dfrac{H}{r} = -4\pi kT + C$
$r = a, \quad T = T_2;$	$r = b, \quad T = T_1$
$H = \dfrac{2\pi k(T_2 - T_1)}{[\ln(b/a)]/L}$	$H = \dfrac{4\pi k(T_2 - T_1)}{(b - a)/ab}$

of insulating material, or a sphere surrounded by a spherical shell of insulation. Let T_2 and T_1 be the temperatures of the inner and outer surfaces of the insulation, and a and b be the inner and outer radii. If $T_2 > T_1$, heat flows outward, and in the steady state the heat current H is the same across all surfaces within the insulation, such as that of radius r, shown by a dashed circle. If A is the area of this surface and dT/dr the corresponding temperature gradient,

$$H = -kA\frac{dT}{dr} = \text{constant.}$$

The heat current can be expressed in terms of T_2, T_1, a, and b, by integrating this equation and inserting the appropriate boundary conditions. The calculations for the cylinder and the sphere have been carried out in parallel in Table 17–2.

17–3 CONVECTION

The term *convection* is applied to the transfer of heat from one place to another by the actual motion of hot material. The hot-air furnace, the hot-water heating system, and the flow of blood in the body are examples. If the heated material is forced to move by a blower or pump, the process is called *forced convection*; if the material flows due to differences in density, the process is called *natural* or *free convection*. To understand the latter, consider a U-tube, as illustrated in Fig. 17–4.

In (a), the water is at the same temperature in both arms of the U and hence stands at the same level in each. In (b), the right side of the U has been heated. The water in this side expands and therefore has smaller density; thus a longer column is needed to balance the pressure produced by the cold water in the left column. When the stopcock is opened, water flows from the top of the warmer column into the colder column. This increases the pressure at the bottom of the U produced by the cold column, and decreases the pressure at this point due to the hot column. Hence at the bottom of the U, water is forced from the cold to the hot side. If heat is continually applied to the hot side and removed from the cold side, the circulation continues of itself. The net result is a continual transfer of heat from the hot to the cold side of the column. In the common household hot-water heating system, the "cold" side corresponds to the radiators and the "hot" side to the furnace.

The anomalous expansion of water which was mentioned in Chapter 15 has an important effect on the way in which lakes and ponds freeze in winter.

(a)   (b)

Fig. 17–4 *Convection is brought about by differences in density.*

Consider a pond at a temperature of, say 20°C throughout, and suppose the air temperature at its surface falls to −10°C. The water at the surface becomes cooled to, say, 19°C. It therefore contracts, becomes more dense than the warmer water below it, and sinks in this less dense water, its place being taken by water at 20°C. The sinking of the cooled water causes a mixing process, which continues until all of the water has been cooled to 4°C. Now, however, when the surface water cools to 3°C, it expands, is less dense than the water below it, and hence floats on the surface. Convection and mixing then cease, and the remainder of the water can lose heat only by *conduction*. Since water is an extremely poor heat conductor, cooling takes place very slowly after 4°C is reached, with the result that the pond freezes first at its surface. Then, since the density of ice is even smaller than that of water at 0°C, the ice floats on the water below it, and further freezing can result only from heat flow upward by conduction.

The mathematical theory of heat convection is quite involved. There is no simple equation for convection, as there is for conduction. This arises from the fact that the heat lost or gained by a surface at one temperature in contact with a fluid at another temperature depends on many circumstances, such as (1) whether the surface is flat or curved, (2) whether the surface is horizontal or vertical, (3) whether the fluid in contact with the surface is a gas or a liquid, (4) the density, viscosity, specific heat, and thermal conductivity of the fluid, (5) whether the velocity of the fluid is small enough to give rise to laminar flow or large enough to cause turbulent flow, (6) whether evaporation, condensation, or formation of scale takes place.

The procedure adopted in practical calculations is first to define a *convection coefficient h* by means of the equation

CONVECTION

$$H = hA \, \Delta T, \tag{17–3}$$

where H is the heat convection current (the heat gained or lost by convection by a surface per unit of time), A is the area of the surface, and ΔT is the temperature difference between the surface and the main body of fluid. The next step is the determination of numerical values of h that are appropriate to a given piece of equipment. Such a determination is accomplished partly by dimensional analysis and partly by an elaborate series of experiments. An enormous amount of research in this field has been done in recent years so that, by now, there are in existence fairly complete tables and graphs from which the physicist or engineer may obtain the convection coefficient appropriate to certain standard types of apparatus.

A case of common occurrence is that of natural convection from a wall or a pipe which is at a constant temperature and is surrounded by air at atmospheric pressure whose temperature differs from that of the wall or pipe by an amount ΔT. The convection coefficients applicable in this situation are given in Table 17–3.

Example The air in a room is at a temperature of 25°C, and the outside air is at −15°C. How much heat is transferred per unit area of a glass windowpane of thermal conductivity 2.5×10^{-3} cgs units and of thickness 2 mm?

To assume that the inner surface of the glass is at 25°C and the outer surface is at −15°C is entirely erroneous, as anyone can verify by touching the inner surface of a glass windowpane on a cold day.

Table 17–3 COEFFICIENTS OF NATURAL CONVECTION IN AIR AT ATMOSPHERIC PRESSURE

Equipment	Convection coefficient h, cal s^{-1} cm^{-2}(C°)$^{-1}$
Horizontal plate, facing ~~downward~~ upward	$0.595 \times 10^{-4}(\Delta T)^{1/4}$
Horizontal plate, facing downward	$0.314 \times 10^{-4}(\Delta T)^{1/4}$
Vertical plate	$0.424 \times 10^{-4}(\Delta T)^{1/4}$
Horizontal or vertical pipe $\left(\dfrac{\text{diameter}}{D}\right)$	$1.00 \times 10^{-4}\left(\dfrac{\Delta T}{D}\right)^{1/4}$

$H = hA\Delta t$ $H = \dfrac{kA\,\Delta T}{L}$

$\dfrac{H}{A} = h\Delta t$ $\Delta T = \dfrac{LH}{kA}$

-15C + 20 = 5C + 20 - 25'

One must expect a much smaller temperature difference across the windowpane, so that in the steady state the rates of transfer of heat (1) by convection in the room, (2) by conduction through the glass, and (3) by convection in the outside air, are all equal.

As a first approximation in the solution of this problem, let us assume that the window is at a uniform temperature T. If $T = 5°C$, then the temperature difference between the inside air and the glass is the same as that between the glass and the outside air, or 20°C. Hence the convection coefficient in both cases is

$$h = 0.424 \times 10^{-4}(20)^{1/4}\ \text{cal s}^{-1}\ \text{cm}^{-2}\ (\text{C}°)^{-1}$$
$$= 0.895 \times 10^{-4}\ \text{cal s}^{-1}\ \text{cm}^{-2}\ (\text{C}°)^{-1},$$

and, from Eq. (17–3), the heat transferred per unit area is

$$\frac{H}{A} = 0.895 \times 10^{-4} \times 20$$
$$= 17.9 \times 10^{-4}\ \text{cal s}^{-1}\ \text{cm}^{-2}.$$

The glass, however, is *not* at uniform temperature; there must be a temperature difference ΔT across the glass sufficient to provide heat conduction at the rate of $17.9 \times 10^{-4}\ \text{cal s}^{-1}\ \text{cm}^{-2}$. Using the conduction equation, Eq. (17–2), we obtain

$$\Delta T = \frac{L}{k} \times \frac{H}{A} = \frac{0.2}{2.5 \times 10^{-3}} \times 17.9 \times 10^{-4}\ \text{C}°$$

$$= 0.14\ \text{C}°.$$

With sufficient accuracy we may therefore say that the inner surface is at 5.07°C and the outer surface is at 4.93°C.

Heat transfer in the human body involves a combination of mechanisms, which together maintain a remarkably constant and uniform temperature despite large changes in environmental conditions. As mentioned above, the chief *internal* mechanism is forced convection, with the heart serving as the pump and the blood as the circulating fluid. Heat transfer with the surroundings involves conduction, convection, and radiation, in proportions depending on circumstances. Total heat loss from the body is on the order of 2000 to 5000 kcal per day, depending on

amount of activity. A dry nude body in still air loses about half its heat by radiation, but under conditions of vigorous activity and copious perspiration, the dominant mechanism is evaporative cooling. Radiation will be discussed in the following sections.

17–4 RADIATION

When one's hand is placed in direct contact with the surface of a hot-water or steam radiator, heat reaches the hand by *conduction* through the radiator walls. If the hand is held above the radiator but not in contact with it, heat reaches the hand by way of the upward-moving *convection* currents of warm air. If the hand is held at one side of the radiator it still becomes warm, even though conduction through the air is negligible and the hand is not in the path of the convection currents. Energy now reaches the hand by *radiation.*

The term radiation refers to the continual emission of energy from the surface of all bodies. This energy is called *radiant energy* and is in the form of electromagnetic waves. These waves travel with the velocity of light and are transmitted through a vacuum as well as through air. (Better, in fact, since they are absorbed by air to some extent.) When they fall on a body which is not transparent to them, such as the surface of one's hand or the walls of the room, they are absorbed, resulting in a transfer of heat to the absorbing material.

The radiant energy emitted by a surface, per unit time and per unit area, depends on the nature of the surface and on its temperature. At low temperatures the rate of radiation is small and the radiant energy is chiefly of relatively long wavelength. As the temperature is increased, the rate of radiation increases very rapidly, in proportion to the fourth power of the absolute temperature. For example, a copper block at a temperature of 100°C (373 K) radiates about 0.03 J s^{-1} or 0.03 W from each square centimeter of its surface. At a temperature of 500°C (773 K), it radiates about 0.54 W from each square centimeter, and at 1000°C (1273 K), it radiates about 4 W per square centimeter. This rate is 130 times as great as that at a temperature of 100°C.

Fig. 17–5 *Rate of emission of radiant energy by a blackbody, per unit range of wavelength, as a function of wavelength. Shaded area indicates visible spectrum.*

At each of these temperatures the radiant energy emitted is a mixture of waves of different wavelengths. At a temperature of 300°C the most intense of these waves has a wavelength of about 5×10^{-4} cm; for wavelengths either greater or less than this value the intensity decreases, as shown by the curve in Fig. 17–5. The corresponding distribution of energy at higher temperatures is also shown in the figure. The area between each curve and the horizontal axis represents the *total* rate of radiation at that temperature. It is evident that this rate increases rapidly with increasing temperature, and also that the wavelength of the most intense wave shifts toward the left, or toward shorter wavelengths, with increasing temperature.

At a temperature of 300°C, practically all of the radiant energy emitted by a body is carried by waves longer than those corresponding to red light. Such waves are called *infrared*, meaning "beyond the red." At a temperature of 800°C a body emits enough visible radiant energy to be self-luminous and appears "red hot." By far the larger part of the energy emitted, however, is still carried by infrared waves. At 3000°C, which is about the temperature of an incandescent lamp filament, the radiant energy contains enough of the shorter wavelengths so that the body appears "white hot."

17–5 STEFAN'S LAW

Experimental measurements of the rate of emission of radiant energy from the surface of a body were made by John Tyndall (1820–1893), and on the basis of these Josef Stefan (1835–1893) concluded in 1879 that the rate of emission could be expressed by the relation

$$R = e\sigma T^4 \qquad (17\text{–}4)$$

which is *Stefan's law*. The quantity R, known as the *radiant emittance*, is equal to the rate of emission of radiant energy per unit area. In the mks system, R is expressed in watts per square meter; in the cgs system it is expressed in ergs per square centimeter per second. The constant σ is a universal physical constant called the *Stefan–Boltzmann constant*; its numerical value in mks units is

$$\sigma = 5.6696 \times 10^{-8} \, \text{W m}^{-2} \, \text{K}^{-4}.$$

The quantity T is the Kelvin temperature of the surface, and e is a quantity called the emissivity of the surface. The emissivity lies between zero and unity, depending on the nature of the surface. The emissivity of copper, for example, is about 0.3. (Strictly speaking, the emissivity varies somewhat with temperature even for the same surface.) In general, the emissivity is larger for rough and smaller for smooth, polished surfaces.   ―OR COLOR―

It may be wondered why it is, if the surfaces of all bodies are continually emitting radiant energy, that all bodies do not eventually radiate away all of their internal energy and cool down to a temperature of absolute zero, where $R = 0$ by Eq. (17–4). The answer is that they *would* do so if energy were not supplied to them in some way. In the case of an electric heater element or the filament of an electric lamp, energy is supplied electrically to make up for the energy radiated. As soon as this energy supply is cut off, these bodies do, in fact, cool down very quickly to room temperature. The reason that they do not cool further is that their surroundings (the walls, and other objects in the room) are also radiating, and some of this radiant energy is intercepted, absorbed, and converted into internal energy. The same thing is true of all other objects in the room: each is both emitting and absorbing radiant energy simultaneously. When any object is hotter than its surroundings, the rate of emission exceeds its rate of absorption. There is thus a net loss of energy and the body cools down, unless heated by some other method. If a body is at a *lower* temperature than its surroundings, its rate of absorption is larger than its rate of emission and its temperature rises. When the body is at the same temperature as its surroundings, the two rates become equal, there is no net gain or loss of energy, and no change in temperature.

If a small body of emissivity e is completely surrounded by walls at a temperature T, the rate of *absorption* of radiant energy per unit area by the body is

$$R = e\sigma T^4.$$

Hence for such a body at a temperature T_1, surrounded by walls at a temperature T_2, the *net* rate of loss (or gain) of energy per unit area by radiation is

$$R_{\text{net}} = e\sigma T_1^4 - e\sigma T_2^4 = e\sigma(T_1^4 - T_2^4). \quad (17\text{–}5)$$

Infrared emission from a body can be studied using a camera equipped with infrared-sensitive film, or better with a special device similar in principle to a television camera, sensitive to infrared radiation. The resulting picture is called a *thermograph*; since emission depends on temperature, thermography permits detailed study of temperature distributions. Some present-day instrumentation is sensitive to differences as small as 0.1 C°.

Thermography has a variety of important medical applications. Local temperature variations in the body are associated with various tumors, such as breast cancers, and growths as small as 1 cm in diameter can be detected. Vascular disorders leading to local temperature anomalies can be studied, and many other examples could be cited.

17–6 THE IDEAL RADIATOR

Suppose that the walls of the enclosure in Fig. 17–6 are kept at temperature T_2 and a number of different bodies having different emissivities are suspended one after another within the enclosure. Regardless of their temperature when they are inserted, it is found that eventually each comes to the same temperature T_2 as that of the walls, even if the air in the enclosure is evacuated. If the bodies are small compared with the size of the enclosure, radiant energy from the walls strikes the surface of each body at the same rate. Of this energy, a part is reflected and the remainder absorbed. In the absence of any other process, the energy absorbed will raise the tempera-

Fig. 17–6 *In thermal equilibrium, the rate of emission of radiant energy equals the rate of absorption. Hence a good absorber is a good emitter.*

ture of the absorbing body, but the temperature is observed *not* to change (once it has reached T_2), and we conclude that each body must *emit* radiant energy at the same rate as it *absorbs* it. Hence a good absorber is a good emitter, and a poor absorber is a poor emitter. But since each body must either absorb or reflect the radiant energy reaching it, a poor absorber must also be a good reflector. Hence a *good reflector* is a *poor emitter.*

This is the reason for silvering the walls of vacuum ("Thermos") bottles. A vacuum bottle is constructed with double glass walls, the space between the walls being evacuated so that heat flow by conduction and convection is practically eliminated. To reduce the radiant emission to as low a value as possible, the walls are covered with a coating of silver, which is highly reflecting and hence is a very poor emitter.

Since a good absorber is a good emitter, the *best* emitter is that surface which is the best absorber. But no surface can absorb more than all of the radiant energy which strikes it. Any surface which *does* absorb all of the incident energy is the best emitting surface possible. Such a surface would reflect no radiant energy, and hence would appear black in color (provided its temperature is not so high that it is self-luminous). It is called an *ideally black surface*, a body having such a surface is called an ideal blackbody, an ideal radiator, or simply a *blackbody.*

No actual surface is ideally black, the closest approach being lampblack, which reflects only about 1%. Blackbody conditions can be closely realized, however, by a small opening in the walls of a closed container. Radiant energy entering the opening is in part absorbed by the interior walls. Of the part reflected, only a very little escapes through the opening, the remainder being eventually absorbed by the walls. Hence the *opening* behaves like an ideal absorber.

Conversely, the radiant energy emitted by the walls or by any body within the enclosure, and escaping through the opening, is, if the walls are of uniform temperature, of the same nature as that emitted by an ideal radiator. This fact is of importance when using an optical pyrometer, described in Section 15–2. The readings of such an instrument are correct only when it is sighted on a blackbody. If used to measure the temperature of a red-hot ingot of iron in the open, its readings will be too low, since iron is a poorer emitter than a blackbody. If, however, the pyrometer is sighted on the iron while still in the furnace, where it is surrounded by walls at the same temperature, "blackbody conditions" are fulfilled and the reading is correct. The failure of the iron to emit as effectively as a blackbody is then just compensated for by the radiant energy which it reflects.

The emissivity e of an ideally black surface is equal to unity. For all real surfaces it is a fraction less than one.

Example   Assuming the total surface of the human body is 1.2 m^2 and the surface temperature is 30°C = 303 K, find the total rate of radiation of energy from the body.

Surprisingly, for infrared radiation the body is a very good approximation to an ideal blackbody, irrespective of skin pigmentation. The rate of energy loss per unit area is given by Eq. (17–4). Taking $e = 1$, we find

$$R = (1)(5.67 \times 10^{-8} \text{ W m}^{-2} \text{ K}^{-4})(303 \text{ K})^4$$

$$= 482 \text{ W m}^{-2}.$$

The total rate of energy loss is then

$$P = (482 \text{ W m}^{-2})(1.2 \text{ m}^2) = 578 \text{ W}.$$

Of course this loss is partially balanced by *absorption* of radiation, which depends on the temperature of the surroundings. The *net* rate of radiative energy transfer is given by Eq. (17–5).

PROBLEMS

17–1 A slab of a thermal insulator is 100 cm^2 in cross section and 2 cm thick. Its thermal conductivity is 0.1 J s^{-1} m^{-1} (C°)$^{-1}$. If the temperature difference between opposite faces is 100 C°, how much heat flows through the slab in one day?

17–2 Suppose that the rod in Fig. 17–1 is of copper, of length 10 cm and of cross-sectional area 1 cm^2. Let T_2 = 100°C and T_1 = 0°C.

a) What is the final steady-state temperature gradient along the rod?

b) What is the heat current in the rod, in the final steady state?

c) What is the final steady-state temperature at a point in the rod 2 cm from its left end?

17-3 A long rod, insulated to prevent heat losses, has one end immersed in boiling water (at atmospheric pressure) and the other end in a water–ice mixture. The rod consists of 100 cm of copper (one end in steam) and a length, L_2, of steel (one end in ice). Both rods are of cross-sectional area 5 cm^2. The temperature of the copper–iron junction is 60°C, after a steady state has been set up.

a) How much heat per second flows from the steam bath to the ice-water mixture?

b) How long is L_2?

17-4 A rod is initially at a uniform temperature of 0°C throughout. One end is kept at 0°C and the other is brought into contact with a steam bath at 100°C. The surface of the rod is insulated so that heat can flow only lengthwise along the rod. The cross-sectional area of the rod is 2 cm^2, its length is 100 cm, its thermal conductivity is 0.8 cgs units, its density is 10 g cm^{-3}, and its specific heat capacity is 0.10 cal gm^{-1} (C°)$^{-1}$. Consider a short cylindrical element of the rod 1 cm in length.

a) If the temperature gradient at one end of this element is 200 C° cm^{-1}, how many calories flow across this end per second?

b) If the average temperature of the element is increasing at the rate of 5 C°s^{-1}, what is the temperature gradient at the other end of the element?

17-5 One experimental method of measuring the thermal conductivity of an insulating material is to construct a box of the material and measure the power input to an electric heater, inside the box, which maintains the interior at a measured temperature above the outside surface. Suppose that in such an apparatus a power input of 120 W is required to keep the interior surface of the box 120 F° above the temperature of the outer surface. The total area of the box is 25 ft^2 and the wall thickness is 1.5 in. Find the thermal conductivity of the material in the commercial system of units.

17-6 A container of wall area 5000 cm^2 and thickness 2 cm is filled with water in which there is a stirrer. The outer surface of the walls is kept at a constant temperature of 0°C. The thermal conductivity of the walls is 0.000478 cgs

units, and the effect of edges and corners can be neglected. The power required to run the stirrer at an angular velocity of 1800 rpm is found to be 100 W. What will be the final steady-state temperature of the water in the container? Assume that the stirrer keeps the entire mass of water at a uniform temperature.

17-7 A boiler with a steel bottom 1.5 cm thick rests on a hot stove. The area of the bottom of the boiler is 1500 cm^2. The water inside the boiler is at 100°C, and 750 g are evaporated every 5 min. Find the temperature of the lower surface of the boiler, which is in contact with the stove.

17-8 A camping icebox, having wall area of 2 m^2 and thickness 5 cm, is constructed of insulating material having a thermal conductivity of 0.05 J s^{-1} m^{-1} (C°)$^{-1}$. The outside temperature is 20° C, and the inside of the box is to be maintained at 5° C by ice. The melted ice leaves the box at a temperature of 15° C. If ice costs 1 cent per kilogram, what will it cost to run the icebox for 1 hr?

17-9 Rods of copper, brass, and steel are welded together to form a Y-shaped figure. The cross-sectional area of each rod is 2 cm^2. The end of the copper rod is maintained at 100°C and the ends of the brass and steel rods at 0°C. Assume there is no heat loss from the surface of the rods. The lengths of the rods are: copper, 46 cm; brass, 13 cm; steel, 12 cm.

a) What is the temperature of the junction point?

b) What is the heat current in the copper rod?

17-10 A compound bar 2 meters long is constructed of a solid steel core 1 cm in diameter surrounded by a copper casing whose outside diameter is 2 cm. The outer surface of the bar is thermally insulated and one end is maintained at 100°C, the other at 0°C.

a) Find the total heat current in the bar.

b) What fraction is carried by each material?

17-11 Heat flows radially outward through a cylindrical insulator of outside radius R_2 surrounding a steam pipe of outside radius R_1. The temperature of the inner surface of the insulator is T_1, that of the outer surface is T_2. (a) At what radial distance from the center of the pipe is the steady-state temperature exactly halfway between temperatures T_1 and T_2? (b) Sketch a graph of T versus r.

17-12 A steam pipe 2 cm in radius is surrounded by a cylindrical jacket of insulating material 2 cm thick. The temperature of the steam pipe is 100° C, and that of the outer surface of the jacket is 20° C. The thermal conductivity of the insulating material is $0.1 \text{ J s}^{-1} \text{ m}^{-1}(\text{C}°)^{-1}$. Compute the temperature gradient, dT/dr, at the inner and outer surfaces of the jacket, and sketch the graph of T versus r.

17–13 A steam pipe of radius 2 cm, carrying steam at 120° C, is surrounded by a cylindrical jacket of cork with inner and outer radii 2 cm and 4 cm, and this in turn is surrounded by a cylindrical jacket of styrofoam having inner and outer radii 4 cm and 6 cm. The thermal conductivity of styrofoam is about $0.01 \text{ J s}^{-1} \text{ m}^{-1} (\text{C}°)^{-1}$ The outer surface of the styrofoam is in contact with air at 20° C.

a) What is the temperature at a radius of 4 cm, where the two insulating layers meet?

b) What is the total rate of transfer of heat out of a 2-m length of pipe?

17–14 A spherical shell has inner and outer radii a and b, respectively, and the temperatures at the inner and outer surfaces are T_a and T_b.

a) Find the temperature in the shell as a function of radius.

b) Find the total rate of heat transfer through the shell.

17–15 An electric transformer is in a cylindrical tank 60 cm in diameter and 1 m high, with flat top and bottom. If the tank transfers heat to the air only by natural convection, and the electrical losses are to be dissipated at the rate of 1 kw, how many degrees will the tank surface rise above room temperature?

17–16

a) What would be the difference in height between the columns in the U-tube in Fig. 17–4 if the liquid is water and the left arm is 1 m high at 4°C while the other is at 75°C?

b) What is the difference between the pressures at the foot of two columns of water each 10 m high if the temperature of one is 4°C and that of the other is 75°C?

17–17 A flat wall is maintained at constant temperature of 100°C, and the air on both sides is at atmospheric pressure and at 20°C. How much heat is lost by natural convection from 1 m^2 of wall (both sides) in 1 hr if (a) the wall is vertical and (b) the wall is horizontal?

17–18 A vertical steam pipe of outside diameter 7.5 cm and height 4 m has its outer surface at the constant temperature of 95°C. The surrounding air is at atmospheric pressure and at 20°C. How much heat is delivered to the air by natural convection in 1 hr?

17–19 What is the radiant emittance of a blackbody at a temperature of (a) 300 K, (b) 3000 K?

17–20 The radiant emittance of tungsten is approximately 0.35 that of a blackbody at the same temperature. A tungsten sphere 1 cm in radius is suspended within a large evacuated enclosure whose walls are at 300 K. What power input is required to maintain the sphere at a temperature of 3000 K if heat conduction along the supports is neglected?

17–21 A small blackened ~~solid copper~~ sphere of radius 2 cm is placed in an evacuated enclosure whose walls are kept at 100°C. At what rate must energy be supplied to the sphere to keep its temperature constant at 127°C?

17–22 A cylindrical metal can 10 cm high and 5 cm in diameter contains liquid helium at 4K, at which temperature its heat of vaporization is 5 cal g^{-1}. Completely surrounding the helium can are walls maintained at the temperature of liquid nitrogen, 80 K, the intervening space being evacuated. How much helium is lost per hour? Assume the radiant emittance of the helium can to be 0.2 that of a blackbody at 4 K.

17–23 A solid cylindrical copper rod 10 cm long has one end maintained at a temperature of 20.00 K. The other end is blackened and exposed to thermal radiation from a body at 300 K, no energy being lost or gained elsewhere. When equilibrium is reached, what is the temperature of the blackened end? [*Hint.* Since copper is a very good conductor of heat at low temperature, $k = 4 \text{ cal s}^{-1} \text{ cm}^{-1} (\text{C}°)^{-1}$, the temperature of the blackened end is only slightly greater than 20 K.]

17–24 The operating temperature of a tungsten filament in an incandescent lamp is 2450 K and its emissivity is 0.30. Find the surface area of the filament of a 25-watt lamp.

17–25 A certain incandescent lightbulb operates at 3000 K. The total surface area of the filament is 0.05 cm^2, and the emissivity is 0.3. What electric power must be supplied to the filament?

17–26 The rate at which radiant energy reaches the surface of the earth from the sun is about 1.4 kW m^{-2}. The distance from earth to sun is about 1.5×10^{11} m, and the radius of the sun is about 0.7×10^9 m.

a) What is the rate of radiation of energy, per unit area, from the sun's surface?

b)   If the sun radiates as an ideal blackbody, what is the temperature of its surface?

17–27 Suppose that both ends of the rod in Fig. 17–1 are kept at a temperature of 0° C, and that the initial temperature distribution along the rod is given by $T = 100 \sin \pi x/L$, where T is in °C. Let the rod be of copper, of length $L = 10$ cm and of cross section 1 cm².

a)   Show the initial temperature distribution in a diagram.

b)   What is the final temperature distribution after a very long time has elapsed?

c)   Sketch curves which you think would represent the temperature distribution at intermediate times.

d)   What is the initial temperature gradient at the ends of the rod?

e)   What is the initial heat current from the ends of the rod into the bodies making contact with its ends?

f)   What is the initial heat current at the center of the rod? Explain. What is the heat current at this point at any later time?

g)   What is the value of $k/\rho C$ for copper, and in what unit is it expressed? (This quantity is called the *diffusivity*.)

h)   What is the initial rate of change of temperature at the center of the rod?

i)   How long a time would be required for the rod to reach its final temperature, if the temperature continued to decrease at this rate? (This time can be described as the *relaxation time* of the rod.)

j)   From the graphs in part (c), would you expect the rate of change of temperature at the midpoint to remain constant, increase, or decrease?

k)   What is the initial rate of change of temperature at a point in the rod, 2.5 cm from its left end?

illumination

$$I_s = \frac{I}{d_o^2}$$

I_s : illumination of surface   ft. candles

I : illumination of source   candles

d^2 : distance between

I

$$\frac{I}{d^2} = \frac{I_s}{d^2} = \frac{I_s}{d^2}$$

Chapter 18

Thermal Properties of Matter

18–1 EQUATIONS OF STATE

The volume V occupied by a definite mass m of any substance depends on the pressure p to which the substance is subjected, and on its temperature T. For every pure substance there is a definite relation between these quantities, called the *equation* of state of the substance. In formal mathematical language, the equation can be written

$$f(m, V, p, T) = 0.$$

The exact form of the function is usually very complicated. It often suffices to know only how some one of the quantities changes when some other is varied, the rest being kept constant. Thus the compressibility k describes the change in volume when the pressure is changed, for a constant mass at a constant temperature, and the coefficient of volume expansion gives the change in volume when the temperature is changed, for a constant mass at a constant pressure.

The term "state" as used here implies an *equilibrium* state. This means that the temperature and pressure are the same at all points. Hence if heat is added at some point to a system in an equilibrium state, we must wait until the processes of heat transfer within the system have brought about a new uniform temperature before the system is again in an equilibrium state.

18–2 THE IDEAL GAS

The simplest equation of state is that of a gas at low pressure. Consider a container whose volume can be varied, such as a cylinder provided with a movable piston. Provision is made for pumping any desired mass of any kind of gas into or out of the cylinder, and the cylinder is provided with a pressure gauge and with a thermometer for determining the Kelvin (absolute) temperature T. Then corresponding values of m, p, V, and T can all be measured. Instead of the mass m, let us express the results in terms of the number of moles, n. Since the molecular mass M is the mass per mole, the total mass m is given by

$$m = nM.$$

In calculations with gases, the number of moles is often the most convenient way to specify the quantity of material.

From measurements of the pressure, volume, temperature, and number of moles, a number of conclusions emerge. First, the volume is proportional to the number of moles. If we double the quantity of gas, keeping pressure and temperature constant, the volume doubles. Second, the volume varies inversely with pressure; if we double the pressure, holding the temperature and quantity of material constant, the gas is compressed to one-half its initial volume. Third, the pressure is proportional to the absolute temperature; if we double the absolute temperature, holding the volume and quantity of material constant, the pressure doubles.

These conclusions can be summarized neatly in a single equation of state:

$$pV = nRT. \qquad (18-1)$$

The constant of proportionality R might be expected to have different values for different gases, but instead it turns out to have the same value for *all* gases, at least at sufficiently high temperature and low pressure. This quantity is called the *universal gas constant*. The numerical value of R depends, of course, on the units in which p, V, n, and T are expressed. The adjective "universal" means that in any one system of units R has the same value for *all* gases. In the mks system, where the unit of p is 1 Pa or 1 N m^{-2} and the unit of V is 1 m^3, the numerical value of R is found to be

$$R = 8.314 \,(\text{N m}^{-2})\text{m}^3\,\text{mol}^{-1}\,\text{K}^{-1}$$
$$= 8.314 \,\text{J mol}^{-1}\,\text{K}^{-1}.$$

In the cgs system, where the unit of p is 1 dyn cm^{-2} and the unit of V is 1 cm^3,

$$R = 8.314 \times 10^7 \,(\text{dyn cm}^{-2})\,\text{cm}^3\,\text{mol}^{-1}\,\text{K}^{-1}$$
$$= 8.314 \times 10^7 \,\text{erg mol}^{-1}\,\text{K}^{-1}.$$

We note that the units of pressure times volume are the same as units of energy, so, in *all* systems of units, R has units of energy per mole, per unit of absolute temperature. In terms of calories,

$$R = 1.99 \,\text{cal mol}^{-1}\,\text{K}^{-1}.$$

In chemical calculations volumes are commonly expressed in liters (ℓ), pressures in atmospheres, and temperatures in kelvins. In this system,

$$R = 0.08207 \,\ell\,\text{atm mol}^{-1}\,\text{K}^{-1}.$$

We now define an *ideal gas* as one for which Eq. (18–1) holds precisely for *all* pressures and temperatures. As the term suggests, the ideal gas is an idealized model which represents the behavior of gases very well in some circumstances, less well in others. Generally, gas behavior approximates the ideal-gas model most closely at very low pressures, when the gas molecules are far apart. However, the deviations are not very great at moderate pressures and at temperatures not too near those at which the gas liquefies.

Deviations from ideal-gas behavior are shown in Fig. 18–1, where the ratio pV/RT is plotted as a function of p and T. For an ideal gas this quantity is constant, but for real gases it varies increasingly at lower and lower temperatures. At sufficiently high temperature and low pressure this ratio approaches the value R for an ideal gas.

It follows from Eq. (18–1), and the value of the universal gas constant R, that one mole of an ideal gas occupies a volume of 0.0224 m^3 or 22.4 ℓ at "standard conditions," or at "normal temperature and pressure" (NTP), that is, at a temperature of 0°C = 273 K, and a pressure of 1 atm = 1.01 × 10^{5}N m^{-2}. Thus, from Eq. (18–1),

$$V = \frac{nRT}{p}$$
$$= \frac{(1 \text{ mol})(8.31 \text{ J mol}^{-1}\,\text{K}^{-1})(273 \text{ K})}{1.01 \times 10^5\,\text{N m}^{-2}}$$
$$= 0.0224 \text{ m}^3 = 22,400 \text{ cm}^3 = 22.4\,\ell.$$

Fig. 18–1 *The limiting value of pV/nT is independent of T for all gases. For an ideal gas, pV/nT is constant.*

For a *fixed mass* (or fixed number of moles) of an ideal gas, the product nR is constant and hence pV/T is constant also. Thus if the subscripts 1 and 2 refer to two states of the same mass of a gas, but at different pressures, volumes, and temperatures,

$$\frac{p_1 V_1}{T_1} = \frac{p_2 V_2}{T_2} = \text{constant.} \qquad (18\text{–}2)$$

If the temperatures T_1 and T_2 are the same, then

$$p_1 V_1 = p_2 V_2 = \text{constant.} \qquad (18\text{–}3)$$

The fact that at constant temperature the product of the pressure and volume of a fixed mass of gas is very nearly constant was discovered experimentally by Robert Boyle in 1660, and Eq. (18–3) is called *Boyle's law*. Although exactly true (by definition) for an *ideal gas*, it is obeyed only approximately by real gases and is not a fundamental law like Newton's laws or the law of conservation of energy.

Example 1 The volume of an oxygen tank is 50 ℓ. As oxygen is withdrawn from the tank, the reading of a pressure gauge drops from 300 lb in^{-2} to 100 lb in^{-2}, and the temperature of the gas remaining in the tank drops from 30° C to 10° C.

a) How many kilograms of oxygen were in the tank originally?

b) How many kilograms were withdrawn?

c) What volume would be occupied by the oxygen withdrawn from the tank at a pressure of 1 atm and a temperature of 20° C?

a) Let us express pressures in pascals, volumes in cubic meters, and temperatures in kelvins. The pressure conversion can be carried out by use of the relations

$$1 \text{ atm} = 14.7 \text{ lb in}^{-2} = 1.01 \times 10^5 \text{ Pa.}$$

A *gauge* pressure of 300 lb in^{-2} is a total pressure of $(300 + 14.7)$ lb in^{-2}, and we have

$$314.7 \text{ lb in}^{-2} = \frac{314.7}{14.7} \text{ atm} = 21.4 \text{ atm}$$

$$= (21.4)(1.01 \times 10^5 \text{ Pa})$$

$$= 2.17 \times 10^6 \text{ Pa.}$$

Similarly, 100 lb in^{-2} (gauge) $= 7.90 \times 10^5$ Pa. We also have 50 ℓ = 0.05 m^3, 30° C = 303 K, and 10° C = 283 K.

The initial number of moles is

$$n_1 = \frac{p_1 V}{RT_1} = \frac{(2.17 \times 10^6 \text{ Pa})(0.05 \text{ m}^3)}{(8.314 \text{ J mol}^{-1} \text{ K}^{-1})(303 \text{ K})}$$

$$= 43.2 \text{ mol.}$$

The original mass was therefore

$$m_1 = (43.1 \text{ mol})(32 \text{ g mol}^{-1})$$

$$= 1378 \text{ g} = 1.378 \text{ kg.}$$

b) The number of moles remaining in the tank is

$$n_2 = \frac{p_2 V}{RT_2} = \frac{(7.90 \times 10^5 \text{ Pa})(0.05 \text{ m}^3)}{(8.314 \text{ J mol}^{-1} \text{ K}^{-1})(283 \text{ K})}$$

$$= 16.8 \text{ mol,}$$

and the mass remaining is

$$(16.8 \text{ mol})(32 \text{ g mol}^{-1}) = 537 \text{ g}$$

$$= 0.537 \text{ kg.}$$

The mass withdrawn is, therefore,

$$1.378 \text{ kg} - 0.537 \text{ kg} = 0.841 \text{ kg.}$$

c) The number of moles withdrawn is

$$43.1 - 16.8 = 26.3,$$

and the volume occupied would be

$$V = \frac{nRT}{p}$$

$$= \frac{(26.3 \text{ mol})(8.314 \text{ J mol}^{-1} \text{ K}^{-1})(293 \text{ K})}{1.01 \times 10^5 \text{ Pa}}$$

$$= 0.634 \text{ m}^3.$$

Example 2 The *McLeod gauge* (pronounced "McLoud"), illustrated in Fig. 18–2, can be used to measure pressures as low as 5×10^{-6} mm of mercury. In part (a), the entire space above point A is occupied by the gas at the low pressure p which is to be measured. When the mercury container B is raised as in (b), the gas in bulb C, whose volume V might be 500 cm^3, is trapped and eventually compressed into a much smaller volume V' above a reference mark on the capillary tube D. Assuming the temperature

Fig. 18–2 The McLeod gauge.

constant, the pressure p' of the compressed gas is given by applying Boyle's law,

$$p' = \frac{pV}{V'}.$$

The pressure at the upper surface of the mercury in capillary E remains at the value p, so that if h is the difference in elevation between the tops of the mercury columns in E and D,

$$p' = p + \rho g h,$$

where ρ is the density of mercury. Elimination of p' between these equations gives

$$p = \frac{\rho g V'}{V - V'} h \approx \frac{V'}{V} \rho g h,$$

to a good approximation, since $V' \ll V$.

For example, if $V'/V = 10^{-4}$ and $h = 4$ mm, $p = 4 \times 10^{-4}$ torr.

Since ρ, g, V, and V' are constants, the pressure p is directly proportional to h, and a uniform pressure scale can be mounted beside tube E. Corrections for capillary depression (due to surface tension) are eliminated if capillaries D and E have the same diameter.

Example 3   In the upper part of the atmosphere (the stratosphere) the temperature varies only slightly with changes in elevation. Find the law of variation of pressure with elevation.

The rate of change of pressure with elevation in a fluid is $dp/dy = -\rho g$.

Let us replace n by m/M in the ideal gas law, where m is the mass and M the molecular weight. Then

$$pV = \frac{m}{M} RT,$$

and the density ρ of an ideal gas is

$$\rho = \frac{m}{V} = \frac{pM}{RT}.$$

Hence

$$\frac{dp}{dy} = -\frac{pMg}{RT}, \quad \frac{dp}{p} = -\frac{Mg}{RT} dy.$$

If g and T are constant, and if p_1 and p_2 are the pressures at two elevations y_1 and y_2, integration of the preceding equation gives

$$\ln \frac{p_2}{p_1} = -\frac{Mg}{RT} (y_2 - y_1). \tag{18–4}$$

This is known as the *barometric equation*.

18–3 pVT-SURFACE FOR AN IDEAL GAS

Since the equation of state for a fixed mass of a substance is a relation among the three variables p, V, and T, it defines a *surface* in a rectangular coordinate system in which p, V, and T are plotted along the three axes. Figure 18–3 shows the pVT-surface of an ideal gas. The solid lines on the surface show the relation between p and V when T is constant (Boyle's law), the dashed lines the relation between V and T when p is constant (Gay-Lussac's law), and the dotted lines the relation between p and T when V is constant. When viewed perpendicular to the pV-plane, the surface appears as in Fig. 18–4(a), and Fig. 18–4(b) is its appearance when viewed perpendicular to the pT-plane.

Every possible equilibrium state of a given quantity of gas corresponds to a point on the surface,

Fig. 18–3 *pVT-surface for an ideal gas.*

and every point of the surface corresponds to a possible equilibrium state. The gas cannot exist in a state that is not on the surface. For example, if the volume and temperature are given, thus locating a point in the base plane of Fig. 18–3, the pressure is then determined by the nature of the gas, and it can have only the value represented by the height of the surface above this point.

In any process in which the gas passes through a succession of equilibrium states, the point representing its state moves along a curve lying in the

pVT-surface. Such a process must be carried out very slowly to give the temperature and pressure time to remain uniform at all points of the gas.

18–4 *pVT*-SURFACE FOR A REAL SUBSTANCE

While all real substances approximate ideal gases at sufficiently low pressures, their behavior departs more and more from that of an ideal gas at high pressures and low temperatures. As the temperature is lowered and the pressure increased, all substances change from the *gas phase* to the *liquid phase* or the *solid phase*. For a fixed mass of a substance, however, there is still a definite relation between the pressure, temperature, and total volume. In other words, the substance has an equation of state in *any* circumstances, and although the general form of the equation is much too complicated to express mathematically, we can represent it *graphically* by a *pVT*-surface. Figure 18–5 is a schematic diagram (to a greatly distorted scale) of the *pVT*-surface of a substance that expands on melting (the most common case) and Fig. 18–6 is the corresponding diagram for a substance which, like water, contracts on melting. We see that the substance can exist in either the solid, liquid, or gas phase, or in two phases simultaneously, or, along the triple line, in all three phases. (The distinction between the terms "gas" and "vapor" will be explained later. For the present they may be considered synonymous.)

(a)

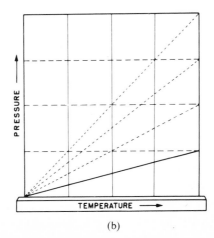

(b)

Fig. 18–4 *Projections of ideal gas pVT-surface on (a) the pV-plane, (b) the pT-plane.*

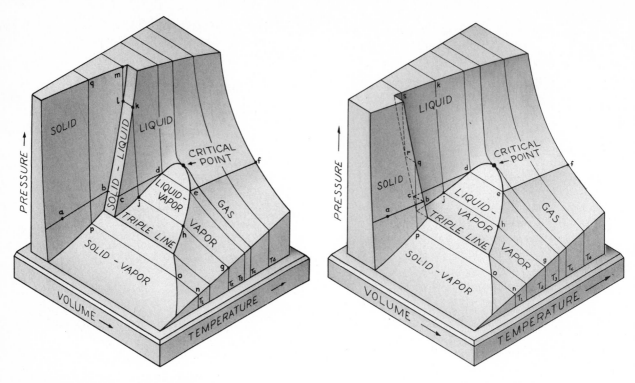

Fig. 18–5 *pVT-surface for a substance that expands on melting.*

Fig. 18–6 *pVT-surface for a substance that contracts on melting.*

In order that the diagram shall represent the properties of a particular *substance*, but not depend on *how much* of the substance is present, we plot along the volume axis not the actual volume V but the *specific volume* v, the volume per unit mass. Thus for a system of mass m,

$$v = \frac{V}{m}.$$

The actual volume of a unit mass of a substance is numerically equal to its specific volume. The specific volume is the reciprocal of the density ρ,

$$\rho = \frac{m}{V} = \frac{1}{v}.$$

To correlate these diagrams with our familiar experiences regarding the behavior of solids, liquids, and gases, let us start with a substance in the solid phase at point a in Fig. 18–5 or 18–6. Let the

substance be contained in a cylinder as in Fig. 18–7, and let a constant force F be exerted on the piston so that the pressure remains constant as the substance expands or contracts. A constant-pressure process is called an *isobaric* process, and lines representing constant-pressure processes are called *isobars*. We now bring the cylinder in contact with some body such as an electric stove, whose temperature can be slowly increased and kept slightly higher than that of the substance, so that heat flows into the substance.

At the start of the process, the temperature of the substance rises as heat is added, at a rate determined by the specific heat capacity of the solid, and the volume increases slightly at a rate determined by its coefficient of expansion. When point b is reached in either diagram the substance starts to melt, that is, to change from the solid to the liquid phase. The temperature ceases to rise even though heat is being continuously supplied. The volume increases in Fig. 18–5 and decreases in Fig. 18–6.

When point *c* is reached, the substance is wholly in the liquid phase. The temperature now starts to rise again, at a rate determined by the specific heat capacity of the liquid, and the volume increases at a rate determined by its coefficient of expansion. (For water at atmospheric pressure, there would be a small *decrease* in volume at first.)

When point *d* is reached, the temperature again ceases to rise, although heat is still supplied, and bubbles of vapor start to form in the cylinder and rise to its upper surface. Since the density of the vapor phase is smaller than that of the liquid, the volume increases.

At point *e*, the substance is wholly in the vapor phase. (The relative volume increase from *d* to *e*, except at very high pressures, is enormously greater than the volume change between liquid and solid, so Figs. 18–5 and 18–6 are not constructed on a uniform scale.) With further addition of heat the temperature rises again, the rate now being determined by the specific heat capacity of the vapor phase (at constant pressure). The volume also increases much more rapidly than did that of the solid or liquid.

If the cylinder is now brought in contact with a body whose temperature is kept slightly *lower* than that of the substance, heat flows out of the substance and all the changes in the original process take place in reversed order.

As a second illustration to show how the *pVT*-surface describes the behavior of a substance, suppose we start with a cylinder containing a gas, at a pressure, volume, and temperature corresponding to point *g* in Fig. 18–5 or 18–6. Let the external pressure be adjusted so that it is always slightly greater than the pressure exerted by the substance, and let the cylinder be kept in contact with a body at a constant

Fig. 18–7

Fig. 18–8 *Portion of the pVT-surface of water at high pressure.*

temperature. A constant-temperature process is called *isothermal*, and lines representing constant-temperature processes are called *isotherms*.

As the pressure increases, the volume decreases along the line *gh*; simultaneously, heat flows from the gas to the body with which it is in contact; otherwise the temperature of the gas would rise. When point *h* is reached, drops of liquid begin to form in the cylinder, and the volume continues to decrease without further increase of pressure.

At point *j*, all of the substance has condensed to the liquid phase. With further increase of pressure the volume decreases, but only slightly because of the small compressibility of liquids. For a substance like that in Fig. 18–6, no other change of phase occurs as the pressure is increased to point *k* and beyond (unless other forms of the solid state exist; see Fig. 18–8 and the accompanying discussion). In Fig. 18–5, however, another break in the curve takes place at point *k*. Crystals of the solid begin to appear, and the volume again decreases without an increase of external pressure. At point ℓ, in Fig. 18–5, the substance has been completely converted to the solid phase. Further increase of pressure reduces the volume of the solid only slightly, and unless the substance can exist in more than one modification of the solid phase, no further changes in phase result.

As a final example, suppose we again start with a gas in the cylinder but at a lower temperature, corresponding to point n in Figs. 18–5 and 18–6. If the pressure is increased isothermally, crystals of the solid begin to appear at point o and the gas changes directly to a solid without passing through the liquid phase. The pressure remains constant along the line op, and at point p the substance is all in the solid phase. In Fig. 18–5 no further phase change takes place but in Fig. 18–6 the solid starts to melt at point q and has completely melted at point r.

In all the processes described above, heat must be removed or added continuously in order to keep the pressure or temperature constant. If at any stage of a process the system is thermally insulated so that there can be no flow of heat in or out, and if the external pressure is kept constant, the system remains in equilibrium. Thus at any point on the surfaces lettered solid–liquid, solid–vapor, or liquid–vapor, *two* phases can coexist in equilibrium, and along the *triple line* all three phases can coexist. A vapor at the pressure and temperature at which it can exist in equilibrium with its liquid is called a *saturated vapor*, and the liquid is called a *saturated liquid*. Thus points e and h represent saturated vapor, and points j and d, saturated liquid. (The term "saturated" is poorly chosen. It does not have the same meaning as a "saturated solution" in chemistry. There is no question here of one substance being dissolved in another.)

18–5 CRITICAL POINT AND TRIPLE POINT

A study of Figs. 18–5 and 18–6 will show that the liquid and gas (or vapor) phases can exist together only if the temperature and pressure are less than those at the point lying at the top of the tongue-shaped surface lettered liquid–vapor. This point is called the *critical point*, and the corresponding values

Table 18–1 CRITICAL CONSTANTS

Substance	Critical temperature, K	Critical pressure, atm	Critical volume, $m^3\ mol^{-1}$	Critical density, $kg\ m^{-3}$
Helium (4)	5.3	2.26	0.057	69.3
Helium (3)	3.34	1.15	0.0726	41.3
Hydrogen (normal)	33.3	12.80	0.065	31.0
Deuterium (normal)	38.4	16.4	0.0603	66.3
Nitrogen	126.2	33.5	0.0901	311
Oxygen	154.8	50.1	0.078	410
Ammonia	405.5	111.3	0.0725	235
Freon 12	384.7	39.6	0.218	555
Carbon dioxide	304.2	72.9	0.094	468
Sulfur dioxide	430.7	77.8	0.122	524
Water	647.4	218.3	0.056	320
Carbon disulfide	552	78	0.170	440

of T, p, and v are the *critical temperature, pressure,* and *specific volume.* A gas at a temperature above the critical temperature, such as T_4, does not separate into two phases when compressed isothermally, but its properties change gradually and continuously from those we ordinarily associate with a gas (low density, large compressibility) to those of a liquid (high density, small compressibility). Table 18–1 lists the critical constants for a few substances. The very low critical temperatures of hydrogen and helium make it evident why these gases defied attempts to liquefy them for many years.

The term *vapor* is sometimes used to mean a gas at any temperature below its critical temperature, and sometimes is restricted to mean a gas in equilibrium with the liquid phase, that is, a saturated vapor. The term is really unnecessary; no sudden change takes place in the properties of a substance when the critical isotherm is crossed either on the portion of the surface "gas and vapor," or on the portion "liquid."

Many substances can exist in more than one modification of the solid phase. Transitions from one modification to another occur at definite temperatures and pressures, like the phase changes from liquid to solid, etc. Water is one such substance, and at least eight types of ice have been observed at very high pressures Figure 18–8 shows a portion of the *pVT*-surface of water at high pressure. Note that ordinary ice (ice I) is the only form whose specific volume is greater than that of the liquid phase.

Because of the difficulty of drawing three-dimensional diagrams, it is customary to represent the *pVT*-surface by its projections onto the *pT*- and *pV*-planes. Figure 18–9 shows the two projections of Fig. 18–5, and Fig. 18–10 shows those of Fig. 18–6. The reader should follow through on these diagrams the isobaric and isothermal processes indicated in Figs. 18–5 and 18–6. The *pT* projection exhibits most clearly the ranges of temperature and pressure in which each phase is stable, and it is often called a *phase diagram*.

The curves in Figs. 18–9(a) and 18–10(b) are loci of corresponding values of pressure and temperature at which the two phases can coexist if a substance is isolated, or at which one phase will transform to the other if heat is supplied or removed. Thus the S-L curve is also a graph of the melting-point temperature or freezing-point temperature of the substance as a function of pressure, the curve S-V is a graph of the *sublimation point* versus *pressure*, and the curve L-V is a graph of the *boiling point* versus *pressure*. The S-V and L-V curves always slope upward to the right. The S-L curve slopes upward to the right for a substance that expands on melting (Fig. 18–9) but upward to the left for a substance such as water that contracts on melting (Fig. 18–10). Thus an increase

(a)

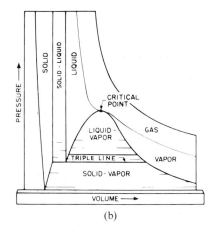

(b)

Fig. 18–9 *Projections of the surface in Fig. 18–5 on the pT- and pV-planes.*

(a)

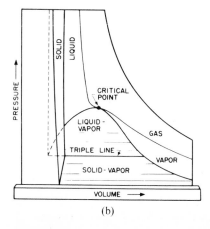

(b)

Fig. 18–10 *Projections of the surface in Fig. 18–6 on the pT- and pV-planes* of pressure always increases the temperature of the sublimation point or boiling point, but the temperature of the freezing point may be raised (Fig. 18–9) or lowered (Fig. 18–10) by an increase in pressure.

The pressure of a vapor in equilibrium with the liquid or solid at any temperature is called the *vapor pressure* of the substance at that temperature. Thus the curves S-V and L-V in Figs. 18–9(a) and 18–10(a) are graphs of vapor pressure versus temperature. The vapor pressure of a substance is a function of *temperature only*, not of volume. That is, in a vessel containing a liquid (or solid) and vapor in equilibrium at a *fixed temperature*, the pressure does not depend on the relative amounts of liquid and vapor present. If the volume is decreased some of the vapor condenses, and vice versa, but if the temperature is kept constant by removing or adding heat the pressure does not change.

The boiling-point temperature of a liquid is the temperature at which its vapor pressure equals the external pressure. Table 18–2 gives the vapor pressure of water as a function of temperature; we see that the vapor pressure is 1 atm at a temperature of 100°C. If the external pressure is reduced to 17.5 mm of mercury, water will boil at room temperature (20°C), while under a pressure of 90 lb in^{-2} (about 6 atm) the boiling point is 160°C.

Table 18–2 VAPOR PRESSURE OF WATER

T, °C	Vapor pressure, torr (mm Hg)	Vapor pressure, lb in^{-2}	T, °F
0	4.58	0.0886	32
5	6.51	0.126	41
10	8.94	0.173	50
15	12.67	0.245	59
20	17.5	0.339	68
40	55.1	1.07	104
60	149	2.89	140
80	355	6.87	176
100	**760**	**14.7**	**212**
120	1490	28.8	248
140	2710	52.4	284
160	4630	89.6	320
180	7510	145	356
200	11650	225	392
220	17390	336	428

Table 18–3 TRIPLE-POINT DATA

Substance	Temperature, K	Pressure, torr (mm Hg)
Helium (4) (λ point)	2.186	38.3
Helium (3)	None	None
Hydrogen (normal)	13.84	52.8
Deuterium (normal)	18.63	128
Neon	24.57	324
Nitrogen	63.18	94
Oxygen	54.36	1.14
Ammonia	195.40	45.57
Carbon dioxide	216.55	3880
Sulfur dioxide	197.68	1.256
Water	**273.16**	**4.58**

The point of intersection of the three equilibrium lines in Figs. 18–9(a) and 18–10(a), which is an end view of the triple line in Figs. 18–9(b) and 18–10(b), is called the *triple point*. There is only one pressure and temperature at which all three phases can coexist. Triple-point data for a few substances are given in Table 18–3.

As numerical examples, consider the pT-diagrams of water and carbon dioxide, in Fig. 18–11. In (a), a horizontal line at a pressure of 1 atm intersects the freezing-point curve at 0°C and the boiling-point curve at 100°C. The boiling point increases with increasing pressure up to the critical temperature of 374°C. Solid, liquid, and vapor can remain in equilibrium only at the triple point, where the vapor pressure is 4.5 mm of mercury and the temperature is 0.01°C.

The freezing point of a substance such as water, which expands on solidifying, is *lowered* by an increase in pressure. The reverse is true for substances which contract on solidifying. The change in the freezing-point temperature is much smaller than is that of the boiling point; an increase of 1 atm lowers the freezing point of water by only about 0.007°C.

The lowering of the freezing point of water (or the melting point of ice) can be demonstrated by passing a loop of fine wire over a block of ice and hanging a weight of a few pounds from each end of the loop. Suppose that the main body of the ice is at 0°C and at atmospheric pressure. The temperature of the small amount of ice directly under the wire

(a) Water

(b) Carbon dioxide

Fig. 18–11 *Pressure-temperature diagrams (not to a uniform scale).*

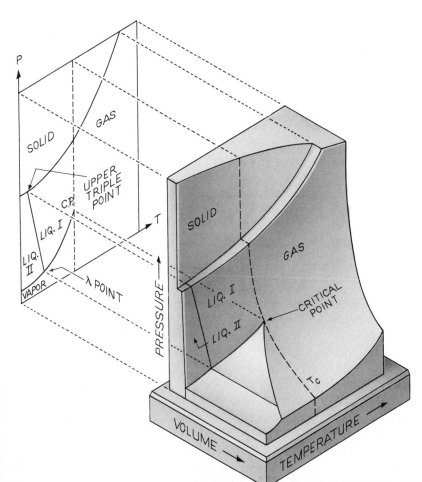

Fig. 18–12 *pVT-surface for helium, and p-T phase diagram.*

decreases until it achieves the melting point appropriate to the pressure under the wire. During this increase of pressure and decrease of temperature, a small amount of melting takes place. The water thus formed is squeezed out from under the wire and coming to the top of the wire where the pressure is atmospheric, it refreezes and liberates heat which passes through the wire and serves to melt the next bit of ice below the wire.

The wire thus sinks farther and farther into the block, eventually cutting its way completely through, but leaving a solid block of ice behind it. This phenomenon is known as *regelation* (refreezing). Since heat is conducted from the top to the bottom of the wire while the wire is cutting through the ice, the greater the thermal conductivity of the wire, the faster will the wire cut through the ice. Even a perfectly conducting wire would not cut through the ice very rapidly, however, because of the very low thermal conductivity of the water film which is always present beneath the wire.

For carbon dioxide, the triple-point temperature is $-56.6°C$ and the corresponding pressure is 5.11 atm. Hence at atmospheric pressure CO_2 can exist only as a solid or vapor. When heat is supplied to solid CO_2, open to the atmosphere, it changes directly to a vapor, without passing through the liquid phase, whence the name "dry ice." Liquid CO_2 can exist only at a pressure greater than 5.11 atm. The steel tanks in which CO_2 is commonly stored contain liquid and vapor (both saturated). The pressure in these tanks is the vapor pressure of CO_2 at the temperature of the tank. If the latter is 20°C the vapor pressure is 56 atm or about 830 lb in^{-2}.

The *pVT*-surface for ordinary helium (of mass number 4) is shown in Fig. 18–12. Two remarkable properties may be seen. First, helium has no triple point at which solid, liquid, and gas coexist in equilibrium. Instead, it has two triple points, the lower one (called the "lambda point") representing the temperature and pressure at which two liquid phases I and II coexist with vapor, and the upper one at which the two liquid phases coexist with solid. The existence of two liquid phases is unique to helium; II is the *superfluid* phase, which has a variety of unusual properties, including very low viscosity, very large thermal conductivity, and the ability of the material

to form a surface film which creeps up the sides of a container. (The isotope of helium with mass number 3 exhibits no superfluid phase.)

Second, as the temperature is lowered, helium does not solidify, but remains liquid all the way to absolute zero. To obtain solid helium, the pressure must be raised to at least 25 atm, at which the helium atoms can coalesce into a crystal lattice.

18–6 EFFECT OF DISSOLVED SUBSTANCES ON FREEZING AND BOILING POINTS

The freezing point of a liquid is lowered when some other substance is dissolved in the liquid. A common example is the use of an "antifreeze" to lower the freezing point of the water in the radiator of an automobile engine.

The freezing point of a saturated solution of common salt in water is about $-20°C$. To understand why a mixture of ice and salt may be used as a freezing mixture we recall that the freezing point is the only temperature at which the liquid and solid phases can exist in equilibrium. When a saturated salt solution is cooled, it freezes at $-20°C$, and crystals of ice (pure H_2O) separate from the solution. In other words, ice crystals and a salt solution can exist in equilibrium only at $-20°C$, just as ice crystals and pure water can exist together only at 0°C.

When ice at 0°C is mixed with a salt solution at 20°C, some of the ice melts, abstracting its heat of fusion from the solution until the temperature falls to 0°C. But ice and salt solution cannot remain in equilibrium at 0°C, so the ice continues to melt. Heat is now supplied by both the ice and the solution, and both cool down until the equilibrium temperature of $-20°C$ is reached. If the melted ice dilutes the salt solution appreciably, the equilibrium temperature rises, but this can be prevented by supplying excess salt so the solution remains saturated. If no heat is supplied from outside, the mixture remains unchanged at $-20°C$. If the mixture is brought into contact with a warmer body, say an ice cream mixture at 20°C, heat flows from the ice cream mixture to the cold salt solution, melting more of the ice but producing no rise in temperature as long as any ice remains. The flow of heat from the ice cream lowers its temperature to its freezing point (below

0°C, since it is itself a solution). Further loss of heat to the ice-salt mixture causes the ice cream to freeze.

The boiling point of a liquid is also affected by dissolved substances but may be either increased or decreased. Thus the boiling point of a water-alcohol solution is *lower* than that of pure water, while the boiling point of a water-salt solution is *higher* than that of pure water.

18–7 HUMIDITY

Atmospheric air is a mixture of gases consisting of about 80 percent nitrogen, 18 percent oxygen, and small amounts of carbon dioxide, water vapor, and other gases. The mass of water vapor per unit volume is called the *absolute humidity*. The total pressure exerted by the atmosphere is the sum of the pressures exerted by its component gases. These pressures are called the *partial pressures* of the components. It is found that the partial pressure of each of the component gases of a gas mixture is very nearly the same as would be the actual pressure of that component alone if it occupied the same volume as does the mixture, a fact known as *Dalton's law*. That is, each of the gases of a gas mixture behaves independently of the others. The partial pressure of water vapor in the atmosphere is ordinarily equal to a few millimeters of mercury.

It should be evident that the partial pressure of water vapor at any given air temperature can never exceed the vapor pressure of water at that particular temperature. Thus at 10°C, from Table 18–2, the partial pressure cannot exceed 8.94 torr, or at 15°C it cannot exceed 12.67 torr. If the concentration of water vapor, or the absolute humidity, is such that the partial pressure equals the vapor pressure, the vapor is *saturated*. If the partial pressure is less than the vapor pressure, the vapor is *unsaturated*. The ratio of the partial pressure to the vapor pressure at the same temperature is called the *relative humidity*, and is usually expressed as a percentage:

Relative humidity (%)

$$= 100 \times \frac{\text{partial pressure of water vapor}}{\text{vapor pressure at same temperature}}.$$

The relative humidity is 100% if the vapor is saturated and zero if no water vapor at all is present.

Example The partial pressure of water vapor in the atmosphere is 10 torr and the temperature is 20°C. Find the relative humidity.

From Table 18–2, the vapor pressure at 20°C is 17.5 torr. Hence,

Relative humidity = (10/17.5)(100%) = 57%.

Since the water vapor in the atmosphere is saturated when its partial pressure equals the vapor pressure at the air temperature, saturation can be brought about either by increasing the water vapor content or by lowering the temperature. For example, let the partial pressure of water vapor be 10 torr when the air temperature is 20°C, as in the preceding example. Saturation or 100% relative humidity could be attained either by introducing enough more water vapor (keeping the temperature constant) to increase the partial pressure to 17.5 torr, *or by lowering the temperature* to 11.4°C, at which, by interpolation from Table 18–2, the vapor pressure is 10 torr.

If the temperature were to be lowered *below* 11.4°C, the vapor pressure would be less than 10 torr. The partial pressure would then be higher than the vapor pressure and enough vapor would condense to reduce the partial pressure to the vapor pressure at the lower temperature. It is this process which brings about the formation of clouds, fog, and rain. The phenomenon is also of frequent occurrence at night when the earth's surface becomes cooled by radiation. The condensed moisture is called *dew*. If the partial pressure is so low that the temperature must fall below 0°C before saturation occurs, the vapor condenses into ice crystals in the form of frost or snow.

The temperature at which the water vapor in a given sample of air becomes saturated is called the *dew point*. Measuring the temperature of the dew point is the most accurate method of determining relative humidity. The usual method is to cool a metal container having a bright polished surface, and to observe its temperature when the surface becomes clouded with condensed moisture. Suppose the dew point is observed in this way to be 10°C, when the air temperature is 20°C. We then know that the water vapor in the air is saturated at 10°C; hence its partial pressure is 8.94 torr, equal to the vapor pressure at 10°C. The pressure necessary for saturation at 20°C is 17.5 torr. The relative humidity is therefore

$$\frac{8.94}{17.5} \times 100 = 51\%.$$

A simpler but less accurate method of determining relative humidity employs a *wet-* and *dry-bulb thermometer.* Two thermometers are placed side by side, the bulb of one being kept moist by a wick dipping in water. The lower the relative humidity, the more rapid will be the evaporation from the wet bulb, and the lower will be its temperature below that of the dry bulb. The relative humidity corresponding to any pair of wet- and dry-bulb temperatures is read from tables.

The rate of evaporation of water from a water–air surface depends on the relative humidity. Evaporation is most rapid when the vapor pressure is low, but when the relative humidity is high evaporation is slower. At 100% relative humidity, *no* further evaporation can take place. The added discomfort which high relative humidity brings to a hot summer day is a familiar effect; the reason is that one of the body's important temperature-control mechanisms, cooling by evaporation of sweat from the skin, is inhibited by high relative humidity.

18–8 THE CLOUD CHAMBER AND THE BUBBLE CHAMBER

The cloud chamber is an apparatus for securing information about elementary particles such as electrons and α-particles. In principle (Fig. 18–13), it consists of a cylindrical enclosure having a glass top B and provided with a movable piston C. The enclosed space contains air, water vapor, and sufficient excess water so that the vapor is saturated. (Other liquids such as alcohol are often used in place of water.) When the piston is suddenly pulled down a short distance, the mixture is lowered in temperature below the dew point. If the air is perfectly clean, the cooled vapor does not immediately condense, and it is said to be *supersaturated.* The point representing its state then lies *above* the pVT-surface, but

Fig. 18–13 *The Wilson cloud chamber.*

this does not contradict a statement made earlier that all equilibrium states of a substance must lie on its pVT-surface, since the supersaturated state is *not* an equilibrium state. It is found that any ions which may be present serve as nuclei upon which the supersaturated vapor condenses to form liquid droplets. Hence if any ions were present just before the expansion, their presence is made evident by the appearance of tiny droplets immediately after the expansion.

Electrons, protons, and α-particles are all capable of traveling several centimeters through air, but as they collide with or pass near the air molecules they may knock off one or more electrons and hence leave behind them a trail of ions. Therefore if such a particle passed through the chamber just before the expansion, a trail of droplets appears after the expansion, indicating the path the particle followed. For photographing the tracks, an intense beam of light is projected transversely through the chamber and a camera mounted above it. Figure 8–10 is a photograph, made with a Wilson cloud chamber, of the paths of the particles resulting from the fission of a uranium nucleus.

The *bubble chamber,* a more recently developed apparatus for studying ionizing particles, makes use of a *superheated liquid* instead of a supersaturated vapor. A superheated liquid is a liquid at a temperature higher than that of its boiling point at the pressure to which it is subjected. It also is a nonequilibrium state. When ions are produced in a superheated liquid, it "boils" in the vicinity of the ion and forms a tiny bubble of vapor. The track of an ionizing particle through it is thus marked by a line of vapor bubbles rather than of liquid droplets. The advantage of the bubble chamber is that the molecules of a liquid are much closer together than those of a gas, so that there is a greater chance that a particle passing through the liquid will collide with a molecule and produce an ion.

PROBLEMS

(*Assume all gases to be ideal.*)

18–1 A tank contains 0.5 m³ of nitrogen at an absolute pressure of 1.5×10^5 N m⁻² and a temperature of 27°C. What will be the pressure if the volume is increased to 5.0 m³ and the temperature is raised to 327°C?

18–2 A tank having a capacity of 2 ft³ is filled with oxygen which has a gauge pressure of 60 lb in⁻² when the temperature is 47°C. At a later time it is found that because of a leak the gauge pressure has dropped to 50 lb in⁻² and the temperature has decreased to 27°C. Find (a) the mass of the oxygen in the tank under the first set of conditions, (b) the amount of oxygen that has leaked out.

18–3 A flask of volume 2 ℓ, provided with a stopcock, contains oxygen at 300 K and atmospheric pressure. The system is heated to a temperature of 400 K, with the stopcock open to the atmosphere. The stopcock is then closed and the flask cooled to its original temperature.

a) What is the final pressure of the oxygen in the flask?
b) How many grams of oxygen remain in the flask?

18–4 A balloon whose volume is 20,000 ft³ is to be filled with hydrogen at atmospheric pressure.

a) If the hydrogen is stored in cylinders of volume 2 ft³ at an absolute pressure of 200 lb in⁻², how many cylinders are required?
b) What is the total weight that can be supported by the balloon, in air at standard conditions?
c) What weight could be supported if the balloon were filled with helium instead of hydrogen?

18–5 Derive from the equation of state of an ideal gas an equation for the density of an ideal gas in terms of pressure, temperature, and appropriate constants.

18–6 At the beginning of the compression stroke, a cylinder of a diesel engine contains 48 in³ of air at atmospheric pressure and a temperature of 27°C. At the end of the stroke, the air has been compressed to a volume of 3 in³ and the gauge pressure has increased to 600 lb in⁻². Compute the temperature.

18–7 A bubble of air rises from the bottom of a lake, where the pressure is 3 atm, to the surface, where the pressure is 1 atm. The temperature at the bottom of the lake is 7°C and the temperature at the surface is 27°C. What is the ratio of the volume of the bubble as it reaches the surface to its volume at the bottom?

18–8 A liter of helium under a pressure of 2 atm and at a temperature of 27°C is heated until both pressure and volume are doubled.

a) What is the final temperature?
b) How many grams of helium are there?

18–9 A flask contains 1 g of oxygen at an absolute pressure of 10 atm and at a temperature of 47°C. At a later time it is found that because of a leak the pressure has dropped to 5/8 of its original value and the temperature has decreased to 27°C.

a) What is the volume of the flask?
b) How many grams of oxygen leaked out between the two observations?

18–10 The submarine *Squalus* sank at a point where the depth of water was 240 ft. The temperature at the surface is 27°C and at the bottom it is 7°C. The density of sea water may be taken as 2 slugs ft⁻³.

a) If a diving bell in the form of a circular cylinder 8 ft high, open at the bottom and closed at the top, is lowered to this depth, to what height will the water rise within it when it reaches the bottom?
b) At what gauge pressure must compressed air be supplied to the bell while on the bottom to expel all the water from it?

18–11 A bicycle pump is full of air at an absolute pressure of 15 lb in⁻². The length of stroke of the pump is 18 in. At what part of the stroke does air begin to enter a tire in which the gauge pressure is 40 lb in⁻²? Assume the compression to be isothermal.

18–12 A vertical cylindrical tank 1 m high has its top end closed by a tightly fitting frictionless piston of negligible weight. The air inside the cylinder is at an absolute pressure of 1 atm. The piston is depressed by pouring mercury on it slowly. How far will the piston descend before mercury spills over the top of the cylinder? The temperature of the air is maintained constant.

18–13 A barometer is made of a tube 90 cm long and of cross section 1.5 cm². Mercury stands in this tube to a height of 75 cm. The room temperature is 27°C. A small amount of nitrogen is introduced into the evacuated space above the mercury and the column drops to a height of 70 cm. How many grams of nitrogen were introduced?

18–14 A large tank of water has a hose connected to it, as shown in Fig. 18–14. The tank is sealed at the top and has compressed air between the water surface and the top. When the water height h_2 is 3 m, the gauge pressure p_1 is $1.0 \times 10^5 \text{N m}^{-2}$. Assume that the air above the water surface expands isothermally.

a) What is the velocity of flow out of the hose when $h_2 = 3$ m?

b) What is the velocity of flow when h_2 has decreased to 2 m? Neglect friction.

$h_1 = 4$ m p_1 h_2 $h_3 = 1$ m   **Fig. 18–14**

18–15 The volume of an ideal gas is 4 ℓ, the pressure is 2 atm, and the temperature is 300 K. The gas first expands at constant pressure to twice its original volume; it is then compressed isothermally to its original volume, and finally cooled at constant volume to its original pressure.

a) Show the process in a pV-diagram.

b) Compute the temperature during the isothermal compression.

c) Compute the maximum pressure.

18–16

a) Show that the density of an ideal gas is given by pM/RT.

b) What is the density of air at normal atmospheric pressure and 20° C?

c) Under these conditions, what is the total mass of air in a dormitory room 14 m × 3 m × 3 m?

18–17 If air pressure at sea level is 1.00×10^5 Pa, what is air pressure at an elevation of 10,000 m, a common altitude for jet aircraft flight, if the temperature of the air is a uniform 0° C? (This is somewhat higher than the summit of Mt. Everest, 8882 m; this effect illustrates why it is difficult to breath on the summit of Everest.)

18–18 Compare the density of water at the critical point with its density at 0° C and 1 atm; what is the ratio of the two densities?

18–19 Compare the density of carbon dioxide at the critical point to the density of solid carbon dioxide ("dry ice") at 1 atm, which is about 1.56 g cm^{-3}.

18–20 Construct two graphs for a real substance, one showing pressure as a function of volume, and the other showing pressure as a function of temperature. Show on each graph the region in which the substance exists as (a) a gas or vapor, (b) a liquid, (c) a solid. Show also the triple point and the critical point.

18–21 A small amount of liquid is introduced into a glass tube, all air is removed, and the tube is sealed off. Describe the behavior of the meniscus when the temperature of the system is raised:

a) If the volume of the tube is much greater than the critical volume.

b) If the volume of the tube is much less than the critical volume.

c) If the volume of the tube is only slightly different from the critical volume.

18–22 A piece of ice at 0°C is placed alongside a beaker of water at 0°C in a glass vessel, from which all air has been removed. If the ice, water, and vessel are all maintained at a temperature of 0°C by a suitable thermostat, describe the final equilibrium state inside the vessel.

18–23

a) What is the relative humidity on a day when the temperature is 68°F and the dew point is 41°F?

b) What is the partial pressure of water vapor in the atmosphere?

c) What is the absolute humidity, in grams per cubic meter?

18–24 The temperature in a room is 40°C. A can is gradually cooled by adding cold water. At 10°C the surface of the can clouds over. What is the relative humidity in the room?

18–25 A pan of water is placed in a sealed room of volume 60 m^3 at a temperature of 27°C and initial relative humidity of 60 percent.

a) How many grams of water will evaporate?

b) What is the absolute humidity in g m^{-3} after equilibrium has been reached?

c) If the temperature of the room is then increased 1 C°, how many more grams of water will evaporate?

18–26

a) What is the dew-point temperature on a day when the air temperature is 20°C and the relative humidity is 60 percent?

b) What is the absolute humidity, expressed in grams per cubic meter?

18–27 An air-conditioning system is required to increase the relative humidity of 10 ft^3 of air per second from 30 percent to 65 percent. The air temperature is 68°F. How many pounds of water are needed by the system per hour?

18–28 In the lower part of the atmosphere (the troposphere) the temperature is not uniform but decreases with increasing elevation. Show that if the temperature variation is approximated by the linear relation

$$T = T_0 - \alpha y,$$

where T_0 is the temperature at the earth's surface and T is the temperature at a height y, the pressure is given by

$$\ln \frac{p_0}{p} = \frac{Mg}{R\alpha} \ln \frac{T_0}{T_0 - \alpha y}$$

where p_0 is the pressure at the earth's surface and M is the molecular mass.

The coefficient α is called the "lapse rate of temperature." While it varies with atmospheric conditions, an average value is about 0.6 C°/100 m.

18–29 The volume of a closed room, kept at a constant temperature of 20° C, is 60 m^3. The relative humidity in the room is initially 10%. If a pan of water is brought into the room, how many grams will evaporate?

Chapter 19

The Laws of Thermodynamics

19–1 ENERGY AND WORK IN THERMODYNAMICS

Thermodynamics is concerned with energy relationships. The principles of thermodynamics are usually stated with reference to some well-defined *system*, usually a specified quantity of matter. A *thermodynamic system* is one that can interact with its surroundings in at least two ways, one of which must be transfer of heat. A familiar example is a quantity of a gas confined in a cylinder with a piston. Energy can be added to the system by conduction of heat, and it is also possible to do mechanical *work* on the system, since the piston exerts a force which can move through a displacement.

Thermodynamics has its roots in intensely practical problems. A steam engine or steam turbine, for example, makes use of the heat of combustion of coal or other fuel to perform mechanical work to drive an electric generator, pull a train, or perform some other useful function. The gasoline engine in an automobile has a similar function. Because of these practical applications, it is customary to discuss energy relations for a thermodynamic system not in terms of the work done *on* the system *by* its surroundings, but rather in terms of the work done *by the system* on its surroundings, which, in view of Newton's third law, is the negative of the work done *on* the system. This convention will be clarified further in the next section.

Heat can be understood on the basis of microscopic mechanical energy, i.e., the kinetic and potential energies of individual molecules in a material; and it is also possible to develop the principles of thermodynamics from a microscopic viewpoint. In the present chapter, that development is deliberately avoided, in order to emphasize that the central principles and concepts of thermodynamics can be treated in a wholly *macroscopic* way, without reference to microscopic models. Indeed, part of the great power and generality of thermodynamics springs from the fact that it is *not* dependent on details of the structure of matter. In Chapter 20 we return to microscopic considerations and examine their relation to the principles of thermodynamics.

19–2 WORK IN VOLUME CHANGES

When a gas expands, it pushes out on its boundary surfaces as they move outward; hence an expanding gas always does a positive quantity of work. To calculate the work done by a thermodynamic system during a volume change, let us consider a solid or fluid contained in a cylinder equipped with a movable piston, as shown in Fig. 19–1(a). Suppose that the cylinder has a cross-sectional area A and that the pressure exerted by the system at the piston face is p. The force exerted by the system is therefore pA. If the piston moves out an infinitesimal distance dx, the

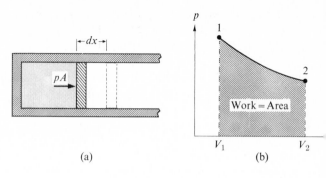

Fig. 19–1 *(a) Force exerted by a system during a small expansion. (b) Work is the area under the curve on a pV-diagram.*

work dW of this force is equal to

$$dW = pA\,dx.$$

But

$$A\,dx = dV,$$

where dV is the change of volume of the system. Therefore

$$dW = p\,dV, \qquad (19\text{–}1)$$

and in a finite change of volume from V_1 to V_2,

$$W = \int_{V_1}^{V_2} p\,dV. \qquad (19\text{–}2)$$

In general the pressure of the system will vary during the volume change, and the integral can be evaluated only if the pressure is known as a function of volume. It is customary to represent this relationship graphically by plotting p as a function of V, as in Fig. 19–1(b). Then Eq. (19–2) may be interpreted graphically as the area under the curve between the limits V_1 and V_2. If the system *expands*, then V_2 is greater than V_1 and the work (and the area) are *positive. Compression* occurs when the final volume is *less* than the initial volume; the work and the area are then *negative.*

If the pressure remains constant while the volume changes, then the work is

$$W = p(V_2 - V_1) \quad \text{(constant pressure only).}$$

Example An ideal gas is kept in thermal contact with a very large body of constant temperature T and undergoes an *isothermal expansion* in which its volume changes from V_1 to V_2. How much work is done?

From Eq. (19–2)

$$W = \int_{V_1}^{V_2} p\,dV.$$

For an ideal gas

$$p = \frac{nRT}{V}.$$

Since n, R, and T are constant,

$$W = nRT\int_{V_1}^{V_2} \frac{dV}{V} = nRT\ln\frac{V_2}{V_1}. \qquad (19\text{–}3)$$

In an expansion, $V_2 > V_1$ and W is positive. At constant T,

$$p_1 V_1 = p_2 V_2, \quad \text{or} \quad \frac{V_2}{V_1} = \frac{p_1}{p_2},$$

and the isothermal work may be expressed also in the form

$$W = nRT\ln\frac{p_1}{p_2}. \qquad (19\text{–}4)$$

On the pV-diagram in Fig. 19–2 an initial state 1 (characterized by pressure p_1 and volume V_1) and a

Fig. 19–2 *Work depends on the path.*

final state 2 (characterized by pressure p_2 and volume V_2) are represented by the two points 1 and 2. There are many ways in which the system may be taken from 1 to 2. For example, the pressure may be kept constant from 1 to 3 (*isobaric* process), and then the volume kept constant from 3 to 2 (*isochoric* process), in which case the work is equal to the area under the line 1→ 3. Another possibility is the path 1 → 4 → 2, in which case the work is the area under the line 4 → 2. The jagged line and the continuous curve from 1 to 2 represent other possibilities, in each of which the work is different. We can see, therefore, that *the work depends not only on the initial and final states but also on the intermediate states, i.e., on the path.*

19–3 HEAT IN VOLUME CHANGES

As we have seen in Chapter 16, heat is energy transferred to or from a system by virtue of a temperature difference between the system and its surroundings. Heat is regarded as positive when it enters a system and negative when it leaves. The performance of work and the transfer of heat are *methods of energy transfer,* that is, methods whereby the energy of a system may be increased or decreased.

An important aspect of energy relations is illustrated by the situation of Fig. 19–3. In Fig.19–3(a), a quantity of gas is contained in a cylinder with a piston, with initial volume of two liters, and maintained at a temperature of 300 K by an electric stove. If the pressure exerted on the piston is only a trifle smaller than the gas pressure, the gas will expand slowly, and heat will flow from the stove to the gas, thereby maintaining the temperature of the gas at the constant value 300 K. Suppose that the gas expands in this slow, controlled isothermal manner until its volume becomes, say, seven liters. A finite amount of heat is absorbed by the gas during this process.

Figure 19–3(b) shows a vessel surrounded by adiabatic walls and divided into two compartments (the lower one of volume two liters, the upper, five liters) by a thin, breakable partition. In the lower compartment the same gas has the same initial volume and temperature as in (a). Suppose the partition is broken and the gas in (b) undergoes a rapid, uncontrolled expansion into the vacuum

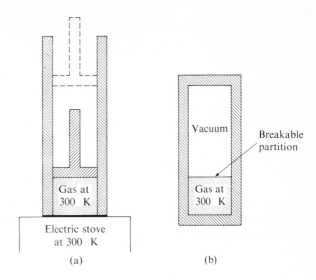

Fig. 19–3 (a) Slow, controlled, isothermal expansion of a gas from an initial state 1 to a final state 2. (b) Rapid, uncontrolled expansion of the same gas starting at the same state 1 and ending at the same state 2.

above, where it encounters no movable piston and therefore exerts no external force. Not only is no work done in this expansion, but also no heat passes through the adiabatic walls. The final volume is seven liters as in (a). The rapid, uncontrolled expansion of a gas into a vacuum is called a *free expansion* and will be discussed later in this chapter. Experiments have indicated that very little temperature change takes place, provided the initial gas pressure is less than a few atmospheres, and therefore the final state of the gas is also the same as the final state in (a). But the states traversed by the gas (the pressures and volumes) while proceeding from state 1 to state 2 are entirely different, so that (a) and (b) represent *two different paths* connecting the *same states* 1 and 2. During path (a) heat is transferred; during path (b) *no* heat is transferred. *Heat*, like work, *depends not only on the initial and final states but also on the path.*

It would be just as incorrect to refer to the "heat in a body" as it would be to speak about the "work in a body." For suppose we assigned an arbitrary value to "the heat in a body" in some standard reference state. The "heat in the body" in some other state would then equal the "heat" in the reference state plus the heat added when the body is carried to the second state. But the heat added depends entirely on the path by which we go from one state to the other, and since there are an infinite number of paths which might be followed, there are an infinite number of values which might equally well be assigned to the "heat in the body" in the second state. Since it is not possible to assign any one value to the "heat in the body," we conclude that this concept is meaningless.

19-4 THE FIRST LAW OF THERMODYNAMICS

The transfer of heat and the performance of work constitute two methods of adding energy to or subtracting energy from a system. Once the transfer of energy has occurred, the system is said to have undergone a change in *internal energy.*

Suppose a system is caused to change from state 1 to state 2 along a definite path and that the heat Q absorbed by the system and the work W done by it are measured. Expressing both Q and W either in thermal units or in mechanical units, we may then calculate the difference $Q - W$. If now we do the same thing over again for many different paths (between the same states 1 and 2), we find experimentally that $Q - W$ *is the same for all paths connecting 1 and 2*. But Q is the energy that has been added to a system by the transfer of heat and W is equal to the energy that has been extracted from the system by the performance of work. The difference $Q - W$, therefore, must represent the internal energy change of the system. It follows that *the internal energy change of a system is independent of the path*, and is therefore equal to the energy of the system in state 2 minus the energy in state 1, or $\Delta U = U_2 - U_1$:

$$\Delta U = U_2 - U_1 = Q - W. \qquad (19\text{–}5)$$

If some arbitrary value is assigned to the internal energy in some standard reference state, its value in any other state is uniquely defined, since $Q - W$ is the same for all processes connecting the states. Equation (19–4) is known as *the first law of thermodynamics.* In applications of the law in this form, it must be remembered that (1) all quantities must be expressed in the same units, (2) Q is positive when heat

goes *into* the system, (3) W is positive when the force exerted *by* the system and the displacement have the same sign, and the system does positive work.

Rewriting Eq. (19–5) in the form $Q = U_2 - U_1 + W$, we may say that when a quantity of heat Q enters a system during a process, some of it ($U_2 - U_1$) remains in the system as increased internal energy, while the remainder (W) leaves the system again in the form of work done by the system against its surroundings.

Internal energy can be interpreted in terms of microscopic mechanical energy, that is, kinetic and potential energies of individual molecules in a material. From the thermodynamic standpoint, however, this is not necessary. Equation (19–5) is the *definition* of the internal energy of a system or, more precisely, of the *change* in its internal energy in any process. As with other forms of energy, only *differences* in internal energy are defined, and not absolute values.

If a system is carried through a process which eventually returns it to its initial state (a *cyclic* process), then

$$U_2 = U_1 \quad \text{and} \quad Q = W.$$

Thus, although net work W may be done by the system in the process, energy has not been created, since an equal amount of energy must have flowed into the system as heat Q.

An *isolated* system is one which does no external work and into which there is no flow of heat. Then, for any process taking place in such a system,

$$W = Q = 0$$

and

$$U_2 - U_1 = 0 \quad \text{or} \quad \Delta U = 0.$$

That is, *the internal energy of an isolated system remains constant.* This is the most general statement of the *principle of conservation of energy.* The internal energy of an *isolated* system cannot be changed by any process (mechanical, electrical, chemical, nuclear, or biological) taking place *within* the system. The energy of a system can be changed only by a flow of heat across its boundary, or by the performance of work. If either of these takes place, the system is no longer isolated. The increase in energy of the system is then equal to the energy flowing in as heat, minus the energy flowing out as work.

19–5 ADIABATIC PROCESS

A process that takes place in such a manner that no heat enters or leaves a system is called an *adiabatic process*. For every adiabatic process, $Q = 0$. This may be accomplished either by surrounding the system with a thick layer of heat-insulating material (such as cork, asbestos, firebrick, or styrofoam), or by performing the process quickly. The flow of heat requires finite time, so any process performed quickly enough will be practically adiabatic. Applying the first law to an adiabatic process, we get

$$U_2 - U_1 = -W \quad \text{(adiabatic process).} \qquad (19\text{–}6)$$

Thus, the change in the internal energy of a system, in an adiabatic process, is equal in absolute magnitude to the work. If the work W is negative, as when a system is compressed, then $-W$ is positive, U_2 is greater than U_1, and the internal energy of the system increases. If W is positive, as when a system expands, the internal energy of the system decreases. An increase of internal energy is usually accompanied by a rise in temperature, and a decrease in internal energy by a temperature drop.

The compression of the mixture of gasoline vapor and air that takes place during the compression stroke of a gasoline engine is an example of an approximately adiabatic process involving a temperature rise. The expansion of the combustion products during the power stroke of the engine is an approximately adiabatic process involving a temperature decrease. Adiabatic processes, therefore, play a very important role in mechanical engineering.

19–6 ISOCHORIC PROCESS

When a substance undergoes a process in which the volume remains unchanged, the process is called *isochoric*. The rise of pressure and temperature produced by a flow of heat into a substance contained in a rigid container of fixed volume is an example of an isochoric process. When the volume does not change, no work is done and therefore, from the first law

$$U_2 - U_1 = Q \quad \text{(isochoric process).} \qquad (19\text{–}7)$$

All the added heat has served to increase the internal energy. The very sudden increase of temperature and pressure accompanying the explosion of gasoline vapor and air in a gasoline engine may be treated mathematically as though it were an isochoric addition of heat.

19–7 ISOTHERMAL PROCESS

A process taking place at constant temperature is said to be *isothermal*. In order for the temperature to remain constant, the changes in the pressure and volume must be carried out very slowly, so that the state approximates thermal equilibrium very closely at every stage of the process. In general, *none* of the quantities Q, W, or $U_2 - U_1$ is constant.

There are a few special cases in which the internal energy of the system depends only on temperature, not on pressure or volume. An ideal gas and an ideal paramagnetic crystal are two examples. When such a system undergoes an isothermal process, its internal energy does not change, and therefore $Q = W$. This relation does *not* hold for systems other than those mentioned.

19–8 ISOBARIC PROCESS

A process taking place at constant pressure is called an *isobaric process*. When water enters the boiler of a steam engine and is heated to its boiling point, vaporized, and then the steam is superheated, all these processes take place isobarically. Such processes play an important role in mechanical engineering and also in chemistry.

As discussed in Section 19–2,

$$W = P(V_2 - V_1). \quad \text{(Isobaric process).} \qquad (19\text{–}8)$$

A simple example is the vaporization of a mass m of liquid to vapor at constant pressure and temperature. If V_L is the volume of liquid and V_V is the volume of vapor, the work done in expanding from V_L to V_V at constant pressure p is

$$W = p(V_V - V_L).$$

The heat absorbed by each unit of mass is the heat of vaporization L. Hence,

$$Q = mL.$$

From the first law,

$$U_V - U_L = mL - p(V_V - V_L). \quad (19\text{-}9)$$

Example One gram of water (1 cm^3) becomes 1671 cm^3 of steam when boiled at a pressure of 1 atm. The heat of vaporization at this pressure is 2256 J g^{-1}. Compute the external work and the increase in internal energy.

$$\begin{aligned} W &= p(V_V - V_L) \\ &= (1.013 \times 10^5\,\text{N m}^{-2})(1671 - 1) \times 10^{-6}\,\text{m}^3 \\ &= 169\,\text{J}. \end{aligned}$$

From Eq. (19-5),

$$\begin{aligned} U_V - U_L &= mL - W \\ &= 2256\,\text{J} - 169\,\text{J} \\ &= 2087\,\text{J}. \end{aligned}$$

Hence the external work, or the external part of the heat of vaporization, equals 169 J, and the increase in internal energy, or the *internal* part of the heat of vaporization, is 2087 J.

19-9 THROTTLING PROCESS

A *throttling process* is one in which a fluid, originally at a constant high pressure, seeps through a porous wall or a narrow opening (needle valve or "throttling" valve) into a region of constant lower pressure without a transfer of heat taking place. In Fig. 19-4(a), a fluid is discharged from a pump at a high pressure, then passes through a throttling valve into a pipe which leads directly to the intake or low-pressure side of the pump. Every successive element of fluid undergoes the throttling process in a continuous stream.

Consider any element of fluid enclosed between the piston and throttling valve of Fig. 19-4(b). Suppose this piston moves toward the right and another piston on the other side of the valve moves to the right also, at such rates that the pressure on the left remains at a constant high value, p_1, and that on the right at a constant lower value p_2. After all the fluid has been forced through the valve, the final state is that of Fig. 19-4(c).

The net work done in this process is the difference between the work done in forcing the righthand piston out and the work done in forcing the lefthand piston in. Let V_1 and V_2 be the initial and final volumes. Since the low-pressure fluid changes in volume from zero to V_2 at the constant pressure p_2,

adiabatic

(a)   Heat insulation   (b)   Fluid at constant high pressure P_1 (c)   Fluid at constant lower pressure P_2

Fig. 19-4 *Throttling process.*

the work is

$$p_2(V_2 - 0),$$

and since the high-pressure fluid changes in volume from V_1 to zero at the constant high pressure p_1, the work is

$$p_1(0 - V_1).$$

The net work W done by the system is, therefore,

$$W = p_2 V_2 - p_1 V_1.$$

Since the process is adiabatic, $Q = 0$; hence, from the first law,

$$U_2 - U_1 = 0 - (p_2 V_2 - p_1 V_1)$$

or

$$\boxed{U_2 + p_2 V_2 = U_1 + p_1 V_1.} \qquad (19\text{–}10)$$

This result is of great importance in steam engineering and in refrigeration. The sum $U + pV$, called the *enthalpy*, is tabulated for steam and for many refrigerants. The throttling process plays the main role in the action of a refrigerator, since this is the process that brings about the drop in temperature needed for refrigeration. Liquids that are about to evaporate (saturated liquids) always undergo a drop in temperature and partial vaporization as a result of a throttling process. Gases, however, may undergo either a temperature rise or a drop, depending on the initial temperature and pressure and on the final pressure.

19–10 DIFFERENTIAL FORM OF THE FIRST LAW

Up to this point we have used the first law of thermodynamics only in its finite form,

$$U_2 - U_1 = Q - W.$$

In this form the equation applies to a process in which states 1 and 2 differ in pressure, volume, and temperature by a finite amount. Suppose states 1 and 2 differ only slightly. Then if only a small amount of heat dQ is transferred, and only a small amount of work dW is done, the energy change dU is also very small. In these circumstances, the first law becomes

$$\boxed{dU = dQ - dW.} \qquad (19\text{–}11)$$

If the system is of such a character that the only work possible is by means of expansion or compression, then $dW = p\,dV$, and

$$\boxed{dU = dQ - p\,dV} \qquad (19\text{–}12)$$

is the *differential form of the first law*, applicable to solids, liquids, and gases.

19–11 INTERNAL ENERGY OF AN IDEAL GAS

Imagine a thermally insulated vessel with rigid walls, divided into two compartments by a partition. Suppose that there is a gas in one compartment and that the other is empty. If the partition is removed, the gas will undergo what is known as a *free expansion* in which no work is done and no heat is transferred. From the first law, since both Q and W are zero, it follows that *the internal energy remains unchanged during a free expansion.* The question as to whether or not the temperature of a gas changes during a free expansion has engaged the attention of physicists for over a hundred years.

This is an important question, for it provides information concerning the dependence of internal energy on other quantities. For if the temperature should change while the internal energy stays the same, one would have to conclude that the internal energy depends on *both* the temperature and the volume, or both the temperature and the pressure, but certainly not on the temperature alone. If, on the other hand, T remains unchanged during a free expansion in which we know U remains unchanged, then the only conclusion that is admissible is that U *is a function of T only.*

Experiment has shown that the internal energy of a real gas does depend to some extent on the pressure or volume as well as on the temperature. At sufficiently low pressures and high temperatures, however, this dependence disappears and U becomes a function only of T. Thus the assumption that U depends only on T becomes an additional part of the definition of an ideal gas, and in a free expansion an ideal gas undergoes no temperature change.

19–12 HEAT CAPACITIES OF AN IDEAL GAS

The temperature of a substance may be changed under a variety of conditions. The volume may be kept constant, or the pressure may be kept constant, or both may be allowed to vary in some definite manner. The amount of heat per mole necessary to cause unit rise of temperature is different in each case. In other words, a substance has many different molar heat capacities. Two, however, are particularly useful, namely, those at *constant volume* and at *constant pressure*. There is a simple and important relation between these two molar heat capacities of an ideal gas.

Figure 19–5 shows two isotherms of an ideal gas at temperatures T and $T + dT$. Since the internal energy of an ideal gas depends only on the temperature, it is constant if the temperature is constant and the isotherms are also curves of *constant internal energy*. The internal energy therefore has a constant value U at every point of the isotherm at temperature T, and a constant value $U + dU$ at every point of the isotherm at $T + dT$. It follows that the *change* in internal energy, dU, is the same in all processes in which the gas is taken from *any* point on one isotherm to *any* point on the other. Thus dU is the same for all the processes *ab*, *ac*, *ad*, and *ef*, in Fig. 19–5.

Fig. 19–5 *The change in internal energy of an ideal gas is the same in all processes between two given temperatures.*

Consider first the process *ab*, at constant volume. To carry out such a process, the gas at temperature T is enclosed in a rigid container and brought in contact with a body at a slightly higher temperature $T + dT$. There will be a flow of heat dQ into the gas, and by definition of the molar heat capacity at constant volume, C_v,

$$dQ = nC_v\, dT. \qquad (19\text{–}13)$$

The pressure of the gas increases during this process, but no work is done, since the volume is constant. Hence, from the first law (in differential form),

$$dU = dQ - dW,$$

we have $nC_v dT - 0$

$$\boxed{dU = nC_v\, dT.} \qquad (19\text{–}14)$$

But dU is the same for *all* processes between the isotherms in Fig. 19–5, so Eq. (19–14) gives the change in internal energy in *all* the processes in Fig. 19–5, *even if they are not at constant volume.*

Now consider the process *ad* in Fig. 19–5, at constant pressure. To carry out such a process the gas might be enclosed in a cylinder with a piston, acted on by a constant external pressure, and brought in contact with a body at a temperature $T + dT$. As heat flows into the gas it expands at constant pressure and does work. By definition of the molar heat capacity at constant pressure, C_p, the heat dQ flowing into the gas is

$$dQ = nC_p\, dT.$$

The work dW is

$$dW = p\, dV.$$

But from the equation of state, since p is constant,

$$pV = nRT, \qquad p\, dV = nR\, dT,$$

so

$$dW = nR\, dT.$$

Then from the first law,

$$dU = dQ - dW,$$
$$nC_v\, dT = nC_p\, dT - nR\, dT,$$

or

$$C_p - C_v = R. \qquad (19\text{–}15)$$

The molar heat capacity of an ideal gas at constant pressure is therefore *greater* than that at constant volume; the difference is the universal gas constant R. Of course, R must be expressed in the same units as C_p and C_v, usually J mol^{-1} (C°)$^{-1}$ or cal mole^{-1} (C°)$^{-1}$. We have shown in Section 18–2 that, in these units, $R = 8.314$ J mol^{-1} (C°)$^{-1}$ = 1.99 cal mole^{-1} (C°)$^{-1}$.

Although Eq. (19–15) was derived for an ideal gas, it is very nearly true for *real* gases at moderate pressures. Measured values of C_p and C_v are given in Table 19–1 for some real gases at low pressures, and the difference is seen to be very nearly 1.99 cal mol^{-1} (C°)$^{-1}$.

In the last column of Table 19–1 are listed the values of the ratio C_p/C_v, denoted by the Greek letter γ (gamma). It is seen that γ is 1.67 for monatomic gases, and is very nearly 1.40 for the so-called permanent diatomic gases. There is no simple regularity for polyatomic gases.

Solids and liquids also expand when heated, if free to do so, and hence perform work. The coefficients of volume expansion of solids and liquids are, however, so much smaller than those of gases that the work is small. The internal energy of a solid or liquid *does* depend on its volume as well as its temperature,

Table 19–1 MOLAR HEAT CAPACITIES OF GASES AT LOW PRESSURE

Type of gas	Gas	C_p, cal mol^{-1} (C°)$^{-1}$	C_v, cal mol^{-1} (C°)$^{-1}$	$C_p - C_v$	$\gamma = \dfrac{C_p}{C_v}$
Monatomic	He	4.97	2.98	1.99	1.67
	A	4.97	2.98	1.99	1.67
Diatomic	H_2	6.87	4.88	1.99	1.41
	N_2	6.95	4.96	1.99	1.40
	O_2	7.03	5.04	1.99	1.40
	CO	6.97	4.98	1.99	1.40
Polyatomic	CO_2	8.83	6.80	2.03	1.30
	SO_2	9.65	7.50	2.15	1.29
	H_2S	8.27	6.2	2.1	1.34

and this must be considered when evaluating the difference between specific heats of solids or liquids. It turns out that here also $C_p > C_v$, but the difference is small and is not expressible as simply as that for a gas. Because of the large stresses set up when solids or liquids are heated and *not* allowed to expand, most heating processes involving them take place at constant pressure, and hence C_p is the quantity usually measured for a solid or liquid.

19–13 ADIABATIC PROCESS OF AN IDEAL GAS

Any process in which there is no flow of heat into or out of a system is called *adiabatic*. To perform a truly adiabatic process it would be necessary that the system be surrounded by a perfect heat insulator, or that the surroundings of the system be kept always at the same temperature as the system. However, if a process such as compression or expansion of a gas is carried out rapidly, it will be nearly adiabatic, since the flow of heat into or out of the system is slow even under favorable conditions. Thus the compression stroke of a gasoline or diesel engine is approximately adiabatic.

Note that external work may be done *on* or *by* a system in an adiabatic process, and that the temperature usually changes in such a process.

Let an ideal gas undergo an infinitesimal adiabatic process. Then $dQ = 0$, $dU = nC_v\, dT$, $dW = p\, dV$, and, from the first law,

$$nC_v\, dT = -p\, dV. \qquad (19\text{–}16)$$

From the equation of state,

$$p\, dV + V\, dp = nR\, dT.$$

Eliminating dT between these equations and making use of the fact that $C_p - C_v = R$, we obtain the relation

$$\frac{dp}{p} + \frac{C_p}{C_v}\frac{dV}{V} = 0,$$

or, if C_p/C_v is denoted by γ,

$$\frac{dp}{p} + \gamma\frac{dV}{V} = 0.$$

To obtain the relation between p and V in a *finite* adiabatic change we may integrate the preceding equation. This gives

$$\ln p + \gamma \ln V = \ln(\text{constant})$$

or

$$\boxed{p V^{\gamma} = \text{constant}} \qquad \text{(Adiabatic process)}.$$

If subscripts 1 and 2 refer to any two points of the process,

$$p_1 V_1^{\gamma} = p_2 V_2^{\gamma}. \qquad (19\text{–}17)$$

It is left as a problem to show that by combining Eq. (19–17) with the equation of state one obtains the alternate forms

$$T_1 V_1^{\gamma-1} = T_2 V_2^{\gamma-1}, \qquad (19\text{–}18)$$

$$T_1 p_1^{(1-\gamma)/\gamma} = T_2 p_2^{(1-\gamma)/\gamma}. \qquad (19\text{–}19)$$

Values of the specific heat ratio γ are listed in Table 19–1 for some common gases.

An adiabatic expansion or compression of an ideal gas may be represented graphically by a plot of Eq. (19–17), as in Fig. 19–6, in which a number of isothermal curves are shown for comparison. The adiabatic curves, at any point, have a somewhat steeper slope than the isothermal curve passing through the same point. That is, as we follow along an adiabatic from right to left (compression process) the curve continually cuts across isotherms of higher and higher temperatures, in agreement with the fact that the

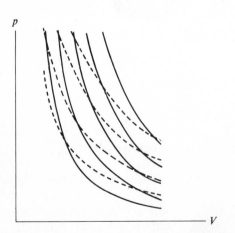

Fig. 19–6 *Adiabatic curves (full lines) versus isothermal curves (dotted lines).*

temperature generally increases in an adiabatic compression.

The work done by an ideal gas in an adiabatic expansion is computed as follows. We have, from Eq. (19–16),

$$W = \int_{V_1}^{V_2} p \, dV = \int_{T_1}^{T_2} - nC_v \, dT. \quad (19\text{–}20)$$

Hence the work may be found from either integral. Consider first $\int p \, dV$. Since $p V^{\gamma} = p_1 V_1^{\gamma} = p_2 V_2^{\gamma} =$ a constant, K, we may write

$$W = \int_{V_1}^{V_2} p \, dV = K \int_{V_1}^{V_2} \frac{dV}{V^{\gamma}}$$

$$= \frac{1}{1-\gamma}(K V_2^{1-\gamma} - K V_1^{1-\gamma}).$$

In the first term let $K = p_2 V_2^{\gamma}$ and in the second term let $K = p_1 V_1^{\gamma}$. Then

$$W = \frac{p_2 V_2 - p_1 V_1}{1 - \gamma}. \qquad (19\text{–}21)$$

This expresses the work in terms of initial and final pressures and volumes. If the initial and final temperatures are known, we may return to Eq. (19–20) and write

$$W = \int_{T_1}^{T_2} - nC_v \, dT = nC_v(T_1 - T_2). \quad (19\text{–}22)$$

Example   The compression ratio of a diesel engine, V_1/V_2, is about 15. If the cylinder contains air at 15 lb in^{-2} (absolute) and 60°F (= 520°R) at the start of the compression stroke, compute the pressure and temperature at the end of this stroke. Assume that air behaves like an ideal gas and that the compression is adiabatic. The value of γ for air is 1.40.

From Eq. (19–17),

$$p_2 = p_1 \left(\frac{V_1}{V_2}\right)^{\gamma}$$

$$= (15 \text{ lb in}^{-2})(15)^{1.4}$$

$$= 665 \text{ lb in}^{-2}.$$

The temperature may now be found from Eqs. (19–18) or (19–19), or by the ideal gas law. Thus, from Eq. (19–18), we have

$$T_2 = T_1 \left(\frac{V_1}{V_2} \right)^{\gamma - 1}$$

$$= 520°R \, (15)^{1.4-1}$$

$$= 1536°R = 1076°F.$$

Or,

$$T_2 = T_1 \frac{p_2 V_2}{p_1 V_1}$$

$$= (520°R) \frac{(665 \text{ lb in}^{-2})(1)}{(15 \text{ lb in}^{-2})(15)}$$

$$= 1536°R.$$

The work done in the compression stroke is found as follows: Let the initial volume V_1 be 60 in³; then $V_2 = 4$ in³. From Eq. (19–21),

$$W = \frac{(665 \text{ lb in}^{-2})(4 \text{ in}^3) - (15 \text{ lb in}^{-2})(60 \text{ in}^3)}{1 - 1.40}$$

$$= -4400 \text{ in. lb}$$

$$= -367 \text{ ft lb.}$$

In the above analysis we have used the ideal-gas equation of state, which holds only when the state changes slowly enough so that, at each step, the pressure and temperature are *uniform* throughout the gas. Thus, the validity of our results is limited to situations where the process is rapid enough to prevent appreciable heat exchange with the surroundings, yet slow enough so the system does not depart very far from thermal and mechanical equilibrium.

19–14 HEAT ENGINES

The dominating feature of an industrial society is its ability to utilize, whether for wise or unwise ends, sources of energy other than the muscles of men or animals. Except for waterpower, where mechanical energy is directly available, most energy supplies are in the form of potential energy of molecular or nuclear aggregations. In chemical or nuclear reactions, some of this potential energy is released and converted to random molecular kinetic energy. Heat can be withdrawn and utilized for heating buildings,

for cooking, or for maintaining a furnace at a high temperature in order to carry out other chemical or physical processes. But to operate a machine, or to propel a vehicle or projectile, *mechanical* energy is required, and one of the problems of the mechanical engineer is to withdraw heat from a high-temperature source and convert as large a fraction as possible to mechanical energy.

This transformation always requires the services of a *heat engine*, such as a steam engine, gasoline engine, diesel engine, or jet engine. We consider for simplicity an engine in which the so-called "working substance" is carried through a *cyclic* process, that is, a sequence of processes in which it eventually returns to its original state. In the condensing type of steam engine used in marine propulsion, the "working substance," in this case pure water, is actually used over and over again. Water is evaporated in the boilers at high pressure and temperature, does work in expanding against a piston or in a turbine, is condensed by cooling water from the ocean, and pumped back into the boilers. The refrigerant in a household refrigerator also undergoes a cyclic process. Internal combustion engines and steam locomotives do not carry a system through a closed cycle, but they can be analyzed in terms of cyclic processes which approximate their actual operations.

All these devices absorb heat from a source at a high temperature, perform some mechanical work, and reject heat at a lower temperature. When a system is carried through a cyclic process, its initial and final internal energies are equal, and from the first law, for any number of complete cycles,

$$U_2 - U_1 = 0 = Q - W,$$

$$Q = W.$$

That is, the **net** *heat flowing into the engine in a cyclic process equals the* **net** *work done by the engine.*

The work is represented by the area enclosed by the curve representing the process in the pV-plane. Thus in Fig. 19–7, for example, where the closed curve shows an arbitrary cyclic process, the area under the upper curve from a to b represents work done *by* the system (positive work) in the expansion from a to b, while the area under the lower curve from b to a

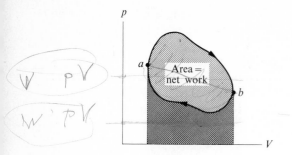

Fig. 19–7 *The area enclosed by the curve representing a cyclic process equals the net work.*

represents work done *on* the system (negative work) in the compression from *b* to *a*. Since the average pressure during the compression process is less than that during the expansion, the positive work exceeds the negative work and the area bounded by the closed curve is the net positive work done by the system. If the same process were traversed counterclockwise the net work done by the system would be negative.

In the operation of a heat engine or of a refrigerator there are always two bodies capable of either providing or absorbing large quantities of heat without appreciable changes of temperature. Thus the flames and hot gases surrounding the boilers of a marine steam installation can give up large quantities of heat at a high temperature, and constitute therefore what may be called the *hot reservoir*. Heat transferred between the hot reservoir and the working substance in a heat engine will be represented by the symbol Q_H, where it is understood that a positive value of Q_H means heat entering the working substance. The ocean water used to cool the condenser of the marine installation constitutes the *cold reservoir*. (The words "hot" and "cold" are, of course, relative.) The heat transferred between the working substance and the cold reservoir will be denoted by Q_C. A negative value of Q_C means heat rejected by the working substance.

The energy transformations in a heat engine are conveniently represented schematically by the *flow diagram* of Fig. 19–8. The engine itself is represented by the circle. The heat Q_H supplied to the engine by the hot reservoir is proportional to the cross section of the incoming "pipeline" at the top of the diagram. The cross section of the outgoing pipeline at the

bottom is proportional to the heat Q_C which is rejected as heat in the exhaust. The branch line to the right represents that portion of the heat supplied which the engine converts to mechanical work, W.

Consider a heat engine operating in a cycle over and over again and let Q_H and Q_C stand for the heats absorbed and rejected by the working substance *per cycle*. The net heat absorbed is

$$Q = Q_H + Q_C, \qquad (19\text{–}23)$$

where Q_C is a negative number. The useful output of the engine is the net work W done by the working substance, and from the first law,

$$W = Q = Q_H + Q_C. \qquad (19\text{–}24)$$

The heat *absorbed* is usually obtained from the combustion of fuel. The heat *rejected* ordinarily has no economic value. The *thermal efficiency* of a cycle is defined as the ratio of the useful work to the heat absorbed ("what you get" divided by "what you pay for"):

$$\boxed{\text{Thermal efficiency} = \frac{W}{Q_H} = \frac{Q_H + Q_C}{Q_H}.} \qquad (19\text{–}25)$$

Because of friction losses, the *useful* work delivered by an engine is less than the work W, and the overall efficiency is less than the thermal efficiency.

Fig. 19–8 *Schematic flow diagram of a heat engine.*

In terms of the flow diagram in Fig. 19–8, the most efficient engine is the one for which the branch pipeline representing the work obtained is as large as possible, and the exhaust pipeline representing the heat rejected is as small as possible, for a given incoming pipeline or quantity of heat supplied.

We shall now consider, without going into the mechanical details of their construction, the gasoline engine, the diesel engine, and the steam engine.

19–15 THE GASOLINE ENGINE

The common gasoline engine is of the four-cycle type, so called because four processes take place in each cycle. Starting with the piston at the top of its stroke, an explosive mixture of air and gasoline vapor is drawn into the cylinder on the downstroke, the intake valve being open and the exhaust valve closed. This is the *intake* stroke. At the end of this stroke the intake valve closes and the piston rises, performing an approximately adiabatic compression of the air–gasoline mixture. This is the *compression* stroke. At or near the top of this stroke a spark ignites the mixture of air and gasoline vapor, and combustion takes place very rapidly. The pressure and temperature increase at nearly constant volume.

The piston is now forced down, the burned gases expanding approximately adiabatically. This is the *power stroke* or *working stroke*. At the end of the power stroke the exhaust valve opens. The pressure in the cylinder drops rapidly to atmospheric, and the rising piston on the *exhaust stroke* forces out most of the remaining gas. The exhaust valve now closes, the intake valve opens, and the cycle is repeated.

For purposes of computation, the gasoline engine cycle is approximated by the *Otto* cycle illustrated in Fig. 19–9. Starting at point *a*, air at atmospheric pressure is compressed adiabatically in a cylinder to point *b*, heated at constant volume to point *c*, allowed to expand adiabatically to point *d*, and cooled at constant volume to point *a*, after which the cycle is repeated. Line *ab* corresponds to the compression stroke, *bc* to the explosion, *cd* to the working stroke, and *da* to the exhaust of a gasoline engine. V_1 and V_2 in Fig. 19–9 are respectively the maximum and minimum volumes of the air in the

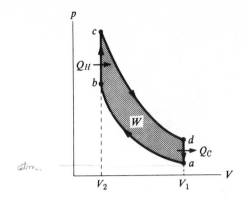

Fig. 19–9 *pV-diagram of the Otto cycle*

cylinder. The ratio V_1/V_2 is called the *compression ratio*, and is about 10 for a modern internal combustion engine.

The work output in Fig. 19–9 is represented by the shaded area enclosed by the figure *abcd*. The heat *input* is the heat supplied at constant volume along the line *bc*. The exhaust heat is removed along *da*. No heat is supplied or removed in the adiabatic processes *ab* and *cd*.

The heat input and the work output can be computed in terms of the compression ratio, assuming air to behave like an ideal gas. The result is

$$\text{Eff}(\%) = 100\left(1 - \frac{1}{(V_1/V_2)^{\gamma-1}}\right),$$

where γ is the ratio of the specific heat capacity at constant pressure to the specific heat capacity at constant volume, C_p/C_v. For a compression ratio of 10 and a value of $\gamma = 1.4$, the efficiency is about 60%. It will be seen that the higher the compression ratio, the higher the efficiency. However, increasing the compression ratio also increases the temperature at the end of the adiabatic compression of the gas–air mixture. If it is too high, the mixture explodes spontaneously and prematurely, instead of burning evenly after ignition by the spark plug. The mechanical strength and wear of engine parts also pose problems; so the maximum practical compression ratio for ordinary gasoline is about 10. Higher ratios can be used with more exotic fuels.

The Otto cycle is of course a highly idealized model; it neglects friction, turbulence, loss of heat to cylinder walls, and many other effects which combine to reduce the efficiency of a real engine. Another source of inefficiency is incomplete combustion. A mixture of gasoline vapor with just enough air for complete combustion of the hydrocarbons to H_2O and CO_2 does not ignite readily. Reliable ignition requires a mixture "richer" in gasoline, but this leads to CO and unburned hydrocarbons in the exhaust. The heat obtained from the gasoline is then less than the total heat of combustion; the difference is wasted, and the exhaust contributes to air pollution. One attack on this problem is the stratified-charge engine, in which the concentration of gasoline vapor near the spark plug is greater than in the remainder of the combustion chamber.

19-16 THE DIESEL ENGINE

In the Diesel cycle, air is drawn into the cylinder on the intake stroke and compressed adiabatically on the compression stroke, to a sufficiently high temperature so that fuel oil injected at the end of this stroke burns in the cylinder without requiring ignition by a spark. The combustion is not as rapid as in the gasoline engine, and the first part of the power stroke proceeds at essentially constant pressure. The remainder of the power stroke is an adiabatic expansion. This is followed by an exhaust stroke which completes the cycle.

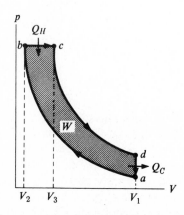

Fig. 19-10 *pV-diagram of the Diesel cycle.*

The idealized air–Diesel cycle is shown in Fig. 19-10. Starting at point a, air is compressed adiabatically to point b, heated at constant pressure to point c, expanded adiabatically to point d and cooled at constant volume to point a.

Since there is no fuel in the cylinder of a diesel engine on the compression stroke, pre-ignition cannot occur and the compression ratio V_1/V_2 may be much higher than that of an internal combustion engine. A value of 15 is typical. The *expansion* ratio V_1/V_3 may be about 5. Using these values, and taking $\gamma = 1.4$, we compute the efficiency of the air–Diesel cycle at about 56%. As with the Otto cycle, this description is an idealized model; efficiencies of real engines are considerably less than this.

19-17 THE STEAM ENGINE

The condensing type of steam engine performs the following sequence of operations. Water is converted to steam in the boiler, and the steam thus formed is superheated above the boiler temperature. Superheated steam is admitted to the cylinder, where it expands against a piston; connection is maintained to the boiler for the first part of the working stroke, which thus takes place at constant pressure. The inlet valve is then closed and the steam expands adiabatically for the rest of the working stroke. The adiabatic cooling causes some of the steam to condense. The mixture of water droplets and steam (known as "wet" steam) is forced out of the cylinder on the return stroke and into the condenser where the remaining steam is condensed into water. This water is forced into the boiler by the feed pump, and the cycle is repeated.

An idealized cycle (called the Rankine cycle) which approximates the actual steam cycle is shown in Fig. 19-11. Starting with liquid water at low pressure and temperature (point a), the water is compressed adiabatically to point b at boiler pressure. It is then heated at constant pressure to its boiling point (line bc), converted to steam (line cd), superheated (line de), expanded adiabatically with some condensation (line ef), and cooled and condensed (along fa) to its initial condition.

The efficiency of such a cycle may be computed in the same way as in the previous examples, by

Fig. 19–11 *The Rankine cycle.*

finding the quantities of heat taken in and rejected along the lines *be* and *fa*. Assuming a boiler temperature of 417°F (corresponding to a pressure of 300 lb in^{-2}), a superheat of 63°F above this temperature (480°F) and a condenser temperature of 102°F, the efficiency of a Rankine cycle is about 32%. Efficiencies of actual steam engines are somewhat lower.

The Rankine cycle may also be applied to analysis of the *steam turbine*, in which the steam expands against rotating turbine blades instead of against a piston moving in a cylinder. Nearly all large electric-power plants that burn coal or oil or use nuclear reactors contain steam turbines. The heat obtained from combustion or from the nuclear reaction is used to boil water; the steam drives a steam turbine connected directly to an electric generator, usually on the same shaft.

19–18 THE SECOND LAW OF THERMODYNAMICS

Although their efficiencies differ from one another, none of the heat engines described above has a thermal efficiency of 100%. That is, none of them absorbs heat and converts it *completely* into work. There is nothing in the *first* law of thermodynamics that precludes this possibility. The first law requires only that the energy output of an engine, in the form of mechanical work, shall equal the difference between the energies absorbed and rejected in the form of heat. An engine which rejected no heat and which converted all the heat absorbed to mechanical work would therefore be perfectly consistent with the first law.

We know now that there is another principle, independent of the first law and not derivable from it, which determines the maximum fraction of the energy absorbed by an engine as heat that can be converted to mechanical work. The basis of this principle lies in the difference between the natures of internal energy and of mechanical energy. The former is the energy of *random* molecular motion, while the latter represents *ordered* molecular motion. Superposed on their random motion, the molecules of a moving body have an ordered motion in the direction of the velocity of the body. The total molecular kinetic energy associated with the ordered motion is what we call in mechanics the kinetic energy of the moving body. The kinetic (and potential) energy associated with the *random* motion constitutes the internal energy. When the moving body makes an inelastic collision and comes to rest, the *ordered* portion of the molecular kinetic energy becomes converted to *random* motion. Since we cannot control the motions of individual molecules, it is impossible to reconvert the random motion *completely* to ordered motion. We can, however, convert *a portion* of it; this is what is accomplished by a heat engine.

The impossibility of converting heat *completely* into mechanical energy forms the basis of one form of the *second law of thermodynamics*. Although we have introduced it with a discussion of molecular motion, the second law (as with all of thermodynamics) deals with directly measurable quantities such as heat and work, and can be stated entirely apart from any molecular theory. One statement of the second law is as follows. *No process is possible whose sole result is the absorption of heat from a reservoir at a single temperature and the conversion of this heat completely into mechanical work.*

If the second law were not true, it would be possible to drive a steamship across the ocean by extracting heat from the ocean or to run a power plant by extracting heat from the surrounding air. It should be noted again that neither of these "impossibilities" violates the first law of thermodynamics. After all, both the ocean and the surrounding air contain an enormous store of internal energy which, in principle, may be extracted in the form of a flow of heat. The second law, therefore, is not a deduction from the first but stands by itself as a separate law of nature, referring to an aspect of nature different from that contemplated by the first law. The first law denies the possibility of creating or destroying energy; the second denies the possibility of utilizing energy in a particular way.

The fact that work may be dissipated completely into heat, whereas heat may *not* be converted entirely into work, expresses an essential one-sidedness of nature. All natural, spontaneous processes may be studied in the light of the second law, and in all such cases, this peculiar one-sidedness is found. Thus, heat always flows spontaneously from a hotter to a colder body; gases always seep through an opening spontaneously from a region of high pressure to a region of low pressure; gases and liquids left by themselves always tend to mix, not to unmix. Salt dissolves in water but a salt solution does not separate by itself into pure salt and pure water. Rocks weather and crumble; iron rusts; people grow old. These are all examples of *irreversible* processes that take place naturally in only one direction and, by their one-sidedness, express the second law of thermodynamics.

19–19 THE REFRIGERATOR

A refrigerator may be considered to be a heat engine operated in reverse. A heat engine takes in heat from a hot reservoir, converts a part of the heat into mechanical work output, and rejects the diference as heat to a cold reservoir. A refrigerator, however, takes in heat from a cold reservoir, the compressor supplies mechanical work *input*, and heat is rejected to a hot reservoir. With reference to the ordinary home refrigerator, the food and ice cubes constitute the cold reservoir, work is done by the electric motor, and the hot reservoir is the air in the kitchen.

The flow diagram of a refrigerator is given in Fig. 19–12. In one cycle, heat Q_C enters the refrigerator at low temperature T_C, work W is done on the refrigerator and heat Q_H leaves at a higher temperature T_H. Both W and Q_H are negative quantities, while Q_C is positive. It follows from the first law that

$$-Q_H = Q_C - W,$$

Fig. 19–12 *Schematic flows diagram of a refrigerator.*

and the heat rejected to the hot reservoir is the sum of the heat taken from the cold reservoir and the heat equivalent of the work done by the motor.

From an economic point of view, the best refrigeration cycle is one that removes the greatest amount of heat Q_C from the refrigerator, for the least expenditure of mechanical work W. We therefore define *the coefficient of performance* (rather than the efficiency) of a refrigerator as the ratio $-Q_C/W$, and since $W = Q_H + Q_C$,

$$\text{Coefficient of performance} = -\frac{Q_C}{Q_H + Q_C}.$$

(19–26)

The principles of the common refrigeration cycle are illustrated schematically in Fig. 19–13. Compressor A delivers gas (CCl_2F_2, NH_3, etc.) at high temperature and pressure to coils B. Heat is removed from the gas in B by water or air cooling, resulting in condensation of the gas to a liquid still under high pressure. The liquid passes through the throttling valve or expansion valve C, emerging as a mixture of liquid and vapor at a lower temperature. In coils D, heat is supplied that converts the remaining liquid into vapor which enters compressor A to repeat the cycle. In a domestic refrigerator, coils D are placed in the ice compartment, where they cool the refrigerator directly. In a larger refrigerating

Low-pressure
gas

High-pressure
gas

Low-pressure
liquid

High-pressure
liquid

Fig. 19–13 *Principle of the mechanical refrigeration cycle.*

plant, these coils are usually immersed in a brine tank and cool the brine, which is then pumped to the refrigerating rooms.

If no work were needed to operate a refrigerator, the coefficient of performance (heat extracted divided by work done) would be infinite. Coefficients of performance of actual refrigerators vary from about 2 to about 6. Experience shows that work is always needed to transfer heat from a colder to a hotter body. This negative statement leads to another statement of the second law of thermodynamics, namely:

No process is possible whose sole result is the transfer of heat from a cooler to a hotter body.

At first sight this and the previous statement of the second law appear to be quite unconnected, but it can be shown that they are in all respects equivalent. Any device that would violate one statement would violate the other. For example, a "workless" refrigerator which violates this "refrigerator" statement of the second law could be used in conjunction with a real heat engine to violate our first statement of the second law by pumping the rejected heat back to the hot reservoir to be reused.

19-20 THE CARNOT CYCLE

Although their efficiencies differ from one another, none of the heat engines which have been described has an efficiency of 100%. An important question remains to /be discussed. What is the maximum attainable efficiency, given a supply of heat at one temperature and a reservoir at a lower temperature

for cooling the exhaust? An idealized engine which can be shown to have the maximum efficiency under these conditions was invented by the French engineer Saki Carnot in 1824 and is called a *Carnot engine.* The *Carnot cycle,* shown in a *p-V* diagram in Fig. 19–14, differs from the Otto and Diesel cycles in that it is bounded by two *isothermals* and two *adiabatics.* Thus all the heat input is supplied at a *single* high temperature and all the heat output is rejected at a *single* lower temperature. (Compare with Figs. 19–9 and 19–10, in which the temperature is different at all points of the lines *bc* and *da.*)

This, however, is not the only feature of the Carnot cycle. There are no "one-way" processes in the Carnot cycle, such as explosions or throttling processes. The isothermal and adiabatic processes of

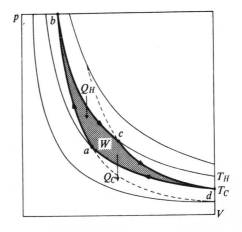

Fig. 19–14 *The Carnot cycle.*

the Carnot cycle are idealizations of actual processes. The direction of either process may be *reversed* by only a slight change in the external pressure; there is no friction, and the working substance is always very close to equilibrium.

   A process that is attended by no equilibrium-disturbing effect (such as an unbalanced force or a finite temperature difference) *and no dissipative effect* (such as friction or electrical resistance) *is said to be* **reversible.** (A reversible process represents the same sort of idealization in thermodynamics that a point mass, a weightless inextensible cord, or a frictionless, massless pulley represents in mechanics. It may be approximated but not perfectly achieved.)

   The isothermal and adiabatic processes of the Carnot cycle may be imagined to proceed in either direction. In the direction shown in Fig. 19–14, heat Q_H goes in, heat Q_C goes out, and work W is done by the engine. If the arrows in the figure are reversed, the cycle becomes a Carnot refrigeration cycle and, *what is most important, the same amount* of heat Q_H which formerly was taken in from the hot reservoir now goes out, *the same amount* of work W, which formerly was delivered to the outside is now required from the surroundings, and *the same amount* of heat Q_C which formerly was rejected to the cold reservoir is now taken in. These numerical equalities would *not* exist if any ordinary engine cycle were reversed.

   Suppose an engine (not a Carnot engine) were to operate between a source of heat at some temperature and a reservoir of heat at a lower temperature, thereby delivering to the outside an amount of work W. Suppose this work W were used to operate a Carnot refrigerator which extracted heat from the colder reservoir and delivered it to the warmer source. It can be shown that if the first engine were more efficient than the engine that would result by operating the Carnot refrigerator backward, then the net effect would be a violation of the second law of thermodynamics. Proceeding along these lines, it has been proved that:

No engine operating between two given temperatures can be more efficient than a Carnot engine operating between the same two temperatures.

The following, known as Carnot's theorem, also follows:

All Carnot engines operating between the same two temperatures have the same efficiency, irrespective of the nature of the working substance.

19–21 THE KELVIN TEMPERATURE SCALE

It follows from Carnot's theorem that the efficiency of a Carnot engine operating between reservoirs at two given temperatures *is independent of the nature of the working substance and is a function only of the temperatures.* If we consider a number of Carnot engines using different working substances and absorbing and rejecting heat to the same two reservoirs, the thermal efficiency is the same for all, or

$$\text{Thermal efficiency} = \frac{Q_H + Q_C}{Q_H} = 1 + \frac{Q_C}{Q_H}$$
$$= \text{constant.}$$

Hence, the ratio Q_H/Q_C is a constant for all the engines. Kelvin proposed that the ratio of the temperatures of the reservoirs be *defined* as equal to this constant ratio of the absolute magnitudes of the quantities of heat absorbed and rejected or, since Q_C is a negative quantity, as equal to the negative of the ratio Q_H/Q_C. Thus,

$$\boxed{\frac{T_H}{T_C} = \frac{|Q_H|}{|Q_C|} = -\frac{Q_H}{Q_C}.} \qquad (19\text{–}27)$$

Since the ratio of temperatures does not involve the properties of any particular substance, the Kelvin temperature scale is truly *absolute.*

   To complete the definition of the Kelvin scale we proceed, as in Chapter 15, to assign the arbitrary value of 273.16 K to the temperature of the triple point of water, T_3. For a Carnot engine operating between reservoirs at the temperatures T and T_3, we have

$$\frac{|Q|}{|Q_3|} = \frac{T}{T_3}$$

or

$$T = 273.16 \text{ K} \frac{|Q|}{|Q_3|}. \qquad (19\text{–}28)$$

Comparing this with the corresponding expression for the ideal gas temperature, namely,

$$273.16 \text{ K} \left[\lim_{p_3 \to 0} \left(\frac{p}{p_3} \right)_{\text{constant volume}} \right],$$

it is seen that, in the Kelvin scale, Q plays the role of a "thermometric property." This does not, however, have the objection attached to a coordinate of an arbitrarily chosen thermometer, inasmuch as the behavior of a Carnot engine is independent of the nature of the working substance.

It is a simple matter to show that if an ideal gas is taken around a Carnot cycle, the ratio of the heats absorbed and rejected, $|Q_H|/|Q_C|$, is equal to the ratio of the temperatures of the reservoirs *as expressed on the gas scale*, defined in Chapter 15. (The proof is somewhat lengthy, and will not be given here.) Since, in both scales, the triple point of water is chosen to be 273.16 K, it follows that *the Kelvin and the ideal gas scales are identical.*

The efficiency of a Carnot engine rejecting heat Q_C to a reservoir at Kelvin temperature T_C and absorbing heat Q_H from a source at Kelvin temperature T_H is, as usual,

$$\text{Efficiency} = 1 + \frac{Q_C}{Q_H} = 1 - \frac{|Q_C|}{|Q_H|}.$$

But, by definition of the Kelvin scale,

$$\frac{|Q_C|}{|Q_H|} = \frac{T_C}{T_H}.$$

Therefore, the efficiency of a Carnot engine is

$$\text{Efficiency (Carnot)} = 1 - \frac{T_C}{T_H}. \qquad (19\text{–}29)$$

Equation (19–29) points the way to the conditions which a real engine, such as a steam engine,

must fulfill to approach as closely as possible the maximum attainable efficiency. These conditions are that the intake temperature T_H must be made as high as possible and the exhaust temperature T_C as low as possible.

The exhaust temperature cannot be lower than the lowest temperature available for cooling the exhaust. This is usually the temperature of the air, or perhaps of river water if this is available at the plant. The only recourse then is to raise the boiler temperature T_H. Since the vapor pressure of all liquids increases rapidly with increasing temperature, a limit is set by the mechanical strength of the boiler. Another possibility is to use, instead of water, some liquid with a lower vapor pressure. Successful experiments in this direction have been made with mercury vapor replacing steam. At a boiler temperature of 200°C, at which the pressure in a steam boiler would be 225 lb in^{-2}, the pressure in a mercury boiler is only 0.35 lb in^{-2}. Liquid sodium is sometimes used for heat transfer in nuclear reactors for similar reasons.

The unavoidable exhaust heat loss in electric power plants creates a serious environmental problem. When a lake or river is used for cooling, the temperature of the body of water may be raised several degrees. Such a temperature change has a severely disruptive effect on the overall ecological balance, inasmuch as relatively small temperature changes can have significant effects on metabolic rates in plants and animals. Since *thermal pollution*, as this effect is called, is an inevitable consequence of the second law of thermodynamics, careful planning is essential to minimize the ecological impact of new power plants.

19–22 ABSOLUTE ZERO

It follows from Eq. (19–28) that the heat transferred isothermally between two given adiabatics decreases as the temperature decreases. Conversely, the smaller the value of Q, the lower the corresponding T. The smallest possible value of Q is zero, and the corresponding T is absolute zero. *Thus, if a system undergoes a reversible isothermal process without transfer of heat, the temperature at which this process takes place is called absolute zero.* In other words, at absolute zero, an isotherm and an adiabatic are identical.

It should be noted that the definition of absolute zero holds for all substances and is therefore independent of the peculiar properties of any one arbitrarily chosen substance. Furthermore, the definition is in terms of purely large-scale concepts. No reference is made to molecules or to molecular energy. Whether absolute zero may be achieved experimentally is a question of some interest and importance. To achieve temperatures below 4.2 K, at which ordinary helium (mass number 4) liquefies, it is necessary to lower the vapor pressure by pumping away the vapor as fast as possible. The lowest temperature that has ever been reached in this way is 0.7 K, and this required larger pumps and larger pumping tubes than are usually employed in low-temperature laboratories. With the aid of the light isotope of helium (mass number 3) which liquefies at 3.2 K, vigorous pumping yields a temperature of about 0.3 K. Still lower temperatures may be achieved magnetically, but it becomes apparent that the closer one approaches absolute zero, the more difficult it is to go further. It is generally accepted as a law of nature that, although absolute zero may be approached as closely as we please, it is impossible actually to reach the zero of temperature This is known as the *"unattainability statement of the third law of thermodynamics."*

19–23 ENTROPY

The first law of thermodynamics is the law of energy, the second law of thermodynamics is the law of entropy, and every process that takes place in nature, whether it be mechanical, electrical, chemical, or biological, must proceed in conformity with these two laws.

A book of this nature is not the place for a thorough exposition of entropy and the second law of thermodynamics. We shall content ourselves with defining entropy, computing its changes in a few instances, and stating some of its properties.

A re-reading of Section 19–4 will recall to mind that when a system is carried from one state to another it is found by experiment that the difference between the heat added and the work done by the system, $Q - W$, has the same value for all paths. The fact that this difference does have the same value makes it possible to introduce the concept of internal energy, the change in internal energy being defined and measured by the quantity $Q - W$.

Entropy, or rather a change in entropy, may be defined in a similar way. Consider two states of a system and a number of *reversible* paths connecting them. (The restriction to reversible paths was not necessary for internal energy changes.) Although the heat added to the system is different along different paths, it may be proved that if the heat ΔQ added in each small interval of the path is divided by the *absolute* temperature T of the system in the interval, and the resulting ratios summed for the entire path, this sum has the same value for all (reversible) paths between the same endpoints. In mathematical symbols,

$$\int_1^2 \frac{dQ}{T} = \begin{array}{l}\text{constant for all reversible} \\ \text{paths between states 1 and 2.}\end{array}$$

It is therefore *possible* (whether it is of any use or not one can only tell later) to introduce a function whose difference between two states, 1 and 2, is defined by the integral above. This function may be assigned any arbitrary value in some standard reference state, and its value in any other state will be a definite quantity. The function is called the *entropy* of the system, and is denoted by S. We then have

$$S_2 - S_1 = \int_1^2 \frac{dQ}{T} \quad \text{(along any reversible path).}$$

(19–30)

If the change is infinitesimal, $dS = dQ/T$. The unit of entropy is 1 J K^{-1}, 1 cal K^{-1}, 1 Btu (°R)$^{-1}$, etc.

Example 1   One kilogram of ice at 0°C is melted and converted to water at 0°C. Compute its change in entropy.

Since the temperature remains constant at 273 K, T may be taken outside the integral sign. Then

$$S_2 - S_1 = \frac{1}{T} \int dQ = \frac{Q}{T}.$$

But Q is simply the total heat which must be supplied to melt the ice, or 80,000 cal. Hence

$$S_2 - S_1 = \frac{80,000}{273} = 293 \text{ cal K}^{-1},$$

and the increase in entropy of the system is 293 cal K^{-1}. In any *isothermal* reversible process, the entropy change equals the heat added divided by the absolute temperature.

Example 2   One kilogram of water at 0°C is heated to 100°C. Compute its change in entropy.

The temperature is not constant and dQ and T must be expressed in terms of a single variable in order to carry out the integration. This may readily be done, since

$$dQ = mc\, dT.$$

Hence

$$S_2 - S_1 = \int_{273K}^{373K} mc \frac{dT}{T} = mc \ln \frac{373 \text{ K}}{273 \text{ K}}$$

$$= 312 \text{ cal K}^{-1}$$

$$= 1306 \text{ J K}^{-1}.$$

Example 3   A gas is allowed to expand adiabatically and reversibly. What is its change in entropy?

In an adiabatic process no heat is allowed to enter or leave the system. Hence $Q = 0$ and there is no change in entropy. It follows that every *reversible* adiabatic process is one of constant entropy, and may be described as *isentropic*.

19–24 THE PRINCIPLE OF THE INCREASE OF ENTROPY

One of the features that distinguish entropy from such concepts as energy, momentum, and angular momentum is that *there is no principle of conservation of entropy.* In fact, the reverse is true. Entropy can be created at will, and there is an increase in entropy in every natural process, if all systems taking part in the process are considered.

Consider the process of mixing 1 kg of water at 100°C with 1 kg of water at 0°C. Let us arbitrarily call the entropy of water zero when it is in the liquid state at 0°C. (This is the reference state often used in engineering work.) Then from Example 2 in the previous section, the entropy of 1 kg of water at 100°C is 312 cal K^{-1} and the entropy of 1 kg at 0°C is zero. The entropy of the system, before mixing, is therefore 312 cal K^{-1}.

After the hot and cold water have been mixed, we have 2 kg of water at a temperature of 50°C or 323 K. From the results of Example 2, we see that the entropy of the system is

$$mc \ln \frac{323}{273} = (2000) \ln \frac{373}{273} = 336 \text{ cal K}^{-1}.$$

There has therefore been an *increase* in entropy of

$$(336 - 312) \text{ cal K}^{-1} = 24 \text{ cal K}^{-1}.$$

Physical mixing of the hot and cold water is, of course, not essential in bringing about the final equilibrium state. We might simply have let heat flow by conduction, or be transferred by radiation, from the hot to the cold water. The same increase in entropy would have resulted.

These examples of the mixing of substances at different temperatures, or the flow of heat from a higher to a lower temperature, are illustrative of all natural (i.e., irreversible) processes. When all the entropy changes in the process are included, the increases in entropy are always greater than the decreases. In the special case of a reversible process, the increases and decreases are equal. Hence we can formulate the general principle, which is considered a part of the second law of thermodynamics, that *when all systems taking part in a process are included, the entropy either remains constant or increases.* In other words, *no process is possible in which the entropy decreases,* when all systems taking part in the process are included.

What is the significance of the increase of entropy that accompanies every natural process? The answer, or one answer, is that it represents the extent to which the Universe "runs down" in that process. Consider again the example of the mixing of hot and cold water. We *might* have used the hot and cold water as the high- and low-temperature reservoirs of a heat engine, and in the course of removing heat from the hot water and giving heat to the cold water we could have obtained some mechanical work. But once the hot and cold water have been mixed and have come to a uniform temperature, this opportunity of converting heat to mechanical work is lost and, moreover, it is lost irretrievably. The lukewarm water will never *unmix* itself and separate into a hotter and a colder portion.* Of course, there is no decrease in *energy* when the hot and cold water are mixed, and what has been "lost" in the mixing process is not *energy*, but *opportunity;* the opportunity to convert a portion of the heat flowing out of the hot water to mechanical work. Hence when entropy increases, energy becomes more unavailable, and we say that the Universe has "run down" to that extent. This is the true significance of the term "irreversible."

The tendency of all natural processes such as heat flow, mixing, diffusion, etc., is to bring about a uniformity of temperature, pressure, composition, etc., at all points. One may visualize a distant future in which, as a consequence of these processes, the entire Universe has attained a state of absolute uniformity throughout. When and if such a state is reached, although there would have been no change in the energy of the Universe, all physical, chemical, and presumably biological processes would have to cease. This goal toward which we appear headed has been described as the "heat death" of the Universe.

19–25 ENERGY CONVERSION

We have seen in this chapter that the laws of thermodynamics place very general limitations on conversion of energy from one form to another. In this day of increasing energy demand and diminishing resources, these matters are of the utmost practical importance. We conclude this chapter with a brief

---

* The branch of physics called "statistical mechanics" would modify this statement to read, "It is highly improbable that the water will separate spontaneously into a hotter and a colder portion, but it is not impossible."

discussion of a few energy-conversion systems, present and proposed.

Most (over 80 percent) of the electric power generated in the United States is obtained from coal-fired steam-turbine generating plants. Modern boilers can transfer about 80 to 90 percent of the heat of combustion of coal into steam. The theoretical thermodynamic efficiency of the turbine, given by Eq. (19–29), is usually limited to about 0.55, and the actual efficiency is typically 90 percent of the theoretical value, or about 0.50. The efficiency of large electrical generators in converting mechanical power to electrical is very large, typically 99 percent. Thus the overall thermal efficiency of such a plant is roughly $(0.85)(0.50)(0.99)$, or about 40 percent.

In 1970, a generator with a capacity of about $1 \text{ GW} \cdot (= 1000 \text{ MW} = 10^9 \text{ W})$ went into operation at the Tennessee Valley Authority's Paradise power plant. The steam is heated to $1003°F$, at a pressure of $3,650 \text{ lb in}^{-2}$. The plant consumes 10,500 tons of coal per day, and the overall thermal efficiency is 39.3 percent.

Nuclear power plants have the same theoretical efficiency limit as coal-fired plants. Because at present it is not practical to run nuclear reactors at as high temperatures as coal boilers, the theoretical thermal efficiency is usually lower. The overall thermal efficiency of a nuclear plant is typically 30 percent.

In both coal-fired and nuclear plants, the energy not converted to electrical energy is wasted and must be disposed of. A common practice is to locate such a plant near a lake or river and use the water for disposal of excess heat; this can raise the water temperature several degrees, often with serious ecological consequences.

Solar energy is an interesting possibility. The power in the sun's radiation is about 1.4 kW per square meter of area at the surface of the earth. This radiation could be collected and focused with mirrors and used to generate steam. A more exotic proposal is to use large banks of photocells for direct conversion to electricity; this conversion, since it does not use heat as an intermediate step, is not limited by the Carnot efficiency but could in principle be carried out with 100-percent efficiency.

An indirect scheme for conversion of solar energy would use the temperature gradient in the ocean. In the Caribbean, for example, the water temperature near the surface is about 25°C, while at a depth of a few hundred meters it may be 10°C. While the Second Law of Thermodynamics forbids taking heat from the ocean and converting it completely into work, there is nothing to forbid running a heat engine between these two temperatures. The thermodynamic efficiency would be very low, but with such a vast reservoir of energy available this would not be a serious problem.

This is only a small sample of present-day activity in energy-conversion research. Many other processes are under discussion or development, and the principles of thermodynamics outlined in this chapter are of central importance in all of them.

PROBLEMS

19–1 A combustion experiment is performed by burning a mixture of fuel and oxygen in a constant-volume "bomb" surrounded by a water bath. During the experiment the temperature of the water is observed to rise. Regarding the mixture of fuel and oxygen as the system; (a) has heat been transferred? (b) has work been done? (c) what is the sign of ΔU?

19–2 A liquid is irregularly stirred in a well-insulated container and thereby undergoes a rise in temperature. Regarding the liquid as the system; (a) has heat been transferred? (b) has work been done? (c) what is the sign of ΔU?

19–3 A resistor immersed in running water carries an electric current. Consider the resistor as the system under consideration.

a) Is there a flow of heat into the resistor?

b) Is there a flow of heat into the water?

c) Is work done?

d) Assuming the state of the resistor to remain unchanged, apply the first law to this process.

19–4 In certain process, 500 cal of heat are supplied to a system, and at the same time 100 J of work are done on the system. What is the increase in the internal energy of the system?

19–5 A piece of ice falls from rest into a lake at 0°C, and one-half of one percent of the ice melts. Compute the minimum height from which the ice falls.

19–6 What must be the initial velocity of a lead bullet at a temperature of 25°C, so that the heat developed when it is brought to rest shall be just sufficient to melt it?

19–7 In a certain process, 200 Btu are supplied to a system and at the same time the system expands against a constant external pressure of 100 lb in^{-2}. The internal energy of the system is the same at the beginning and end of the process. Find the increase in volume of the system.

19–8 An inventor claims to have developed an engine which takes in 100,000 Btu from its fuel supply, rejects 25,000 Btu in the exhaust, and delivers 25 kWh of mechanical work. Do you advise investing money to put this engine on the market?

19–9 A vessel with rigid walls and covered with asbestos is divided into two parts by an insulating partition. One part contains a gas at temperature T and pressure p. The other part contains a gas at temperature T' and pressure p'. The partition is removed. What conclusion may be drawn by applying the first law of thermodynamics?

19–10 A mixture of hydrogen and oxygen is enclosed in a rigid insulating container and exploded by a spark. The temperature and pressure both increase considerably. Neglecting the small amount of energy provided by the spark itself, what conclusion may be drawn by applying the first law of thermodynamics?

19–11 When water is boiled under a pressure of 2 atm, the heat of vaporization is 2.20×10^6 J kg^{-1} and the boiling point is 120°C. At this pressure, one kilogram of water has a volume of 10^{-3} m^3, and one kilogram of steam a volume of 0.824 m^3.

a) Compute the work done when one kilogram of steam is formed at this temperature.

b) Compute the increase in internal energy.

19–12 When a system is taken from state a to state b, in Fig. 19–15, along the path acb, 80 J of heat flow into the system, and 30 J of work are done.

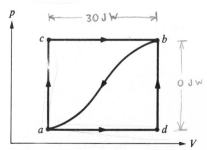

Fig. 19–15

a) How much heat flows into the system along path adb if the work is 10 J?

b) When the system is returned from b to a along the curved path, the work is 20 J. Does the system absorb or liberate heat, and how much?

c) If $U_a = 0$ and $U_d = 40$ J, find the heat adsorbed in the processes ad and db.

19–13 A steel cylinder of cross-sectional area 0.1 ft² contains 0.4 ft³ of glycerin. The cylinder is equipped with a tightly fitting piston which supports a load of 6000 lb. The temperature of the system is increased from 60°F to 160°F. Neglect the expansion of the steel cylinder. Find (a) the increase in volume of the glycerin, (b) the mechanical work of the 6000 lb force, (c) the amount of heat added to the glycerin [specific heat of glycerin = 0.58 Btu lb⁻¹(F°)⁻¹], (d) the change in internal energy of the glycerin.

19–14 The volume of 1 mole of an ideal gas is increased isothermally (T = constant) from 1 to 20 liters at 0°C. The pressure of the gas at any moment is given by the equation $pV = RT$, where $R = 8.31$ J mole⁻¹ K⁻¹ and T is the kelvin temperature. How many joules of work are done?

19–15 Calculate the work done when a gas expands from volume V_1 to V_2, the relation between pressure and volume being

$$\left(p + \frac{a}{V^2}\right)(V - b) = K,$$

where a, b, and K are constants.

19–16

a) The change in internal energy, dU, of a system consisting of n moles of a pure substance, in an infinitesimal process at constant volume, is equal to $nC_v\, dT$. Explain why the internal energy change in a process at constant pressure is *not* equal to $nC_p\, dT$.

b) Explain why the change in internal energy of an ideal gas, in *any* infinitesimal process, is given by $nC_v\, dT$.

19–17 A cylinder contains 1 mole of oxygen gas at a temperature of 27°C. The cylinder is provided with a frictionless piston which maintains a constant pressure of 1 atm on the gas. The gas is heated until its temperature increases to 127°C.

a) Draw a diagram representing the process in the pV-plane.

b) How much work is done by the gas in this process?

c) On what is this work done?

d) What is the change in internal energy of the gas?

e) How much heat was supplied to the gas?

f) How much work would have been done if the pressure had been 0.5 atm?

19–18

a) Compare the quantity of heat required to raise the temperature of 1 g of hydrogen through 1 C°, at constant pressure, with that required to raise the temperature of 1 g of water by the same amount.

b) Of the substances listed in Table 19–1, which has the largest specific heat capacity in cal g⁻¹ (C°)⁻¹?

19–19 Ten liters of air at atmospheric pressure are compressed isothermally to a volume of 2 ℓ, and are then allowed to expand adiabatically to a volume of 10 ℓ. show the process in a pV-diagram.

19–20 An ideal gas is contained in a cylinder closed with a movable piston. The initial pressure is 1 atm and the initial volume is 1 ℓ. The gas is heated at constant pressure until the volume is doubled, then heated at constant volume until the pressure is doubled, and finally expanded adiabatically until the temperature drops to its initial value. Show the process in a pV-diagram.

19–21 An ideal gas initially at 10 atm and 300 K is permitted to expand adiabatically until its volume doubles. Find the final pressure and temperature if the gas is (a) monatomic; (b) diatomic.

19–22 A gasoline engine takes in air at 20° C and 1 atm, and compresses it adiabatically to 1/10 the original volume. Find the final temperature and pressure.

19–23 An ideal gas undergoes an adiabatic expansion, during which its temperature drops from T_1 to T_2. Show that the work done by the gas is given by $nC_v(T_1 - T_2)$, where n is the number of moles.

19–24 A certain ideal gas has $\gamma = 1.33$. Determine the molar heat capacities at constant volume and at constant pressure.

19–25 Compressed air at a pressure of 2×10^6 Pa is used to drive an air engine which exhausts at a pressure of 2×10^5 Pa. What must be the temperature of the compressed air in order that there may be no possibility of frost forming in the exhaust ports of the engine? Assume the expansion to be adiabatic. (*Note:* Frost frequently forms in the exhaust ports of an air-driven engine. This happens when the moist air is cooled below 0°C by the expansion which takes place in the engine.)

19–26 Initially at a temperature of 140°F, 10 ft³ of air expands at a constant gauge pressure of 20 lb in⁻² to a volume of 50 ft³, and then expands further adiabatically to a final volume of 80 ft³ and a final gauge pressure of 3 lb in⁻². Sketch the process in the pV-plane and compute the work done by the air.

19–27 The cylinder of a pump compressing air from atmospheric pressure into a very large tank at 60 lb in⁻² gauge pressure is 10 in. long.

a) At what position in the stroke will air begin to enter the tank? Assume the compression to be adiabatic.

b) If the air is taken into the pump at 27°C, what is the temperature of the compressed air?

19–28 Two moles of helium are initially at a temperature of 27° C and occupy a volume of 20 ℓ. The helium is first expanded at constant pressure until the volume has doubled, and then adiabatically until the temperature returns to its initial value.

a) Draw a diagram of the process in the pV-plane.

b) What is the total heat supplied in the process?

c) What is the total change in internal energy of the helium?

d) What is the total work done by the helium?

e) What is the final volume?

19–29 A heat engine carries 0.1 mole of an ideal gas around the cycle shown in the pV diagram of Fig. 19–16. Process 1–2 is at constant volume, process 2–3 is adiabatic, and process 3–1 is at a constant pressure of 1 atm. The value of γ for this gas is 5/3.

a) Find the pressure and volume at points 1, 2, and 3.

b) Find the net work done by the gas in the cycle.

Fig. 19–16

19–30 A cylinder contains oxygen at a pressure of 2 atm. The volume is 3 ℓ and the temperature is 300 K. The oxygen is carried through the following processes:

1) Heated at constant pressure to 500 K.

2) Cooled at constant volume to 250 K.

3) Cooled at constant pressure to 150 K.

4) Heated at constant volume to 300 K.

a) Show the four processes above in a pV-diagram, giving the numerical values of p and V at the end of each process.

b) Calculate the net work done by the oxygen.

c) Find the net heat flowing into the oxygen.

d) What is the efficiency of this device as a heat engine?

19–31 What is the thermal efficiency of an engine which operates by taking an ideal gas through the following cycle? Let $C_v = 3$ cal mole⁻¹(C°)⁻¹.

1) Start with n moles at P_0, V_0, T_0.

2) Change to $2P_0$, V_0, at constant volume.

3) Change to $2P_0$, $2V_0$, at constant pressure.

4) Change to P_0, $2V_0$, at constant volume.

5) Change to P_0, V_0, at constant pressure.

19–32 A Carnot engine whose low-temperature reservoir is at 280 K has an efficiency of 40%. It is desired to increase this to 50%.

a) By how many degrees must the temperature of the high-temperature reservoir be increased if the temperature of the low-temperature reservoir remains constant?

b) By how many degrees must the temperature of the low-temperature reservoir be decreased if that of the high-temperature reservoir remains constant?

19–33 A Carnot engine whose high-temperature reservoir is at 400 K takes in 100 cal of heat at this temperature in each cycle, and gives up 80 cal to the low-temperature reservoir.

a) What is the temperature of the latter reservoir?

b) What is the thermal efficiency of the cycle?

19–34 A Carnot refrigerator takes heat from water at 0°C and rejects heat to a room at 27°C. Suppose that 50 kg of water at 0°C are converted to ice at 0°C.

a) How much heat is rejected to the room?

b) How much energy must be supplied to the refrigerator?

19–35 A Carnot engine is operated between two heat reservoirs at temperatures of 400 K and 300 K.

a) If the engine receives 1200 cal from the reservoir at 400 K in each cycle, how many calories does it reject to the reservoir at 300 K?

b) If the engine is operated in reverse, as a refrigerator, and receives 1200 cal from the reservoir at 300 K, how many calories does it deliver to the reservoir at 400 K?

c) How many calories would be produced if the mechanical work required to operate the refrigerator in part (b) were converted directly to heat?

19–36 Is it possible to cool a closed, insulated room by operating an electric refrigerator in the room, leaving the refrigerator door open?

19–37 Consider the entropy of water to be zero when it is in the liquid phase at 0°C and atmospheric pressure.

a) How much heat must be supplied to 1 kg of water to raise its temperature from 0°C to 100°C?

b) What is the entropy of 1 kg of water at 100°C?

c) 1 kg of water at 0°C is mixed with 1 kg at 100°C. What is the final temperature?

d) Find the entropy of the hot and cold water before mixing, and the entropy of the system after mixing. (This is another example of the increase of entropy in an irreversible process.)

19–38

a) Draw a graph of a Carnot cycle, plotting Kelvin temperature vertically and entropy horizontally (a temperature–entropy diagram).

b) Show that the area under any curve in a temperature–entropy diagram represents the heat absorbed by the system.

c) Derive from your diagram the expression for the thermal efficiency of a Carnot cycle.

19–39 Heat is added to 0.5 kg of ice at 0° C until it is all melted.

a) What is the change in entropy of the water?

b) If the source of heat is a very massive body at a temperature of 20° C, what is the change in entropy of this body?

c) What is the total change in entropy of the water and the heat source?

19–40 A block of aluminum of mass 1 kg, initially at 100° C, is dropped into 1 kg of water initially at 0° C.

a) What is the final temperature?

b) What is the total change in entropy of the system?

19–41 Two moles of an ideal gas undergo a reversible isothermal expansion from 0.02 m^3 to 0.04 m^3 at a temperature of 300 K. What is the change in entropy of the gas?

19–42 A solar power plant is to be built with a power output capacity of 1000 MW. What land area must the solar energy collectors occupy if they are: (a) photocells with 90-percent efficiency? (b) mirrors that generate steam for a turbine-generator unit with overall efficiency of 30 percent?

19–43 An engine is to be built to extract power from the temperature gradient of the ocean. If the surface and deep-water temperatures are 25° C and 10° C, respectively, what is the maximum theoretical efficiency of such an engine?

19–44 A coal-fired steam-turbine power plant that produces 1000 MW of electrical power is located beside a river. The overall thermal efficiency of the plant is 40 percent.

a) What is the total thermal power input to the plant?

b) At what rate is waste heat discharged from the plant?

c) If the waste heat is delivered to the river, and if the temperature rise must be no greater than 5 C°, how much water must be available per second?

d) In part (c), if the river is 100 m wide and 5 m deep, what must be the minimum flow velocity of the water?

Chapter 20

Molecular Properties of Matter

20–1 MOLECULAR THEORY OF MATTER

Thousands of physical and chemical observations support the hypothesis that matter in all phases is composed of tiny particles called *molecules*. The molecules of any one substance are identical. They have the same structure, the same mass, and the same mechanical and electrical properties. Many of the large-scale properties of matter which have been discussed heretofore, such as elasticity, surface tension, condensation, and vaporization, can be comprehended with deeper understanding in terms of the molecular theory.

The simplest model of a molecule is that of a rigid sphere, like a small billiard ball, capable of moving, of colliding with other molecules or with a wall, and of exerting attractive or repulsive forces on neighboring molecules. In other parts of physics and chemistry it is important to consider the structure of the molecule, but it is not necessary at this point. The smallest molecules are of the order of 10^{-10} m in size; the largest are at least 10,000 times this large.

One essential characteristic of a molecule is the force that exists between it and a neighboring molecule. There is, of course, a force of *gravitational* attraction between every pair of molecules, but it turns out that this is negligible in comparison with

the forces to be considered now. The forces that hold the molecules of a liquid (or solid) together are chiefly of *electrical* origin and do not follow a simple inverse-square law.

When the separation of the molecules is large, as in a gas, the force is extremely small and attractive. The attractive force increases as a gas is compressed and its molecules are brought closer together. But since tremendous pressures are needed to compress a liquid (i.e., to force its molecules closer together than their normal spacing in the liquid state), we conclude that at separations only slightly less than the dimensions of a molecule the force is repulsive and relatively large. Thus the force must vary with separation in somewhat the fashion shown in Fig. 20–1. At large separations the force is small and attractive. As the molecules are brought closer together, the force of attraction becomes larger, passes through a maximum, and then decreases to zero at a separation r_0. When the distance between the molecules is less than r_0, the force becomes repulsive.

A single pair of molecules could remain in equilibrium at a center-to-center spacing equal to r_0 in Fig. 20–1. If they were separated slightly, the force between them would be attractive and they would be drawn together. If they were forced closer together than the distance r_0, the force would be one of

Fig. 20–1 *The force between two molecules changes from an attraction when the separation is large, to a repulsion when the separation is small.*

repulsion and they would spring apart. If they were either pulled apart or pushed together, and then released, they would oscillate about their equilibrium separation r_0.

If the only property of a molecule were the force of attraction between it and its neighbors, all matter would eventually coalesce into the liquid or solid phase. The existence of gases points to another property which enables molecules to stay apart, namely, molecular motion. The more vigorous the motion, the less chance there is for condensation into the liquid or solid phase. In solids the molecules execute vibratory motion about more or less fixed centers. The vibratory motion is relatively weak, and the centers remain fixed at regularly spaced positions which compare a *space lattice.* This gives rise to the extraordinary regularity and symmetry of crystals. A photograph of the individual, very large molecules of *necrosis virus protein* is shown in Fig. 20–2. It was taken with an electron microscope at a magnification of about 80,000. The molecules look like neatly stacked oranges.

In liquids the intermolecular distances are usually only a trifle greater than those of a solid. The molecules execute vibratory motion of greater energy about centers which are free to move but which remain at approximately the same distances from one another. Liquids show a certain regularity of structure only in the immediate neighborhood of a few molecules. This is called *short-range order*, in contrast with the *long-range order* of a solid crystal.

The molecules of a gas have the greatest kinetic energy, so great in fact that they are, on the average, far away from one another, where only very small attractive forces exist. A molecule of a gas therefore moves with linear motion until a collision takes place either with another molecule or with a wall. In molecular terms, an *ideal gas* is a gas whose molecules exert no forces of attraction at all. The mathematical analysis of a collection of such idealized molecules in random linear motion between collisions is called the *kinetic theory of gases*, and was perfected by Clausius, Maxwell, and Boltzmann in the latter part of the nineteenth century. The analysis of an ideal gas given in this chapter is the simplest example of kinetic theory.

Most common substances exist in the solid phase at low temperatures. When the temperature is raised beyond a definite value, the liquid phase results, and when the temperature of the liquid is

Fig. 20–2 *Crystal of necrosis virus protein. The actual size of the entire crystal is about two thousandths of a millimeter (0.002 mm). By permission of Dr. Ralph Wyckoff, University of Arizona.*

raised further, the substance exists in the gaseous phase; i.e., from a large-scale or *macroscopic* point of view, the transition from solid to liquid to gas is in the direction of increasing temperature. From a *molecular* point of view, this transition is in the direction of increasing molecular kinetic energy. Evidently temperature and molecular kinetic energy are related.

20-2 AVOGADRO'S NUMBER

Surprising as it may seem, it was not until 1811 that the suggestion was first made, by Dalton, that the molecules of a given chemical substance are all alike. Before this, even by those who accepted a molecular theory, it had been assumed that the ultimate particles of the same substance varied in size and shape, just as one might find pebbles of different sizes, all made of the same material.

We now know that a mole of any pure substance contains a definite number of identical molecules. The number of molecules in a mole is called *Avogadro's number*, denoted N_A. This number can be determined in a variety of ways; the first reasonably precise determination was made by the French physicist Jean Perrin in the early 1900's. Before describing Perrin's experiment, we first recall an expression previously derived for the variation of atmospheric pressure with elevation. In Example 3 at the end of Section 18–2 we derived from the laws of hydrostatics and the equation of state of an ideal gas an expression for the ratio of the pressures at two different elevations in an ideal gas at a uniform temperature:

$$\ln \frac{p_2}{p_1} = -\frac{Mg}{RT}(y_2 - y_1). \qquad (20\text{-}1)$$

The molecular mass M equals the mass of one mole, and can be expressed as the product of Avogadro's number N_A and the mass m of a single molecule:

$$M = N_A m \qquad (20\text{-}2)$$

Also, the number of moles n in a given sample of a pure substance equals the total number of molecules

N divided by Avogadro's number N_A:

$$n = \frac{N}{N_A}.$$

The ideal gas law can therefore be written

$$pV = nRT = \frac{N}{N_A} RT,$$

or

$$\boldsymbol{n} = \frac{N}{V} = p \frac{N_A}{RT}, \qquad (20\text{-}3)$$

where $\boldsymbol{n}$ is the number of molecules per unit volume, N/V. Since N_A and R are universal constants, it follows that at constant temperature T the number of molecules per unit volume is directly proportional to the pressure. Combining Eqs. (20–2) and (20–3) with Eq. (20–1), we get

$$\ln \frac{\boldsymbol{n}_2}{\boldsymbol{n}_1} = -\frac{N_A mg}{RT}(y_2 - y_1), \qquad (20\text{-}4)$$

where $\boldsymbol{n}_2$ and $\boldsymbol{n}_1$ are the numbers of molecules per unit volume at elevations y_2 and y_1.

One might well ask the following question: If the earth's atmosphere consists of widely separated molecules, not in contact with each other, and if each one is attracted by the earth, then why do the molecules not all settle down to the earth's surface like drops of rain? The explanation is found in the random motions of the molecules, which, in the absence of any gravitational force, would eventually result in their dispersal throughout interstellar space. There are thus two competing tendencies, the gravitational force drawing all molecules toward the earth and their random motions tending to disperse them. The actual distribution represents the compromise between these two tendencies.

It is well known that extremely small particles, called *colloidal* particles, in a liquid of density less than that of the particles, do not all settle to the bottom but remain permanently suspended. These particles are large enough to be observed with a microscope, and the number per unit volume at different heights in the liquid can be counted. Perrin found that the number per unit volume decreased with height, following the same law as predicted by Eq. (20–4) for the number of molecules per unit volume in the atmosphere. Furthermore, the particles are observed to be in a state of continuous motion, called "Brownian motion" after the English botanist Robert Brown, who first observed this effect in 1809 when studying with a microscope small pollen grains suspended in a liquid.

Perrin reasoned that the colloidal particles were analogous in their behavior to the molecules of a "gas," molecules which were actually large enough to be seen and counted. If a colloidal suspension could be made in which all the particles were exactly alike, and if the mass m of each could be determined, and if Eq. (20–4) really describes the vertical distribution of particles, then Avogadro's number could be found by counting the number of particles per unit volume at two known heights in the suspension.

It would occupy too much space to describe the ingenious methods used by Perrin to obtain uniform particles, to count the number per unit volume, and to measure their masses. Suffice it to say that the difficulties were surmounted and that Perrin finally obtained for N_A a value between 6.5 and 7.2×10^{23} molecules per mole. The most precise value of N_A to date, obtained by using x-rays to measure the distance between the layers of molecules in a crystal, is

$$N_A = 6.02217 \times 10^{23} \text{ molecules mol}^{-1}.$$

Once Avogadro's number has been determined, it can be used to compute the mass of a molecule. The mass of 1 mole of atomic hydrogen is 1.008 g. Since by definition the number of molecules in 1 mole is Avogadro's number, it follows that the mass of a single atom of hydrogen is

$$m_H = \frac{1.008 \text{ g mol}^{-1}}{6.022 \times 10^{23}} = 1.673 \times 10^{-24} \text{ g}.$$

The mass of an atom of atomic mass 1 is $1/N_A$ or 1.660×10^{-24} g. Thus, for an oxygen molecule of molecular mass 32,

$$m_{O_2} = (32)(1.660 \times 10^{-24} \text{ g}) = 53.12 \times 10^{-24} \text{ g}.$$

At standard conditions, 1 mole of an ideal gas occupies 22,400 cm³. The number of molecules per cubic centimeter in an ideal gas at standard conditions is therefore

$$\frac{6.022 \times 10^{23}}{22,400 \text{ cm}^3} = 2.69 \times 10^{19} \text{ molecules cm}^{-3}.$$

This is known as *Loschmidt's number*, and of course it is the same for *all* gases. The number of molecules per unit volume can also be computed from Eq. (20–3), at any pressure and temperature.

Numbers of this enormous magnitude are very difficult to comprehend. Sir James Jeans, the eminent British astrophysicist, offered a striking illustration. Each human breath contains the order of 400 cm³, and thus the order of 10^{22} molecules. Let us assume that George Washington's dying breath has by now become uniformly distributed through the atmosphere, which Jeans estimated to contain 10^{44} molecules. Thus the number of molecules in a breath is roughly equal to the number of breaths in the atmosphere, and since the total volume of the lungs (typically 2000 to 3000 cm³) is several times the volume of a single breath, each of us at this moment probably has several molecules of Washington's last breath in his lungs!

20–3 PROPERTIES OF MATTER

A molecular theory of matter obviously accomplishes nothing if it simply endows the molecules of a substance with all the properties of that substance. The blue color of copper sulfate is not "explained" by postulating that it consists of molecules, each of which is blue. The fact that a gas is capable of expanding indefinitely is not explained by assuming that it consists of molecules each one of which can expand indefinitely.

Instead, what we must do is to attempt to account for the *complex* properties of matter in bulk as a consequence of *simple* properties ascribed to its molecules. Thus by assuming that a monatomic gas consists of a large number of particles having no properties other than mass and velocity, we can explain the observed equation of state of a gas at low pressure and can show that C_V for the gas should equal 2.99 cal mol^{-1}(C°)$^{-1}$ (see Table 19–1). If we add the hypothesis that the "particles" are not simply geometrical points but have a finite size, then the general features of the viscosity, thermal conductivity, and coefficient of diffusion of a gas can be understood, as well as the fact that polyatomic gases have larger values of C_V than monatomic gases. By assuming that there are forces between the particles, the equation of state can be brought into better agreement with that of a real gas, and the phenomena of liquefaction and solidification at low temperatures can be explained.

The electrical and magnetic properties of matter, and the emission and absorption of light by matter, call for a molecular model that is itself an aggregate of subatomic particles. Some of these are electrically charged, and the forces between molecules originate in these electric charges. To pursue this development of molecular theory further would take us into the area of atomic and nuclear structure. In the next section we return to the starting point and see what properties of a gas at low pressure can be explained by the simplest possible molecular model, an aggregate of particles having mass and velocity.

20–4 KINETIC THEORY OF AN IDEAL GAS

We know that a gas in a cylinder such as that in Fig. 20–3 exerts a pressure against the piston. The molecular theory interprets the pressure not as a static push but as the average effect of many tiny impulsive blows resulting from the collisions of the molecules with the piston. Thus if the piston is forced to the left by a spring, we would expect it not to remain absolutely at rest but to oscillate slightly about some average position, as if it were being bombarded by a helter-skelter rain of shot fired against it. (It is just

these random molecular impacts, which do not exactly balance at all instants, that are responsible for the Brownian motion of suspended particles.)

We assume that the gas is composed of a large number of molecules, all having the same mass m and flying about in the cylinder with randomly directed velocities having magnitudes that differ from one molecule to another. Consider first the collision of a single molecule with the piston. Let the mass of the molecule be m, and let the x-component of its velocity be v_{x1}. Figure 20–3 illustrates the collision process. When all the molecules are considered, the *average y-components* of their velocities are not altered by collisions with the piston, since on the average the gas does not gain or lose any vertical motion as a whole. We shall therefore assume that the y-component of the velocity of each molecule is unchanged in the collision. Also, we identify the internal energy of the gas as the sum of the kinetic energies of its molecules. Since the internal energy remains constant if the gas is isolated, the average kinetic energy of the molecules is constant. We therefore assume also that the kinetic energy of any one molecule is constant in a collision with the walls, that is, the collisions are perfectly elastic. Therefore the magnitude of the velocity v is the same before and after the collision, and since the y-component is unchanged, the *magnitude* of the x-component is unchanged also, although it is reversed in direction.

The average force exerted on the piston by one molecule can now be computed by setting the average force equal to the average rate of change of momentum. Since the x-velocity of the molecule reverses from $+v_{x1}$ to $-v_{x1}$ in each collision, the magnitude of the change in momentum in each

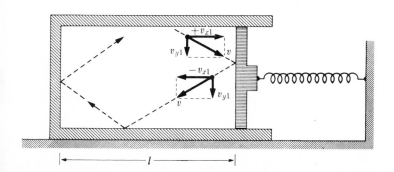

Fig. 20–3 *Collision of a single molecule with a piston.*

collision is $2mv_{x1}$. To find the average force *during the time of one collision* we would divide the change of momentum by this time. However, after having made one collision with the piston, the molecule must travel to the other end of the cylinder and back before colliding again, and during all this time it exerts no force on the piston. To get the correct time average force we must average over the *total* time between collisions. As the molecule travels to the other end of the cylinder and back (see Fig. 20–3), the magnitude of its x-velocity remains constant. If ℓ is the length of the cylinder, the time for the round trip is

$$t = \frac{2\ell}{v_{x1}},$$

and the average force exerted on the piston by one molecule is

$$\text{Average force} = \text{Average rate of change of momentum}$$
$$= \frac{2mv_{x1}}{2\ell/v_{x1}} = \frac{mv_{x1}^2}{\ell}. \qquad (20-5)$$

Now consider other molecules 2, 3, etc., with x-velocities v_{x2}, v_{x3}, etc. The average force exerted by each is given by an expression like that in Eq. (20–5), and the *total* average force is

$$\text{Average force} = \frac{m}{\ell}(v_{x1}^2 + v_{x2}^2 + v_{x3}^2 + \cdots).$$

The average pressure p equals the total average force divided by the piston area A, so

$$p = \frac{m}{V}(v_{x1}^2 + v_{x2}^2 + v_{x3}^2 + \cdots), \qquad (20-6)$$

where $V = \ell A$ is the total volume of the cylinder.

Let N be the total number of molecules. The average value of the square of the x-velocity of all molecules is

$$\overline{v_x^2} = \frac{v_{x1}^2 + v_{x2}^2 + v_{x3}^2 + \cdots}{N}. \qquad (20-7)$$

Hence, Eq. (20–6) can be written

$$p = \frac{Nm\overline{v_x^2}}{V}. \qquad (20-8)$$

The *magnitude* of the resultant velocity v, or the *speed* of any molecule, is given by

$$v^2 = v_x^2 + v_y^2 + v_z^2,$$

and averaging over all molecules,

$$\overline{v^2} = \overline{v_x^2} + \overline{v_y^2} + \overline{v_z^2}.$$

But since the x-, y-, and z-directions are all equivalent,

$$\overline{v_x^2} = \overline{v_y^2} = \overline{v_z^2}.$$

Hence

$$\overline{v^2} = 3\overline{v_x^2},$$

and Eq. (20–8) becomes $N\left(\frac{1}{2}mv^2\right) = U$

$$pV = \tfrac{1}{3}Nm\overline{v^2} = (\tfrac{2}{3}N)(\tfrac{1}{2}m\overline{v^2}) \qquad (20-9)$$

But $\tfrac{1}{2}m\overline{v^2}$ is the average kinetic energy of a single molecule, and the product of this and the total number of molecules N equals the total random kinetic energy, or internal energy U. Hence the product pV equals two-thirds of the internal energy:

$$pV = \tfrac{2}{3}U. \qquad (20-10)$$

By experiment, at low pressures, the equation of state of a gas is

$$pV = nRT.$$

The theoretical and experimental laws will therefore be in complete agreement if we set

$$U = \tfrac{3}{2}nRT. \qquad (20-11)$$

The average kinetic energy of a single molecule is then

$$\frac{U}{N} = \tfrac{1}{2}m\overline{v^2} = \frac{3nRT}{2N}.$$

But the number of moles, n, equals the total number of molecules, N, divided by Avogadro's number, N_A, the number of molecules per mole:

$$n = \frac{N}{N_A}, \qquad \frac{n}{N} = \frac{1}{N_A}.$$

Hence

$$\tfrac{1}{2}m\overline{v^2} = \frac{3}{2}\frac{R}{N_A} T.$$

The ratio R/N_A occurs frequently in molecular theory. It is called the *Boltzmann constant, k*:

$$k = \frac{R}{N_A} = \frac{8.31 \text{ J mol}^{-1}\text{ K}^{-1}}{6.02 \times 10^{23} \text{ molecules mol}^{-1}}$$

$$= 1.38 \times 10^{-23} \text{ J molecule}^{-1}\text{ K}^{-1}.$$

Since R and N_A are universal constants, the same is true of k. Then

$$\boxed{\tfrac{1}{2}m\overline{v^2} = \tfrac{3}{2}kT.} \qquad (20\text{–}12)$$

The average kinetic energy per molecule, therefore, *depends only on the temperature* and not on the pressure, volume, or species of molecule.

The average value of the square of the speed is

$$\overline{v^2} = \frac{3kT}{m},$$

and the square root of this, or the root-mean-square speed v_{rms}, is

$$v_{\text{rms}} = \sqrt{\overline{v^2}} = \sqrt{\frac{3kT}{m}} = \sqrt{\frac{3RT}{M}}. \qquad (20\text{–}13)$$

Example 1 What is the average kinetic energy of a molecule of a gas at a temperature of 300 K?

$$\tfrac{1}{2}m\overline{v^2} = \tfrac{3}{2}kT = (\tfrac{3}{2})(1.38 \times 10^{-23} \text{ J K}^{-1})(300 \text{ K})$$

$$= 6.20 \times 10^{-21} \text{ J}.$$

Example 2 What is the total random kinetic energy of the molecules in 1 mole of a gas at a temperature of 300 K?

$$U = (N_A)(\tfrac{3}{2}kT) = \tfrac{3}{2}RT$$

$$= (\tfrac{3}{2})(8.31 \text{ J mol}^{-1}\text{ K}^{-1})(1 \text{ mol})(300 \text{ K})$$

$$= 3750 \text{ J} = 900 \text{ cal}.$$

Example 3 What is the root-mean-square speed of a hydrogen molecule at 300 K?

The mass of a hydrogen molecule (see Section 20–2) is

$$m_{\text{H}_2} = (2)(1.66 \times 10^{-27} \text{ kg}) = 3.32 \times 10^{-27} \text{ kg}.$$

Hence

$$v_{\text{rms}} = \sqrt{\frac{3kT}{m}} = \sqrt{\frac{3(1.38 \times 10^{-23} \text{ J K}^{-1})(300 \text{ K})}{3.32 \times 10^{-27} \text{ kg}}}$$

$$= 1934 \text{ m s}^{-1}.$$

Alternatively,

$$v_{\text{rms}} = \sqrt{\frac{3RT}{M}} = \sqrt{\frac{(3)(8.31 \text{ J mol}^{-1}\text{ K}^{-1})(300 \text{ K})}{2 \times 10^{-3} \text{ kg mol}^{-1}}}$$

$$= 1934 \text{ m s}^{-1}.$$

Example 4 What is the root-mean-square speed of a molecule of mercury vapor at 300 K?

The *kinetic energy* of a mercury molecule (or atom, since mercury vapor is monatomic) is the same as that of a hydrogen molecule at the same temperature. The mass of a mercury atom is

$$m_{\text{Hg}} = (201)(1.66 \times 10^{-27} \text{ kg}) = 334 \times 10^{-27} \text{ kg}.$$

Therefore,

$$v_{\text{rms}} = \sqrt{\frac{3(1.38 \times 10^{-23} \text{ J K}^{-1})(300 \text{ K})}{334 \times 10^{-27} \text{ kg}}}$$

$$= 193 \text{ m s}^{-1}.$$

When a gas expands against a moving piston it does work. The source of this work is the random kinetic energy of the gas molecules. Conversely, when work is done in compressing a gas, the random kinetic energy of its molecules increases. But if the collisions with the walls are perfectly elastic, as we have assumed, how can a molecule gain or lose energy in a collision with a piston? To understand this, we must consider the collision of a molecule with a *moving* wall.

When a molecule makes a collision with a *stationary* wall, it exerts a momentary force on the wall but does no work, since the wall does not move. But if the wall is in motion, work *is* done in a collision. Thus, if the piston in Fig. 20–3 is moving to the right, work is done on it by the molecules that strike it, and their velocities (and kinetic energies) after colliding are smaller than they were before a collision. The collision is still completely elastic, since the work done on the moving piston is just equal to the decrease in the kinetic energy of the molecule. Similarly, if the piston is moving toward the left, the kinetic energy of a colliding molecule is *increased*, the increase in kinetic energy being equal to the work done on the molecule.

20–5 MOLAR HEAT CAPACITY OF A GAS

When heat flows into a system in a process at constant volume, no work is done and all of the energy inflow goes into an increase in the internal energy U of the system. From the molecular viewpoint, the internal energy of a system is the sum of the kinetic and potential energies of its molecules. If this sum can be computed as a function of temperature, then, from its rate of change with temperature, we can derive a theoretical expression for the molar heat capacity. \

The simplest system is a monatomic ideal gas, for which the molecular energy is wholly kinetic and is given by Eq. (20–11):

$$\text{Random kinetic energy} = U = \tfrac{3}{2}nRT.$$

If the temperature is increased by dT, the random kinetic energy increases by

$$dU = \tfrac{3}{2}nR\,dT.$$

In a process in which the temperature increases by dT at constant volume, the energy flowing into a system is, by definition of C_v,

$$dQ = nC_v\,dT = dU.$$

Hence,

$$\boxed{C_v = \tfrac{3}{2}R.}$$

We showed in Section 19–12 and Table 19–1 that the experimental values of C_v for monatomic gases are, in fact, almost exactly equal to $\tfrac{3}{2}R$. This agreement is a striking confirmation of the basic correctness of the kinetic model of a gas and did much to establish this theory at a time when many scientists still refused to accept it.

Since $C_p = C_v + R$, it follows that the theoretical ratio of molar heat capacities for a monatomic ideal gas is

$$\frac{C_p}{C_v} = \gamma = \frac{\tfrac{3}{2}R + R}{\tfrac{3}{2}R} = \frac{5}{3} = 1.67.$$

This also is in good agreement with the experimental values in Table 19–1.

Polyatomic gases present a more complicated problem. We might expect that a molecule composed of two or more atoms could have additional "internal" energy associated with vibratory motion of the atoms relative to each other and of rotation of the molecule about its center of mass. Also, since every atom consists of a positive nucleus and one or more electrons, there may be additional energy associated with position and motion of these electric charges.

The *temperature* of a gas depends on the average random *translational* kinetic energy of its molecules. When heat flows into a *monatomic* gas, at constant volume, all of this energy goes into an increase in random translational molecular kinetic energy, as evidenced by the agreement between the measured values of C_v and the values computed from the increase in translational kinetic energy. But when heat flows into a *polyatomic* gas, we would expect that a part of the energy would go toward increasing "molecular internal energy." Hence, when equal amounts of heat flow into a monatomic and a polyatomic gas containing the same number of molecules, it is to be expected that the temperature rise in the polyatomic gas will be *less* than in the monatomic, because only a part of the energy is available for increasing *translational* molecular kinetic energy. In other words, to produce *equal* increases of temperature, *more* heat would have to be supplied to the polyatomic gas. This is a qualitative explanation, then, of the larger values of C_v for polyatomic gases listed in Table 19–1.

20–6 MEASUREMENT OF MOLECULAR SPEEDS

Direct measurements of the distribution of molecular speeds have been made by a number of methods. Figure 20–4 is a diagram of the apparatus used by Zartman and Ko in 1930–1934, a modification of a technique developed by Stern in 1920. Metallic silver is melted and evaporated in the oven O. A beam of silver atoms escapes through a small opening in the oven and passes through the slits S_1 and S_2 into an evacuated region. The cylinder C can be rotated at approximately 6000 rpm about the axis A. If the cylinder is at rest, the molecular beam enters the cylinder through a slit S_3 and strikes a curved glass plate G. The molecules stick to the glass plate, and the number arriving at any portion can be determined by removing the plate and measuring with a recording microphotometer the darkening that has resulted.

Now suppose the cylinder is rotated. Molecules can enter it only during the short time intervals

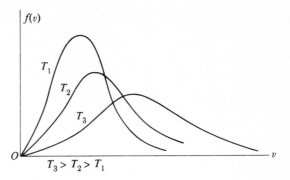

Fig. 20–5 Maxwell–Boltzmann distribution curves for various temperatures. As the temperature increases, the curve becomes flatter, and its maximum shifts to higher temperature.

during which the slit S_3 crosses the molecular beam. If the rotation is clockwise, as indicated, the glass plate moves toward the right while the molecules cross the diameter of the cylinder. They therefore strike the plate to the left of the point of impact when the cylinder is at rest, and the more slowly they travel, the farther to the left is this point of impact. The blackening of the plate is therefore a measure of the "velocity spectrum" of the molecular beam.

Several other techniques have been devised in more recent years for measuring distributions of molecular speeds with greater precision. Results of these experiments can be compared with a predicted speed distribution derived from a more detailed kinetic theory of gases and called the *Maxwell–Boltzmann distribution*. This theoretical distribution is shown in Fig. 20–5 for several different temperatures. At any temperature the curve shows a peak corresponding to the most probable speed; very small and very large values of v occur infrequently. As the temperature increases, the peak shifts to higher and higher speeds, corresponding to the increase in average molecular kinetic energy with temperature. Measured molecular speed distributions are generally in excellent agreement with the theoretical prediction of the Maxwell–Boltzmann distribution.

20–7 CRYSTALS

Many materials can exist in a variety of solid forms; a familiar example is the element *carbon*. The black

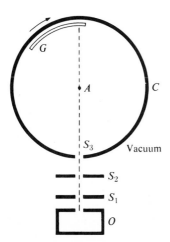

Fig. 20–4 Apparatus used by Zartman and Ko in studying distribution of velocities.

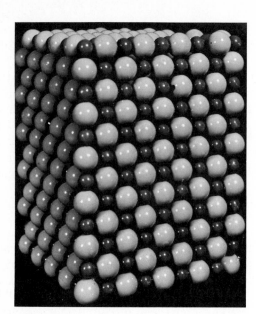

Fig. 20–6 *Close packing of the sodium ions (black spheres) and chlorine ions (white spheres) in a sodium chloride crystal. (Courtesy of Alan Holden; reprinted with permission of Educational Services, Inc., from* Crystals and Crystal Growing, *Doubleday and Co., New York, 1960.)*

soot deposited on a kettle by a smoky campfire is nearly pure carbon. The so-called "lead" in a pencil is not lead at all, but chiefly a different form of carbon called graphite. Diamond is a third form of solid carbon.

The remarkably dissimilar mechanical, thermal, electrical, and optical properties of these three forms of carbon can be understood in terms of the arrangement of the carbon atoms in the solid structure. In soot the atoms have no regular arrangement and are said to constitute an *amorphous* solid. Graphite and diamond are both crystals, but the orderly arrangement of atoms is different in the two crystal lattices. The study of the spatial arrangements of atoms (or molecules or ions) in the various types of crystal lattices is a fascinating area of physics. We shall be concerned here with only a few of the fundamental ideas needed to understand some of the mechanical and thermal properties of single crystals or solids composed of an aggregate of crystals, that is, *poly-crystalline solids.*

The particles comprising a crystal lattice are *closely packed,* as seen in Fig. 20–2 and as suggested by the model in Fig. 20–6, where the sodium and chlorine ions of a sodium chloride crystal are represented as spheres touching one another. Strictly speaking, each ion consists of a nucleus surrounded by electrons in motion, so that the boundary of an ion must be thought of as the average positions of the outermost electrons. It follows that the outer electrons of one particle in a crystal lattice come close to and even at times interpenetrate with the outer electrons of a neighboring particle.

To make crystal structures simpler to comprehend, it is customary to exaggerate the distances between neighboring particles, as shown in Fig. 20–7, where a *face-centered cubic* lattice structure is depicted. Much can be learned by representing crystal lattices in this way, with the aid of a supply of gum drops and toothpicks.

Suppose that the chemical composition and density of a very small volume element of a substance are measured at many different places. If the composition and density are the same at all points, the substance is said to be *homogeneous.* If the physical properties determining the transport of heat, electricity, light, etc., are the *same in all directions,* the substance is said to be *isotropic.* Cubic crystals are both homogeneous and isotropic, whereas all other crystals, although homogeneous, are not isotropic (i.e., they are *anisotropic*).

The *anisotropy* (accent on the syllable "ot") of noncubic crystals can be understood when the forces

Fig. 20–7 *Symbolic representation of a sodium chloride crystal, with exaggerated distances between ions.*

Fig. 20–8 *The forces between neighboring particles in a crystal may be visualized by imagining every particle to be connected to its neighbors by springs. In the case of a cubic crystal, all springs are assumed to have the same spring constant. Anisotropy is connected with differing spring constants in different directions.*

between neighboring particles in a lattice are taken into account. A convenient way of visualizing these forces is shown in Fig. 20–8, an idealized model which represents a crystal lattice as a large number of spheres, each connected to its neighbors by springs. The force constants of parallel springs are imagined equal, but the force constant of each spring pointing in one direction may be different from that of springs pointing in different directions. If each of the particles represented by spheres in Fig. 20–8 is imagined to be vibrating about its equilibrium position, a fairly accurate picture of the dynamic character of a crystal lattice will result. The mathematical calculations of the vibration frequencies of such a complex mechanical system was first attempted by Born and Von Karman early in the twentieth century. It is, of course, a very laborious and complicated calculation, but it has been carried out in detail for sodium chloride, potassium chloride, diamond, and silver.

The elastic properties of crystals can be partially understood from a diagram such as that in Fig. 20–8. Young's modulus is simply a measure of the stiffness of the springs in the direction in which the crystal is pulled or pushed. The shear modulus depends on the springs which are stretched and those which are compressed when the crystal is twisted. One of the first tests of the correctness of a lattice structure representing a particular material is to compare a calculated value of an elastic modulus with a measured value.

20–8 HEAT CAPACITY OF A CRYSTAL

In Section 20–4 a gas was assumed to consist of N particles in rapid linear motion, so far apart on the average that no force of interaction existed except upon collision with one another or with a wall. By applying simple mechanical principles, it was shown that the product of the pressure p and the volume V is equal to

$$pV = (\tfrac{2}{3}N)(\tfrac{1}{2}m\overline{v^2}) = \tfrac{2}{3}U,$$

where U is the total energy of the particles, in this case, entirely translational kinetic energy. In order for this result to agree with the equation of state of an ideal gas,

$$pV = nRT,$$

the average total energy per particle U/N, under equilibrium conditions, must be equal to

$$\frac{U}{N} = \tfrac{3}{2}kT.$$

Since there are three translational degrees of freedom (i.e., three components of velocity), $\tfrac{1}{2}kT$ of energy on the average is attributed to each degree of freedom. With some limitations, this is true also of rotational degrees of freedom. The statement that with every degree of freedom of motion of a molecule, there is associated an average energy, at equilibrium, equal to $\tfrac{1}{2}kT$, is termed the principle of *equipartition of energy.*

The methods of calculation just referred to represent a simplified version of the branch of physics called the *kinetic theory of gases.* There is, however, another way in which the behavior of large numbers of particles may be described and calculated, which, in some respects, is more general and more powerful. This more general method is called *statistical mechanics*. In statistical mechanics it is not necessary to specify whether the particles are molecules of a gas, ions occupying sites in a crystal lattice, electrons, or other entities. The important thing is whether they are almost independent, and if so, whether the weak interaction among the particles is sufficient to enable the particles to exchange energy and finally come to equilibrium. If this is the case, the methods of statistical mechanics enable us to find the equilibrium value of the *average* energy per particle from a knowledge of the types of energy possessed by each individual particle. If the energy of an individual particle is expressed as the sum of a number of squared terms, such as $\frac{1}{2}mv_x^2$ or $\frac{1}{2}I\omega^2$, or $\frac{1}{2}kx^2$, then it can be proved *rigorously* that, at equilibrium, the average energy per particle *associated with each squared* term is $\frac{1}{2}kT$. This is the rigorous statement of the principle of the equipartition of energy.

If we consider an assemblage of N vibrators, each vibrating along the x-axis with very small amplitude almost independently (but not enough to prevent a weak interaction with neighbors), we may consider the energy of each vibrator to consist of two squared terms,

$$\tfrac{1}{2}mv_x^2 + \tfrac{1}{2}kx^2.$$

The principle of equipartition then gives us, for the average energy per particle at equilibrium, the sum of two terms each equal to $\frac{1}{2}kT$, or a total of kT.

Now a crystal is composed of N particles, each of which interacts strongly with its neighbors and vibrates in a very complicated way. In spite of the complications, it may be shown quite generally that the actual situation is *equivalent* to $3N$ oscillators, almost independent of one another, each vibrating with its own frequency. It follows therefore that the total energy at equilibrium is

$$U = 3NkT,$$

or, in terms of the number of moles n and the universal gas constant R,

$$U = 3nRT,$$

and finally

$$C_v = \frac{1}{n}\frac{dU}{dT} = 3R = 25.0\,\text{J mol}^{-1}\,\text{K}^{-1}.$$

This is the law of Dulong and Petit, which we encountered in Chapter 16. As Fig. 16–5 shows, many solids approach this value of C_v at high temperatures, but at low temperatures C_v invariably decreases; it is *not* constant, as the law of Dulong and Petit suggests, but approaches zero as T approaches zero.

A very careful analysis shows that there is nothing wrong with the methods of statistical mechanics or with the equivalence between the actual vibrations of a crystal lattice and $3N$ almost independent oscillators. What is wrong is the assumption that a vibrator can assume *any value* of the energy given by $\frac{1}{2}mv_x^2 + \frac{1}{2}kx^2$. In classical newtonian mechanics, this is certainly correct; but any treatment of mechanical systems on an atomic scale should include considerations of *quantum mechanics*, which we shall discuss in later chapters. A fundamental result of quantum mechanics is that a vibrator of frequency f may *not* take on *any* value of the energy, but only *discrete energy values* given by whole-number multiples of the product hf:

$$1hf, 2hf, 3hf, \ldots,$$

where h is a universal constant of nature, called *Planck's constant*. When this postulate is used in conjunction with the classical methods of statistical mechanics, it is found that the average energy per particle at equilibrium is always *less* than kT, but that at high temperatures it is approximately equal to kT.

This behavior may be understood qualitatively by noting that at very low temperatures the quantity kT is much *smaller* than the smallest energy increment hf that the vibrators can accommodate. As a result, at low T most of the systems remain in their lowest energy states because the energy hf required to make transitions to the next higher energy levels is not available. Thus, at low temperatures the average energy per vibrator is less than kT, and correspondingly the heat capacity per molecule is less than k. But in the other limit, where kT is *large* compared to hf, the classical equipartition theorem holds, and the total heat capacity is $3k$ per molecule, or $3R$ per mole, as the Dulong and Petit relation predicts. This behavior is illustrated by Fig. 16–3. Quantitative understanding of the temperature variation of specific heats was one of the triumphs of quantum mechanics during the early years of its development, early in the twentieth century.

PROBLEMS

Molecular Data

N_A = Avogadro's number
    = 6.02×10^{23} molecules mol^{-1},
Mass of a hydrogen atom = 1.67×10^{-27} kg,
Mass of a nitrogen molecule = $28 \times 1.66 \times 10^{-27}$ kg,
Mass of an oxygen molecule = $32 \times 1.66 \times 10^{-27}$ kg.

20–1 Consider an ideal gas at 0°C and at 1 atm pressure. Imagine each molecule to be, on the average, at the center of a small cube.
a) What is the length of an edge of this small cube?
b) How does this distance compare with the diameter of a molecule?

20–2 A mole of liquid water occupies a volume of 18 cm³. Imagine each molecule to be, on the average, at the center of a small cube.
a) What is the length of an edge of this small cube?
b) How does this distance compare with the diameter of a molecule?

20–3 What is the length of the side of a cube, in a gas at standard conditions, that contains a number of molecules equal to the population of the United States (about 200 million)?

20–4 The lowest pressures readily attainable in the laboratory are the order of 10^{-10} torr (the order of 10^{-13} atm). At this pressure and ordinary temperature (say $T = 300$ K) how many molecules are present in a volume of 1 cm³?

20–5 Experiment shows that the size of an oxygen molecule is the order of 2×10^{-10} m. Make a rough estimate of the pressure at which the finite volume of the molecules should cause noticeable deviations from ideal-gas behavior at ordinary temperatures ($T = 300$ K).

20–6
a) What is the average translational kinetic energy of a molecule of oxygen at a temperature of 300 K?
b) What is the average value of the square of its speed?
c) What is the root-mean-square speed?
d) What is the momentum of an oxygen molecule traveling at this speed?
e) Suppose a molecule traveling at this speed bounces back and forth between opposite sides of a cubical vessel 10 cm on a side. What is the average force it exerts on the walls of the container?
f) What is the average force per unit area?
g) How many molecules traveling at this speed are necessary to produce an average pressure of 1 atm?
h) Compare with the number of oxygen molecules actually contained in a vessel of this size, at 300 K and atmospheric pressure.

20–7 The speed of propagation of a sound wave in air at 27°C is about 350 m s^{-1}. Compare this with the root-mean-square speed of nitrogen molecules at this temperature.

20–8 At what temperature is the rms speed of oxygen molecules equal to the rms speed of hydrogen molecules at 0° C?

20–9 Isotopes of uranium are sometimes separated by gaseous diffusion, using the fact that the rms speeds of their molecules in vapor are slightly different, and hence the vapors diffuse at slightly different rates. Assuming the atomic masses for ^{235}U and ^{238}U are 235 g mol^{-1} and 238 g mol^{-1}, respectively, what is the ratio of the rms speed of the ^{235}U atoms in the vapor to that of the ^{238}U atoms, assuming the temperature is uniform?

20–10 Compute the specific heat capacity at constant volume of hydrogen gas, and compare it with the specific heat capacity of water.

20–11 The molar heat capacities of many diatomic gases can be understood on the assumption that there is an additional average kinetic energy of kT per molecule associated with rotational motion. On this assumption, calculate the molar heat capacity at constant volume (C_v) for an ideal diatomic gas, and compare with the experimental values in Table 19–1.

20–12 What is the total random kinetic energy of the molecules in 1 mole of helium at a temperature of (a) 300 K? (b) 301 K? (c) Compare the difference between these with the change in internal energy of 1 mole of helium when its temperature is increased by 1 K, as computed from the relation $\Delta U = nC_V \Delta T$.

20–13 A flask contains a mixture of mercury vapor, neon, and helium. Compare (a) the average kinetic energies of the three types of atoms, and (b) the root-mean-square speeds.

20–14

a) At what temperature is the rms speed of hydrogen molecules equal to the speed of the first earth satellite (about 18,000 mi hr^{-1})?

b) At what temperature is the rms speed equal to the escape speed from the gravitational field of the earth?

20–15 Smoke particles in the air typically have masses the order of 10^{-16} kg. The Brownian motion of these particles resulting from collisions with air molecules can be observed with a microscope.

a) Find the root-mean-square speed of Brownian motion for such a particle in air at 300 K.

b) Would the speed be different if the particle were in hydrogen gas at the same temperature? Explain.

20–16 In the kinetic–molecular model of an ideal gas discussed in Section 20–4, the effect of gravity on molecular motion was neglected. How can this omission be justified?

20–17 The oven in Fig. 20–4 contains bismuth at a temperature of 840 K; the drum is 10 cm in diameter and rotates at 6000 rev min^{-1}. Find the displacement on the glass plate G, measured from a point directly opposite the slit, of the points of impact of the molecules Bi and Bi$_2$. Assume that all molecules of each species travel with the rms speed appropriate to that species.

20–18 Suppose the two atoms in an oxygen molecule are represented as two point masses 2×10^{-10} m apart.

a) Calculate the moment of inertia of a molecule about an axis through the midpoint, perpendicular to the line joining the atoms.

b) If the rotational kinetic energy of the molecule is $\frac{1}{2}kT$, what is the root-mean-square angular velocity of rotation at $T = 300$ K?

20–19 The density of crystalline sodium chloride is 2.16 g cm^{-3}, and its crystal structure is shown in Fig. 20–7. Find the spacing of adjacent atoms in the crystal lattice. The molecular "weight" of sodium is 23.0 g mol^{-1} and that of chlorine is 35.5 g mol^{-1}.

Chapter 21

Traveling Waves

21–1 INTRODUCTION

A wave is any disturbance from an equilibrium condition which travels or *propagates* with time from one region of space to another. There are many examples of waves in everyday experience, and examples of wave phenomena are found in all branches of physics. The concept of waves, in fact, is one of the most important unifying elements in all of physics.

Waves on the surface of a body of water, produced by the wind or by some other disturbance, are a familiar sight. A sound is carried by means of traveling waves in the intervening atmosphere between source and listener. Many of the observed properties of light are best explained by a wave theory, and we believe that light waves are of the same fundamental nature as radio waves, infrared and ultraviolet waves, x-rays, and gamma rays. One of the outstanding developments of twentieth-century physics has been the discovery that all matter is endowed with wave properties and that a beam of electrons, for example, is reflected by a crystal in much the same way as is a beam of x-rays.

The waves most easily comprehended are *mechanical waves*, in which the wave travels through some material medium as it is displaced from an equilibrium state. Imagine a medium consisting of a large number of particles, each connected or coupled to its neighbors by *elastic* material. If one end of the medium is disturbed or displaced in any way, the displacement does not occur immediately at all other parts of the medium. The original displacement gives rise to an elastic force in the material adjacent to it; then the next particle is displaced, and then the next, and so on. In other words, *the displacement is propagated along the medium with a definite speed.*

In Fig. 21–1(a) the medium is a spring, or even just a wire under tension. If the left end is given a small displacement in a direction perpendicular to the medium, this transverse displacement will occur at successive intervals of time at each coil of the spring, resulting in the propagation of a *transverse pulse* along the spring.

In Fig. 21–1(b) the medium is a liquid or a gas contained in a tube closed at the right end with a rigid wall and at the left end with a movable piston. If the piston is moved slightly toward the right, a *longitudinal pulse* is propagated through the medium in the tube. Longitudinal waves can also travel in solid materials.

In Fig. 21–1(c) the medium is a liquid contained in a trough. The horizontal motion of a flat piece of wood at the left end will create a displacement of the liquid which is *both* longitudinal and transverse, and this disturbance travels along the medium.

(a)   Transverse displacement

(b)   Longitudinal displacement

Fig. 21–1 *Propagation of disturbances.*

(c)   Longitudinal and transverse displacement

(d)   Torsional displacement

In Fig. 21–1(d) the medium is a set of "dumb-bells" connected to a steel strip. A slight rotation of the lowest dumbbell constitutes a *torsional displacement* which is propagated up the medium with a finite speed.

21–2 PERIODIC WAVES

When the motions of the particles are at right angles to the direction of travel of the wave, the wave is called *transverse.* If the particles move in the direction of propagation, the wave is called *longitudinal.* Suppose that one end of a medium is forced to vibrate periodically, the displacement y (either transverse or longitudinal) varying with time according to the equation of simple harmonic motion:

$$y = \begin{cases} Y \sin \omega t, \\ \text{or} \\ Y \cos \omega t. \end{cases}$$

The resulting continuous train of disturbances, trav-

eling with a speed depending on the properties of the medium, is called a *wave.*

To fix our ideas, suppose that one end of a stretched string is forced to vibrate periodically in a transverse direction with simple harmonic motion of amplitude Y, frequency f, and period $\tau = 1/f$. For the present we shall assume the string to be long enough so that any effects at the far end need not be considered. A *continuous train* of transverse sinusoidal waves then advances along the string. The shape of a portion of the string near the end, at intervals of $\frac{1}{8}$ of a period, is shown in Fig. 21–2 for a total time of one period. The string is assumed to have been vibrating for a sufficiently long time so that the shape of the string is sinusoidal for an indefinite distance from the driven end. It will be seen from the figure that the waveform advances steadily toward the right, as indicated by the short arrow pointing to one particular wave crest, while any one point on the string (see the black dot) oscillates about its equilibrium position with simple harmonic motion. It is important to distinguish between the motion of the

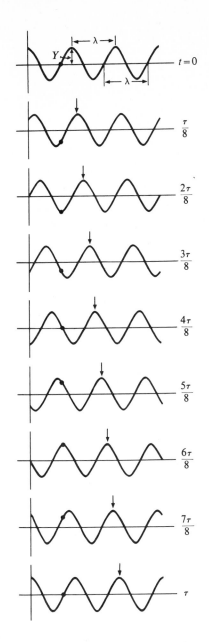

Fig. 21–2 *A sinusoidal transverse wave traveling toward the right, shown at intervals of one-eighth of a period.*

waveform, which moves with constant velocity c along the string, and the motion of *a particle of the string*, which is simple harmonic and transverse to the string.

The distance between two successive maxima (or between any two successive points in the same phase) is the *wavelength* of the wave and is denoted by λ. Since the waveform, traveling with constant velocity c, advances a distance of one wavelength in a time interval of one period, it follows that $c = \lambda/\tau$, or

$$c = f\lambda. \qquad (21\text{–}1)$$

That is, *the velocity of propagation equals the product of frequency and wavelength.*

To understand the mechanics of a *longitudinal* wave, consider a long tube filled with a fluid and provided with a plunger at the left end, as shown in Fig. 21–3. The dots represent particles of the fluid. Suppose the plunger is forced to undergo a simple harmonic vibration parallel to the direction of the tube. During a part of each oscillation, a region whose pressure is above the equilibrium pressure is formed. Such a region is called a *condensation* and is represented by closely spaced dots. Following the production of a condensation, a region is formed in which the pressure is lower than the equilibrium value. This is called a *rarefaction* and is represented by widely spaced dots. The condensations and rarefactions move to the right with constant velocity c, as indicated by successive positions of the small vertical arrow. The velocity of longitudinal waves in air (sound) at 20°C is 344 m s^{-1} or 1130 ft s^{-1}. The motion of a single particle of the medium, shown by a heavy black dot, is simple harmonic, parallel to the direction of propagation.

The wavelength is the distance between two successive condensations or two successive rarefactions, and the same fundamental equation, $c = f\lambda$, holds in this, as in all types of periodic or repetitive waves.

21–3 MATHEMATICAL REPRESENTATION OF A TRAVELING WAVE

Suppose that a wave of any sort travels from left to right in a medium. Let us compare the motion of any one particle of the medium with that of a second particle to the right of the first. We find that the

Fig. 21–3 *A sinusoidal longitudinal wave traveling toward the right, shown at intervals of one-eighth of a period.*

second particle moves in the same manner as the first, but after a lapse of time that is proportional to the distance of the second particle from the first. Hence if one end of a stretched string oscillates with simple harmonic motion, all other points oscillate with simple harmonic motion of the same amplitude and frequency. The *phase angle* of the motion, however, is different for different points.

Let the displacement of a particle at the origin ($x = 0$) be given by

$$y = Y \sin \omega t = Y \sin 2\pi ft. \qquad (21\text{–}2)$$

The time required for the wave disturbance to travel from $x = 0$ to some point x to the right of the origin is given by x/c, where c is the wave speed, as usual. The motion of point x at time t is the same as the motion of point $x = 0$ at the earlier time $t - x/c$. Thus the displacement of point x at time t is obtained simply by replacing t in Eq. (21–2) by $(t - x/c)$, and we find

$$
\begin{aligned}
y(x, t) &= Y \sin \omega \left(t - \frac{x}{c} \right) \\
&= Y \sin 2\pi f \left(t - \frac{x}{c} \right).
\end{aligned}
\qquad (21\text{–}3)
$$

The notation $y(x,t)$ is a reminder that the displacement y is a function of two variables x and t, corresponding to the fact that the displacement of a point depends on both the location of the point and the time.

Equation (21–3) can be rewritten in several alternate forms, conveying the same information in different ways. In terms of the period τ and wavelength λ, we find

$$y(x, t) = Y \sin 2\pi \left(\frac{t}{\tau} - \frac{x}{\lambda} \right). \qquad (21\text{–}4)$$

Another convenient form is obtained by defining a quantity k, called the *propagation constant* or the *wave number*:

$$k = \frac{2\pi}{\lambda}. \qquad (21\text{–}5)$$

In terms of k and ω, the wavelength–frequency relation $c = \lambda f$ becomes

$$\omega = ck, \qquad (21\text{–}6)$$

and Eq. (21–4) can be written

$$y(x, t) = Y \sin(\omega t - kx). \qquad (21\text{–}7)$$

Which of these various forms is used is a matter of convenience in a specific problem; fundamentally they all say the same thing.

At any given *time t*, Eq. (21–4) or (21–7) gives the displacement y of a particle from its equilibrium position, as a function of the *coordinate x* of the particle. If the wave is a transverse wave in a string, the equation represents the *shape* of the string at that instant, as if we had taken a photograph of the string. Thus at time $t = 0$,

$$y = Y \sin(-kx) = -Y \sin kx = -Y \sin 2\pi \frac{x}{\lambda}.$$

This curve is plotted in Fig. 21–4.

At any given *coordinate x*, Eq. (21–4) or (21–7) gives the displacement y of the particle at that coordinate, as a function of *time*. Thus at the position $x = 0$,

$$y = Y \sin \omega t = Y \sin 2\pi \frac{t}{\tau}.$$

This curve is plotted in Fig. 21–5.

The above formulas may be used to represent a wave traveling in the negative x-direction by making a simple modification. In this case the displacement of point x at time t is the same as the motion of point $x = 0$ at the *later* time $(t + x/c)$, and in Eq. (21–2) we must replace t by $(t + x/c)$. Thus, for a wave traveling in the x-direction,

$$y = Y \sin 2\pi f\left(t + \frac{x}{c}\right)$$

$$= Y \sin 2\pi\left(\frac{t}{\tau} + \frac{x}{\lambda}\right)$$

$$= Y \sin(\omega t + kx). \qquad (21–8)$$

Fig. 21–4

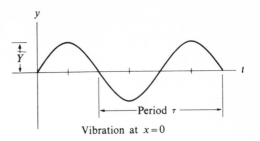

Vibration at $x=0$

Fig. 21–5

It is essential to distinguish carefully between the *speed of propagation c* of the waveform and the *particle speed v* of a particle of the medium in which the wave is traveling. The wave speed c is given by

$$c = \lambda f = \frac{\omega}{k}. \qquad (21–9)$$

The particle speed v for any point in a transverse wave, that is, at a fixed value of x, is obtained by taking the derivative of y with respect to t, holding x constant. Such a derivative is called a *partial derivative* and is written $\partial y/\partial t$. Thus for a sinusoidal wave given by

$$y = Y \sin(\omega t - kx), \qquad (21–10)$$

we have

$$v = \frac{\partial y}{\partial t} = \omega Y \cos(\omega t - kx). \qquad (21–11)$$

The *acceleration* of the point is the *second* partial derivative:

$$a = \frac{\partial^2 y}{\partial t^2} = -\omega^2 Y \sin(\omega t - kx). \qquad (21–12)$$

One may also compute partial derivatives with respect to x, holding t constant. The first derivative $\partial y/\partial x$ is the *slope* of the string at any point. The second partial derivative with respect to x is

$$\frac{\partial^2 y}{\partial x^2} = -k^2 Y \sin(\omega t - kx). \qquad (21–13)$$

It follows from Eqs. (21–12) and (21–13) that

$$\frac{\partial^2 y/\partial t^2}{\partial^2 y/\partial x^2} = \frac{\omega^2}{k^2} = c^2,$$

since $c = \omega/k$. The *partial differential equation*

$$\frac{\partial^2 y}{\partial t^2} = c^2 \frac{\partial^2 y}{\partial x^2} \qquad (21\text{–}14)$$

is one of the most important in all of physics. It is called the *wave equation*, and whenever it occurs, the conclusion is made immediately that y is propagated as a traveling wave along the x-axis with a wave speed c.

21–4 SPEED OF A TRANSVERSE WAVE

We shall now derive a relation between the speed of propagation c of a wave on a stretched string and its mechanical properties, *mass per unit length* and *tension*. We consider a particularly simple wave motion, a transverse *pulse*. If a string stretched between two fixed supports is struck a transverse blow at some point, or if a small portion is displaced sideways and released, disturbances will be observed to travel outward in both directions from the displaced portion. Each of these is called a *transverse pulse*, the direction of motion of particles of the string being at right angles to the direction of propagation of the pulse. Each pulse retains its shape as it travels, and each travels with a constant speed which we shall represent by c.

Consider the string depicted in Fig. 21–6 under a tension S and with linear density (mass per unit length) μ. In Fig. 21–6(a) the string is at rest. At time $t = 0$, a constant transverse force F is applied at the left end of the string. As a result, this end moves up with a constant transverse speed v. Figure 21–6(b) shows the shape of the string after a time t has elapsed. All points of the string at the left of the point P are moving with speed v, whereas all points at the right of P are still at rest. The boundary between the moving and the stationary portions is traveling to the right with the *speed of propagation c*. The left end of the string has moved up a distance vt and the boundary point P has advanced a distance ct.

The speed of propagation c can be calculated by setting the transverse impulse (transverse force × time) equal to the change of transverse momentum of the moving portion (mass × transverse velocity). The impulse of the transverse force F in time t is Ft. By similar triangles,

$$\frac{F}{S} = \frac{vt}{ct}, \qquad F = S\frac{v}{c}.$$

Hence

$$\text{Transverse impulse} = S\frac{v}{c}t.$$

The mass of the moving portion is the product of the mass per unit length μ and the length ct. Hence

$$\text{Transverse momentum} = \mu ctv.$$

Applying the impulse–momentum theorem, we obtain

$$S\frac{v}{c}t = \mu ctv,$$

and therefore

$$\boxed{c = \sqrt{S/\mu} \qquad \text{(transverse).}} \qquad (21\text{–}15)$$

Thus it is seen that the velocity of propagation of a transverse pulse in a string depends only on the tension (a force) and the mass per unit length. Although the above calculation of wave speed considered only a very special kind of pulse, a little thought shows that *any* shape of wave disturbance can be considered as a series of pulses with different rates of transverse displacement. Thus, although derived for a special case, Eq. (21–15) is valid for *any* transverse wave motion on a string, including, in

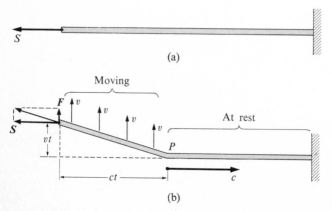

Fig. 21–6 *Propagation of a transverse disturbance in a string.*

particular, the sinusoidal and other periodic waves discussed in Section 21–3.

Example 1   One end of a rubber tube is fastened to a fixed support. The other end passes over a pulley at a distance of 8 m from the fixed end, and carries a load of 2 kg. The mass of the tube, between the fixed end and the pulley, is 600 g. What is the speed of a transverse wave in the tube?

The tension in the tube equals the weight of the 2-kg load, or

$$S = (2 \text{ kg})(9.8 \text{ m s}^{-2}) = 19.6 \text{ N}.$$

The mass of the tube, per unit length, is

$$\mu = \frac{m}{L} = \frac{0.60 \text{ kg}}{8 \text{ m}} = 0.075 \text{ kg m}^{-1}.$$

The speed of propagation is therefore

$$c = \sqrt{\frac{S}{\mu}} = \sqrt{\frac{19.6 \text{ N}}{0.075 \text{ kg m}^{-1}}} = 16 \text{ m s}^{-1}.$$

Example 2   Suppose that a sine wave of amplitude $Y = 10$ cm and wavelength $\lambda = 3$ m travels along the tube from left to right. What is the maximum *transverse* speed of a point of the tube?

The equation of the wave is

$$y = Y \sin (\omega t - kx).$$

The transverse speed is

$$v = \frac{\partial y}{\partial t} = \omega Y \cos (\omega t - kx).$$

The maximum transverse speed is

$$v_{\max} = \omega Y = 2\pi f Y = 2\pi \frac{c}{\lambda} Y$$

$$= (2\pi) \left(\frac{16 \text{ m s}^{-1}}{3 \text{ m}} \right) (0.10 \text{ m})$$

$$= 4.9 \text{ m s}^{-1}.$$

21–5 SPEED OF A LONGITUDINAL WAVE

The propagation speed of longitudinal as well as transverse waves is determined by the mechanical properties of the medium. Figure 21–7 shows a fluid (liquid or gas) of density ρ in a tube of cross-sectional area A, under a pressure p. In Fig. 21–7(a) the fluid is at rest. At time $t = 0$, the piston at the left end of the tube is set in motion toward the right with a speed v. Figure 21–7(b) shows the fluid after a time t has elapsed. All portions of the fluid at the left of point P are moving with speed v, whereas all portions at the right of P are still at rest. The boundary between the moving and the stationary portions travels to the right with the speed of propagation c. The piston has moved a distance vt and the boundary has advanced a distance ct. As in the case of a transverse disturbance in a string, the speed of propagation can be computed from the impulse–momentum theorem.

The quantity of fluid set in motion in time t is the amount that originally occupied a volume of length ct and of cross-sectional area A. The mass of this fluid is therefore ρctA and the longitudinal momentum it has acquired is

$$\text{Longitudinal momentum} = \rho ctAv.$$

We next compute the increase of pressure, Δp, in the moving fluid. The original volume of the moving fluid, Act, has been decreased by an amount Avt. From the definition of bulk modulus B (see Chapter 10),

$$B = \frac{\text{change in pressure}}{\text{fractional change in volume}} = \frac{\Delta p}{Avt/Act}.$$

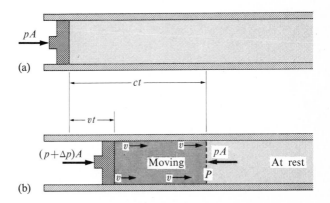

(a)

(b)

Fig. 21–7 *Propagation of a longitudinal disturbance in a fluid confined in a tube.*

Therefore,

$$\Delta p = B\frac{v}{c}.$$

The pressure in the moving fluid is therefore $p + \Delta p$, and the force exerted on it by the piston is $(p + \Delta p)A$. The net force on the moving fluid (see Fig. 21–7(b)) is ΔpA, and the longitudinal impulse is

$$\text{Longitudinal impulse} = \Delta pAt = B\frac{v}{c}At.$$

Applying the impulse-momentum theorem, we obtain

$$B\frac{v}{c}At = \rho ctAv,$$

and therefore

$$\boxed{c = \sqrt{B/\rho} \quad \text{(longitudinal).}} \quad (21\text{–}16)$$

The speed of propagation of a longitudinal pulse in a fluid therefore depends only on the bulk modulus and the density of the medium.

When a solid bar is struck a blow at one end, the situation is somewhat different from that of a fluid confined in a tube of constant cross section, since the bar expands slightly sidewise when it is compressed longitudinally. It can be shown, by the same type of reasoning as that just given, that the velocity of a longitudinal pulse in the bar is given by

$$\boxed{c = \sqrt{Y/\rho} \quad \text{(longitudinal),}} \quad (21\text{–}17)$$

where Y is Young's modulus, defined in Chapter 10.

As with the calculation for the transverse wave on a string, Eqs. (21–16) and (21–17) are valid for all wave motions, not just the special case discussed here. In particular, they are valid for sinusoidal and other periodic waves, discussed in Section 21–3.

21–6 ADIABATIC CHARACTER OF A LONGITUDINAL WAVE

It is a familiar fact that compression of a fluid causes a rise in its temperature unless heat is withdrawn in some way. Conversely, an expansion is accompanied by a temperature decrease unless heat is added. As a longitudinal wave advances through a fluid, the regions which are compressed at any instant are slightly warmer than those that are expanded. The condition is present, therefore, for the conduction of heat from a condensation to a rarefaction. The quantity of heat conducted per unit time and per unit area depends on the thermal conductivity of the fluid and upon the distance between a condensation and its adjacent rarefaction (half a wavelength). Now for ordinary frequencies, say from 20 vibrations per second to 20,000 vibrations per second (the frequency range in which the human ear is sensitive) and for even the best known heat conductors, the wavelength is too large and the thermal conductivity too small for an appreciable amount of heat to flow. The compressions and rarefactions are therefore *adiabatic* rather than isothermal.

In the expression for the speed of a longitudinal wave in a fluid, $c = \sqrt{B/\rho}$, the bulk modulus B is defined by the relation

$$B = \frac{\text{change in pressure}}{\text{fractional change in volume}}.$$

The change in volume produced by a given change of pressure depends upon whether the compression (or expansion) is adiabatic or isothermal. Thus there are two bulk moduli, the adiabatic bulk modulus B_{ad} and the isothermal bulk modulus. The expression for the speed of a longitudinal wave should therefore be written

$$\boxed{c = \sqrt{\frac{B_{ad}}{\rho}}.} \quad (21\text{–}18)$$

In the case of an ideal gas, the relation between pressure p and volume V during an adiabatic process is given by

$$pV^{\gamma} = \text{constant}, \quad (21\text{–}19)$$

where γ is the ratio of the heat capacity at constant pressure to the heat capacity at constant volume.

The definition of the adiabatic bulk modulus is

$$B_{ad} = -\left(\frac{dp}{dV/V}\right)_{ad} = -V\left(\frac{dp}{dV}\right)_{ad}.$$

To calculate the adiabatic bulk modulus B_{ad}, we must evaluate the derivative $(dp/dV)_{ad}$ with the aid of the adiabatic equation. Taking logarithms of both sides of Eq. (21–19), we obtain

$$\ln p + \gamma \ln V = \ln \text{constant},$$

and, taking the differential of this equation, we get

$$\frac{dp}{p} + \gamma \frac{dV}{V} = 0,$$

whence

$$\left(\frac{dp}{dV}\right)_{ad} = -\gamma \frac{p}{V}.$$

Thus for an ideal gas the adiabatic bulk modulus is given by

$$B_{ad} = \gamma p, \tag{21–20}$$

while the isothermal bulk modulus B is given simply by

$$B = p. \tag{21–21}$$

In each case the bulk modulus, characterizing the material's resistance to being compressed, is proportional to the pressure, but the adiabatic modulus is larger than the isothermal by a factor $\gamma = C_p/C_v$, the ratio of the heat capacities at constant pressure and constant volume.

Combining Eqs. (21–18) and (21–20), we obtain

$$c = \sqrt{\gamma p/\rho} \quad \text{(ideal gas)}. \tag{21–22}$$

But, for an ideal gas,

$$p/\rho = RT/M,$$

where R is the universal gas constant, M the molecular mass, and T the Kelvin temperature. Therefore

$$\boxed{c = \sqrt{\gamma RT/M} \quad \text{(ideal gas)}, \tag{21–23}}$$

and since for a given gas, γ, R, and M are constants, we see that the velocity of propagation is proportional to the square root of the Kelvin temperature.

Let us use Eq. (21–23) to compute the velocity of longitudinal waves in air. The mean molecular mass of air is 28.8 g mol^{-1} = 28.8×10^{-3} kg mol^{-1}. Also, $\gamma = 1.40$ and $R = 8.31$ J mol^{-1} K^{-1}. At $T = 300$ K we obtain

$$c = \sqrt{\frac{(1.40)(8.31 \text{ J mol}^{-1}\text{ K}^{-1})(300 \text{ K})}{28.8 \times 10^{-3} \text{ kg mol}^{-1}}}$$

$$\doteq 348 \text{ m s}^{-1} = 1140 \text{ ft s}^{-1} = 778 \text{ mi hr}^{-1}.$$

This agrees with the measured speed at this temperature to within 0.3 percent.

Longitudinal waves in air give rise to the sensation of *sound*. The ear is sensitive to a range of sound frequencies from about 20 to about 20,00 hertz (one hertz = 1 Hz = 1 cycle s^{-1}). From the relation $c = f\lambda$, the corresponding wavelength range is from about 17 m, corresponding to a 20-Hz note, to about 1.7 cm, corresponding to 20,000 Hz.

The *molecular* nature of a gas has been ignored in the preceding discussion, and a gas has been treated as though it were a continuous medium. Actually, we know that a gas is composed of molecules in random motion, separated by distances which are large compared with their diameters. The vibrations which constitute a wave in a gas are superposed on the random thermal motion. At atmospheric pressure, the mean free path is about 10^{-5} cm, while the displacement amplitude of a faint sound may be only one ten-thousandth of this amount. An element of gas in which a sound wave is traveling can be compared to a swarm of gnats, where the swarm as a whole can be seen to oscillate slightly while individual insects move about through the swarm, apparently at random.

Since the *shape* of a fluid does not change when a longitudinal wave passes through it, it is not as easy to visualize the relation between particle motion and wave motion as it is for transverse waves in a string. Figure 21–8 may help in correlating these motions To use this figure, cut a slit about $\frac{1}{16}$ in. wide and $3\frac{1}{4}$ in. long in a 3-in. × 5-in. card (or fasten two cards edge-to-edge with a $\frac{1}{16}$-in. gap), place the card over

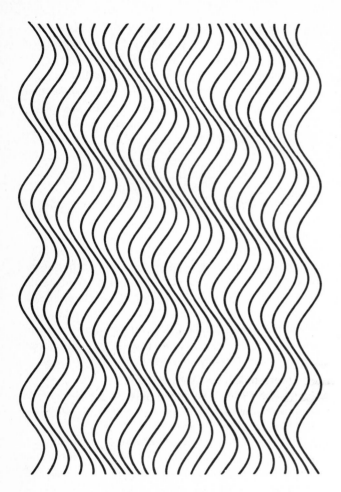

Fig. 21–8 Diagram for illustrating longitudinal traveling waves.

21–7 WATER WAVES

Probably the most familiar type of wave motion is that observed at the surface of a body of water and produced by the winds or some other disturbance. The oscillations of the water particles in these waves are not confined to the surface, however, but extend with diminishing amplitude to the very bottom. Furthermore, the oscillations have both a longitudinal and a transverse component. For this reason, and also because the motion is governed by the laws of hydrodynamics, a complete analysis of this wave motion calls for mathematical methods beyond the scope of this book. The basic physical approach, however, is perfectly straightforward.

Consider a long canal of rectangular cross section and with frictionless walls, containing a depth h of an ideal incompressible liquid of density ρ and surface tension γ. When a train of waves travels along the canal, each element of the liquid is displaced from its equilibrium position both horizontally and vertically. The restoring force on the element results in part from pressure differences brought about by the variations in depth from point to point, and in part from surface tension effects arising from the curvature of the free surface. The solution of the resulting equations of motion must satisfy the conditions that the pressure at the upper surface is constant and equal to atmospheric pressure, and that the vertical displacement at the bottom of the canal is always zero. In addition, the motion of the fluid must satisfy the equation of continuity.

For simple harmonic waves, in which the x- and y-components of displacement are sine or cosine functions, the speed c is found to be given by

$$c^2 = \left(\frac{g\lambda}{2\pi} + \frac{2\pi\gamma}{\rho\lambda}\right) \tanh \frac{2\pi h}{\lambda}, \quad (21\text{-}24)$$

where g is the acceleration of gravity and λ is the wavelength.

We immediately recognize one very significant difference between this expression and that for the speed of propagation of transverse waves in a string, or compressional waves in a fluid. Here, the speed depends not only on the properties of the medium *but on the wavelength λ as well*. When this is the case,

the figure with the slit at the top of the diagram, and move the card downward with constant velocity. The portions of the sine curves that are visible through the slit correspond to a row of particles along which there is traveling a longitudinal, sinusoidal wave. Notice that each particle executes simple harmonic motion about its equilibrium position, with a phase that increases continuously along the slit, while the regions of maximum condensation and rarefraction move from left to right with constant speed. Moving the card upward produces a wave traveling from right to left.

the medium is said to be *dispersive*. Such a dependence of speed on wavelength occurs in many other types of waves. For example, a prism separates a parallel beam of white light into a spectrum, with different wavelengths emerging in different directions, because glass is a dispersive medium for light waves.

In general, the particles of liquid move in ellipses in vertical planes parallel to the length of the canal, the long axis of the ellipse being horizontal. This motion can be considered as the superposition of two simple harmonic oscillations of the same frequency and different amplitudes, one in the horizontal and one in the vertical direction, and with a phase difference of 90°. The wave can therefore be considered as the superposition of a longitudinal and a transverse wave 90° out of phase with each other and having different amplitudes. If the wavelength is as small or smaller than the depth, the two amplitudes are very nearly equal at the surface and the surface particles move in circles. The amplitudes of both components decrease with increasing depth, but the vertical component decreases more rapidly than does the horizontal. At the bottom, the vertical component becomes zero and the oscillation is wholly longitudinal.

Figure 21–9 illustrates the paths of motion of particles in the surface layer and at some depth below the surface. The upper horizontal dotted line represents the free surface of the liquid at rest. The circles are the paths of particles whose equilibrium position is at the center of the circle. When a train of

waves travels from left to right, the particles revolve clockwise in these paths. The full line gives the shape of the surface at the instant shown.

The lower dotted line passes through the equilibrium positions of particles at some depth below the surface. Their paths are elliptical, as indicated, and the dashed line is their locus at the instant when the free surface has the shape shown by the full line.

Similar waves can occur at the surfaces of elastic solids. The rigidity of the material is an added complication in this case, with the result that the phase difference between the longitudinal and transverse components of displacement changes with depth beneath the surface. Such waves, called *Rayleigh waves*, promise to become important in connection with mechanical devices for processing radar, television, and radio signals. For example, it is possible to make *mechanical* band-pass filters for the megahertz frequency range which are simpler and more compact than their electrical analogs.

There are certain properties possessed by all waves. Propagation, reflection, refraction, absorption, interference, and diffraction are the most common. These wave phenomena may be demonstrated effectively by means of a "ripple tank" consisting of a horizontal sheet of glass or plastic forming the bottom of a shallow tank, filled with water about 1 cm deep. A strong source of light beneath the tank sends light through the water waves, which cast a shadow on a translucent screen (tracing cloth) above the tank. A convenient source of water waves is provided by a smooth sphere about the size of a

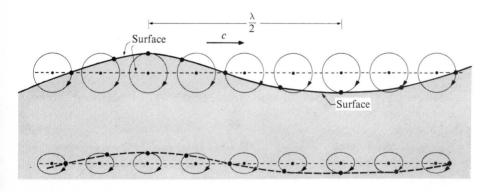

Fig. 21–9 *Particle motion and wave shape for waves in a canal.*

pencil eraser attached to a rod and periodically dipped into the water. At low frequencies the wavelength λ may be made as large as, say, 10 cm. The surface tension of water is about 0.070 N m^{-1}, and the density of water is 10^3 kg m^{-3}. Under these conditions, the second term in the parentheses of Eq. (21-24) turns out to be negligible compared with the first; also, tanh $2\pi h/\lambda$ is very nearly equal to $2\pi h/\lambda$. Then

$$c \approx \sqrt{\frac{g\lambda}{2\pi} \cdot \frac{2\pi h}{\lambda}} \approx \sqrt{gh}. \qquad (21\text{-}25)$$

Thus for *long* waves, the speed depends only on the acceleration due to gravity, g, and the depth h, and is independent of the wavelength (no dispersion). Since, according to Eq. (21-25), the shallower the water, the slower the speed of the waves, it is possible to eliminate reflections from the edges of the ripple tank by beveling the edges of the tank. A wave approaching a "shore" is therefore slowed down to zero speed.

PROBLEMS

21-1 The equation of a certain traveling transverse wave is

$$y = 2 \sin 2\pi \left(\frac{t}{0.01} - \frac{x}{30} \right),$$

where x and y are in centimeters and t is in seconds. What are (a) the amplitude, (b) the wavelength, (c) the frequency, and (d) the speed of propagation of the wave?

21-2 Show that Eq. (21-4) may be written

$$y = -Y \sin \frac{2\pi}{\lambda} (x - ct).$$

21-3 A traveling transverse wave on a string is represented by the equation in Problem 21-2. Let Y = 8 cm, λ = 16 cm, and c = 2 cm s^{-1}.

a) At time t = 0, compute the transverse displacement y at 2-cm intervals of x (i.e., at x = 0, x = 2 cm, x = 4 cm, etc.) from x = 0 to x = 32 cm. Show the results in a graph. This is the shape of the string at time t = 0.

b) Repeat the calculations, for the same values of x, at times t = 1 s, t = 2 s, t = 3 s, and t = 4 s. Show on

the same graph the shape of the string at these instants. In what direction is the wave traveling?

21-4 The equation of a transverse traveling wave on a string is

$$y = 2 \cos \pi (0.05x - 200t),$$

where x and y are in cm and t is in seconds.

a) Find the amplitude, wavelength, frequency, period, and velocity of propagation.

b) Sketch the shape of the string at the following values of t: 0, 0.0025 s, and 0.005 s.

c) If the mass per unit length of the string is 5 g cm^{-1}, find the tension.

21-5 A steel wire 6 m long has a mass of 60 g and is stretched with a tension of 1000 N. What is the speed of propagation of a transverse wave in the wire?

21-6 One end of a horizontal string is attached to a prong of an electrically driven tuning fork whose frequency of vibration is 240 Hz. The other end passes over a pulley and supports a mass of 5 kg. The linear mass density of the string is 0.02 kg m^{-1}.

a) What is the speed of a transverse wave in the string?

b) What is the wavelength?

21-7 One end of a stretched rope is given a periodic transverse motion with a frequency of 10 Hz. The rope is 50 m long, has total mass 0.5 kg, and is stretched with a tension of 400 N.

a) Find the wave speed and the wavelength.

b) If the tension is doubled, how must the frequency be changed to maintain the same wavelength?

21-8 One end of a rubber tube 20 m long, of total mass 1 kg, is fastened to a fixed support. A cord attached to the other end passes over a pulley and supports a body of mass 10 kg. The tube is struck a transverse blow at one end. Find the time required for the pulse to reach the other end.

21-9 A metal wire has the properties: coefficient of linear expansion = 1.5 × 10^{-5}(C°)$^{-1}$, Young's modulus = 2.0 × 10^{11} N m^{-2}, density = 9.0 × 10^3 kg m^{-3}. At each end are rigid supports. If the tension is zero at 20°C, what will be the speed of a transverse wave at 8°C?

21-10 A metal wire, of density 5 × 10^3 kg m^{-3}, with a Young's modulus equal to 2.0 × 10^{11} N m^{-1}, is stretched between rigid supports. At one temperature the speed of a transverse wave is found to be 200 m s^{-1}. When the temperature is raised 100 C°, the speed decreases to 160 m s^{-1}. What is the coefficient of linear expansion?

21–11 A transverse sine wave of amplitude 10 cm and wavelength 200 cm travels from left to right along a long horizontal stretched string with a speed of 100 cm s^{-1}. Take the origin at the left end of the undisturbed string. At time $t = 0$, the left end of the string is at the origin and is moving downward.

a) What is the frequency of the wave?

b) What is the angular frequency?

c) What is the propagation constant?

d) What is the equation of the wave?

e) What is the equation of motion of the left end of the string?

f) What is the equation of motion of a particle 150 cm to the right of the origin?

g) What is the (absolute) maximum transverse velocity of any particle of the string?

h) Find the transverse displacement and the transverse velocity of a particle 150 cm to the right of the origin, at time $t = 3.25$ s.

i) Make a sketch of the shape of the string, for a length of 400 cm, at time $t = 3.25$ s.

21–12 What must be the stress (F/A) in a stretched wire of a material whose Young's modulus is Y, in order that the speed of longitudinal waves shall equal 10 times the speed of transverse waves?

21–13 The speed of longitudinal waves in water is approximately 1450 m s^{-1} at 20°C. Compute the adiabatic compressibility ($1/B_{ad}$) of water and compare with the isothermal compressibility in Table 10–2.

21–14 Provided the amplitude is sufficiently great, the human ear can respond to longitudinal waves over a range of frequencies from about 20 Hz to about 20,000 Hz. Compute the wavelengths corresponding to these frequencies (a) for waves in air, (b) for waves in water. (See Problem 21–13).

21–15 Middle C on the piano corresponds to a frequency of 262 Hz = 262 s^{-1}.

a) Find the wavelength and wave number of the corresponding sound wave in air at 20° C.

b) What is the angular frequency?

c) What is the wavelength of the wave in water at 20° C. if the wavespeed is 1450 m s^{-1}?

21–16 What is the ratio of the speed of sound in a diatomic gas to the rms speed of gas molecules at the same temperature?

21–17

a) If the propagation of sound waves in gases were characterized by isothermal rather than adiabatic expansions and compressions, show that the speed of sound would be given by $(RT/M)^{1/2}$.

b) What would be the speed of sound in air at 20° C in this case?

c) Under what circumstances might the wave propagation be expected to be isothermal?

21–18 A longitudinal wave propagates in a steel bar having density 7.0 g cm^{-3} and Young's modulus 2×10^{11} N m^{-2}.

a) What is the wave speed?

b) By what factor is this speed greater than the speed of sound in air at 20° C?

21–19 At a temperature of 27°C, what is the speed of longitudinal waves in (a) argon, (b) hydrogen? (c) Compare with the speed in air at the same temperature.

21–20 What is the difference between the speeds of longitudinal waves in air at −3°C and at 57°C?

21–21 The sound waves from a loudspeaker spread out nearly uniformly in all directions when their wavelength is large compared with the diameter of the speaker. When the wavelength is small compared with the diameter of the speaker, much of the sound energy is concentrated in the forward direction. For a speaker of diameter 25 cm, compute the frequency for which the wavelength of the sound waves, in air, is (a) 10 times the diameter of the speaker, (b) equal to the diameter of the speaker, (c) 1/10 the diameter of the speaker.

21–22 A steel pipe 100 m long is struck at one end. A person at the other end hears two sounds as a result of two longitudinal waves, one in the pipe and the other in the air. What is the time interval between the two sounds? Take Young's modulus of steel to be 2×10^{11} N m^{-2}.

21–23 Explain why the breakers on a sloping beach are always parallel to the shore line, whatever may be the direction of the waves far from the shore.

21–24 (a) By how many meters per second, at a temperature of 27°C, does the speed of sound in air increase per Celsius degree rise in temperature? [*Hint:* compute dc in terms of dT, and approximate finite changes by differentials.] (b) Is the rate of change of speed with temperature the same at all temperatures?

21–25 A loose coil of flexible rope of length L and mass M rests on a frictionless table. A force S is applied at one end of the rope and more and more rope is pulled from the coil with a constant velocity v.

a) What is the relation between S and v?

b) A transverse pulse is produced while the rope is being pulled from the coil. How does the pulse behave?

c) What is the work of the force S at the moment when the last bit of rope leaves the coil?

d) What is the kinetic energy of the rope?

e) How come?

21–26 The ends of a flexible rope of length L and mass M are joined so as to form a circular loop of radius R. The loop rotates about an axis through its center on a horizontal frictionless surface with a constant linear speed v.

a) Apply Newton's second law to a short piece of rope and calculate the tension S in the rope.

b) Explain the effect of imparting to the rope, while it is spinning, a small rapid radial displacement.

Chapter 22

Vibrating Bodies

22–1 BOUNDARY CONDITIONS FOR A STRING

Let us consider what happens when a wave pulse or wave train, advancing along a stretched string, arrives at the end of the string. If the end is fastened to a rigid support, it just remains at rest. The arriving wave exerts a force on the support, and the reaction to this force "kicks back" on the string and sets up a *reflected* pulse or wave train traveling in the reverse direction.

At the opposite extreme from a rigidly fixed end is one which is perfectly free to move in the direction transverse to the length of the string. For example, the string might be tied to a light ring which slides on a smooth rod perpendicular to the length of the string; the ring and rod maintain the tension of the string but exert no transverse force. At a free end the arriving pulse or wave train causes the end to "overshoot," and again a reflected wave is set up. The conditions imposed on the motion of the end of the string, such as attachment to a rigid support or the complete absence of transverse force, are called *boundary conditions*.

The multiflash photograph of Fig. 22–1 shows the reflection of a pulse at a fixed end of a string. (The camera was tipped vertically while the photographs were taken so that successive images lie one under the other. The "string" is a rubber tube and it

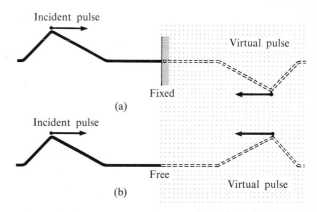

Fig. 22–1 *A pulse starts in the upper left corner and is reflected from the fixed end of the string at the right.*

Fig. 22–2 *Description of reflection of a pulse (a) at a fixed end of a string, and (b) at a free end, in terms of an imaginary "virtual" pulse.*

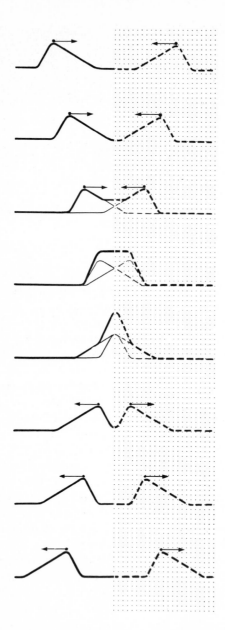

Fig. 22-3 *Reflection at a free end.*

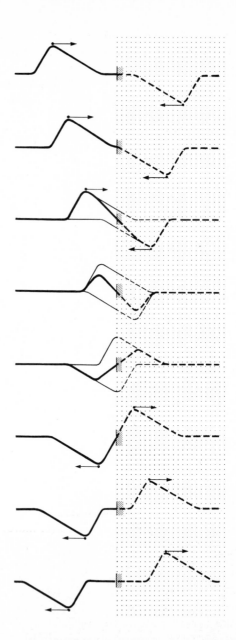

Fig. 22-4 *Reflection at a fixed end.*

sags somewhat.) It will be seen that the pulse is reflected with its displacement and its direction of propagation both reversed. When reflection takes place at a free end, the direction of propagation is reversed but the direction of the displacement is unchanged.

It is helpful to think of the process of reflection in the following way. Imagine the string to be extended indefinitely beyond its actual terminus. The actual pulse can be considered to continue on into the imaginary portion as though the support were not there, while at the same time a "virtual" pulse, which has been traveling in the imaginary portion, moves out into the real string and forms the reflected pulse. The nature of the reflected pulse depends on whether the end is fixed or free. The two cases are shown in Fig. 22–2.

The displacement at a point where the actual and virtual pulses cross each other is the algebraic sum of the displacements in the individual pulses. Figures 22–3 and 22–4 show the shape of the region near the end of the string for both types of reflected pulses. It will be seen that Fig. 22–3 corresponds to a free end and Fig. 22–4 to a fixed end. In the latter case, the incident and reflected pulses combine in such a way that the displacement of the end of the string is always zero.

22–2 STANDING WAVES IN A STRING

When a continuous train of waves arrives at a fixed end of a string, a continuous train of reflected waves appears to originate at the end and travel in the opposite direction. Provided the elastic limit of the string is not exceeded and the displacements are sufficiently small, the actual displacement of any point of the string is the algebraic sum of the displacements of the incident and reflected waves, a fact called the *principle of superposition*. This principle is extremely important in all types of wave motion and applies not only to waves in a string but to sound waves in air, to light waves, and, in fact, to wave motion of any sort. The general term *interference* is applied to the effect produced by two (or more) wave trains which are simultaneously passing through a given region.

The appearance of the string in these circumstances gives no evidence that two waves are traversing it in opposite directions. If the frequency is sufficiently great so that the eye cannot follow the motion, the string appears subdivided into a number of segments, as in the time exposure photograph of Fig. 22–5(a). A multiflash photograph of the same string, in Fig. 22–5(b), indicates a few of the instantaneous shapes of the string. At any instant (except those when the string is straight) its shape is a sine curve, but whereas in a traveling wave the amplitude remains constant while the wave progresses, here the

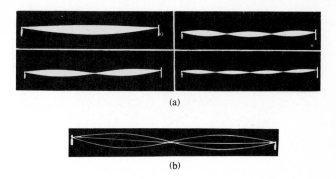

(a)

(b)

Fig. 22–5 *(a) Standing waves in a stretched string (time exposure). (b) Multiflash photograph of a standing wave, with nodes at the center and at the ends.*

waveform remains fixed in position (longitudinally) while the amplitude fluctuates. Certain points known as the *nodes* remain always at rest. Midway between these points, at the *loops* or *antinodes*, the fluctuations are a maximum. The vibration as a whole is called a *standing* wave.

To understand the formation of a standing wave, consider the separate graphs of the waveform S at four instants one-tenth of a period apart, shown in Fig. 22–6. The dotted curves represent a wave traveling to the right. The dashed curves represent a wave of the same propagation speed, wavelength, and amplitude traveling to the *left*. The heavy curves represent the resultant waveform, obtained by applying the principle of superposition, that is, by adding displacements. At those places marked N at the bottom of Fig. 22–6, the resultant displacements are *always* zero. These are the **nodes**. Midway between the nodes, the vibrations have the *largest* amplitude. These are the **antinodes**. It is evident from the figure that

$$\left. \begin{array}{l} \text{Distance between adjacent nodes} \\ \qquad\qquad\qquad \text{or} \\ \text{Distance between adjacent antinodes} \end{array} \right\} = \frac{\lambda}{2}.$$

The equation of a standing wave may be obtained by adding the displacements of two waves of

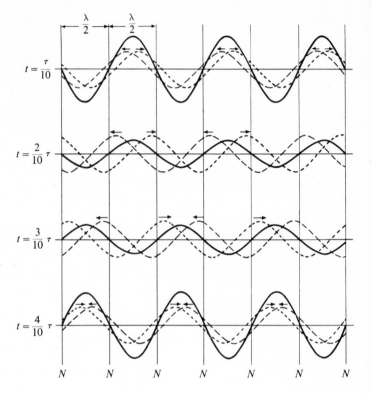

Fig. 22–6 *The formation of a standing wave.*

equal amplitude, period, and wavelength, but travel-ing in opposite directions. Thus, if

$$y_1 = A \sin(\omega t - kx) \qquad \text{(positive } x\text{-direction)},$$
$$y_2 = -A \sin(\omega t + kx) \qquad \text{(negative } x\text{-direction)},$$

then

$$y_1 + y_2 = A[\sin(\omega t - kx) - \sin(\omega t + kx)].$$

Introducing the expressions for the sine of the sum and difference of two angles and combining terms, we obtain

$$y_1 + y_2 = -[2A \cos \omega t]\sin kx. \qquad (22\text{–}1)$$

The shape of the string at each instant is, therefore, a sine curve whose amplitude (the expression in brackets) varies with time.

22–3 VIBRATION OF A STRING FIXED AT BOTH ENDS

Thus far we have discussed a long string fixed at one end and have considered the standing waves set up near that end by interference between the incident and reflected waves. Let us next consider the more usual case, that of a string fixed at *both* ends. A continuous train of sine or cosine waves is reflected and re-reflected; and since the string is fixed at both ends, both ends must be nodes. Since adjacent nodes are one-half wavelength apart, the length of the string may be

$$\frac{\lambda}{2}, \quad \frac{2\lambda}{2}, \quad \frac{3\lambda}{2},$$

or, in general, *any* integer number of half-wave-lengths. Or, to put it differently, if one considers a particular string of length L, standing waves may be set up in the string by vibrations of a number of different frequencies, namely, those which give rise to waves of wavelengths

$$\lambda = 2L, \frac{2L}{2}, \frac{2L}{3}, \cdots = \frac{2L}{n} \qquad (n = 1, 2, 3, \ldots).$$

$$(22\text{–}2)$$

From the relation $f = c/\lambda$, and since c is the same for all frequencies, the possible frequencies are

$$f = \frac{c}{2L}, \frac{2c}{2L}, \frac{3c}{2L}, \cdots = \frac{nc}{2L} \qquad (n = 1, 2, 3, \ldots).$$

$$(22\text{–}3)$$

The lowest frequency, $c/2L$, is called the *fundamental* frequency f_1 and the others are the *overtones*. The frequencies of the latter are therefore $2f_1$, $3f_1$, $4f_1$, and so on. Overtones whose frequencies are integral multiples of the fundamental are said to form a *harmonic series*. The fundamental is the *first harmonic*. The frequency $2f_1$ is the *first overtone* or the *second harmonic*, the frequency $3f_1$ is the *second overtone* or the *third harmonic*, and so on.

The above results may also be obtained directly from Eq. (22–1). The boundary conditions require that $y_1 + y_2 = 0$ at the ends of the string, that is at $x = 0$ and $x = L$. Since the sine of zero is zero, the first condition is satisfied automatically. The second requires that $\sin kL = 0$, and this is true only when k has certain special values. The sine of an angle is zero only when the angle is zero or an integer multiple of π (180°). Thus we must have

$$kL = n\pi \qquad (n = 1, 2, 3, \ldots). \qquad (22\text{–}4)$$

We do not include the possibility $n = 0$ because that gives $k = 0$, i.e., a wave with zero displacement *everywhere* (a possible case, to be sure, but not a very interesting one!).

Replacing k above by $2\pi/\lambda$, we obtain

$$\frac{2\pi L}{\lambda} = n\pi \qquad \text{or} \qquad \lambda = \frac{2L}{n},$$

in agreement with Eq. (22–2).

Each of the frequencies given by Eq. (22–3) corresponds to a possible *normal mode* of motion,

that is, a motion in which each particle of the string moves sinusoidally, all with the same frequency. As the above analysis shows, there is an infinite number of normal modes, each with its characteristic frequency. This situation is in striking contrast with the simpler harmonic oscillator system, a single mass and a spring. The harmonic oscillator has only one normal mode and one characteristic frequency, while the vibrating string has an infinite number. When a body suspended from a spring is pulled down and released, only one frequency of vibration will ensue. If a string is initially distorted so that its shape is the same as *any one* of the possible harmonics, it will vibrate, when released, at the frequency of that particular harmonic. But when a piano string is struck, not only the fundamental, but many of the overtones are present in the resulting vibration. This motion is therefore a combination or *superposition* of normal modes. Several frequencies and motions are present simultaneously, and the displacement of any point on the string is the sum or superposition of displacements associated with the individual modes.

The fundamental frequency of the vibrating string is $f_1 = c/2L$, where $c = \sqrt{S/\mu}$. It follows that

$$f_1 = \frac{1}{2L}\sqrt{\frac{S}{\mu}}. \qquad (22\text{–}5)$$

Stringed instruments afford many examples of the implications of this equation. For example, all such instruments are "tuned" by varying the tension S, an increase of tension increasing the frequency or pitch, and vice versa. The inverse dependence of frequency on length L is illustrated by the long strings of the bass section of the piano or the bass viol compared with the shorter strings of the piano treble or the violin. Pressing the strings against the fingerboard of a violin or guitar with the fingers to change the length of the vibrating portion of the string is the usual means of varying pitch when playing these instruments. One reason for winding the bass strings of a piano with wire is to increase the mass per unit length μ, so as to obtain the desired low frequency without resorting to a string which is inconveniently long.

22-4 RESONANCE

Whenever a body capable of oscillating is acted on by a periodic series of impulses having a frequency equal to one of the natural frequencies of oscillation of the body, the body is set into vibration with a relatively large amplitude. This phenomenon is called *resonance*, and the body is said to *resonate* with the applied impulses.

A common example of mechanical resonance is provided by pushing a swing. The swing is a pendulum with a single natural frequency depending on its length. If a series of regularly spaced pushes is given to the swing, with a frequency equal to that of the swing, the motion may be made quite large. If the frequency of the pushes differs from the natural frequency of the swing, or if the pushes occur at irregular intervals, the swing will hardly execute any vibration at all.

Unlike a simple pendulum, which has only one natural frequency, a stretched string (and other systems to be discussed later in this chapter) has a large number of natural frequencies. Suppose that one end of a stretched string is fixed while the other is moved back and forth in a transverse direction. The amplitude at the driven end is fixed by the driving mechanism. Standing waves will be set up in the string, whatever the value of the frequency f. If the frequency is not equal to one of the natural frequencies of the string, the amplitude at the antinodes will be fairly small. However, if the frequency is equal to *any one* of the natural frequencies, the string is in resonance and the amplitude at the antinodes will be very much larger than that at the driven end. In other words, although the driven end is not a node, it lies much closer to a node than to an antinode when the string is in resonance. In Fig. 22-5(a), the right end of the string was held fixed and the left end was forced to oscillate vertically with small amplitude. The photographs show the standing waves of relatively large amplitude which resulted when the frequency of oscillation of the left end was equal to the fundamental frequency or to one of the first three overtones.

A steel bridge or, for that matter, any elastic structure, is capable of vibrating with certain natural frequencies. If the regular footsteps of a column of soldiers were to have a frequency equal to one of the natural frequencies of a bridge which the soldiers are crossing, a vibration of dangerously large amplitude might result. Therefore, in crossing a bridge, a column of soldiers is always ordered to break step.

Tuning a radio or television receiver is an example of electrical resonance. By turning a dial, the natural frequency of an alternating current in the receiving circuit is made equal to the frequency of the waves broadcast by the desired station. Optical resonance may also take place between atoms in a gas at low pressure and light waves from a lamp containing the same atoms. Thus, light from a sodium lamp may cause the sodium atoms in a glass bulb to glow with characteristic yellow sodium light.

The phenomenon of resonance may be demonstrated with the aid of the longitudinal waves set up in air by a vibrating plate or tuning fork. If two identical tuning forks are placed some distance apart and one is struck, the other will be heard when the first is suddenly damped. Should a small piece of wax or modeling clay be put on one of the forks, the frequency of that fork will be altered enough to destroy the resonance.

A similar phenomenon can be demonstrated with a piano. With the damper pedal depressed so the dampers are lifted and the strings free to vibrate, one sings a steady tone into the piano. When the singing stops the piano seems to continue to sing the same note. The sound waves from the voice excite vibrations in those strings with natural frequencies close to those (fundamental or harmonics) of the note sung initially.

22-5 INTERFERENCE OF LONGITUDINAL WAVES

The phenomenon of interference between two longitudinal waves in air may be demonstrated with the aid of the apparatus depicted in Fig. 22-7. A wave emitted by a source S is sent into a metal tube, where it divides into two waves, one following the constant path SAR, the other the path SBR, which may be varied by sliding the tube B to the right. Suppose the frequency of the source is 350 Hz. Then the wavelength is $\lambda = c/f = 1$ m. If both paths are of equal length, the two waves arrive at R at the same time

Source
S

A

Sliding tube

B

R
Receiver

Fig. 22–7 *Apparatus for demonstrating interference of longitudinal waves.*

and the vibrations set up by both waves are in phase. The resulting vibration has an amplitude equal to the sum of the two individual amplitudes and the phenomenon of *reinforcement* may be detected either with the ear at R or with the aid of a microphone, amplifier, and loudspeaker.

Now suppose the tube B is moved out a distance of 0.25 m, thereby making the path SBR 0.5 m longer than the path SAR. The righthand wave travels a distance $\lambda/2$ greater than the lefthand wave and the vibration set up at R by the righthand wave is therefore in opposite phase to that set up by the lefthand wave. The consequent interference is shown by the marked reduction in sound at R.

If the tube B is now pulled out another 0.25 m, so that the *path difference, SBR* minus *SAR*, is 1 m (1 wavelength), the two vibrations at R again reinforce each other. Thus

$$\left\{\begin{matrix}\text{Reinforcement takes place} \\ \text{when the path difference}\end{matrix}\right\} = 0, \lambda, 2\lambda, \text{etc.}$$

$$\left\{\begin{matrix}\text{Interference takes place} \\ \text{when the path difference}\end{matrix}\right\} = \frac{\lambda}{2}, \frac{3\lambda}{2}, \frac{5\lambda}{2}, \text{etc.}$$

A similar phenomenon can be demonstrated with two loudspeakers driven by the same amplifier, as shown in Fig. 22–8. Suppose the speakers both emit a pure sinusoidal sound wave of constant frequency. When a microphone is placed at point A in the figure, equidistant from the speakers, it receives a strong acoustic signal, corresponding to the fact that two waves emitted from the speakers in phase also arrive at point A in phase and add to each other. But at point B the signal is much *weaker* than when only one speaker is present, corresponding to the fact that

the wave from one speaker travels a half-wavelength farther than that from the other, the two arrive a half-cycle out of phase and cancel each other out almost completely. An experiment closely analogous to this one but using light waves, provided conclusive evidence of the wave nature of light.

22–6 LONGITUDINAL STANDING WAVES

Longitudinal waves traveling along a tube of finite length are reflected at the ends of the tube in much the same way that transverse waves in a string are reflected at its ends. Interference between the waves traveling in opposite directions again gives rise to standing waves.

When reflection takes place at a closed end, the displacement of the particles there must always be

B

A

Amplifier

Fig. 22–8

zero. Hence a closed end is a *displacement node*. If the end of the tube is open, the nature of the reflection is more complex and depends on whether the tube is wide or narrow compared with the wavelength. If the tube is narrow compared with the wavelength, which is the case in most musical instruments, the reflection is such that the open end is a *displacement antinode*. Thus the longitudinal waves in a column of fluid are reflected at the closed and open ends of a tube in the same way that transverse waves in a string are reflected at fixed and free ends, respectively.

The reflections at the openings where the instrument is blown are usually characterized by an antinode located at or near the opening. The effective length of the air column of a wind instrument is thus less definite than the length of a string fixed at its ends.

Standing longitudinal waves in a column of gas may be demonstrated conveniently with the aid of the apparatus shown in Fig. 22–9, known as Kundt's tube. A glass tube a few feet long is closed at one end with glass and at the other with a flexible diaphragm. The gas to be studied is admitted to the tube at a known temperature and at atmospheric pressure. A source of longitudinal waves, S, whose frequency may be varied, causes vibration of the flexible diaphragm. A small amount of light powder or cork dust is sprinkled uniformly along the tube.

When a frequency is found at which the air column is in resonance, the amplitude of the standing waves becomes large enough for the gas particles to sweep the cork dust along the tube, at all points where the gas is in motion. The powder therefore collects at the displacement nodes, where the gas remains at rest. Sometimes a wire, running along the axis of the tube, is maintained at a dull red heat by an electric current. The nodes show themselves as

hot points, compared with the antinodes, which are cooled by moving gas.

With careful manipulation and a good variable-frequency source, a fair determination of the velocity of the wave may be obtained with Kundt's tube. Since, in a standing wave, the distance between two adjacent nodes is one-half (not one!) wavelength, the wavelength λ is obtained by measuring the distance between alternate clumps of powder. When the frequency f is known, the velocity c is then

$$c = f\lambda.$$

At a displacement node, the pressure variations above and below the average are *maximum*, whereas at a displacement antinode, pressure does not vary. This may be understood easily when it is realized that two small masses of gas on opposite sides of a displacement node vibrate in *opposite phase*. When they approach each other, the pressure at the node is maximum, and when they recede from each other, the pressure at the node is minimum. Two small masses of gas, however, on opposite sides of a displacement *antinode* vibrate *in phase*, and hence cause no pressure variations at the antinode.

Figure 22–10 may be helpful in visualizing a longitudinal standing wave. To use this figure, cut a slit about $\frac{1}{16}$ in. wide and $3\frac{1}{2}$ in. long in a card. Place the card over the diagram with the slit horizontal and move it vertically with constant velocity. The portions of the curves that appear in the slit will correspond to the oscillations of the particles in a longitudinal standing wave.

22–7 VIBRATIONS OF ORGAN PIPES

If one end of a pipe is open and a stream of air is directed against an edge, vibrations are set up and

Fig. 22–9 *Kundt's tube for determining the velocity of sound in a gas. The dots represent the density of the gas molecules at an instant when the pressure at the displacement nodes is a maximum or a minimum.*

Fig. 22–10

be $c/4L$, which is one-half that of an open pipe of the same length. In the language of music, the *pitch* of a closed pipe is one octave lower (a factor of two in frequency) than that of an open pipe of equal length. From the remaining diagrams of Fig. 22–12, it may be seen that the second, fourth, etc., harmonics are missing. Hence, *in a closed pipe, the fundamental frequency is $c/4L$ and only the odd harmonics are present.*

Almost always in an organ pipe several modes are present at once. The extent to which modes higher than the fundamental are present depends on the cross section of the pipe, the ratio of length to width, whether the pipe is straight or tapered, and to a slight extent on the material. The harmonic content of the tone is an important factor in determining the "tone quality" or timbre. A very thin pipe produces

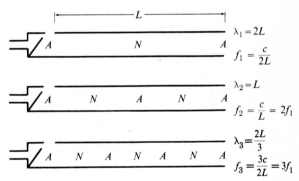

Fig. 22–11 *Modes of vibration of an open organ pipe.*

the tube resonates at its natural frequencies. As in the case of a plucked string, the fundamental and overtones exist at the same time. In the case of a pipe open at both ends, the fundamental frequency f_1 corresponds to an antinode at each end and a node in the middle, as shown at the top of Fig. 22–11. Succeeding diagrams of Fig. 22–11 show two of the overtones, which are seen to be the second and third harmonics. *In an open pipe the fundamental frequency is $c/2L$ and all harmonics are present.*

The properties of a *closed* pipe (open at one end but closed at the other) are shown in the diagrams of Fig. 22–12. The fundamental frequency is seen to

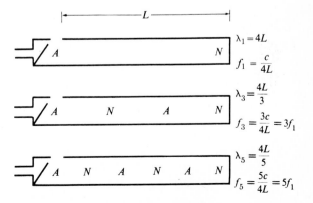

Fig. 22–12 *Modes of vibration of a closed organ pipe.*

$$\lambda_1 = 2L$$
$$f_1 = \frac{c}{2L}$$

$$\lambda_2 = L$$
$$f_2 = \frac{c}{L} = 2f_1$$

Fig. 22–13 *Modes of vibration of a rod.*   (a)   (b)

a tone rich in higher harmonics which the ear perceives as thin and "stringy," while a fatter pipe produces principally the fundamental tone, perceived as a softer, more flute-like tone.

22–8 VIBRATIONS OF RODS AND PLATES

A rod may be set in longitudinal vibration by clamping it at some point and stroking it with a chamois skin that has been sprinkled with rosin. In Fig. 22–13(a) the rod is clamped in the middle; consequently, when it is stroked near the end, a standing wave is set up with a node in the middle and antinodes at each end, exactly the same as the fundamental mode of an open organ pipe. The fundamental frequency of the rod is then $c/2L$, where c is the velocity of a longitudinal wave in the rod. Since the velocity of a longitudinal wave in a solid is much greater than that in air, a rod has a higher fundamental frequency than an open organ pipe of the same length.

By clamping the rod at a point $\frac{1}{4}$ of its length from one end, as shown in Fig. 22–13(b), the second harmonic may be produced.

When a stretched flexible membrane, such as a drumhead, is struck a blow, a two-dimensional pulse travels outward from the struck point and is reflected and re-reflected at the boundary of the membrane. If some point of the membrane is forced to vibrate periodically, continuous trains of waves travel along the membrane. Just as with the stretched string, standing waves can be set up in the membrane; each of these waves has a certain natural frequency. The lowest frequency is the fundamental and the others are overtones. In general, when the membrane is vibrating, a number of overtones are present.

The nodes of a vibrating membrane are *lines* (nodal lines) rather than points. The boundary of the membrane is evidently one such line. Some of the other possible nodal lines of a circular membrane are indicated by arrows in Fig. 22–14. The natural

frequency of each mode is given in terms of the fundamental f_1. It will be noted that the frequencies of the overtones are *not* integer multiples of f_1. That is, they are not harmonics.

The restoring force in a vibrating flexible membrane arises from the tension with which it is stretched. A metal plate, if sufficiently thick, vibrates in a similar way, the restoring force being produced by bending stresses in the plate. The study of vibrations of membranes and plates is of importance in connection with the design of loudspeaker diaphragms and the diaphragms of telephone receivers and microphones.

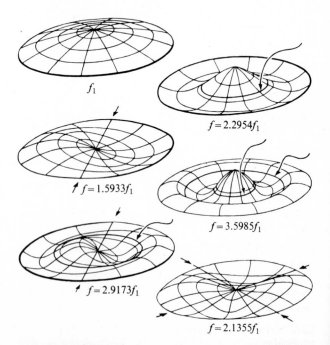

f_1

$f = 2.2954f_1$

$f = 1.5933f_1$

$f = 3.5985f_1$

$f = 2.9173f_1$

$f = 2.1355f_1$

Fig. 22–14 *Modes of vibration of a membrane, showing nodal lines. The frequency of each mode is given in terms of the fundamental frequency, f_1. (Adapted from* Vibration and Sound, *by Philip M. Morse, 2nd Edition, McGraw-Hill Book Company, Inc., 1948. By permission of the publishers.)*

PROBLEMS

22-1 A steel piano wire 50 cm long, of mass 5 g, is stretched with a tension of 400 N.

a) What is the frequency of its fundamental mode of vibration?

b) What is the number of the highest overtone that could be heard by a person who is capable of hearing frequencies up to 10,000 Hz?

22-2 A steel wire of length $L = 100$ cm and density $\rho = 8\,\text{g cm}^{-3}$ is stretched tightly between two rigid supports. Vibrating in its fundamental mode, the frequency is $f = 200$ Hz.

a) What is the speed of transverse waves on this wire?

b) What is the longitudinal stress (force per unit area) in the wire?

c) If the maximum acceleration at the midpoint of the wire is 80,000 cm s^{-2}, what is the amplitude of vibration at the midpoint?

22-3 A stretched string is observed to vibrate with a frequency of 30 Hz in its fundamental mode when the supports are 60 cm apart. The amplitude at the antinode is 3 cm. The string has a mass of 30 g.

a) What is the speed of propagation of a transverse wave in the string?

b) Compute the tension in the string.

22-4 The fundamental frequency of the A-string on a cello is 220 Hz. The vibrating portion of the string is 68 cm long and has a mass of 1.29 g. With what tension must it be stretched?

22-5 A standing wave of frequency 1100 Hz in a column of methane at 20°C produces nodes that are 20 cm apart. What is the ratio of the heat capacity at constant pressure to that at constant volume?

22-6 An aluminum object is hung from a steel wire. The fundamental frequency for transverse standing waves on the wire is 300 Hz. The object is then immersed in water so that one-half of its volume is submerged. What is the new fundamental frequency?

22-7 Standing waves are set up in a Kundt's tube by the longitudinal vibration of an iron rod 1 m long, clamped at the center. If the frequency of the iron rod is 2480 Hz and the powder heaps within the tube are 6.9 cm apart, what is the speed of the waves (a) in the iron rod, and (b) in the gas?

22-8 The speed of a longitudinal wave in a mixture of helium and neon at 300 K was found to be 758 m s^{-1}. What is the composition of the mixture?

22-9 The atomic mass of iodine is 127 g mol^{-1}. A standing wave in iodine vapor at 400 K produces nodes that are 6.77 cm apart when the frequency is 1000 Hz. Is iodine vapor monatomic or diatomic?

22-10 A copper rod 1 m long, clamped at the $\frac{1}{4}$ point, is set in longitudinal vibration and is used to produce standing waves in a Kundt's tube containing air at 300 K. Heaps of cork dust within the tube are found to be 4.95 cm apart. What is the speed of the longitudinal waves in copper?

22-11 Find the fundamental frequency and the first four overtones of a 6-in. pipe (a) if the pipe is open at both ends, (b) if the pipe is closed at one end. (c) How many overtones may be heard by a person having normal hearing for each of the above cases?

22-12 A long tube contains air at a pressure of 1 atm and temperature 77°C. The tube is open at one end and closed at the other by a movable piston. A tuning fork near the open end is vibrating with a frequency of 500 Hz. Resonance is produced when the piston is at distances 18.0, 55.5, and 93.0 cm from the open end.

a) From these measurements, what is the speed of sound in air at 77°C?

b) From the above result, what is the ratio of the specific heats γ for air?

22-13 An organ pipe A of length 2 ft, closed at one end, is vibrating in the first overtone. Another organ pipe B of length 1.35 ft, open at both ends, is vibrating in its fundamental mode. Take the speed of sound in air as 1120 ft s^{-1}. Neglect end corrections. (a) What is the frequency of the tone from A? (b) What is the frequency of the tone from B?

22-14 A plate cut from a quartz crystal is often used to control the frequency of an oscillating electrical circuit. Longitudinal standing waves are set up in the plate with displacement antinodes at opposite faces. The fundamental frequency of vibration is given by the equation

$$f_1 = \frac{2.87 \times 10^5}{s},$$

where f_1 is in hertz and s in the thickness of the plate in centimeters.

a) Compute Young's modulus for the quartz plate.

b) Compute the thickness of plate required for a frequency of 1200 kHz. (1 kHz = 1000 Hz.) The density of quartz is 2.66 g cm^{-3}.

22–15 In Problem 22–4, what percent increase in tension is needed to increase the frequency from 220 Hz to 233 Hz, corresponding to a rise in pitch from A to A-sharp?

22–16 The longest pipe found in most medium-size pipe organs is 16 feet long. What is the frequency of the corresponding note if the pipe is (a) open at both ends; (b) open at one end, closed at the other?

22–17 The frequency of middle C is 262 Hz.

a) If an organ pipe is open at both ends, what length must it have to produce this note at 20° C?

b) At what temperature will the frequency be 6 percent higher, corresponding to a rise in pitch from C to C-sharp?

22–18 A certain organ pipe produces a frequency of 440 Hz in air. If the pipe is filled with helium at the same temperature, what frequency does it produce?

Chapter 23

Acoustic Phenomena

23–1 SOUND WAVES

In this chapter we are concerned primarily with longitudinal waves in air which, when striking the ear, give rise to the sensation of *sound*. The human ear is sensitive to waves in the frequency range from about 20 to 20,000 Hz, although the term *sound wave* is sometimes applied also to similar waves with frequencies outside the range of human audibility.

The simplest sound waves are sinusoidal waves with definite frequency, amplitude, and wavelength. When such a wave arrives at the ear, it causes a vibration of the air particles at the eardrum with a definite frequency and amplitude. This vibration may also be described in terms of the variation of *air pressure* at the same point. The air pressure rises above atmospheric pressure, and then drops below atmospheric pressure, with simple harmonic motion of the same frequency as that of an air particle.

The simplest sinusoidal sound wave is described by the wave function

$$y = Y \sin (\omega t - kx), \qquad (23\text{–}1)$$

where the symbols have the same meaning as in Section 21–3. It is usually much easier to measure the *pressure* variations in sound waves than the displacements, so it is worthwhile to develop a relation

between the two. Let p be the instantaneous pressure fluctuation at any point, that is, the amount by which the pressure differs from normal atmospheric pressure. If the displacements of two neighboring points x and $x + \Delta x$ are the same, the gas between these points is neither compressed nor rarefied, there is no volume change, and consequently $p = 0$. Only when y varies from one point to a neighboring point is there a change of volume and therefore of pressure.

The fractional volume change $\Delta V/V$ in an element near point x turns out to be given simply by $\partial y/\partial x$, which is the rate of change of y with x as we go from one point to a neighboring point. From the definition of the bulk modulus B, Eq. (10–11), $p = -B\Delta V/V$, and we find

$$p = -B\left(\frac{\partial y}{\partial x}\right). \qquad (23\text{–}2)$$

The negative sign arises because, when $\partial y/\partial x$ is positive, the displacement is greater at $x + \Delta x$ than at x, corresponding to an increase in volume and a *decrease* in pressure. For the sinusoidal wave of Eq. (23–1) we find

$$p = -BkY \cos (\omega t - kx). \qquad (23\text{–}3)$$

The maximum amount by which the pressure differs from atmospheric, that is, the maximum value of p,

389

is called the *pressure amplitude*, denoted P. Equation (23–3) shows that

$$P = BkY. \qquad (23\text{–}4)$$

We may also develop a relation between p and the particle velocity v of points in the medium, not to be confused with the wave propagation velocity c. The particle velocity is given by $v = \partial y/\partial t$, and for Eq. (23–1) this is

$$v = \frac{\partial y}{\partial t} = -\omega Y \cos(\omega t - kx). \qquad (23\text{–}5)$$

The maximum velocity, called the velocity amplitude V, is given by $V = \omega Y$. Combining this with Eq. (23–4), we find

$$P = \frac{BkV}{\omega} = \frac{BV}{c}. \qquad (23\text{–}6)$$

Thus the relation between pressure amplitude and velocity amplitude does not depend on frequency. Recalling that the propagation velocity c is given by

$$c = \sqrt{B/\rho}, \qquad (23\text{–}7)$$

we may rewrite Eq. (23–6) as follows:

$$P = \sqrt{B\rho}\,V. \qquad (23\text{–}8)$$

The quantity $(B\rho)^{\frac{1}{2}}$ is called the *mechanical impedance* of the system, in reference to an analogy with electric circuits, in which pressure is analogous to voltage, velocity to current, and mechanical impedance to resistance or impedance. In terms of this analogy, Eq. (23–8) is the mechanical analog of Ohm's law.

Example Measurements of sound waves show that the maximum pressure variations in the loudest sounds that the ear can tolerate without pain are of the order of 30 N m^{-2} (above and below atmospheric pressure, which is about 100,000 N m^{-2}). Find the corresponding maximum displacement, if the frequency is 100 Hz and $c = 350$ m s^{-1}.

We have $\omega = (2\pi)(1000\text{ Hz}) = 6283$ s^{-1} and

$$k = \frac{\omega}{c} = \frac{6283\text{ s}^{-1}}{350\text{ m s}^{-1}} = 18.0\text{ m}^{-1}.$$

The adiabatic bulk modulus for air is

$$B = (1.4)(1.01 \times 10^5\text{ N m}^{-2}) = 1.42 \times 10^5\text{ N m}^{-2}.$$

From Eq. (23–4) we find

$$Y = \frac{P}{Bk} = \frac{(30\text{ N m}^{-2})}{(1.42 \times 10^5\text{ N m}^{-2})(18.0\text{ m}^{-1})}$$

$$= 1.17 \times 10^{-5}\text{ m}$$

$$= 0.017\text{ mm}.$$

Thus the displacement amplitude of even the loudest sound is *extremely small*. The maximum pressure variation in the *faintest* audible sound of frequency 1000 Hz is only about 3×10^{-5} N m^{-2}. The corresponding displacement amplitude is about 10^{-9} cm. By way of comparison, the wavelength of yellow light is 6×10^{-5} cm, and the diameter of a molecule is about 10^{-8} cm. The ear is an extremely sensitive organ!

23–2 INTENSITY

An essential aspect of wave propagation of all sorts is transfer of *energy*. A familiar example is the energy supply of the earth, which reaches us from the sun via electromagnetic waves. The intensity I of a traveling wave is defined as *the time average rate at which energy is transported by the wave, per unit area*, across a surface perpendicular to the direction of propagation. More briefly, the intensity is the average *power* transported per unit area.

We have seen that the power developed by a force equals the product of force and velocity. Hence the power per unit area in a sound wave equals the product of the excess pressure (force per unit area) and the *particle* velocity, given by Eqs. (23–3) and (23–5), respectively. We find

$$pv = \omega BkY^2 \cos^2(\omega t - kx). \qquad (23\text{–}9)$$

The intensity is, by definition, the average value of this quantity. The average value of the function $\cos^2 z$ is $\frac{1}{2}$, so we find

$$I = \tfrac{1}{2}\omega BkY^2. \qquad (23\text{–}10)$$

It is usually more convenient to express I in terms of

the pressure amplitude P. Using Eq. (23–4), we find

$$I = \frac{P^2}{2\rho c} = \frac{P^2}{2\sqrt{\rho B}}. \qquad (23\text{–}11)$$

The fact that the intensity is proportional to the *square* of the amplitude is a characteristic of all kinds of wave motion.

The intensity of a sound wave of the largest amplitude tolerable to the human ear (about $P = 30$ N m^{-2}) is

$$I = \frac{(30\,\text{N m}^{-2})^2}{2(1.22\,\text{kg m}^{-3})(346\,\text{m s}^{-1})}$$

$$= 1.06\,\text{J s}^{-1}\,\text{m}^{-1} = 1.06\,\text{W m}^{-2}$$

$$= 1.06 \times 10^{-4}\,\text{W cm}^{-2}.$$

The unit 1 W cm^{-2} is a mixed unit, neither cgs nor mks. We mention it here because it is unfortunately in general use among acousticians.

The pressure amplitude of the faintest sound wave which can be heard is about 3×10^{-5} N m^{-2} and the corresponding intensity is about 10^{-12} W m^{-2} or 10^{-16} W cm^{-2}.

The *total* power carried across a surface by a sound wave equals the product of the intensity at the surface and the surface area, if the intensity over the surface is uniform. The average power developed as sound waves by a person speaking in an ordinary conversational tone is about 10^{-5} W, while a loud shout corresponds to about 3×10^{-2} W. Since the population of the city of New York is about eight million persons, the acoustical power developed, if all were to speak at the same time, would be about 60 W, or enough to operate a moderate-sized electric light. On the other hand, the power required to fill a large auditorium with loud sound is considerable. Suppose the intensity over the surface of a hemisphere 20 m in radius is 10^{-4} W cm^{-2}. The area of the surface is about 25×10^6 cm^2. Hence the acoustic power output of a speaker at the center of the sphere would have to be

$$(10^{-4}\,\text{W cm}^{-2})(25 \times 10^6\,\text{cm}^2) = 2500\,\text{W},$$

Table 23–1 NOISE LEVELS DUE TO VARIOUS SOURCES (representative values)

Source or description of noise	Noise level, db	Intensity, W m^{-2}
Threshold of pain	120	1
Riveter	95	3.2×10^3
Elevated train	90	10^{-3}
Busy street traffic	70	10^{-5}
Ordinary conversation	65	3.2×10^{-6}
Quiet automobile	50	10^{-7}
Quiet radio in home	40	10^{-8}
Average whisper	20	10^{-10}
Rustle of leaves	10	10^{-11}
Threshold of hearing	0	10^{-12}

or 2.5 kW. The electrical power input to the speaker would need to be considerably larger, since the efficiency of such devices is not very high.

23–3 INTENSITY LEVEL AND LOUDNESS

Because of the large range of intensities over which the ear is sensitive, a logarithmic rather than an arithmetic intensity scale is convenient. Accordingly, the *intensity level* β of a sound wave is defined by the equation

$$\beta = 10 \log \frac{I}{I_0}, \qquad (23\text{–}12)$$

where I_0 is an arbitrary reference intensity, taken as 10^{-12} W m^{-2}, corresponding roughly to the faintest sound which can be heard. Intensity levels are expressed in *decibels*, abbreviated db.*

If the intensity of a sound wave equals I_0 or 10^{-12} W m^{-2}, its intensity level is 0 db. The maximum intensity which the ear can tolerate, about 1 W m^{-2}, corresponds to an intensity level of 120 db. Table 23–1 gives the intensity levels in decibels of a number

---

* Originally, a scale of intensity levels in *bels* was defined by the relation

$$\text{Intensity level} = \log I/I_0.$$

This unit proved rather large, and hence the *decibel*, one-tenth of a bel, has come into general use. The unit is named in honor of Alexander Graham Bell.

Fig. 23–1 *Auditory area between threshold of hearing and threshold of feeling (light lines), and spectrogram of street noise (heavy line). (Courtesy of Dr. Harvey Fletcher.)*

of familiar noises. It is taken from a survey made by the New York City Noise Abatement Commission.

The range of frequencies and intensities to which the ear is sensitive is conveniently represented by a diagram such as that of Fig. 23–1, which is a graph of the *auditory area* of a person of good hearing. The height of the lower curve at any frequency represents the intensity level of the faintest pure (sinusoidal) tone of that frequency which can be heard. It will be seen from the diagram that the ear is most sensitive to frequencies between 2000 and 3000 Hz where the *threshold of hearing*, as it is called, is about −5 db. The upper curve represents the intensity level of the loudest pure tone which can be tolerated. At intensities above this curve, which is called the *threshold of feeling or pain*, the sensation changes from one of hearing to discomfort or even pain. The height of the upper curve is approximately constant at a level of about 120 db for all frequencies. Every pure tone which can be heard may be represented by a point lying somewhere in the area between these two curves.

Only about 1 percent of the population has a threshold of hearing as low as the bottom curve in Fig. 23–1; 50 percent of the population can hear pure tones of a frequency of 2500 cycles when the intensity level is about 8 db, and 90 percent when the level is 20 db. For a loud tone of intensity level 80 db, the range of audibility is from 20 to 20,000 Hz, but at a level of 20 db it is only from about 200 to

about 15,000 Hz. At a frequency of 1000 Hz the range of intensity level is from about 3 db to about 120 db, whereas at 100 Hz it is only from 30 db to 120 db.

The term *loudness* refers to a sensation in the consciousness of a human observer. It is purely subjective, as contrasted with the objective quantity *intensity*, and is not directly measurable with instruments. Loudness increases with intensity but there is no simple linear relationship. Pure tones of the same intensity but different frequencies do not necessarily produce sensations of equal loudness. Thus, for a listener whose auditory area is represented in Fig. 23–1, a pure tone of intensity level 30 db and frequency 60 Hz is completely inaudible, while one of the same intensity level but of frequency 1000 Hz is well above the threshold of audibility. In order for the first tone to appear as loud as the second, its intensity level would have to be raised to about 65 db.

Sounds which are not pure tones do not have a single frequency and hence cannot be represented by a single point on the diagram. A sound such as that from a musical instrument, consisting of a mixture of a relatively few frequencies (the fundamental and overtones), can be represented by a set of points, each point giving the intensity and frequency of one particular overtone. A sound such as street noise, while it cannot be considered as made up of a fundamental and overtones, can still be represented on the diagram. The sound is picked up by a microphone and sent through an electrical network which selects a narrow range of frequencies and measures the average intensity within this range. By repeating the process at a large number of frequencies throughout the audible range, a series of points is obtained which can be plotted. A continuous curve drawn through them is called the *spectrogram* of the sound. A typical spectrogram of street noise is shown in Fig. 23–1.

The term "spectrogram" is borrowed from optics. The process just described is entirely analogous to the optical one of dispersing a beam of light waves into a spectrum by means of a prism and measuring the intensity at a number of wavelengths throughout the spectrum. The light emitted by a gas in an electrical discharge is a mixture of waves of a number of definite frequencies and corresponds to the sound emitted by a musical instrument. Most light beams, however, are a mixture of all frequencies and are therefore the optical analog of noise.

The *total* intensity level of a noise can be found from its spectrogram by an integration process. There are also instruments known as noise meters which measure this level directly. The level of the street noise in Fig. 23–1 is about 85 db and is shown by the short heavy line.

23–4 QUALITY AND PITCH

A string that has been plucked or a plate that has been struck, if allowed to vibrate freely, will vibrate with many frequencies at the same time. It is a rare

occurrence for a body to vibrate with only one frequency. A carefully made tuning fork struck lightly on a rubber block may vibrate with only one frequency, but in the case of musical instruments, the fundamental and many harmonics are usually present at the same time. The impulses that are sent from the ear to the brain give rise to one net effect which is characteristic of the instrument. Suppose, for example, the sound spectrum of a tone consisted of a fundamental of 200 Hz and harmonics 2, 3, 4 and 5, all of different intensity, whereas the sound spectrum of another tone consisted of exactly the same frequencies but with a different intensity distribution. The two tones would sound different; they are said to differ in *quality* or *timbre*.

Adjectives used to describe the quality of musical tones are purely subjective in character, such as reedy, golden, round, mellow, tinny, etc. The quality of a sound is determined in part by the number of overtones present and their respective intensity-versus-time curves. The sound spectra of several musical instruments are shown in Fig. 23–2.

Another important factor in determining tone quality is the behavior at the beginning and end of a tone. A piano tone begins percussively with a thump and then dies away gradually; a harpsichord tone, in addition to having different harmonic content, begins much more quickly and incisively with a click, the higher harmonics beginning before the lower ones. The cessation of tone when the key is released is also much prompter with the harpsichord than the piano. Similar effects are present in other musical instru-

Fig. 23–2 *Sound spectra of some musical instruments (Courtesy of Dr. Harvey Fletcher.)*

ments; with wind and string instruments the player has considerable control over the attack and decay of the tone, and these characteristics help define the unique characteristics of each instrument.

The term *pitch* refers to the attribute of a sound sensation that enables one to classify a note as "high" or "low." Like loudness, it is a subjective quantity and cannot be measured with instruments. Pitch is related to the objective quantity *frequency*, but there is no one-to-one correspondence. For a pure tone of constant intensity, the pitch becomes higher as the frequency is increased, but the pitch of a pure tone of constant frequency becomes lower as the intensity level is raised.

Many of the notes played on musical instruments are rich in harmonics, some of which may be more prominent than the fundamental. Presented with an array of frequencies constituting a harmonic series, the ear still assigns a characteristic pitch to the combination, this pitch being that associated with the fundamental frequency of the series. So definite is this pitch sensation that it is possible to eliminate the fundamental frequency entirely, by means of filters, without changing the pitch! The ear apparently will supply the fundamental, provided the correct harmonics are present. It is this rather surprising property of the ear that enables a small loudspeaker which does not radiate low frequencies well to give nevertheless the impression of good radiation in the low-frequency region. Because the speaker is a fairly efficient radiator for the frequencies of the harmonics, the listener believes he is actually hearing the low frequencies, when instead he is hearing only multiples of these frequencies and his ear is supplying the fundamental. It is possible, by deliberate distortion of the harmonics associated with low musical notes, to make a very small radio set, totally inadequate in the low-frequency range, sound somewhat like a larger, acoustically superior console set. Such synthetic bass is, to the critical ear, inferior to true bass reproduction, where the harmonic content is closer to that of the original sound.

23–5 MUSICAL INTERVALS AND SCALES

When certain musical tones are produced in succession or together, even an untrained listener recog-

nizes a relationship among them. Such relations are described in musical language by words such as octave, major third, minor third, etc. The listener recognizes something basic in these combinations of tones, and experimental measurement of the fundamental frequencies of the separate tones discloses that their relationships, one to the other, may be approximated by simple whole-number ratios. For example, the fundamental frequencies of middle C of the piano and its *octave* above, C′, have a ratio of 1 to 2. Another basic set of tones is obtained by playing C, E, G. These constitute what is known as a *major triad*, and the frequencies are found to be approximately proportional to 4, 5, and 6.

Starting at middle C of the piano, and playing only the white keys toward the right, it is possible to find, within only nine notes, three major triads. These are shown in Table 23–2; they enable one to calculate the frequencies of all the notes, once one of them has been chosen arbitrarily. By international agreement, the frequency of the A above middle C is chosen to be 440 vibrations per second. The frequencies of all the notes, shown in row 5 of Table 23–2, are those that correspond to the *just diatonic scale* of the key of C. The frequency ratios of adjacent notes are seen to be either 9/8, 10/9, or 16/15. The interval between two tones whose frequencies bear the ratio 9/8 or 10/9 are called *whole tones*, whereas the interval between two notes of frequency ratio 16/15 is a *half tone*.

If a major diatonic scale were constructed starting at D instead of at C, four new notes would be needed for a perfect diatonic scale. If all possible musical keys were to be provided for, many additional notes would be needed for each octave. To avoid this tremendous complication, what is known as the *equally tempered scale* has been devised. In this scheme there are 12 half-tone intervals in every range of an octave, adjacent notes a half tone apart bearing the constant ratio of the twelfth root of 2, i.e., 1.05946.

Simple as this scheme is, it results in no one scale being exactly diatonic. Since the ratios of the diatonic scale were originally selected to suit the preferences of the ear (being made up of three sets of major triads, each of which constitutes a harmonious combination), this means that an instrument tuned to

Table 23–2 FREQUENCY RELATIONS IN THE JUST DIATONIC SCALE AND IN THE EQUALLY TEMPERED SCALE

	Do	Re	Mi	Fa	Sol	La	Ti	Do'	Re'
Frequency relations	Middle C	D	E	F	G	A	B	C'	D'
Octave, key of C	1							2	
Major triad, key of C	4		5		6			(8)	
Major triad, key of F				4		5		6	
Major triad, key of G		(3)			4		5		6
Just scale, key of C	264	297	330	352	396	440	495	528	594
Intervals in just scale	9/8 whole	10/9 whole	16/15 half	9/8 whole	10/9 whole	9/8 whole	16/15 half	9/8 whole	
Equally tempered scale suitable for all keys	261.6	293.7	329.6	349.2	392.0	440	493.9	523.3	587.4
Intervals in equally tempered scale	$\sqrt[6]{2}$	$\sqrt[6]{2}$	$\sqrt[12]{2}$	$\sqrt[6]{2}$	$\sqrt[6]{2}$	$\sqrt[6]{2}$	$\sqrt[12]{2}$	$\sqrt[6]{2}$	

the equally tempered scale is not quite so pleasant to the ear. Thus a piano whose white keys are tuned to the "just" scale sounds better (i.e., more "in tune") than one tuned to equal temperament when played in the key of C, but in some other keys it sounds worse. Hence an instrument to be used for compositions in all keys (as, for example, the *Well-Tempered Clavier* of Bach, containing compositions in all the major and minor keys) is almost always tuned to the equally tempered scale.

23–6 RADIATION FROM A PISTON

As we have seen, sound waves can be produced in a wide variety of ways. One of the simplest to understand is the moving cone of a loudspeaker, and it is instructive to examine the action of such a system.

For simplicity, we consider an idealized model in which the cone is represented as a plane moving perpendicularly to its surface. A single surface oscillating this way produces sound waves radiating away from both sides. To simplify the problem further, let us assume that we have an oscillating piston fitting closely in a large wall or *baffle*, as in Fig. 23–3(a). Then only the waves radiated from the right side of the piston need be considered in the space to the right of the baffle.

The directional distribution of radiated energy depends on the shape of the piston and on its dimensions relative to the wavelength of the emitted waves. The mathematics becomes very complicated for any shapes except circles and rectangles. The simplest case is that of a very long rectangle like a long plank, and Fig. 23–3(b) is an end view of such

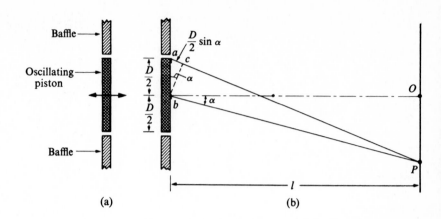

Fig. 23–3 *(a) An oscillating piston set in a baffle. (b) Construction for locating diffraction minima in radiation pattern from a long rectangular source.*

a piston. Imagine the right face of the piston, of width D, to be subdivided into a large number of very narrow strips parallel to the length of the piston and perpendicular to the plane of the diagram. Each of these strips can be treated as a *line source*, sending out waves having *cylindrical* wave surfaces coaxial with the source. We now apply the principle of superposition to these cylindrical waves.

Point P is a point in space at a distance ℓ large compared with the width D of the source, and with the wavelength λ. Consider the two elementary line sources at a and b, one at the top edge of the piston and the other just below its centerline. With P as a center and Pb as radius, strike the arc shown by the dotted line, intersecting the line Pa at c. The distance ac is then the path difference between the waves reaching P from a and b. Since ℓ is large compared with D, the arc bc is very nearly a straight line and abc is very nearly a right triangle with the angle α equal to the angle between Pb and the normal Ob. The path difference ac is then

$$ac = \frac{D}{2}\sin\alpha. \qquad (23\text{–}13)$$

If the angle α has such a value that ac equals a half-wavelength, the waves from a and b will reach P 180° out of phase and will very nearly cancel one another. The cancellation will not be complete because (1) the waves have slightly different distances to travel, and the amplitude decreases with distance, and (2) the directions to P are not exactly the same for both sources, and a line source, unlike a spherical

source, does not radiate uniformly in all directions. However, both of these effects will be small if ℓ is large compared with D.

Consider next the pair of line sources just below a and b. Except for the small differences mentioned, the same diagram as in Fig. 23–3(b) can be constructed for them, and the waves from these two sources will *also* cancel at P. Proceeding in this way over the entire surface of the piston, we see that very nearly complete cancellation results in a direction making an angle α with the normal, provided that

$$ac = \frac{\lambda}{2},$$

or, from Eq. (23–5), that

$$\sin\alpha = \frac{\lambda}{D}. \qquad (23\text{–}14)$$

As an example, if $D = 30$ cm and $\lambda = 15$ cm (corresponding to a frequency of about 2000 Hz),

$$\sin\alpha = \frac{15}{30} = 0.50,$$

$$\alpha = 30°,$$

and no energy is radiated at an angle of 30° on either side of the normal.

Other minima occur in directions for which $\sin\alpha = 2\lambda/D$, $3\lambda/D$, etc. This can be shown by dividing the surface of the piston into quarters, sixths, etc., and pairing off one element against another, as in Fig. 23–3(b).

The relatively simple discussion above, while it gives the angular positions of the *minima*, does not give those of the *maxima* nor does it give the relative intensities in other directions. The complete analysis is too lengthy to give here and we shall only state the results. There is a maximum intensity of radiated energy along the normal to the piston ($\alpha = 0$) and in other directions approximately halfway between the minima. However, the greatest amount of energy is concentrated in the region between the first two minima on either side of the normal, and for most purposes the energy radiated in other directions can be neglected.

The analysis of the radiation pattern from a *circular* piston can be carried out in the same way as that for a long rectangle. The piston is subdivided into narrow circular zones instead of long strips, and the effect of the waves from all the zones is summed at a distant point. There is a maximum of intensity along the axis of the piston, as would be expected. The angle α at which the first minimum occurs is given by

$$\sin \alpha = 1.22 \frac{\lambda}{D}, \qquad (23\text{–}15)$$

where D is now the piston diameter, and about 85 percent of the radiated energy is concentrated within a cone of this half-angle. If $D = 12$ in., $\lambda = 6$ in.,

$$\sin \alpha = 1.22 \times \frac{6}{12} = 0.61,$$

$$\alpha \approx 37°.$$

Other minima and maxima of rapidly decreasing intensity surround the central maximum. Figure 23–4 is a space diagram of the intensity distribution in front of an oscillating circular piston set in a baffle.

This *directivity* of the sound radiated by a piston has a number of important applications. In a motion picture theater, for example, we wish the sound waves radiated by a speaker behind the screen to spread out over a large angle. If a loudspeaker 12 in. in diameter, approximated by a plane circular piston, is the sound source, then at a frequency of 2000 Hz the directly radiated wave is concentrated mainly in a "beam" of half-angle 37°, centered on the speaker

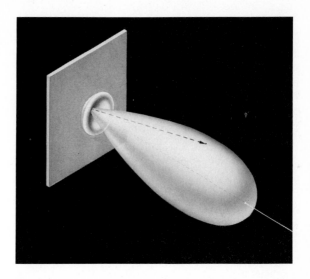

Fig. 23–4 *Space diagram of intensity distribution in front of an oscillating circular piston set in a baffle. A vector drawn from the center of the piston to any point on the surface has a length proportional to the sound intensity in that direction, as observed at a large fixed distance from the piston.*

axis. For a frequency of 10,000 Hz, or a wavelength of about 0.1 ft (about the upper frequency limit for such sound systems), the half-angle is

$$\sin \alpha = 1.22\left(\frac{0.1}{1.0}\right) = 0.122, \qquad \alpha \approx 7°,$$

while at a frequency of 1200 Hz, or a wavelength of about 0.8 ft,

$$\sin \alpha = 1.22\left(\frac{0.8}{1.0}\right) \approx 1.0, \qquad \alpha \approx 90°.$$

This means that for frequencies of 1200 Hz or less the radiated energy spreads out fairly uniformly, while at 10,000 Hz it is concentrated in a narrow beam, like a searchlight beam.

Since the intelligibility of speech depends largely on the high-frequency components, these are usually channeled into number of speakers directed toward different parts of the auditorium, while the low frequencies can be handled by a single speaker, since the low-frequency radiation pattern has a much wider angular spread. On the other hand, it is sometimes desirable, as in underwater sound signaling or when leading the cheering section, to produce a beam having only a small angular divergence. To accomplish this, the diameter of the source (approximating it by a circular piston) must be large compared with the wavelength of the radiated sound.

If instead of a vibrating piston set in a wall there is merely an aperture in the wall and a train of plane waves is incident from the left, the waves transmitted through the aperture will propagate beyond it in the same way as waves originating at a piston set in the aperture. If the wavelength is small compared with the dimensions of the aperture, the spreading of the waves is small, while if the wavelength is large, the waves spread out in all directions.

When there is an obstacle in the path of a train of waves, the resultant effect at the far side of the obstacle is due to those portions of the advancing wave surface that are *not* obstructed. In very general terms, if the wavelength is relatively small, the spreading is small and the obstacle casts a sharp "shadow." The larger the wavelength, the greater the spreading or bending of the waves. Thus one *can* hear around the corner of a wall but the effect is greater for waves of long wavelength (or low frequency) than for waves of short wavelength (or high frequency). The general term for the phenomena described above, in which one is concerned with the resultant effect of a large number of waves from different parts of a source, is *diffraction*.

23–7 APPLICATIONS OF ACOUSTIC PHENOMENA

The relevance of acoustical principles is by no means limited to sound and hearing. Analysis of mechanical waves is a powerful investigative tool in a wide variety of fields. Elastic waves in water are reflected by solid bodies immersed in the water, such as schools of fish, submarines, or submerged wrecked ships. *Sonar* systems make use of reflections of sound waves in water to determine position and motion of such bodies.

Analysis of elastic waves in the earth provides important information about its structure. The interior of the earth may be pictured crudely as made of concentric spherical shells, with mechanical properties such as density and elastic moduli that are different in different shells. Waves produced by explosions or earthquakes are reflected and refracted at the interfaces between these shells, and analysis of these waves permits partial determination of the dimensions and properties of the shells.

Some of the most interesting recent applications of acoustics lie in the field of medicine, where sound waves are used for both diagnosis and therapy. In diagnosis, *ultrasonic* (i.e., above 20 kHz) frequencies are usually used because their short wavelengths permit study of smaller-scale phenomena than with audible sound. Reflection of such waves from regions in the interior of the body can be used to detect a wide variety of anomalous conditions such as tumors, and to study various phenomena such as heart-valve action. Ultrasound is more sensitive than x-rays in distinguishing various kinds of tissues; it is believed to be less hazardous than x-rays, although possible hazards of ultrasound have not yet been thoroughly explored. At much higher power levels, ultrasound appears to have promise as a selective destroyer of pathological tissue, and may find usefulness in the treatment of certain cancers as well as arthritis and related diseases.

In medical and other applications of ultrasonics, a variety of electronic instrumentation is used. The sound is always produced by first generating an electrical wave, and using this to drive a *transducer* which converts electrical to mechanical (i.e., sound) waves. Thus the *transducer* is similar in action to a loudspeaker except that the frequency ranges may be different; frequencies as high as several megahertz (i.e., several million cycles per second) are routinely used. The detecting instruments include transducers, functioning in the role of microphones, as well as amplifiers and devices for displaying reflected signals. A common technique is to send out pulses, and observe the transmitted and reflected pulse "blips" on an oscilloscope, which permits a direct measurement of the time interval between transmission and reflection.

The applications of acoustic principles to environmental problems are obvious. It is now recognized that noise is an important aspect of environmental degradation. The design of quiet mass-transit vehicles, for example, involves detailed study of

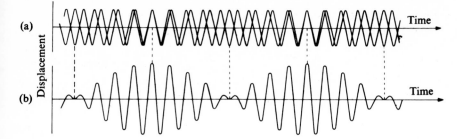

Fig. 23–5 *Beats are fluctuations in amplitude produced by two sound waves of slightly different frequency.*

sound generation and propagation in the motors, wheels, and supporting structures of vehicles. Excessive noise levels can lead to permanent hearing impairment; recent studies have shown that many young "rock" musicians have suffered hearing losses typical of persons 65 years of age, and even prolonged listening to high-level rock music (90 to 100 db) can lead to permanent damage.

23–8 BEATS

Standing waves in an air column have been cited as one example of *interference*. They arise when two wave trains of the same amplitude and frequency are traveling through the same region in opposite directions. We now wish to consider another type of interference which results when two wave trains of equal amplitude but slightly different frequency travel through the same region. Such a condition exists when two tuning forks of slightly different frequency are sounded simultaneously or when two piano strings struck by the same key are slightly "out of tune."

Let us consider some one point of space through which the waves are simultaneously passing. The displacements due to the two waves separately are plotted as a function of time on graph (a) in Fig. 23–5. If the total extent of the time axis represents one second, the graphs correspond to frequencies of 16 Hz and of 18 Hz. Applying the principle of superposition to find the resultant vibration, we get graph (b), where it is seen that the amplitude varies with time. These variations of amplitude give rise to variations of loudness which are called *beats*. Two strings may be tuned to the same frequency by

tightening one of them while sounding both until the beats disappear.

The production of beats may be treated mathematically as follows. The displacements due to the two waves passing simultaneously through some one point of space may be written

$$y_1 = A \cos 2\pi f_1 t, \qquad y_2 = A \cos 2\pi f_2 t.$$

(The amplitudes are assumed equal.)

By the principle of superposition, the resultant displacement is

$$y = y_1 + y_2 = A[\cos 2\pi f_1 t + \cos 2\pi f_2 t],$$

and, since

$$\cos a + \cos b = 2 \cos\left(\frac{a+b}{2}\right) \cos\left(\frac{a-b}{2}\right),$$

this may be written

$$y = \left[2A \cos 2\pi\left(\frac{f_1 - f_2}{2}\right)t\right] \cos 2\pi\left(\frac{f_1 + f_2}{2}\right)t.$$

$$(23–16)$$

The resulting vibration can then be considered to be of frequency $(f_1 + f_2)/2$, or the average frequency of the two tones, and of amplitude given by the expression in brackets. The amplitude therefore varies with time at a frequency $(f_1 - f_2)/2$. If f_1 and f_2 are nearly equal, this term is small and the amplitude fluctuates very slowly. When the amplitude is large, the sound is loud, and vice versa. A beat, or a maximum of amplitude, will occur when cos $2\pi t(f_1 - f_2)/2$ equals 1 or -1. Since *each* of these values occurs once in each cycle, the number of beats per second is *twice* the frequency $(f_1 - f_2)/2$; *the number of beats per second equals the difference of the frequencies.*

Beats between two tones can be detected by the ear up to a beat frequency of 6 or 7 per second. At higher frequencies, individual beats can no longer be distinguished and the sensation merges into one of *consonance* or *dissonance* depending on the frequency ratio of the tones. A beat frequency, even though it lies within the frequency range of the ear, is not necessarily interpreted by the ear as a tone of that frequency. Nevertheless, a tone can be heard of frequency equal to the frequency difference between two others sounded simultaneously. Such a tone is called a *difference* tone. Although not as easy to recognize, a frequency called a *summation* tone and equal to the sum of the frequencies of the two tones can also be heard. The general term applied to both difference tones and summation tones is *combination* tones.

23-9 THE DOPPLER EFFECT

When a source of sound, or a listener, or both, are in motion relative to the air, the pitch of the sound, as heard by the listener, is in general not the same as when source and listener are at rest. The most common example is the sudden drop in pitch of the sound from an automobile horn as one meets and passes a car proceeding in the opposite direction. This phenomenon is called the *Doppler effect.*

Let v_L and v_S represent the velocities of a listener and a source, relative to the air. We shall consider only the special case in which the velocities lie along the line joining listener and source. Since these velocities may be in the same or opposite directions, and the listener may be either ahead of or behind the source, a convention of signs is required. We shall take the positive directions of v_L and v_S as that *from* the position of the listener, toward the position of the source. The velocity of propagation of sound waves, c, will always be considered positive.

In Fig. 23-6, a listener L is at the left of a source S. The positive direction is then from left to right and

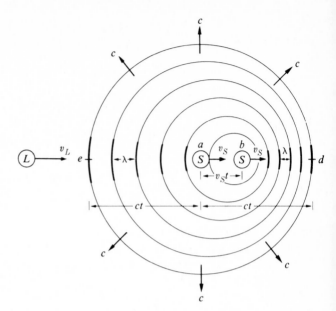

Fig. 23-6 *Wave surfaces emitted by a moving source.*

both v_L and v_S are positive in the diagram. The sound source is at point a at time $t = 0$ and at point b at a later time t. The outer circle represents the wave surface emitted at time $t = 0$. This surface (in free space) is a sphere with center at a, and is traveling radially outward at all points with speed c. (The fact that the wave originated at a *moving* source does not affect its speed after leaving the source. The wave speed c is a property of the *medium* only; the waves forget about the source as soon as they leave it.) The radius of this sphere (the distance ea or ad) is therefore ct. The distance ab equals $v_S t$, so

$$eb = (c + v_S)t, \qquad bd = (c - v_S)t.$$

In the time interval between $t = 0$ and $t = t$, the number of waves emitted by the source is $f_S t$, where f_S is the frequency of the source. In front of the source these waves are crowded into the distance bd, while behind the source they are spread out over the distance eb. The wavelength in front of the source is therefore

$$\lambda = \frac{(c - v_S)t}{f_S t} = \frac{c - v_S}{f_S}, \qquad (23\text{-}17)$$

while the wavelength behind the source is

$$\lambda = \frac{(c + v_S)t}{f_S t} = \frac{c + v_S}{f_S}. \qquad (23\text{–}18)$$

The waves approaching the moving listener L have a speed of propagation *relative to him*, given by $c + v_L$. The frequency f_L at which the listener encounters these waves is

$$f_L = \frac{c + v_L}{\lambda} = \frac{c + v_L}{(c + v_S)/f_S},$$

or

$$\boxed{\frac{f_L}{c + v_L} = \frac{f_S}{c + v_S},} \qquad (23\text{–}19)$$

which expresses the frequency f_L as heard by the listener in terms of the frequency f_S of the source.

This general relation includes all possibilities for collinear motion of source and listener relative to the medium. If one happens to be at rest in the medium, the corresponding velocity is zero, and of course when both are at rest or have the same velocity relative to the medium, then $f_L = f_S$. Whenever source or listener is moving in the opposite direction to that which we have designated as positive, the corresponding velocity to be used in Eq. (23–19) is negative. The examples illustrate these sign conventions.

Examples   Let $f_S = 1000$ Hz, and $c = 1000$ ft s^{-1}. The wavelength of the waves emitted by a stationary source is then $c/f_S = 1.00$ ft.

a) What are the wavelengths ahead of and behind the moving source in Fig. 23–6 if its velocity is 100 ft^{-1}?

In front of the source,

$$\lambda = \frac{c - v_S}{f_S} = \frac{1000 - 100}{1000} = 0.90 \text{ ft.}$$

Behind the source.

$$\lambda = \frac{c + v_S}{f_S} = \frac{1000 + 100}{1000} = 1.10 \text{ ft.}$$

b) If the listener L in Fig. 23–6 is at rest and the source is moving away from him at 100 ft s^{-1}, what

is the frequency as heard by the listener?

Since

$$v_L = 0 \qquad \text{and} \qquad v_S = 100 \text{ ft s}^{-1},$$

we have

$$f_L = f_S \frac{c}{c + v_S}$$

$$= 1000 \left(\frac{1000}{1000 + 100} \right) = 909 \text{ Hz.}$$

c) If the source in Fig. 23–6 is at rest and the listener is moving toward the left at 100 ft s^{-1}, what is the frequency as heard by the listener?

The positive direction (from listener to source) is still from left to right, so

$$v_L = -100 \text{ ft s}^{-1}, \qquad v_S = 0,$$

$$f_L = f_S \frac{c + v_L}{c}$$

$$= 1000 \left(\frac{1000 - 100}{1000} \right) = 900 \text{ Hz.}$$

Thus, while the frequency f_L as heard by the listener is less than the frequency f_S both when the source moves away from the listener and when the listener moves away from the source, the decrease in frequency is not the same for the same speed of recession.

In the preceding equations, the velocities v_L, v_S and c are all *relative to the air*, or more generally, to the medium in which the waves are traveling. The Doppler effect exists also for electromagnetic waves in empty space, such as light waves or radio waves. In this case, there is no "medium" relative to which a velocity can be defined, and we can speak only of the *relative* velocity v of source and receiver.

In deriving the Doppler relations for light, we must use the relativistic kinematic relations developed in Chapter 14. The wave velocity c is the velocity of light and is the same for both source and listener. In the frame of reference in which the listener is at rest, the source is moving away from the listener with velocity v. (If, instead, the source is *approaching* the listener, v is negative.) The source frequency is again f_S, but this is measured in the rest frame of the source; in the rest frame of the *listener* the corresponding frequency f'_S is less than this by the time-dilation factor $(1 - v^2/c^2)^{1/2}$.

The Doppler relation for light analogous to Eq. (23–19) for sound is obtained by using the time-dilated frequency $f'_S = f_S\sqrt{1 - v^2/c^2}$, in place of f_S. That is, in the rest frame of L,

$$\lambda = \frac{c + v}{f'_S} = \frac{c + v}{f_S\sqrt{1 - v^2/c^2}}.$$

The frequency f_L measured by the listener (i.e., the frequency of arrival of the waves at L) is then given by

$$f_L = \frac{c}{\lambda} = \frac{cf_S\sqrt{1 - v^2/c^2}}{c + v} = \frac{f_S\sqrt{c^2 - v^2}}{c + v}.$$

This can be simplified further by noting that

$$\sqrt{c^2 - v^2} = \sqrt{c - v} \cdot \sqrt{c + v}.$$

The final result is

$$\boxed{f_L = \left(\sqrt{\frac{c - v}{c + v}}\right)f_S.} \qquad (23\text{–}20)$$

When v is positive, the source moves *away* from the listener and f_L is always *less* than f_S; when v is negative, the source moves *toward* the listener and f_L is *greater* than f_S. Thus, the qualitative effect is the same as for sound, although the quantitative relationship is different.

The Doppler effect provides a convenient means of tracking an artificial satellite which is emitting a radio signal of constant frequency f_S. The frequency f_L of the signal that is received on the earth decreases as the satellite is passing, since the velocity component *toward* the earth decreases from position 1 to position 2 in Fig. 23–7, and then points *away* from the earth from 2 to 3. If the received signal is combined with a constant signal generated in the receiver to give rise to *beats*, then the beat frequency may be such as to produce an audible note whose pitch decreases as the satellite passes overhead.

A similar technique is used by law-enforcement officers to measure automobile speeds. An electro-

Fig. 23–7 *Change of velocity component along the line of sight of a satellite passing a tracking station.*

magnetic wave is emitted by a source at the side of the road, typically attached to a police car. The wave is reflected from a moving car, which thus acts as a moving source; the reflected wave is Doppler-shifted in frequency. Measurement of the frequency shift using beats, as with satellite tracking, permits simple measurement of the speed.

The Doppler effect for *light* is important in astronomy. Analysis of the spectra of light from distant stars shows shifts in wavelength compared to spectra from the same elements on earth. These can be interpreted as Doppler shifts due to motion of the stars. The shift is nearly always toward the longer wavelength or red end of the spectrum, and is therefore called the *red shift*. Such observations have provided practically all the evidence for the "exploding universe" cosmological theories, which represent the universe as having evolved from a great explosion several billion years ago in a relatively small region of space.

PROBLEMS

23–1

a) If the pressure amplitude in a sound wave is tripled, by how many times is the intensity of the wave increased?

b) By how many times must the pressure amplitude of a sound wave be increased in order to increase the intensity by a factor of 16 times?

23–2

a) Two sound waves of the same frequency, one in air and one in water, are equal in intensity. What is the ratio of the pressure amplitude of the wave in water to that of the wave in air?

b) If the pressure amplitudes of the waves are equal, what is the ratio of their intensities?

c) What is the difference between their intensity levels? (The speed of sound in water may be taken as 1490 m s^{-1}.)

23–3

a) Relative to the arbitrary reference intensity of 10^{-16} W cm^{-2}, what is the intensity level in decibels of a sound wave whose intensity is 10^{-10} W cm^{-2}?

b) What is the intensity level of a sound wave in air whose pressure amplitude is 0.2 N m^{-2}?

23–4

a) Show that if β_1 and β_2 are the intensity levels in db of sounds of intensities I_1 and I_2 respectively, the difference in intensity levels of the sound is

$$\beta_2 - \beta_1 = 10 \log \frac{I_2}{I_1}.$$

b) Show that if P_1 and P_2 are the pressure amplitudes of two sound waves, the difference in intensity levels of the waves is

$$\beta_2 - \beta_1 = 20 \log \frac{P_2}{P_1}.$$

23–5

a) Show that if the reference level of intensity is $I_0 = 10^{-16}$ W cm^{-2}, the intensity level of a sound of intensity I (in W cm^{-2}) is

$$\beta = 160 + 10 \log I.$$

b) Is this relation valid if I is expressed in W m^{-2} instead of W cm^{-2}? If not, how should it be modified?

23–6 The intensity due to a number of independent sound sources is the sum of the individual intensities. How many decibels greater is the intensity level when all five quintuplets cry simultaneously than when a single one cries? How many more crying babies would be required to produce a further increase in the intensity level of the same number of decibels?

23–7 A window whose area is 1 m^2 opens on a street where the street noises result in an intensity level, at the window, of 60 db. How much "acoustic power" enters the window via the sound waves?

23–8 A source of sound emits a total power of 10 W, uniformly in all directions. At what distance from the source is the sound level (a) 100 db? (b) 60 db?

23–9 A certain sound source radiates uniformly in all directions in air. At a distance of 5 m the sound level is 80 db. The frequency is 440 Hz.

a) What is the displacement amplitude at this distance?

b) What is the pressure amplitude?

c) At what distance is the sound level 60 db?

23–10

a) What are the upper and lower limits of intensity level of a person whose auditory area is represented by the graph of Fig. 23–3?

b) What are the highest and lowest frequencies he can hear when the intensity level is 40 db?

23–11 Two loudspeakers, A and B, radiate sound uniformly in all directions. The output of acoustic power from A is 8×10^{-4} W, and from B it is 13.5×10^{-4} W. Both loudspeakers are vibrating in phase at a frequency of 173 Hz.

a) Determine the difference in phase of the two signals at a point C along the line joining A and B, 3 m from B and 4 m from A.

b) Determine the intensity at C from speaker A if speaker B is turned off, and the intensity at C from speaker B if speaker A is turned off.

c) With both speakers on, what is the intensity and intensity level at C?

23–12 What should be the diameter of a sound source in the form of a circular piston set in a wall, if the central lobe of the diffraction pattern is to have a half-angle of 45°, for a frequency of 10,000 Hz?

23–13 The sound source of a sonar system operates at a frequency of 50,000 Hz. Approximate the source by a circular disk set in the hull of a destroyer. The velocity of sound in water can be taken as 1450 m s^{-1}.

a) What is the wavelength of the waves emitted by the source?

b) What must be the diameter of the source if the half-angular divergence of the main beam is not to be more than 10°?

c) What is the difference in frequency between the directly radiated waves and the waves reflected from the hull of a submarine traveling directly away from the destroyer at 15 mi hr^{-1}?

23–14 Two identical piano strings when stretched with the same tension have a fundamental frequency of 400 Hz. By what fractional amount must the tension in one string be increased in order that 4 beats s^{-1} shall occur when both strings vibrate simultaneously?

23–15 In Table 23–2, the interval from C to G is called a *fifth*. Compare the frequency ratio for this interval in the just scale to that in the equally tempered scale. Which is larger, and by what percent?

23–16 In Table 23–2, compare the frequency ratio for the interval C to G with the ratio for the interval E to B, in the just scale.

a)   By what percent do the two ratios differ?

b)   Show that the two ratios are equal in the equally tempered scale.

23–17 The frequency ratio of a half-tone interval on the equally tempered scale is 1.059. Find the velocity of an automobile passing a listener at rest in still air, if the pitch of the car's horn drops a half-tone between the times when the car is coming directly toward him and when it is moving directly away from him.

23–18

a)   Refer to Fig. 23–9 and the examples in Section 23–9. Suppose that a wind of velocity 50 ft s^{-1} is blowing in the same direction as that in which the source is moving. Find the wavelengths ahead of and behind the source.

b)   Find the frequency heard by a listener at rest when the source is moving away from him.

23–19 A railroad train is traveling at 100 ft s^{-1} in still air. The frequency of the note emitted by the locomotive whistle is 500 Hz. What is the wavelength of the sound waves (a) in front of and (b) behind the locomotive? What would be the frequency of the sound heard by a stationary listener (c) in front of and (d) behind the locomotive? What frequency would be heard by a passenger on a train traveling at 50 ft s^{-1} and (e) approaching the first, (f) receding from the first? (g) How would each of the preceding answers be altered if a wind of velocity 30 ft s^{-1} were blowing in the same direction as that in which the locomotive was traveling?

23–20 A train of plane sound waves of frequency f_0 and wavelength λ_0 travels horizontally toward the right. It strikes and is reflected from a large, rigid, vertical plane

surface, perpendicular to the direction of propagation of the wave train and moving toward the left with a velocity v.

a)   How many waves strike the surface in a time interval t?

b)   At the end of this time interval, how far to the left of the surface is the wave that was reflected at the beginning of the time interval?

c)   What is the wavelength of the reflected waves, in terms of λ_0?

d)   What is the frequency, in terms of f_0?

e)   A listener is at rest at the left of the moving surface. Describe the sensation of sound which he hears as a result of the combined effect of the incident and reflected wave trains.

23–21 Two whistles, A and B, each have a frequency of 500 Hz. A is stationary and B is moving toward the right (away from A) at a velocity of 200 ft s^{-1}. An observer is between the two whistles, moving toward the right with a velocity of 100 ft s^{-1}. Take the velocity of sound in air as 1100 ft s^{-1}.

a)   What is the frequency from A as heard by the observer?

b)   What is the frequency from B as heard by the observer?

c)   What is the beat frequency heard by the observer?

23–22 A man stands at rest in front of a large smooth wall. Directly in front of him, between him and the wall, he holds a vibrating tuning fork of frequency 400 Hz. He now moves the fork toward the wall with a velocity of 4 ft s^{-1}. How many beats per second will he hear between the sound waves reaching him directly from the fork, and those reaching him after being reflected from the wall?

23–23 A source of sound waves, S, emitting waves of frequency 1000 Hz, is traveling toward the right in still air with a velocity of 100 ft s^{-1}. At the right of the source is a large, smooth, reflecting surface moving toward the left with a velocity of 400 ft s^{-1}.

a)   How far does an emitted wave travel in 0.01 s?

b)   What is the wavelength of the emitted waves in front of (i.e., at the right of) the source?

c)   How many waves strike the reflecting surface in 0.01 s?

d)   What is the velocity of the reflected waves?

e)   What is the wavelength of the reflected waves?

23–24

a)   Show that Eq. (23–11) can be written

$$f_L = f_s\left(1 - \frac{v}{c}\right)^{1/2}\left(1 + \frac{v}{c}\right)^{-1/2}.$$

b)   Use the binomial theorem to show that if $v \ll c$, then, to a good approximation,

$$f_L = f_s\left(1 + \frac{v}{c}\right).$$

c)   An earth satellite emits a radio signal of frequency 10^8 Hz. An observer on the ground detects beats between the received signal and a local signal also of frequency 10^8 Hz. At a particular moment, the beat frequency is 2400 Hz. What is the component of the satellite's velocity directed toward the earth at this moment?

$f = 1000 \, Hz \qquad \longrightarrow V = 100 \, ^{ft}/_s$

$s = v \, t$

$s = (100)(.01)$

$V = 400$

$c = \lambda f$

$\lambda = \frac{c}{f} = \frac{1020}{1000} = 1.1$

Appendix

Natural Trigonometric Functions   **T-1**

Common Logarithms   **T-2**

Periodic Table of the Elements   **T-4**

The International System of Units   **T-5**

Physical Constants   **T-8**

Unit Conversion Factors   **T-9**

NATURAL TRIGONOMETRIC FUNCTIONS

De-gree	Ra-dian	Sine	Co-sine	Tan-gent	De-gree	Ra-dian	Sine	Co-sine	Tan-gent
0°	0.000	0.000	1.000	0.000					
1°	0.017	0.017	1.000	0.017	46°	0.803	0.719	0.695	1.036
2°	0.035	0.035	0.999	0.035	47°	0.820	0.731	0.682	1.072
3°	0.052	0.052	0.999	0.052	48°	0.838	0.743	0.669	1.111
4°	0.070	0.070	0.998	0.070	49°	0.855	0.755	0.656	1.150
5°	0.087	0.087	0.996	0.087	50°	0.873	0.766	0.643	1.192
6°	0.105	0.105	0.995	0.105	51°	0.890	0.777	0.629	1.235
7°	0.122	0.122	0.993	0.123	52°	0.908	0.788	0.616	1.280
8°	0.140	0.139	0.990	0.141	53°	0.925	0.799	0.602	1.327
9°	0.157	0.156	0.988	0.158	54°	0.942	0.809	0.588	1.376
10°	0.175	0.174	0.985	0.176	55°	0.960	0.819	0.574	1.428
11°	0.192	0.191	0.982	0.194	56°	0.977	0.829	0.559	1.483
12°	0.209	0.208	0.978	0.213	57°	0.995	0.839	0.545	1.540
13°	0.227	0.225	0.974	0.231	58°	1.012	0.848	0.530	1.600
14°	0.244	0.242	0.970	0.249	59°	1.030	0.857	0.515	1.664
15°	0.262	0.259	0.966	0.268	60°	1.047	0.866	0.500	1.732
16°	0.279	0.276	0.961	0.287	61°	1.065	0.875	0.485	1.804
17°	0.297	0.292	0.956	0.306	62°	1.082	0.883	0.469	1.881
18°	0.314	0.309	0.951	0.325	63°	1.100	0.891	0.454	1.963
19°	0.332	0.326	0.946	0.344	64°	1.117	0.899	0.438	2.050
20°	0.349	0.342	0.940	0.364	65°	1.134	0.906	0.423	2.145
21°	0.367	0.358	0.934	0.384	66°	1.152	0.914	0.407	2.246
22°	0.384	0.375	0.927	0.404	67°	1.169	0.921	0.391	2.356
23°	0.401	0.391	0.921	0.424	68°	1.187	0.927	0.375	2.475
24°	0.419	0.407	0.914	0.445	69°	1.204	0.934	0.358	2.605
25°	0.436	0.423	0.906	0.466	70°	1.222	0.940	0.342	2.748
26°	0.454	0.438	0.899	0.488	71°	1.239	0.946	0.326	2.904
27°	0.471	0.454	0.891	0.510	72°	1.257	0.951	0.309	3.078
28°	0.489	0.469	0.883	0.532	73°	1.274	0.956	0.292	3.271
29°	0.506	0.485	0.875	0.554	74°	1.292	0.961	0.276	3.487
30°	0.524	0.500	0.866	0.577	75°	1.309	0.966	0.259	3.732
31°	0.541	0.515	0.857	0.601	76°	1.326	0.970	0.242	4.011
32°	0.559	0.530	0.848	0.625	77°	1.344	0.974	0.225	4.332
33°	0.576	0.545	0.839	0.649	78°	1.361	0.978	0.208	4.705
34°	0.593	0.559	0.829	0.675	79°	1.379	0.982	0.191	5.145
35°	0.611	0.574	0.819	0.700	80°	1.396	0.985	0.174	5.671
36°	0.628	0.588	0.809	0.727	81°	1.414	0.988	0.156	6.314
37°	0.646	0.602	0.799	0.754	82°	1.431	0.990	0.139	7.115
38°	0.663	0.616	0.788	0.781	83°	1.449	0.993	0.122	8.144
39°	0.681	0.629	0.777	0.810	84°	1.466	0.995	0.105	9.514
40°	0.698	0.643	0.766	0.839	85°	1.484	0.996	0.087	11.43
41°	0.716	0.656	0.755	0.869	86°	1.501	0.998	0.070	14.30
42°	0.733	0.669	0.743	0.900	87°	1.518	0.999	0.052	19.08
43°	0.750	0.682	0.731	0.933	88°	1.536	0.999	0.035	28.64
44°	0.768	0.695	0.719	0.966	89°	1.553	1.000	0.017	57.29
45°	0.785	0.707	0.707	1.000	90°	1.571	1.000	0.000	

COMMON LOGARITHMS

N	0	1	2	3	4	5	6	7	8	9
10	0000	0043	0086	0128	0170	0212	0253	0294	0334	0374
11	0414	0453	0492	0531	0569	0607	0645	0682	0719	0755
12	0792	0828	0864	0899	0934	0969	1004	1038	1072	1106
13	1139	1173	1206	1239	1271	1303	1335	1367	1399	1430
14	1461	1492	1523	1553	1584	1614	1644	1673	1703	1732
15	1761	1790	1818	1847	1875	1903	1931	1959	1987	2014
16	2041	2068	2095	2122	2148	2175	2201	2227	2253	2279
17	2304	2330	2355	2380	2405	2430	2455	2480	2504	2529
18	2553	2577	2601	2625	2648	2672	2695	2718	2742	2765
19	2788	2810	2833	2856	2878	2900	2923	2945	2967	2989
20	3010	3032	3054	3075	3096	3118	3139	3160	3181	3201
21	3222	3243	3263	3284	3304	3324	3345	3365	3385	3404
22	3424	3444	3464	3483	3502	3522	3541	3560	3579	3598
23	3617	3636	3655	3674	3692	3711	3729	3747	3766	3784
24	3802	3820	3838	3856	3874	3892	3909	3927	3945	3962
25	3979	3997	4014	4031	4048	4065	4082	4099	4116	4133
26	4150	4166	4183	4200	4216	4232	4249	4265	4281	4298
27	4314	4330	4346	4362	4378	4393	4409	4425	4440	4456
28	4472	4487	4502	4518	4533	4548	4564	4579	4594	4609
29	4624	4639	4654	4669	4683	4698	4713	4728	4742	4757
30	4771	4786	4800	4814	4829	4843	4857	4871	4886	4900
31	4914	4928	4942	4955	4969	4983	4997	5011	5024	5038
32	5051	5065	5079	5092	5105	5119	5132	5145	5159	5172
33	5185	5198	5211	5224	5237	5250	5263	5276	5289	5302
34	5315	5328	5340	5353	5366	5378	5391	5403	5416	5428
35	5441	5453	5465	5478	5490	5502	5514	5527	5539	5551
36	5563	5575	5587	5599	5611	5623	5635	5647	5658	5670
37	5682	5694	5705	5717	5729	5740	5752	5763	5775	5786
38	5798	5809	5821	5832	5843	5855	5866	5877	5888	5899
39	5911	5922	5933	5944	5955	5966	5977	5988	5999	6010
40	6021	6031	6042	6053	6064	6075	6085	6096	6107	6117
41	6128	6138	6149	6160	6170	6180	6191	6201	6212	6222
42	6232	6243	6253	6263	6274	6284	6294	6304	6314	6325
43	6335	6345	6355	6365	6375	6385	6395	6405	6415	6425
44	6435	6444	6454	6464	6474	6484	6493	6503	6513	6522
45	6532	6542	6551	6561	6571	6580	6590	6599	6609	6618
46	6628	6637	6646	6656	6665	6675	6684	6693	6702	6712
47	6721	6730	6739	6749	6758	6767	6776	6785	6794	6803
48	6812	6821	6830	6839	6848	6857	6866	6875	6884	6893
49	6902	6911	6920	6928	6937	6946	6955	6964	6972	6981
50	6990	6998	7007	7016	7024	7033	7042	7050	7059	7067
51	7076	7084	7093	7101	7110	7118	7126	7135	7143	7152
52	7160	7168	7177	7185	7193	7202	7210	7218	7226	7235
53	7243	7251	7259	7267	7275	7284	7292	7300	7308	7316
54	7324	7332	7340	7348	7356	7364	7372	7380	7388	7396

N	0	1	2	3	4	5	6	7	8	9
55	7404	7412	7419	7427	7435	7443	7451	7459	7466	7474
56	7482	7490	7497	7505	7513	7520	7528	7536	7543	7551
57	7559	7566	7574	7582	7589	7597	7604	7612	7619	7627
58	7634	7642	7649	7657	7664	7672	7679	7686	7694	7701
59	7709	7716	7723	7731	7738	7745	7752	7760	7767	7774
60	7782	7789	7796	7803	7810	7818	7825	7832	7839	7846
61	7853	7860	7868	7875	7882	7889	7896	7903	7910	7917
62	7924	7931	7938	7945	7952	7959	7966	7973	7980	7987
63	7993	8000	8007	8014	8021	8028	8035	8041	8048	8055
64	8062	8069	8075	8082	8089	8096	8102	8109	8116	8122
65	8129	8136	8142	8149	8156	8162	8169	8176	8182	8189
66	8195	8202	8209	8215	8222	8228	8235	8241	8248	8254
67	8261	8267	8274	8280	8287	8293	8299	8306	8312	8319
68	8325	8331	8338	8344	8351	8357	8363	8370	8376	8382
69	8388	8395	8401	8407	8414	8420	8426	8432	8439	8445
70	8451	8457	8463	8470	8476	8482	8488	8494	8500	8506
71	8513	8519	8525	8531	8537	8543	8549	8555	8561	8567
72	8573	8579	8585	8591	8597	8603	8609	8615	8621	8627
73	8633	8639	8645	8651	8657	8663	8669	8675	8681	8686
74	8692	8698	8704	8710	8716	8722	8727	8733	8739	8745
75	8751	8756	8762	8768	8774	8779	8785	8791	8797	8802
76	8808	8814	8820	8825	8831	8837	8842	8848	8854	8859
77	8865	8871	8876	8882	8887	8893	8899	8904	8910	8915
78	8921	8927	8932	8938	8943	8949	8954	8960	8965	8971
79	8976	8982	8987	8993	8998	9004	9009	9015	9020	9025
80	9031	9036	9042	9047	9053	9058	9063	9069	9074	9079
81	9085	9090	9096	9101	9106	9112	9117	9122	9128	9133
82	9138	9143	9149	9154	9159	9165	9170	9175	9180	9186
83	9191	9196	9201	9206	9212	9217	9222	9227	9232	9238
84	9243	9248	9253	9258	9263	9269	9274	9279	9284	9289
85	9294	9299	9304	9309	9315	9320	9325	9330	9335	9340
86	9345	9350	9355	9360	9365	9370	9375	9380	9385	9390
87	9395	9400	9405	9410	9415	9420	9425	9430	9435	9440
88	9445	9450	9455	9460	9465	9469	9474	9479	9484	9489
89	9494	9499	9504	9509	9513	9518	9523	9528	9533	9538
90	9542	9547	9552	9557	9652	9566	9571	9576	9581	9586
91	9590	9595	9600	9605	9609	9614	9619	9624	9628	9633
92	9638	9643	9647	9652	9657	9661	9666	9671	9675	9680
93	9685	9689	9694	9699	9703	9708	9713	9717	9722	9727
94	9731	9736	9741	9745	9750	9754	9759	9763	9768	9773
95	9777	9782	9786	9791	9795	9800	9805	9809	9814	9818
96	9823	9827	9832	9836	9841	9845	9850	9854	9859	9863
97	9868	9872	9877	9881	9886	9890	9894	9899	9903	9908
98	9912	9917	9921	9926	9930	9934	9939	9943	9948	9952
99	9956	9961	9965	9969	9974	9978	9983	9987	9991	9996

PERIODIC TABLE OF THE ELEMENTS

Period	IA	IIA	IIIB	IVB	VB	VIB	VIIB	VIIIB	VIIIB	VIIIB	IB	IIB	IIIA	IVA	VA	VIA	VIIA	Noble gases
1	1 H 1.008																	2 He 4.003
2	3 Li 6.939	4 Be 9.012											5 B 10.811	6 C 12.011	7 N 14.007	8 O 15.999	9 F 18.998	10 Ne 20.183
3	11 Na 22.990	12 Mg 24.312											13 Al 26.982	14 Si 28.086	15 P 30.974	16 S 32.064	17 Cl 35.453	18 Ar 39.948
4	19 K 39.102	20 Ca 40.08	21 Sc 44.956	22 Ti 47.90	23 V 50.942	24 Cr 51.996	25 Mn 54.938	26 Fe 55.847	27 Co 58.933	28 Ni 58.71	29 Cu 63.54	30 Zn 65.37	31 Ga 69.72	32 Ge 72.59	33 As 74.922	34 Se 78.96	35 Br 79.909	36 Kr 83.80
5	37 Rb 85.47	38 Sr 87.62	39 Y 88.905	40 Zr 91.22	41 Nb 92.906	42 Mo 95.94	43 Tc (99)	44 Ru 101.07	45 Rh 102.91	46 Pd 106.4	47 Ag 107.87	48 Cd 112.40	49 In 114.82	50 Sn 118.69	51 Sb 121.75	52 Te 127.60	53 I 126.90	54 Xe 131.30
6	55 Cs 132.91	56 Ba 137.34	57 La 138.91	72 Hf 178.49	73 Ta 180.95	74 W 183.85	75 Re 186.2	76 Os 190.2	77 Ir 192.2	78 Pt 195.09	79 Au 196.97	80 Hg 200.59	81 Tl 204.37	82 Pb 207.19	83 Bi 208.98	84 Po (210)	85 At (210)	86 Rn (222)
7	87 Fr (223)	88 Ra (226)	89 Ac (227)	104 Rf(?) (259)	105 Ha(?) (260)													

58 Ce 140.12	59 Pr 140.91	60 Nd 144.24	61 Pm (145)	62 Sm 150.35	63 Eu 151.96	64 Gd 157.25	65 Tb 158.92	66 Dy 162.50	67 Ho 164.93	68 Er 167.26	69 Tm 168.93	70 Yb 173.04	71 Lu 174.97
90 Th 232.04	91 Pa (231)	92 U 238.03	93 Np (237)	94 Pu (242)	95 Am (243)	96 Cm (247)	97 Bk (249)	98 Cf (251)	99 Es (254)	100 Fm (253)	101 Md (256)	102 No (253)	103 Lr (257)

THE INTERNATIONAL SYSTEM OF UNITS

The Système International d'Unites, abbreviated SI, is the system developed by the General Conference on Weights and Measures and adopted by nearly all the industrial nations of the world. It is based on the mksa (meter-kilogram-second-ampere) system. The following material is reproduced from Publication SP–7012 of the Scientific and Technical Information Office of the National Aeronautics and Space Administration.

Names and Symbols of SI Units

Quantity	Name of Unit	Symbol	
SI Base Units			
length	meter	m	
mass	kilogram	kg	
time	second	s	
electric current	ampere	A	
thermodynamic temperature	kelvin	K	
luminous intensity	candela	cd	
amount of substance	mole	mol	
SI Derived Units			
area	square meter	m^2	
volume	cubic meter	m^3	
frequency	hertz	Hz	s^{-1}
mass density (density)	kilogram per cubic meter	kg/m^3	
speed, velocity	meter per second	m/s	
angular velocity	radian per second	rad/s	
acceleration	meter per second squared	m/s^2	
angular acceleration	radian per second squared	rad/s^2	
force	newton	N	$kg \cdot m/s^2$
pressure (mechanical stress)	pascal	Pa	N/m^2
kinematic viscosity	square meter per second	m^2/s	
dynamic viscosity	newton-second per square meter	$N \cdot s/m^2$	
work, energy, quantity of heat	joule	J	$N \cdot m$
power	watt	W	J/s
quantity of electricity	coulomb	C	$A \cdot s$
potential difference, electromotive force	volt	V	W/A
electric field strength	volt per meter	V/m	
electric resistance	ohm	Ω	V/A
capacitance	farad	F	$A \cdot s/V$
magnetic flux	weber	Wb	$V \cdot s$
inductance	henry	H	$V \cdot s/A$
magnetic flux density	tesla	T	Wb/m^2
magnetic field strength	ampere per meter	A/m	
magnetomotive force	ampere	A	
luminous flux	lumen	lm	$cd \cdot sr$
luminance	candela per square meter	cd/m^2	

Quantity	Name of Unit	Symbol	
illuminance	lux	lx	lm/m²
wave number	1 per meter	m⁻¹	
entropy	joule per kelvin	J/K	
specific heat capacity	joule per kilogram kelvin	J/(kg · K)	
thermal conductivity	watt per meter kelvin	W/(m · K)	
radiant intensity	watt per steradian	W/sr	
activity (of a radioactive source)	1 per second	s⁻¹	

SI Supplementary Units

plane angle	radian	rad
solid angle	steradian	sr

Definitions of SI Units

meter (m)

The *meter* is the length equal to 1,650,763.73 wavelengths in vacuum of the radiation corresponding to the transition between the levels $2 p_{10}$ and $5 d_5$ of the krypton-86 atom.

kilogram (kg)

The *kilogram* is the unit of mass; it is equal to the mass of the international prototype of the kilogram. (The international prototype of the kilogram is a particular cylinder of platinum-iridium alloy which is preserved in a vault at Sèvres, France, by the International Bureau of Weights and Measures.)

second (s)

The *second* is the duration of 9,192,631,770 periods of the radiation corresponding to the transition between the two hyperfine levels of the ground state of the cesium-133 atom.

ampere (A)

The *ampere* is that constant current which, if maintained in two straight parallel conductors of infinite length, of negligible circular cross section, and placed 1 meter apart in vacuum, would produce between these conductors a force equal to 2×10^{-7} newton per meter of length.

kelvin (K)

The *kelvin*, unit of thermodynamic temperature, is the fraction 1/273.16 of the thermodynamic temperature of the triple point of water.

candela (cd)

The *candela* is the luminous intensity, in the perpendicular direction, of a surface of 1/600,000 square meter of a blackbody at the temperature of freezing platinum under a pressure of 101,325 newtons per square meter.

mole (mol)

The *mole* is the amount of substance of a system which contains as many elementary entities as there are carbon atoms in 0.012 kg of carbon 12. The elementary entities must be specified and may be atoms, molecules, ions, electrons, other particles, or specified groups of such particles.

newton (N)

The *newton* is that force which gives to a mass of 1 kilogram an acceleration of 1 meter per second per second.

joule (J)

The *joule* is the work done when the point of application of 1 newton is displaced a distance of 1 meter in the direction of the force.

watt (W)

The *watt* is the power which gives rise to the production of energy at the rate of 1 joule per second.

volt (V)

The *volt* is the difference of electric potential between two points of a conducting wire carrying a constant current of 1 ampere, when the power dissipated between these points is equal to 1 watt.

ohm (Ω)

The *ohm* is the electric resistance between two points of a conductor when a constant difference of potential of 1 volt, applied between these two points, produces in this conductor a current of 1 ampere, this conductor not being the source of any electromotive force.

coulomb (C)

The *coulomb* is the quantity of electricity transported in 1 second by a current of 1 ampere.

farad (F)

The *farad* is the capacitance of a capacitor between the plates of which there appears a difference of potential of 1 volt when it is charged by a quantity of electricity equal to 1 coulomb.

henry (H)

The *henry* is the inductance of a closed circuit in which an electromotive force of 1 volt is produced when the electric current in the circuit varies uniformly at a rate of 1 ampere per second.

weber (Wb)

The *weber* is the magnetic flux which, linking a circuit of one turn, produces in it an electromotive force of 1 volt as it is reduced to zero at a uniform rate in 1 second.

lumen (lm)

The *lumen* is the luminous flux emitted in a solid angle of 1 steradian by a uniform point source having an intensity of 1 candela.

radian (rad)

The *radian* is the plane angle between two radii of a circle which cut off on the circumference an arc equal in length to the radius.

steradian (sr)

The *steradian* is the solid angle which, having its vertex in the center of a sphere, cuts off an area of the surface of the sphere equal to that of a square with sides of length equal to the radius of the sphere.

SI Prefixes

The names of multiples and submultiples of SI units may be formed by application of the prefixes:

Factor by which unit is multiplied	Prefix	Symbol
10^{12}	tera	T
10^9	giga	G
10^6	mega	M
10^3	kilo	k
10^2	hecto	h
10	deka	da
10^{-1}	deci	d
10^{-2}	centi	c
10^{-3}	milli	m
10^{-6}	micro	μ
10^{-9}	nano	n
10^{-12}	pico	p
10^{-15}	femto	f
10^{-18}	atto	a

PHYSICAL CONSTANTS

Based on values in the *American Institute of Physics Handbook* (1957), except where superseded by the determination of Taylor, Parker, and Langenberg, *Revs. Mod. Phys.* **41**, 375 (1969). The probable error for each value, properly considered part of the datum, has been omitted.

Name of Quantity	Symbol	Value
Speed of light in vacuum	c	2.9979×10^8 m s^{-1}
Charge of electron	q_e	-1.602×10^{-19} C
Rest mass of electron	m_e	9.10×10^{-31} kg
Ratio of charge to mass of electron	q_e/m_e	1.759×10^{11} C kg^{-1}
Planck's constant	h	6.626×10^{-34} J s
Boltzmann's constant	k	1.381×10^{-23} J K^{-1}
Avogadro's number (chemical scale)	N_0	6.023×10^{23} molecules mole^{-1}
Universal gas constant (chemical scale)	R	8.314 J mole^{-1} K^{-1}
Mechanical equivalent of heat	J	4.185×10^3 J kcal^{-1}
Standard atmospheric pressure	1 atm	1.013×10^5 N m^{-2}
Volume of ideal gas at 0° C and 1 atm (chemical scale)		22.415 liters mole^{-1}
Absolute zero of temperature	0 K	$-273.15°$ C
Acceleration due to gravity (sea level, at equator)		9.78049 m s^{-2}
Universal gravitational constant	G	6.673×10^{-11} N · m^2 kg^{-2}
Mass of earth	m_E	5.975×10^{24} kg
Mean radius of earth		6.371×10^6 m $= 3959$ mi
Equatorial radius of earth		6.378×10^6 m $= 3963$ mi
Mean distance from earth to sun	1 AU	1.49×10^{11} m $= 9.29 \times 10^7$ mi
Eccentricity of earth's orbit		0.0167
Mean distance from earth to moon		3.84×10^8 m $= 60$ earth radii
Diameter of sun		1.39×10^9 m $= 8.64 \times 10^5$ mi
Mass of sun	m_S	1.99×10^{30} kg $= 333{,}000 \times$ mass of earth
Coulomb's law constant	$k = 1/4 \pi \epsilon_0$	8.9874×10^9 N · m^2 C^{-2}
Faraday's constant (1 faraday)	F	96,487 C mole^{-1}
Mass of neutral hydrogen atom	$m_H{}^1$	1.007825 u
Mass of proton	m_p	1.007277 u
Mass of neutron	m_n	1.008665 u
Mass of electron	m_e	5.486×10^{-4} u
Ratio of mass of proton to mass of electron	m_p/m_e	1836.11
Rydberg constant for nucleus of infinite mass	R_∞	109,737 cm^{-1}
Rydberg constant for hydrogen	R_H	109,678 cm^{-1}
Wien displacement law constant		0.2898 cm K^{-1}

Numerical constants: $\pi = 3.142$; $e = 2.718$; $\sqrt{2} = 1.414$; $\sqrt{3} = 1.732$.

UNIT CONVERSION FACTORS

Length

1 m = 100 cm = 1000 mm = 10^6 μm = 10^9 nm

1 km = 1000 m = 0.6214 mi

1 m = 3.281 ft = 39.37 in.; 1 cm = 0.3937 in.

1 ft = 30.48 cm; 1 in. = 2.540 cm

1 mi = 5280 ft = 1.609 km

1 Å = 10^{-10} m = 10^{-8} cm = 10^{-1} nm

Area

1 cm² = 0.155 in²;   1 m² = 10^4 cm² = 10.76 ft²

1 in² = 6.452 cm²;   1 ft² = 144 in² = 0.0929 m²

Volume

1 liter = 1000 cm³ = 10^{-3} m³ = 0.0351 ft³ = 61.02 in.³

1 ft³ = 0.02832 m³ = 28.32 liters = 7.477 gallons

Time

1 min = 60 s; 1 hr = 3600 s

1 da = 86,400 s; 1 yr = 3.156×10^7 s

Velocity

1 cm s^{-1} = 0.03281 ft s^{-1}; 1 ft s^{-1} = 30.48 cm s^{-1}

1 mi min^{-1} = 60 mi hr^{-1} = 88 ft s^{-1}

1 km hr^{-1} = 0.2778 m s^{-1}; 1 mi hr^{-1} = 0.4470 m s^{-1}

Acceleration

1 m s^{-2} = 100 cm s^{-2} = 3.281 ft s^{-2}

1 cm s^{-2} = 0.01 m s^{-2} = 0.03281 ft s^{-2}

1 ft s^{-2} = 0.3048 m s^{-2} = 30.48 cm s^{-2}

1 mi hr^{-1} s^{-1} = 1.467 ft s^{-2}

Mass

1 kg = 10^3 g = 0.0685 slug

1 g = 6.85×10^{-5} slug

1 slug = 14.59 kg

1 u = 1.661×10^{-27} kg

Force

1 N = 10^5 dyn = 0.2247 lb

1 lb = 4.45 N = 4.45×10^5 dyn

Pressure

1 Pa = 1 N m^{-2} = 1.451×10^{-4} lb in^{-2} = 0.209 lb ft^{-2}

1 lb in^{-2} = 6891 Pa; 1 lb ft^{-2} = 47.85 Pa

1 atm = 1.013×10^5 Pa = 14.7 lb in^{-2} = 2117 lb ft^{-2}

Energy

1 J = 10^7 ergs = 0.239 cal; 1 cal = 4.186 J
     (based on 15° calorie)

1 ft lb = 1.356 J; 1 Btu = 1055 J = 252 cal

1 eV = 1.602×10^{-19} J

Mass–Energy Equivalence

1 kg $\leftrightarrow$ 8.988×10^{16} J

1 u $\leftrightarrow$ 931.5 MeV

1 eV $\leftrightarrow$ 1.073×10^{-9} u

Power

1 W = 1 J s^{-1}

1 hp = 746 W = 550 ft lb s^{-1}

1 Btu hr^{-1} = 0.293 W

Answers to Odd-Numbered Problems

PART I

Chapter 1

1-1 230 N, 55° above negative x-axis

1-3 a) 19.3 N in a direction midway between the 10-N forces

b) 8.46 N in a direction midway between the 10-N forces

1-5 15 N, 53° above the x-axis

26 N, 28° below the x-axis

1-7 25.7 N, 30.6 N

1-9 a) 18.5 N   b) 9.2 N

1-11 308 N, 25° above the x-axis

1-13 a) 7 N, 2.9 N   b) 7.6 N   c) 11 N

1-15 $F_x = 380$ N, west; $F_y = 960$ N, north

1-17 5 mi, 53.1° south of east

1-19 a) $N_x = -1$ cm, $N_y = 1$ cm

b) 1.41 cm, 135° counterclockwise from $+x$-axis

Chapter 2

2-1 a) The earth   b) 4 N, the book   c) No

d) 4 N, the earth, the book, upward

e) 4 N, the hand, the book, downward

f) First   g) Third   h) No   i) No   j) Yes

k) Yes   l) One   m) No   n) Nothing

2-3 a) 10 N   b) 20 N

2-5 a) 150 N in A, 180 N in B, 200 N in C

b) 200 N in A, 280 N in B, 200 N in C

c) 550 N in A, 670 N in B, 200 N in C

d) 167 N in A, 58 N in B, 125 N in C

2-7 a) Parts (b) and (c) can be solved.

b) In part (a), another side or angle is needed.

2-9 3150 N

2-11 a) 20 N   b) 30 N

2-13 a) $w/(2 \sin \theta)$ b) $w/(2 \tan \theta)$

2-17 22 N

2-19 a) 76 N   b) 24 N   c) From 15.4 to 84. 6 N

2-21 $P = mg(\sin \phi + \mu \cos \phi)/(\cos \phi - \mu \sin \phi)$,

$\phi = $ arc ctn μ

2-23 1077 N

2-25 a) Held back   b) 480 N

2-27 a) 3 N   b) 4 N   c) 5 N

2-29 b) 10 N   c) 30 N

Chapter 3

3-3 b) 2.5 m   c) 2.87 m

3-5 125 N, 100 N to the right, 25 N up

3-7 a) −30 N, 50 N   b) 5/3   c) 60 N

d) 1 m from right end

3-9 a) 270 N   b) 120 N

3-11 a) 325 N, 375 N   b) 225 N   c) 235 N   d) 864 N

3-13 a) arctan 0.3 or 16.7°   b) 24 cm   c) 0.53

3-17 a) 215 N   b) 185 N   c) 295 N
3-19 a) 10 N   b) 5 cm to right of c.g.   c) 31 cm
3-23 $3\frac{1}{3}$ ft from end
3-25 a) Top: 100 N up, 66.7 N away from door;
        Bottom: 100 N up, 66.7 N toward door
     b) 120 N, 56° up from horizontal
3-27 4.6×10^6 m from center of earth

Chapter 4

4-1 a) 15 mi hr^{-1} b) 22 ft s^{-1} c) 671 cm s^{-1}
4-3 61 cm s^{-1}, 60.1 cm s^{-1}, 60.01 cm s^{-1}; 60 cm s^{-1}
4-5 a) 0, 6.3 m s^{-2}, -11.2 m s^{-2}
     b) 100 m, 230 m, 320 m
4-9 a) 3.95 ft s^{-2} b) 620 ft
4-11 a) 2.67 km hr^{-1} s^{-1} b) 7.5 s   c) 83.4 m, 104.3 m
4-13 a) 12.5 cm s^{-2} b) 7840 cm
4-15 a) 100 m   b) 20 m s^{-1}
4-17 a) 24 ft s^{-1}, 29 ft s^{-1}, 34 ft s^{-1} b) 2.5 ft s^{-1}
     c) 21.5 ft s^{-1} d) 8.6 s   e) 23 ft   f) 1 s
     g) 24 ft s^{-1}
4-19 a) 32 ft s^{-1} b) 2.83 s   c) 16 ft   d) 22.6 ft s^{-1}
     e) 16 ft s^{-1}
4-21 a) 8.7 s   b) 37.5 m   c) $v_A = 26$ m s^{-1},
        $v_T = 17.5$ m s^{-1}
4-23 a) 94 ft s^{-1} b) 124 ft   c) 53 ft s^{-1}
     d) 150 ft s^{-2} e) 1.96 s   f) 93 ft s^{-1}
4-25 10.74 m
4-27 a) 48 ft s^{-1} b) 36 ft   c) zero
     d) 32 ft s^{-2}, downward   e) 80 ft s^{-1}
4-29 a) 32 ft   b) 13 ft s^{-1}, -32 ft s^{-2}
     c) -51 ft s^{-1}, -32 ft s^{-2} d) 37 ft s^{-1}
4-31 a) 11.3 m   b) 0.51 s   c) 10 m s^{-1}, -9.8 m s^{-2}
4-33 a) 4 m s^{-2} b) 6 m s^{-1} c) 4.5 m
4-35 a) 64.8 s   b) 101 mi
4-37 d) $k = 3.33 \times 10^{-3}$ ft^{-1} e) -1.33 ft s^{-2}
4-39 $v^2 = v_0^2 + K(y_0^2 - y^2)$
4-41 a) car A b) $t = 0, 2, 1.19$s, -1.69s   c) $t = 1$s,
     .667s   d) $t = 1,21, (-0.55)$
4-43 a) 50 s   b) 150 s
4-45 40 min, 30 min
4-47 N 19.5° E, 170 mi hr^{-1}
4-49 a) Zero, 10 m s^{-1} westward   b) 17.3 m s^{-1},
        20 m s^{-1}
4-51 a) E 41.7° S   b) 2.24 m s^{-1} c) 446 s

Chapter 5

5-1 a) 0.102 kg   b) 0.00102 g   c) 0.0311 slug
5-3 The two forces in an action-reaction pair never
     act on the same body.
5-5 a) 0.5 slug   b) 500 ft

5-7 a) 2 m s^{-2} b) 100 m   c) 20 m s^{-1}
5-9 a) 19.6 N   b) 2.55 s
5-11 a) 1.62×10^{-15} N   b) 3.33×10^{-9} s
     c) 1.8×10^{15} m s^{-2}
5-13 6.2×10^{24} kg
5-15 1.9 m s^{-2}
5-17 21,600 N
5-19 a) 59.0 N   b) Constant velocity   c) 6.25 m s^{-1}
5-21 a) 40 lb   b) 4 ft s^{-2} downward   c) Zero
5-23 240 ft
5-25 $W = 2wa/(g + a)$
5-27 a) 37°   b) 6.4 ft s^{-2} c) 2.5 s
5-29 3.13 s, 9.8 m
5-31 a) 2000 N, 1000 N   b) 5000 N
5-33 a) 36.8 N   b) 2.45 m s^{-2}
5-35 a) 2 lb   b) 1.9 lb
5-37 a) 1.96 m s^{-2} b) 3.14 N
5-39 a) To the left   b) 0.65 m s^{-2} c) 424 N
5-41 $2\, m_2g/(4m_1 + m_2)$, $m_2g/(4m_1 + m_2)$
5-43 a) 2.7 m s^{-2} b) 112.5 N   c) 87.5 N
5-45 2.43 m
5-47 g/μ
5-49 a) Left   b) Right   c) Right   d) Left
     e) Down slope   f) Always perpendicular to plane
5-51 a) 133 lb   b) 33 lb   c) 33 lb
5-53 c) $v_T = (mg/k)^{1/2}$ d) $t_R = (m/kg)^{1/2}$
     e) $v = v_T(e^{2t/t_R} - 1)/(e^{2t/t_R} + 1)$
5-55 a) Up   b) Constant   c) Constant   d) Stop

Chapter 6

6-1 a) 0.5 s   b) 12 ft s^{-1} c) 20 ft s^{-1}, 53° below
     horizontal
6-3 a) 20 s   b) 6000 ft   c) $v_x = 300$ ft s^{-1}.
     $v_y = 640$ ft s^{-1}
6-5 a) 30.6 m   b) 62.5 m   c) 25 m s^{-1}, 24.5 m s^{-1};
     35 m s^{-1}, 44.4° below horizontal
6-7 a) 0.022 m   b) 0.087 m   c) 0.196 m   d) 4.9 m
6-9 a) 121 ft s^{-1} b) 57.2 ft   c) 3.8 s
6-11 a) 1.01 m   b) -3.91 m   c) Yes
6-13 17 ft s^{-1}
6-15 a) 21.8 m s^{-1} b) 13.5 m s^{-1}, 36.2° below
     horizontal
6-17 2 s or 1.57 s
6-19 $R = 2v_0^2 \sin(\theta_0 - \alpha)\cos\theta_0/g\cos^2\alpha$
6-21 Straight line, 27° to vertical
6-23 a) 25° above horizontal   b) 650 mi
     c) 10,000 mi hr^{-1} d) 875 mi   e) 5 min
6-25 a) $v_x = 2 + 3t^2$, $v_y = -2t$
     b) 14.6 m s^{-1}, $-15.9°$ c) 24 N, -4N
     d) 24 N, $-9.46°$

6–27 a) 6.5 cm s^{-1}, 25° E of N   b) 3.25 cm s^{-2}, 25° E
of N

6–29 a) 66,700 mi hr^{-1} b) 0.0193 ft s^{-2}

6–31 5.97°

6–33 22 m s^{-1}

6–35 a) .127   b) To the very edge

6–37 a) 6.73 s   b) No

6–39 a) 52°   b) No   c) The bead remains at the
bottom

6–41 a) 28°   b) 1910 m   c) 0.533 W

6–43 a) 1530 ft   b) 1440 lb

6–45 a) 0.40 W   b) 4.49 s   c) 2 W   d) He would strike
the ground 9.9 m from the center of the wheel.

6–49 36,000 km above earth

6–51 a) 98 min   b) 26.5 ft s^{-2}

6–53 9.8 oz

Chapter 7

7–1 6.0 × 10^7 J

7–3 9850 ft lb

7–5 a) 6 lb   b) 48 lb   c) 22.5 ft lb

7–7 a) 55,000 ft lb   b) 4 times

7–9 4.5 × 10^{-17} J

7–11 9.8 J

7–13 a) 10 N, 20 N, 40 N   b) 0.5 J, 2 J, 8 J

7–15 a) 33 J   b) 13 J

7–17 a) 160 ft lb   b) 160 ft lb

7–19 a) 2400 J   b) 518 J   c) 1412 J   d) 470 J
e) $b + c + d = a$

7–21 a) 240 N   b) 235 J

7–23 a) 11.2   b) 5.0 m s^{-1}

7–25 0.1 m

7–27 a) 0.27   b) 3.6 J

7–31 1.81 m s^{-1}

7–33 8.02 m s^{-1}

7–35 a) 5.51 m s^{-1} b) 0.0091 m

7–37 a) 4.42 m s^{-1} b) 14.7 N

7–39 a) 24.6 ft s^{-1} b) 27.2 ft s^{-1} c) 22.5 lb
d) 10 ft s^{-1} e) 29° with vertical
f) 11.8 ft s^{-1} g) Parabolic trajectory
h) 12.64 ft s^{-1}

7–43 470 W, 0.47 kW

7–45 154 hp

7–47 $1.06

7–49 a) 1070 N   b) 5 hp, 80 hp

7–51 a) 8200 J   b) 8200 J   c) 0.55 hp

7–53 a) 27,600 N   b) 10.9 m s^{-1} c) 144,000 J
d) 353,000 J   e) 295 hp

7–55 a) 1070 N   b) 41.9 hp   c) 15.6 hp   d) 9.1%

7–57 227 km^2

7–59 a) 1.8 × 10^{14} J   b) 1.8 × 10^{20} W
c) 1.83 × 10^{10} kg

7–61 a) 4.55 × 10^{-17} J   b) 1.82 × 10^{-14} J,
2.79 × 10^{-14} J

Chapter 8

8–1 a) 2 × 10^5 kg m s^{-1} b) 40 m s^{-1}
c) 28.2 m s^{-1}

8–3 a) 8 × 10^5 m s^{-2} b) 4 × 10^4 N   c) 5 × 10^{-4} s
d) 20 N s

8–5 No; No; transferred to earth

8–7 a) 0.67 m s^{-1} b) 13,333 J   c) 1 m s^{-1}

8–9 780 ft s^{-1}

8–11 26 cm

8–13 a) 0.39 ft   b) 1950 ft lb   c) 3.9 ft lb

8–15 3.3 cm

8–17 17 mi hr^{-1}, 53° E of S

8–19 a) 10 cm s^{-1} b) 0.14 J   c) −70 cm s^{-1},
80 cm s^{-1}

8–21 5 cm s^{-1}, −25 cm s^{-1}

8–23 a) 50 cm s^{-1}, 53° below x-axis   b) 0.038 J

8–25 d) $\rho A(v_J - v_B)$ e) $\rho A(v_J - v_B)^2$
f) $\rho A v_B(v_J - v_B)^2$

8–27 a) $v_A = 22.0$ m s^{-1}, $v_B = 15.6$ m s^{-1} b) No; 0.19

8–29 a) 21,400 m

8–31 a) 7.20 ft s^{-1} b) 375

8–33 a) 8.85 kg m s^{-1} b) 4430 N

8–35 $v_A = 26.0$ m s^{-1}, $v_B = 15$ m s^{-1}; 60° from initial
direction of A.

8–37 0.91 m s^{-1}

8–39 a) 149° from direction of electron
b) 1.06 × 10^{-20} kg m s^{-1} c) 1.45 × 10^{-16} J

8–41 16 N

8–43 25 × 10^6 dyn, 250 N   b) Yes

8–45 15.9

8–47 4.23 × 10^7 m s^{-1}; greater

8–49 c) Measure mc^2 along the hypotenuse; excess-
length is E_k.

8–51 7.13 × 10^3 m s^{-1}; no

Chapter 9

9–1 a) 1.5 rad   b) 1.57 rad, 90°   c) 120 cm, 120 ft

9–3 471, 942, 1414 cm s^{-1}; 600, 1800, 5400 rev min^{-1}

9–5 −13 rad s^{-2}, 57 rev, 3.3 s

9–7 7.5 s

9–9 a) 20 rev s^{-1} b) 66 rev   c) 63 m s^{-1}
d) 7940 m s^{-2}

9–11 a) 15 cm s^{-2} b) 65 cm s^{-2} c) 126 cm s^{-2}

9–13 164,000 rev min^{-1}
9–15 b) $1/(12)^{1/2}$ rad
9–17 a) 2.67 kg m^2 b) 10.7 kg m^2
c) 1.6×10^{-3} kg m^2
9–19 $11m\ell^2/16$, 0.478 ℓ
9–21 a) $mb^2/12$ b) $mb^2/3$
9–23 a) 50.5 N   b) 40 N   c) 9.42 s
9–25 0.59
9–27 a) 6.25 rad s^{-2} b) 250 J   c) 4.74 rad s^{-2}; tension
is less than weight
9–29 a) 21.3 lb   b) 52.3 ft s^{-1} c) 2.45 s
9–31 a) 3.27 m s^{-2}, 0 rad s^{-2}, 26.1 N, 26.1 N
b) 0.754 m s^{-2}, 7.54 rad s^{-2}, 21.1 N, 36.2 N
9–33 a) 2.4 m s^{-1}, 3.2 m s^{-1} b) 0.05 kg m^2
c) 16 m s^{-2} d) 7.8 m s^{-2} e) 0.51 N, tension
9–35 a) 2 m s^{-2} b) 9.8 N
9–37 a) 1066 J   b) 36.3 m
9–39 1.01 slug ft^2
9–41 a) 2×10^7 J   b) 18 min
9–43 a) 2.96 rad s^{-2} b) 12.2 rad s^{-2} c) 50,200 J
9–45 1.45 J
9–47 a) 12 rad s^{-1} b) 0.027 J
9–49 $(mg \pm \mu Mg)/M\omega^2$
9–51 0.079 rev s^{-1}
9–53 a) -0.05 rad s^{-1} b) 32.7°   c) 36.2°
9–55 a) 2 slug ft^2 b) 2620 ft lb
9–57 a) 1.8 N   b) 4300 rev min^{-1}
9–59 a) 26,180 rad s^{-2} b) 1.00 N-m   c) Clockwise,
looking into page   d) 3.5×10^{-3} N m

Chapter 10

10–1 4.8×10^{11} N m^{-2}
10–3 b) 14×10^6 lb in^{-2} c) 0.016×10^6 lb in^{-2}
10–5 4.1×10^8 N m^{-2}, 0.054 mm
10–7 62.8 N   b) 5.7 mm   c) 1.82×10^{-5} cm
10–9 1.66 mm
10–11 a) 1.82 m   b) 1.5×10^8 N m^{-2}, 3.0×10^8 N m^{-2}
c) 1.36×10^{-3}, 1.49×10^{-3}
10–13 $\frac{1}{2}QP = P^2Y/2 = Q^2/2Y$
10–15 10^8 N m^{-2}, 5.0×10^{-4}, 2.5 mm, 0.95×10^{-4}
10–17 4.0×10^9 N m^{-2}, 2.5×10^{-10} m^2 N^{-1}
10–19 4.7×10^4 N
10–21 Steel: 0.64×10^{-6} atm^{-1}; water: 50×10^{-6} atm^{-1};
water is 78 times more compressible.
10–23 0.13 m

Chapter 11

11–1 $B = A \cos \theta_0$, $C = A \sin \theta_0$
11–3 a) 9470 cm s^{-2}, 377 cm s^{-1}
b) 5680 cm s^{-2}, 301 cm s^{-1} c) 0.0368 s

11–5 a) 2.37×10^4 ft s^{-2} b) 740 lb   c) 43 mi hr^{-1}
11–7 22.4 kg
11–9 a) 7.8 Hz   b) 9.8×10^{10} N m^{-2}
11–11 a) 3.12 kg   b) 4.31 cm below equilibrium,
moving upward   c) 36 N
11–13 a) 31 cm s^{-1} b) 49 cm s^{-2} c) 0.33 s
d) 100 cm
11–15 0.785 s
11–17 a) 5.6 lb   b) 13.6 lb, 8 lb, 2.4 lb
c) 0.62 ft lb, 0.077 ft lb
11–19 a) $k = k_1 + k_2$ b) $k = k_1 + k_2$
c) $1/k = (1/k_1) + (1/k_2)$ d) $2^{1/2}$
11–21 a) 1.4 s, 3.5 cm   b) Yes; friction   c) No
11–23 0.248 m
11–25 4.8 s
11–27 a) 284 dyn cm rad^{-1} b) 5.8 C°
11–31 a) 0.21 m   b) 0.70 rad s^{-1}
11–33 a) 81 ft
11–35 67 cm

Chapter 12

12–1 2.08×10^4 N m^{-2}, .206 atm
12–5 a) 1.077×10^5 N m^{-2} b) 1.037×10^5 N m^{-2}
c) 1.037×10^5 N m^{-2} d) 5 cm Hg
e) 68 cm water
12–7 a) 889 lb in^{-2} b) 25,132 lb
12–9 a) $\frac{1}{4}$ b) $\frac{3}{4}$
12–11 a) 98 N   b) 1.95 N
12–13 3.62 m^2
12–15 8330 m^3, 9250 kg
12–17 a) 5 cm   b) 490 N m^{-2}
12–19 a) 100 lb ft^{-3} b) E: 5 lb; D: 15 lb
12–23 100.87 g
12–25 2.12 W lb in
12–29 6.25×10^4 lb ft
12–31 a) $x = 30$ ft   b) Yes
12–33 a) Yes   b) For rubber, not for soap   c) No
12–35 1372 N m^{-2}
12–37 a) 70.7 cm Hg   b) 71.2 cm Hg   c) 11 cm
12–39 4.3 cm
12–43 3.6×10^{-4} mm

Chapter 13

13–1 a) 14 m s^{-1} b) 4.4×10^{-3} m^3 s^{-1}
13–3 10 cm
13–5 39.6 ft s^{-1}
13–7 a) 1.21×10^{-3} m^3 s^{-1} b) 0.90 m
13–9 a) 16 ft s^{-1} b) 0.79 ft^3 s^{-1}
13–11 5.0×10^5 Pa

13–13 a) 6.4×10^5 Pa   b) 0.265 m³ s⁻¹

13–13 a) 6.4×10^5 Pa   b) 0.265 m^3 s^{-1}
13–15 0.0268 m^3 min^{-1}
13–17 12 ft^3 min^{-1}
13–19 $.816$ m^3 s^{-1}
13–21 38.8 m s^{-1}
13–23 4.0×10^4 N
13–25 8 Pa
13–29 a) 0.77 cm s^{-1} b) 1.89 cm s^{-1}
13–31 325 ft s^{-1}
13–33 a) Turbulent   b) 5.3 liter s^{-1}
13–35 a) Yes   b) No

Chapter 14

14–1 a) 4.3×10^{-8} s   b) 10.4 m
14–3 The bolt at A appears to come first.
14–7 $\Delta s' = [(\Delta s)^2 - c^2 (\Delta t)^2]^{1/2}$
14–9 5.77×10^{-9} s
14–11 $v/c = E/mc^2$
14–13 2.60×10^8 m s^{-1}
14–15 a) 0.5×10^{-4} b) 15%   c) Factor of 2.30
14–17 1.18×10^{-18} m s^{-2}
14–19 2.988×10^8 m s^{-1}
14–21 1.64×10^{-13} J
14–23 a) 4.55×10^{-15} J, 4.92×10^{-15} J
b) 18.2×10^{-15} J, 27.9×10^{-15} J
14–25 a) 1.8×10^{14} J   b) 1.8×10^{20} W
c) 1.84×10^{10} kg

Chapter 15

15–1 600.45 K
15–3 a) 20°C   b) 35°C   c) −15°C
15–5 6.83 cm Hg
15–9 about 7%
15–11 235 cm
15–13 20 cm, 10 cm
15–15 250 ft
15–17 a) About 1 C°   b) Lose
15–19 2.61×10^5 lb
15–21 9.35×10^7 N m^{-2}
15–23 a) 3×10^{-4} b) 9000 lb in^{-2} c) 4.5×10^5 lb
d) 21,000 lb in^{-2} e) 105° F
15–25 a) 893 atm   b) 36.2° C
15–27 494 atm

Chapter 16

16–1 a) Yes   b) No
16–3 4470 cal
16–5 370 years
16–7 80.7

16–9 a) $1:0.093:0.031$ b) $1:0.83:0.35$
16–11 0.1 Btu lb^{-1}(F°)$^{-1}$
16–13 0.092 cal g^{-1}(C°)$^{-1}$
16–15 a) 72°C   b) Temperature of apparatus does not change.
16–17 a) 0.806   b) 84,280 cal
16–19 a) 273 J   b) 3.42 J mol^{-1} K^{-1}
c) 10.9 J mol^{-1} K^{-1}
16–21 b) 3.01 J g^{-1} (C°)$^{-1}$
16–23 106 g
16–25 0° C with 0.2 g of ice left
16–27 539 cal g^{-1}
16–29 24° C
16–31 40°C
16–33 1.84 kg
16–35 a) 1602 ft^3 b) 310 ft^3
16–37 a) 12,000 Btu hr^{-1} b) 3515 W
16–39 a) 8.05×10^6 J   b) 805 N

Chapter 17

17–1 4.32×10^5 J
17–3 a) 1.8 cal s^{-1} b) 20 cm
17–5 0.20 Btu in hr^{-1} ft^{-2} (F°)$^{-1}$
17–7 110° C
17–9 a) 40° C   b) 2.4 cal s^{-1}
17–11 $(R_1 R_2)^{1/2}$
17–13 a) 90.0° C   b) 21.7 J s^{-1}
17–15 88.5 C°
17–17 a) 7.32×10^5 cal   b) 8.05×10^5 cal
17–19 a) 460 W m^{-2} b) 4.6×10^6 W m^{-2}
17–21 1.78 W
17–23 20.03 K
17–25 6.9 W
17–27 d) 31.4 C° cm^{-1} e) 29 cal s^{-1} f) Zero
g) 1.11×10^{-4} m^2 s^{-1} h) 10.9 C° s^{-1}
i) 3.33 s   j) Decrease   k) 5.4 C° s^{-1}

Chapter 18

18–1 0.3×10^5 Pa
18–3 a) 0.75 atm   b) 2 g
18–5 $\rho = pM/RT$
18–7 3.25
18–9 a) 82 cm^3 b) 0.33 g
18–11 When the piston has moved 13.1 inches.
18–13 0.0023 g
18–15 b) 600 K   c) 4 atm
18–17 0.286×10^5 Pa
18–19 Dry ice is 3.3 times as dense.
18–21 a) Rises   b) Falls   c) Approximately stationary

18-23 a) 37.2% b) 865 Pa or .0086 atm
c) 6.4 g m^3
18-25 a) 11.6 g b) 29 g m^{-3} c) 110 g
18-27 1.3 lb hr^{-1}
18-29 0.929 kg

Chapter 19

19-1 a) Yes b) No c) Negative
19-3 a) No b) Yes c) Yes d) Work done on the
resistor equals heat transferred to water.
19-5 171 m
19-7 10.8 ft^3
19-9 $U_1 = U_2$
19-11 a) 0.167×10^6 J b) 2.03×10^6 J
19-13 a) 1.1×10^{-2} ft^3 b) 650 ft lb c) 1800 Btu
d) 1.4×10^6 ft lb
19-15 $K \ln[(V_2 - b)/(V_1 - b)] - [(a/V_1) - (a/V_2)]$
19-17 b) 830 J c) On the atmosphere d) 2110 J
e) 2940 J f) 830 J (Same as (b))
19-21 a) 3.15 atm, 189 K b) 3.78 atm, 227 K
19-25 267° C
19-27 a) When piston is 3.13 in. from bottom b) 477 K
19-29 a) $p_1 = 1$ atm, $V_1 = 2.46$ l;
$p_2 = 2$ atm, $V_2 = 2.46$ l;
$p_3 = 1$ atm, $V_3 = 3.74$ l
b) 49.4 J
19-31 18%
19-33 a) 320 K b) 20%
19-35 a) 900 cal b) 1600 cal c) 400 cal
19-37 a) 4.19×10^5 J b) 1306 J K^{-1} c) 50° C
d) 1306 J K^{-1} before, 1408 J K^{-1} after
19-39 a) 613 J K^{-1} b) -571 J K^{-1} c) 42 J K^{-1}
19-41 11.5 J K^{-1}
19-43 0.0503

Chapter 20

20-1 a) 3×10^{-7} cm b) About 10 times as great
20-3 1.95×10^{-4} cm
20-5 1000 atm
20-7 508 m s^{-1}; smaller by factor 1.45
20-9 1.00636
20-11 20.8 J mol^{-1} K^{-1} or 4.97 cal mol^{-1} K^{-1}
20-13 a) All equal b) 1:3.15:7.03
20-15 a) 1.1 cm s^{-1} b) No
20-17 Bi: 9.9 mm; Bi$_2$: 14 mm
20-19 2.82×10^{-10} m

Chapter 21

21-1 a) 2 cm b) 30 cm c) 100 Hz d) 3000 cm s^{-1}
21-5 320 m s^{-1}

21-7 a) 200 m s^{-1}, 20 m b) Increase by $2^{1/2}$
21-9 6320 cm s^{-1}
21-11 a) 0.5 Hz b) π rad s^{-1} c) $\pi/100$ cm^{-1}
d) $y = 10 \sin[(\pi x/100) - \pi t]$ e) $y = -10 \sin \pi t$
f) $y = 10 \sin[(3\pi/2) - \pi t)]$ g) 10π cm s^{-1}
h) $y = 5 \sqrt{2}$ cm, $v = -5\pi\sqrt{2}$ cm s^{-1}
21-13 48×10^{-6} atm^{-1}
21-15 a) 1.31 m, 4.80 m^{-1} b) 1644 s^{-1} c) 5.54 m
21-17 b) 290 m s^{-1}
21-19 a) 323 m s^{-1}, 1316 m s^{-1} b) 347 m s^{-1}
21-21 a) 135 Hz b) 1350 Hz c) 13,500 Hz
21-25 a) $F = Mv^2/L$ b) Stationary with respect to
table or speed $2v$ c) Mv^2 d) $\frac{1}{2}Mv^2$
e) Sudden accelerations of successive elements
of rope are analogous to a series of inelastic
collisions; energy is not conserved.

Chapter 22

22-1 a) 200 Hz b) 49 (50th harmonic)
22-3 a) 36 m s^{-1} b) 65 N
22-5 1.28
22-7 a) 5000 m s^{-1} b) 340 m s^{-1}
22-9 Diatomic
22-11 a) 1140, 2280, 3420, 4560, 5700 Hz
b) 570, 1710, 2850, 3990, 5130 Hz
c) 16, 17
22-13 a) 420 Hz b) 415 Hz
22-15 2.9%
22-17 a) 0.655 m b) 56° C

Chapter 23

23-1 a) 9 times b) 4 times
23-3 a) 60 db b) 77 db
23-5 b) No; $\beta = 120 + 10 \log I$
23-7 10^{-6} W
23-9 a) 0.256 μm b) 0.293 N m^{-2} c) 50 m
23-11 a) π b) $I_A = 4 \times 10^{-6}$ W m^{-2},
$I_B = 12 \times 10^{-6}$ W m^{-2}
c) 2.1×10^{-6} W m^{-2}, 63.2 db
23-13 a) 0.029 m b) 21 cm c) 455 Hz
23-15 1.5000, 1.4983; just interval is 0.11% larger.
23-17 24 mi hr^{-1} or 10.7 m s^{-1}
23-19 a) 2.06 ft b) 2.46 ft c) 548 Hz d) 460 Hz
e) 572 Hz f) 439 Hz g) 2.12 ft, 2.40 ft, 547,
458, 572, and 437 Hz
23-21 a) 454 Hz b) 462 Hz c) 8 beats s^{-1}
23-23 a) 11 ft b) 1 ft c) 15 waves
d) 1100 ft s^{-1} e) 0.49 ft

Index

Index

Absolute temperature, 340
Absolute zero, 270, 341
Acceleration, 51–53
   angular, 158–160
   average, 52, 90–91
   centripetal, 98
   components, 70, 91–93
   due to gravity: table, 107
   instantaneous, 52–53, 91
   variable, 56–60
Accelerometer, 79–80
Acoustic phenomena, 389–402
Action and reaction, 17–20
Addition, of vectors, 6–11
Adiabatic process, 327
   in longitudinal wave, 370–371
Adiabatic wall, 266
Air table, 28
Air track, 28
Amorphous solid, 358
Amplitude, 197
   of wave, 364
Angular acceleration, 158–160
Angular and linear motion, relation,
   160–161
Angular harmonic motion, 204
Angular momentum, 167–170
Angular quantities, vector
   representation, 170–173
Angular velocity, 157–158

Antinode, 379
Archimedes' principle, 219–221
Atmospheric pressure, 217, 269
Auditory area, 392
Average velocity, 89–90
Avogadro's number, 351–352

Backspin, 241
Baffle, 395
Ballistic pendulum, 144–145
Ballistic trajectories, 93–97
Balloon, 219
Banked curve, 101–102
Barometer, 216–217
Barometric equation, 308
Beats, 399–400
Bel, 391
Bernoulli's equation, 235–241
Binding energy, 130–131
Blackbody, 301
Body temperature, 270
Boiling point, 313
Boltzmann constant, 355
Boundary conditions, for string, 377
Bourdon gauge, 216–217
Boyle's law, 307
Breaking stress, 188
British engineering system, 71
British thermal unit, 282
Bubble, 224

Bubble chamber, 318–319
Bulk modulus, 188–191
Buoyancy, 218–221

Caloric, 279
Calorie, 282–284
   food-value, 288
Calorimetry, examples, 288–289
Capillarity, 225–227
Carnot cycle, 339–340
Cartesian diver, 227
Cavendish balance, 73
Celsius temperature scale, 271–272
Center, of oscillation, 205–207
   of percussion, 206–207
Center of gravity, 21, 39–42
Centigrade temperature scale,
   271–272
Centrifuge, 98–99
Centripetal acceleration, 98
Centripetal force, 99–102, 118
Change of phase, 286–288
Circular motion, 97–104
Close packing, 358
Cloud chamber, 287–288, 318
Coefficient of thermal expansion,
   272–274
Coefficient of friction, 27–29
   table, 28
Coefficient of performance, 338

Coefficient of viscosity, 242
Collisions, 143–146
    elastic, 145–146
    inelastic, 143–145
Commutative law, for vector sum, 7
Compass needle, 42
Components, of acceleration, 91
    of vector, 8–11
Compressibility, 189
Compression ratio, 335
Compressive strain, 186
Compressive stress, 184
Concurrent forces, 15
Condensation, in wave, 365
Conduction, of heat, 293–296
Conductivity, thermal, 293–295
Conductor, thermal, 294
Conic section, 104
Conical pendulum, 101
Conservation, of angular
        momentum, 168–170
    of energy, 120–126
        in thermodynamics, 326–329
    of mass and energy, 260
    of momentum, 142–150
Conservative force, 125–126
Constant acceleration, motion with,
        53–59
Contact angle, 225–227
Contact force, 4
Continuity equation, 234
Contraction, of length, 256
Convection, 296–298
Convection coefficient, 297–298
Coordinate transformation, galilean,
        252
    Lorentz, 257–258
Correspondence principle, 261
Couple, 16, 42
Critical point, 312–316
Cross product, 171
Crystals, 357–361
    heat capacity, 359–361
Cubic crystal, 358
Curve, banked, 101–102
Curved flight, of spinning ball, 240
Cyclic motion, 196
Cyclic process, in engine, 333

Dam, force and torque on, 221
Deceleration, 53

Decibel, 390–393
Definition, operational, 1
Density, 213–214
    of water, table, 273
Derivative, 50
Dew point, 317–318
Diathermic wall, 266
Difference, of vectors, 7, 11
Difference tone, 400
Diffraction, of sound, 398
Dilation, of time, 254
Displacement vector, 49
Dissipative force, 125–126
Doppler effect, 400–402
    relativistic, 402
Doppler radar, 402
Dot product, 116
Ductility, 187
Dulong and Petit, law of, 285, 360
Dynamics, 69
Dyne, 71

Efficiency, thermal, 334–335
Efflux, speed of, 236–237
Elastic collision, 145–146
Elastic force, 124
Elastic limit, 187
Elastic modulus, 188–191
    table, 188
Elasticity, 183–191
    and plasticity, 186–188
Emissivity, 301
Energy, conservation of, 120–126
    equipartition, 359
    in harmonic motion, 197–198
    internal, 326–329
    kinetic, 117
        of molecule, 355–357
        in rotation, 166–167
    and mass, 130
    mechanical, 117–128
    potential, elastic, 124–125
        gravitational, 119–124
    radiant, 298
    relativistic, 259–261
    in thermodynamics, 323–330
Energy conversion, 344–345
Engine, 333–337
    diesel, 336
    gasoline, 335–336
    steam, 336–337

Enthalpy, 329
Entropy, 342–344
Equal temperament, 394–395
Equation of state, 305–312
Equilibrium, 15–47
    phase, 288
    of a point, 15–34
    of a rigid body, 35–47
Equipartition of energy, 359
Ether, 251
Expansion, thermal, 272–274
External force, 77

Face-centered cubic crystal, 358
Fahrenheit temperature scale,
        271–272
First law of thermodynamics,
        326–329
Fixed points, on temperature scale,
        271
Floating, 219
Flow tube, 233–236
Fluid, 213
Force, 4–5, 70–71
    centripetal, 99–102
    conservative, 125–126
    dissipative, 125–126
    internal, 77
    measurement of, 5
    viscous, 81
Force arm, 35
Force constant, 191
Force diagram, 20
Fracture, 187
Free-body diagram, 20
Free fall, 57–59
Frequency, 196
    of wave, 364
Friction, 26–29
    kinetic, 27–29
    static, 26–27

Galilean transformation, 252
Galilei, Galileo, 15
Gas, kinetic theory, 350–357
Gas constant, 306
Gas thermometer, 268–271
Gasoline engine, 335–336
Gauges, pressure, 216
Gradient, temperature, 294
Gravitation, Newton's law, 72–73

Gravitational constant, 73
Gravitational mass, 75
Gravity, acceleration due to, 57
  table, 107
  center of, 21, 39–42
  specific, 213
Gyroscope, 173–175

Half-tone, 394–395
Harmonic motion, 195–206
  angular, 204
Harmonics, 381
Hearing, 392
  loss of, 398–399
Heat, 279–289
  in volume change, 325–326
Heat capacity, 282–285
  constant pressure, 284
  constant volume, 284
  of gas, 356–357
  ideal gas, 330–331
  measurement of, 283–285
  of solid, 359–361
  variation with temperature, 283–284
Heat current, 294
Heat death, 344
Heat engine, 333–337
Heat flow, 293–296
  radial, 295
Heat of combustion, 288
Heat of fusion, 287
Heat of transformation, 288
Heat of vaporization, 287
Heat phenomena, 265–274
Heat transfer, 279, 293–301
  in human body, 298
Hooke's law, 115
Horsepower, 129
Hovercraft, 28
Human arm, 38
Humidity, 317–318
Hydrodynamics, 233–247
Hydrometer, 219–220
Hydrostatic paradox, 215–216
Hydrostatic pressure, 185
Hydrostatics, 213–227
Hydrotherapy, 221
Hygrometer, 317–318
Hysteresis, 187–188

Ideal fluid, 233
Ideal gas, 305–309
  adiabatic process, 331–333
  heat capacity, 330–331
  internal energy, 329–330
  kinetic theory, 350–357
Idealized model, 20
Impedance, mechanical, 390
Impulse, 139–142
Indefinables, 1
Inelastic collision, 143–145
Inertia, moment of, 161–166
Inertial frame of reference, 251
Inertial mass, 75
Inertial reference system, 17–18
Instantaneous acceleration, 90–91
Instantaneous velocity, 90
Insulator, thermal, 294
Integration, in kinematics, 55–57
Intensity, of sound, 390–393
Interference, 379, 383–384
Interferometer, Michelson, 2
Internal energy, 326–329
  of ideal gas, 329–330
Internal force, 77
Internal potential energy, 127–128
Internal work, 126–127
Interval, musical, 394–395
Invariance, 251
Isentropic process, 343
Isobar, 310
Isobaric process, 327
Isochoric process, 327
Isochronous motion, 199
Isolated system, 326
Isotherm, 311
Isothermal process, 324, 327
Isotropic solid, 358

Kelvin temperature scale, 269–272, 340–342
Kilocalorie, 288
Kinematics, 49
Kinetic energy, 117
  of molecule, 355–357
  relativistic, 130, 259–261
  in rotation, 166–167
Kinetic friction, 27
Kinetic theory of gases, 350–357

Laminar flow, 234, 241
Latent heat, 287

Length, relativity of, 255–257
  standard of, 1–4
Length contraction, 256
Lift, on aircraft wing, 240
Light, speed of, 251
Line of flow, 233–234
Linear and angular motion, relation, 160–161
Linear momentum, 139
Liquid, structure of, 350
Longitudinal wave, 363–372
  adiabatic character, 370–371
Loop, 379
Loop-the-loop, 103
Lorentz transformation, 257–258
Loschmidt's number, 352
Loudness, 391–393

Macroscopic description, 323
Manometer, 216–217
Mass, 69–70
  and energy, 130
  gravitational, 75
  inertial, 175
  relativistic, 130–131
  standard of, 2–4
  and weight, 74–75
Massless particle, 261
McLeod gauge, 307–308
Mechanical impedance, 390
Meniscus, 226
Metric system, 1–4
Michelson interferometer, 2
Mode, normal, 381–386
Model, idealized, 20
Moderator, 146
Modulus, elastic, 188–191
  table of, 188
Modulus of rigidity, 189
Molar heat capacity, 282–285
  of gas, 356–357
  of solid, 359–361
Molar specific heat, 282–285
Molecular forces, nature of, 349
Molecular speed, measurement, 357
Molecular theory of matter, 349–360
Molecule, 349
  speed of, 354–357
Moment, 35–36
  and angular acceleration, 161–163
Moment arm, 35

Moment of inertia, 161–166
Momentum, 139–150
    angular, 167–170
    conservation of, 142–150
    relativistic, 149
Monatomic gas, 356
Mössbauer effect, 155
Motion, in a plane, 89–107
    in straight line, 49–63

Newton, Isaac, 15
Newtonian fluid, 243
Newton's first law, 15–17
Newton's law of gravitation, 72–73
Newton's second law, 69–71
Newton's third law, 17–20
Node, 379
Noise level, 391
Normal boiling point, 269
Normal component, 92
Normal force, 26
Normal melting point, 269
Normal mode, 381–386
Normal sublimation point, 269
Nuclear reactor, 146
Nutation, 175

Optical pyrometer, 268
Organ pipe, 384–386
Oscillation, 195–204
    center of, 205–207
Oscillatory motion, 195
Otto cycle, 335–336
Overtones, 381

Parallelogram, in vector addition, 7
Parametric equations, 89
Particle, massless, 261
Pascal, pressure unit, 217
Pascal's law, 215
Pendulum, ballistic, 144–145
    conical, 101
    physical, 204–205
    simple, 203–204
Percussion, center of, 206–207
Period, 196
    of wave, 364
Periodic motion, 195
Periodic waves, 364–368
Permanent set, 187

Phase angle, 198–199, 366
Phase change, 286–288
Phase diagram, 313
Phase equilibrium, 269, 288
Photon, 149–150
Pitch, 394
Pitot tube, 239–240
Plasticity, 186–188
Plate, vibration of, 386
Poise, 242
Poiseuille's law, 243–245
Poisson's ratio, 188–191
Polygon, for vector addition, 7
Potential energy, 119–125
    internal, 127–128
Pound, 71
Power, 128–129
    in rotation, 166–167
Powers of ten, prefixes, 3
Prandtl tube, 239–240
Precession, 174–175
Prefixes, for units, 3
Pressure, 185
    absolute, 216
    in fluid, 214–217
    gauge, 216
Pressure gauges, 216–217
Pressure gradient, 244
Principle of superposition, 379–386
Problems, solving, 12
Projectile motion, 93–97
Propagation constant, 366
Proper length, 256
Proper time, 255
Proportional limit, 187
Pump, diaphragm, 218–219
    diffusion, 217–218
    rotary, 218
pVT surface, for ideal gas, 308–309
    for real substance, 309–312
Pyrometer, 268

Quantity of heat, 281–282
Quantum, 149–150
Quantum mechanics, and specific
    heats, 360–361

Radial heat flow, 295
Radian, 158

Radiant emittance, 299
Radiant energy, 298
Radiation, acoustic, from piston,
    395–398
    of heat, 298–301
Radiator, ideal, 300–301
Range, of projectile, 95
Rankine cycle, 336
Rankine temperature scale, 271–272
Rarefaction, 365
Reactor, nuclear, 146
Recoil, 146–147
Rectangular components, 9–11
Rectilinear motion, 49–63
Reference temperature, 267
Reflection, of wave pulse, 377–379
Refrigerator, 338–339
Regelation, 316
Relative humidity, 317–318
Relative velocity, 60–63
Relativistic mass, 130–131
Relativity, 251–262
    general theory, 261–262
    of length, 255–257
    and newtonian mechanics, 261–262
    of simultaneity, 253
    of time, 253–255
Relaxation time, 82
Reservoir, heat, 334
Resistance thermometer, 268
Resolution, of vector, 8–10
Resonance, 382
Rest energy, 130–131, 260
Restoring force, 124, 195–196
Resultant, 8
Reversible process, 340
Reynolds number, 245–247
Righthand rule, 170–171
Rigid body, 157
Rocket propulsion, 148–149
Rod, vibration of, 386
Rotation, 157–175
    of earth, 106–107
Rotational motion, 15

Satellite, 104–106
Saturated liquid, 312
Saturated vapor, 312
Scalar product, 116
Scalar quantity, 6

Scale, musical, 394–395
  just, 394–395
  tempered, 394–395
Second law of thermodynamics, 337–345
Sedimentation velocity, 245
Shear modulus, 188–191
Shear strain, 186
Shear stress, 184
Simple harmonic motion, 195–206
Simultaneity, relative nature of, 253
Slug, 71
Soap film, 222, 224
Solar energy, 344
Sonar, 398
Sound, 371, 389–402
Specific gravity, 213
Specific heat capacity, 282–285
Specific volume, 310
Speed, 51
  of light, 251–252
  of longitudinal wave, 369–370
  molecular, 354–357
  of propagation, 367–374
  root-mean-square, 354
  of wave on string, 368–369
Sphere, gravitational interaction, 73
Sphygmomanometer, 217
Stagnation point, 234
Standards, 1–4
Standing wave, 379–386
State coordinate, 265–266
Static friction, 27
Steady state, in heat flow, 294–295
Steam turbine, 337
Stefan–Boltzmann constant, 299
Stefan's law, 299–301
Stokes' law, 81, 245
Strain, 186
Streamline, 233
Stress, 183–186
  thermal, 274
Structure, of gases, 350–357
  of matter, 349–360
  of solids, 349–351, 357–360
Sublimation, 288
Sublimation point, 313
Supercooling, 287
Superfluid, 316
Superheated liquid, 318

Superposition principle, 379–386
Supersaturated vapor, 318
Surface tension, 221–227
  table, 223
Symbols, for physical quantities, 4

Tangential component, 92
Temperature, 265–267
Temperature gradient, 294
Temperature interval, 272
Temperature scale, 269–272, 394–395
Tensile strain, 186
Tensile stress, 183–184
Tensor, 184
Terminal velocity, 245
Thermal conductivity, 293–295
Thermal efficiency, 334–335
Thermal equilibrium, 266–267
Thermal expansion, 272–274
Thermal phenomena, 265–274
Thermal pollution, 341
Thermal properties, of matter, 305
Thermal stress, 274
Thermocouple, 267
Thermodynamics, first law, 326
  second law, 337–345
  third law, 342
  zeroth law, 267
Thermograph, 300
Thermometer, 267–272
Thermometric property, 268–269
Third law of thermodynamics, 342
Threshold, of hearing, 392
  of pain, 392
Throttling process, 328
Timbre, 385, 393–395
Time, relativity of, 255–257
  standard of, 2–4
Time dilation, 254
Tone quality, 385, 393–395
Top, 173–175
Torr, 217
Torricelli's theorem, 236–237
Torque, 35
  and angular acceleration, 161–163
Torque converter, 167
Torsion modulus, 189
Traction, 41
Transducer, 398
Transfer of heat, 293–301

Transformation, galilean, 252
  Lorentz, 257–258
Transient heat flow, 295
Translational motion, 15
Transverse wave, 363–369
Triad, 394
Triple line, 312
Triple point, 269, 312–316
Tube of flow, 233–236
Turbine, steam, 337
Turbulence, 234, 245–247

Ultimate strength, 188
Ultrasonics, 398
Units, conversion, 3–4
  standard, 2–4
  systems of, 71–72
Universal gas constant, 306
Urine, 214

Vapor, 312
  saturated, 312
Vapor pressure, 314
Vector addition, 6–11
Vector difference, 7, 11
Vector product, 171
Vector quantity, 6
Vectors, 5–11
Velocity, angular, 157–158
  average, 49–50, 89–90
  components of, 60–63
  instantaneous, 50–51, 90
  relative, 60–63
Velocity gradient, 242
Velocity transformation, galilean, 252
  relativistic, 258
Vena contracta, 237
Venturi tube, 238–239
Vibration, 377–386
Viscosimeter, 241
Viscosity, 27–28, 241–245
Viscous force, 81
Volume, specific, 310
Volume strain, 186

Water wave, 372–374
Wave, 363–373
  longitudinal, adiabatic character, 370–371
  in water, 372–374

Wave motion, 195
Wave number, 366
Wave pulse, 363, 377–379
Wave speed, 367–374
Wavelength, 365
Weight, 4, 21
    buoyancy correction, 219–220
    and mass, 74–75

Whole tone, 394–395
Work, 113–117
    as integral, 115
    internal, 126–127
    and kinetic energy, 117–119
    in rotation, 166–167
    in thermodynamics, 323–330
    in volume change, 323–325

Work and energy, relativistic,
    259–261

Yield point, 187
Young's modulus, 188

Zero, absolute, 341–342
Zeroth law of thermodynamics, 267